S0-AJV-917

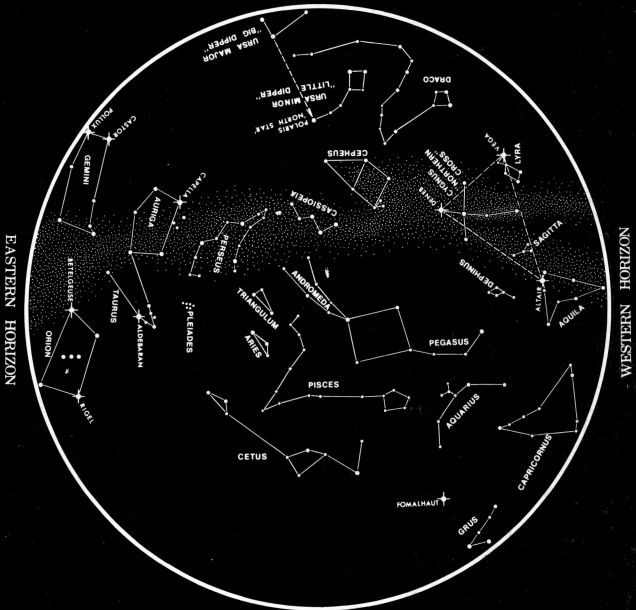

NORTHERN HORIZON

URSA MAJOR "BIG DIPPER"

DRACO

URSA MINOR "LITTLE DIPPER"

POLARIS "NORTH STAR"

POLLUX

CASTOR

GEMINI

CAPELLA

CEPHEUS

VEGA

LYRA

CYGNUS "NORTHERN CROSS"

DENEB

AURIGA

CASSIOPEIA

SAGITTA

ALDEBARAN

PERSEUS

DELPHINUS

ALTAIR

AQUILA

TAURUS

BETELGEUSE

ANDROMEDA

TRIANGULUM

PLEIADES

PEGASUS

ARIES

ORION

PISCES

AQUARIUS

RIGEL

CAPRICORNUS

CETUS

FOMALHAUT

GRUS

EASTERN HORIZON

WESTERN HORIZON

SOUTHERN HORIZON

THE NIGHT SKY IN NOVEMBER

Chart time (Local Standard Time):

10 pm...First of November
9 pm...Middle of November
8 pm...Last of November

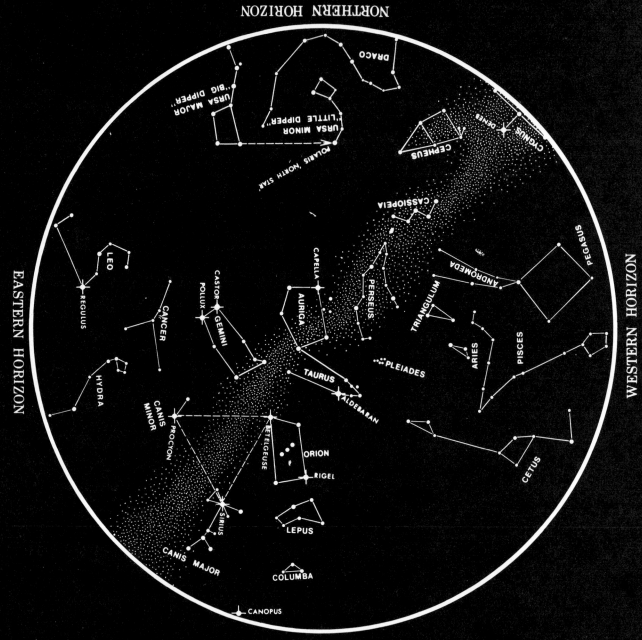

THE NIGHT SKY IN JANUARY

Chart time (Local Standard Time):

10 pm...First of January
9 pm...Middle of January
8 pm...Last of January

DISCOVERING THE UNIVERSE

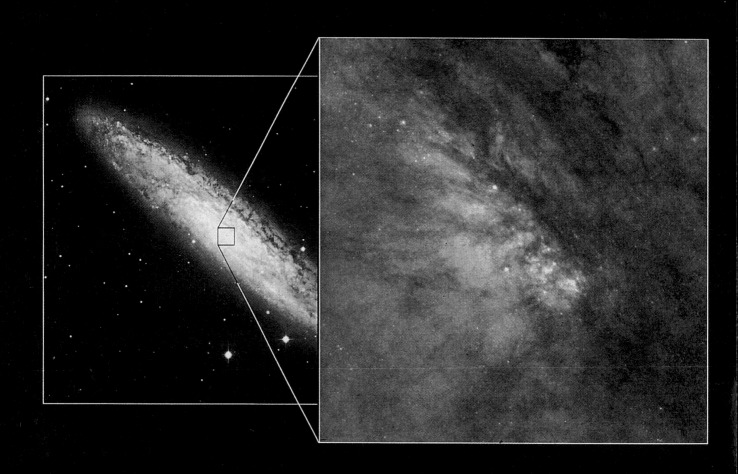

*Fourth
Edition*

DISCOVERING
THE UNIVERSE

William J. Kaufmann III
Late of San Diego State University

Neil F. Comins
University of Maine

W. H. Freeman and Company
New York

To the memory of William J. Kaufmann III

Cover Image and Frontispiece: A Burst of Star Formation
Stars continue to form throughout the universe. On the cover, a Hubble Space Telescope image shows a small part of an enormous spiral galaxy named NGC 253. (NGC stands for the *New General Catalogue* of galaxies.) Called a starburst galaxy, NGC 253 is located 8 million light-years from the Earth in the constellation Sculptor. It is notable because of the tremendous amount of star formation presently occurring in it. In this image, colorful clouds of interstellar gas and dust surround a cluster of bright, newly formed stars. Facing the title page is a ground-based telescopic view of NGC 253 (left) and the 1000-light-year-wide region from which the cover image was taken. (Ground-based view: Carnegie Institutes of Washington; HST view: Jay Gallagher/University of Wisconsin-Madison, Alan Watson/Lowell Observatory, Flagstaff, Arizona, and NASA.)

Contents Images: p. ix:NASA; p. x:A. S. McEwen, USGS; p. xi:Gary McDonald; p. xii:NASA; p. xiii:Jeff Hester and Paul Scowen (Arizona State University) and NASA; p. xiv:Dennis di Cicco; p. xv:R. C. Krann-Korteweg/Photo Researchers; p. xvi:Drawing by Joh Kagaya for Makoto Inoue, National Observatory of Japan.

Library of Congress Cataloging-in-Publication Data
Kaufmann, William J.
    Discovering the Universe / William J. Kaufmann III, Neil F. Comins. —4th ed.
        p.   cm.
    Includes bibliographical references and index.
    ISBN 0-7167-2646-7 (softcover)
    1. Astronomy.   2. Cosmology.   I. Comins, Neil F. Title.
    QB43.2.K376   1996
520—dc20   95-38765

Illustration credits are listed on pages 425–426.

Copyright © 1987, 1990, 1993, 1996 by Neil F. Comins

No part of this book may be reproduced by any mechanical, photographic, or electronic process, or in the form of a phonographic recording, nor may it be stored in a retrieval system, transmitted, or otherwise copied for public or private use, without written permission from the publisher.

Printed in the United States of America

Fourth printing 1997, RRD

Contents Overview

Contents

Contents

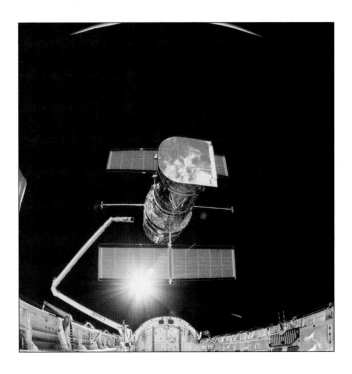

Contents

FOUNDATIONS III
THE STARS

Preface

A course in introductory astronomy attracts different students for different reasons. Some of you have a lifelong passion for the beauty of the sky; you long to understand how the Earth and universe got here. Others of you have only a passing interest and a free course opening, or you heard that this is a great course. Or you have a science requirement to fill and a long-lived dread of math and science. You've all come to the right place; this book was written to accommodate all of you. This edition of *Discovering the Universe* pays special attention to the way you learn. It evolved from my perceptions of the backgrounds and needs of over 6000 students to whom I have taught introductory astronomy.

Furthermore, you've come to astronomy at the right time. The range of topics that astronomers understand has skyrocketed in the last few decades. What you will be learning in this book is far more accurate—and breathtaking—than the astronomy that your parents, your grade school teachers, or even your older siblings learned.

The first edition of this textbook changed how introductory astronomy is taught. Most paperback introductory textbooks are just pared down versions of an author's longer text, but not this one. In it, William J. Kaufmann III introduced hundreds of color photographs, wonderful color drawings and paintings, clear and compelling prose, the full breadth and currency of a rapidly changing field, and many important learning aids. You will find all those features here. Yet I have made many revisions in this new edition, including an improved order of topics, a refined structure, and greater sensitivity to the beliefs that you bring to this course. This is my first textbook, and while it builds on Professor Kaufmann's work, you will find much that is fresh and new.

This book provides a broad overview of the cosmos, from our nearest neighbors in the solar system to the secrets in the cores of distant galaxies to the origin and evolution of the universe. Moreover, it is designed to show how the physical laws governing distant stars and planets are the same as those that affect our lives on Earth. You will see the same principles of gravity and light at work on Earth and in the stars; you will see convection, the process that naturally stirs a pot of boiling water, occurring inside Jupiter, the Sun, and the other stars. You will discover that nuclear fusion, a potentially limitless, safe source of energy on Earth, occurs throughout the richly varied life histories of stars.

Above all, you will see how discoveries in science are made—using careful observations, increasingly sensitive instruments, and constantly refined scientific models. The result is almost daily discoveries about the universe. And you will participate in the process of discovery by learning to question what you think you know. That, too, is part of *Discovering the Universe.*

Each part now lays the foundations for understanding the facts

The book is divided into four parts: modern astronomy, the solar system, the stars, and the universe. Your study of astronomy will begin with what people have seen with the naked eye for millennia. You will then move into a realm visible only through telescopes as we go from the familiar to the more mysterious.

Part I begins with naked-eye astronomy, the nature of gravity, and its effects on the planets and other bodies. Then we turn to the nature of light and how telescopes work, along with how stars and other astronomical bodies emit light. Part II covers the solar system, starting with the Earth and Moon. We explore each of the planets and their moons, the smaller bodies in the solar system, and finally the Sun. Part III describes the nature and evolution of stars, and it concludes with a discus-

sion of black holes. *Discovering the Universe* has been written so that stars can be studied before planets, as your instructor may prefer. In Part IV we turn to the Milky Way Galaxy, other galaxies, quasars, and the entire universe. The book ends by exploring the age-old question of whether there is other life in the universe.

But presenting facts isn't enough. A book has to guide you through the facts to the ideas behind them. Therefore, each of the four parts of the book begins with a **Foundations** section. These new "mini-chapters" give you an overview of upcoming chapters *and* present the essential physics you will use. Presenting this basic material just when it is needed will help prevent it from disrupting your sense of discovery. Read it with special care: This material must not be missed!

New questions address misconceptions that block real understanding

While teaching thousands of students, it became abundantly clear to me that everyone develops ideas about astronomy that are *wrong*. Errors are inevitable as we each struggle to understand our complex environment. Misconceptions are compounded by all the incorrect impressions fed to us by television and the movies. Raise your hand if you believe that asteroids are crowded together as they were depicted in the movie *The Empire Strikes Back* from the *Star Wars* trilogy.

Memorizing for an exam is one thing; true understanding is something else altogether. Extensive research has shown that it is virtually impossible for anyone to accept new information that conflicts with established beliefs unless we first understand why our old ideas are wrong. I have written this book expressly to help you identify your own misconceptions and ponder alternatives.

Near the beginning of each chapter are numbered questions under the heading **What Do You Think?** Each question addresses a very common misconception. I urge you to honestly consider your own answers before reading the text. When the correct science is presented in the chapter, an icon with the corresponding number appears in the margin. The same questions and their answers are finally listed at the end of each chapter. If you were wrong, I urge you to think about where your ideas came from and how the incorrect information has distorted your understanding of the natural world.

Questioning old beliefs is disconcerting, but it prepares us to deal with the world far more rationally.

New essays use what you have learned to raise further questions

One of the most effective ways to learn is to consider a single situation from different perspectives. In particular, to help you better understand the Earth as an astronomical body, I have written four essays that explore what the world would be like if something familiar to us changed. What if there were no Moon? What if we had infrared-sensitive eyes? How would our view of the universe change? Thinking about alternative scenarios can shed light on the universe we really do have.

One **What If . . . ?** essay follows each of the book's four major parts, pulling together the ideas and techniques of the preceding chapters. The essays grew from fuller discussions in my book *What If the Moon Didn't Exist?* Use them to test your understanding and broaden your view of the universe.

New boxes help you do astronomy by yourself

This edition again includes some essential equations used in astronomy. Additional explanations, and practice with unfamiliar equations, are set off in numbered boxes, each called **An Astronomer's Toolbox.** Each box also has a worked example so those of you uncomfortable with mathematics can follow step by step. The boxes should help you see for yourself how key results are derived, whatever your math background.

Other new boxes, called **Eyes on . . . ,** give "hands-on" information. For example, they will help you buy your first telescope and orient yourself among the unfamiliar names on star charts.

Not every box involves a project that requires a clear sky devoid of city lights: "Eyes on . . . Getting Connected," for example, will walk you through your first on-line session, because the resources you need are no longer all in the library.

Chapters preview, summarize, and review the main ideas

People learn in different ways. Some of you are visual learners, while others rely more heavily on text; Professor Kaufmann recognized that when he introduced large, full-color illustrations to his astronomy text.

Still, *everyone* needs frequent reinforcement of key ideas, and this text incorporates a variety of features to help you study and review:

- Each chapter and foundations section starts with an **overview**, preceding the "What Do You Think?" questions. This introductory paragraph summarizes the main points of the chapter.

- **Full-sentence headings** introduce each topic. Taken together later, they will also provide an outline for reviewing what you have learned. In addition, numbered, full-sentence headings are now grouped beneath a broader main heading of only a word or two.

- When important astronomical terms are introduced, they appear in boldface type. These **Key Terms** also appear at the end of the chapter, followed by a full summary list of **Key Ideas**. Key terms are again defined in the **Glossary**.

- **Review** and **Advanced Questions** at the end of each chapter will stimulate you to explore the ideas in the text. **Discussion Questions** require even more thought, more physics and chemistry, and sometimes even outside reading. Any question relying on mathematics is flagged with an asterisk and answered briefly at the back of the book. The question sets also include new **Observing Projects**, for use in conjunction with the "Eyes on . . ." boxes and the **star charts** located inside the book covers.

Color illustrations show astronomy happening today

As Professor Kaufmann always did, I have made every effort to bring the book up-to-date, adding changes up to the very last minute, much to the chagrin of my editors. For example, dozens of photographs are new, including remarkable successes of the repaired Hubble Space Telescope.

I have included results from the Magellan spacecraft orbiting Venus, the impact on Jupiter of Comet Shoemaker-Levy 9, and the visit of spacecraft Galileo to Jupiter. I review critical new visual evidence for black holes and for emerging planetary systems. I discuss how the gorgeous new images of recent star births clarify our understanding of the life cycles of stars. I have also included recent discoveries of new moons of Saturn, an upcoming comet's passage near the Sun, a galaxy close to our own, galaxies deep in space that may

hold the key to how star systems evolve, and many other new discoveries.

This edition makes it easier to study related topics

Discovering the Universe can be covered in a single semester or quarter course. Teachers of introductory courses are often torn between a desire to cover a broad range of topics and the restrictions of time. The four parts of this edition will make it easier for your instructor to choose among topics when needed. I have tried to ensure that concepts and relationships do not get lost in a staggering amount of detail. To that end, I have made changes to the order of presentation.

First, each planet is explored separately, followed by a discussion of its moons. Professor Kaufmann had himself planned to alter the sequence of the third edition, which was based on comparative planetology. My experience agrees: Unless you encounter the planets one by one, you may easily become confused as to which planet's surface goes with which atmosphere and which interior. You will study each planet from its atmosphere to the properties of its interior—from what we can see to what our theories tell us. You can rely on a consistent organization as you explore the universe.

Second, expanded coverage of the stars and galaxies has resulted in a more balanced presentation. I have also consolidated information about black holes from several chapters of the previous edition into one chapter, and I have revised the chapter on stars so that their evolution unfolds chronologically. What used to be an afterword on the search for extraterrestrial life is now a full chapter.

Third, I have carefully adjusted the level of presentation. I have drawn on analogies that I know from classroom experience can make your job of understanding the unfamiliar that much easier. Finally, the index has been enlarged.

This textbook remains Bill Kaufmann's legacy

It is with great sadness and sense of loss that I note the death of William J. Kaufmann III, author of the first three editions of this book. Bill died unexpectedly in 1994. We in the astronomy community, as well as the thousands he touched through his writings, will miss him. Bill pioneered the short, paperback astronomy text. His revolutionary use of full color added a depth to the course that we take for granted today. His prose gave each edition a clarity and sense of discovery that I sincerely hope to continue. This fourth edition is dedicated to his memory.

It has been an exciting challenge for me, after writing a book for the layperson, to enter the exacting world of textbook writing—let alone to fill Bill Kaufmann's shoes. I believe that the book you are holding is worthy of its heritage. While I wrote it, I wore three hats: one as a very active teacher of introductory astronomy, another as a writer, and a third as a student. For several years I have been relearning how students learn; whatever I wrote, I read with a student's eye to see if it made sense. Please let me know if the book "worked" for you.

Acknowledgments

I am deeply grateful to the astronomers and teachers who reviewed the manuscript of this edition:

William P. Beres, *Wayne State University*
Gerald Cecil, *University of North Carolina*
George V. Coyne, S.J., *University of Arizona*
Stephen C. Danford, *University of North Carolina*
Richard D. Dietz, *University of Northern Colorado*
Dennis Hibbert, *North Seattle Community College*
Terry Jay Jones, *University of Minnesota, Twin Cities*
Thomas M. Jordan, *Ball State University*
Jerry LaSala, *University of Southern Maine*
Michael Lieber, *University of Arkansas*
Anthony Lomazzo, *University of Bridgeport*
Thomas E. Lutz, *late of Washington State University*
James A. Morgan, *University of Maryland*
Charles J. Peterson, *University of Missouri*
Steve Shore, *Indiana University*
Edward M. Sion, *Villanova University*
Mike Skrutskie, *University of Massachusetts, Amherst*
Alex G. Smith, *University of Florida*
Ronald Stoner, *Bowling Green State University*
Richard Ward, *Las Positas College*
William J. F. Wilson, *University of Calgary*
Robert L. Zimmerman, *University of Oregon*

I would like to express my appreciation to Robert L. Biewen, president of W. H. Freeman and Company, who gave me the opportunity and responsibility of revising this landmark text. My sincerest thanks go also to others at Freeman: Holly Hodder, acquisitions editor, who shepherded the book so efficiently; John Haber, development editor, whose ability to refine a raw manuscript is breathtaking; Georgia Lee Hadler, senior project editor, who handled the monumental task of transforming the manuscript into a finished book; and Paul Monsour, whose copyediting put the final polish on the text.

I also thank Blake Logan, who created the attractive design; Travis Amos and Larry Marcus, who obtained many new photographs; Susan Wein, who oversaw the illustration program; Kenny Beck, who did the page layout; Tomo Narashima, who created wonderful new airbrush art; Network

Graphics, which produced all the other drawings; Sheila Anderson who coordinated the production; and Patrick Shriner, who oversaw the supplements available with this book.

Special thanks to my friends Craig Bohren, for his comments about light and optics, and Steven Stahler, for his insights and explanations of early star formation. And most of all I thank my wife, Suzanne, who caught syntax errors the rest of us missed, and my sons, James and Joshua, who had to put up with daddy hard at work on "the book."

Every effort has been made to make this book as error-free as possible. Nevertheless, some inaccuracies have inevitably crept in. I would appreciate hearing from you if you find an error or wish to comment on the text.

Neil F. Comins

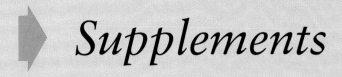 # *Supplements*

This book is even better on CD-ROM!

In 1994 Freeman released the CD-ROM version of William J. Kaufmann's hardcover textbook, *Universe.* It was the first multimedia science textbook. ***Discovering the Universe 4.0,*** the CD-ROM version of this textbook, continues our tradition of innovation. It is the first CD-ROM for a science textbook fully integrated to the Internet. Through Internet browsers, its WebNotes™ feature links the topical coverage of the text to the vast amounts of astronomical resources available on the Internet. A brief tutorial describes how the Internet can serve as an almost limitless resource for students and instructors, as well as for amateur and professional astronomers.

The Internet features are only one part of this truly splendid CD-ROM. It includes all the text and illustrations from the book. A wealth of additional images—including animations and videos—provides added dimensions to the text's topic coverage. Students will benefit from an interactive electronic study guide called Q & A. Each chapter contains a selection of practice multiple-choice questions that provide built-in feedback for each possible answer. A scorekeeper at the bottom of the screen keeps a running record of how the student is doing.

There are numerous other features on this supercharged CD-ROM that will make the study of astronomy more exciting and rewarding.

Other outstanding supplements also accompany the fourth edition:

The **Instructor's Resource Guide** by George A. Carlson of Citrus College contains teaching strategies, suggestions for class discussions and projects, solutions to computational problems and other end-of-chapter questions, additional questions, and a comprehensive list of outside resources.

Both full-color **overhead transparencies** and **slides,** in sets containing a selection of the most important images from the textbook, are available.

A large number of test questions are available in printed, Macintosh, and IBM versions. The **test bank** has been revised and expanded by T. Alan Clark and William J. F. Wilson of the University of Calgary.

Discovering the Universe at a Glance

 FOUNDATIONS

Foundations preview the upcoming chapters and introduce important background. Like each chapter, they begin with a stunning photograph. Their careful overview will motivate students and set the stage for each of the book's four parts.

Modern Astronomy

IN THE PAGES AHEAD

You will learn how astronomers study the uni... they feel justified in believing the theories that ar... explain what they see. You will learn what m... "scientific" and how scientific theories are con... modified, and sometimes rejected. You will di... tronomy today is done as much with pencil, p... puters as with telescopes. You will also learn th... essential for understanding what you see in th... the book focuses on the "everyday" aspect... from naked-eye observations of the sky to... planets and the use of telescopes to wider... vi...

The Hubble Space Telescope Our quest for knowledge about the universe is epitomized by the building of such technological marvels as the Hubble Space Telescope (HST), shown here in 1990 as it was being launched. Since the installation of corrective optics, in 1993, this telescope, now fully functional, is providing new insights into the nature of everything in space.

What Do You Think? questions are based on common misconceptions that students have shared with me over the years. Numbered icons in the text will help students find the answers for themselves.

The Nature of Stars

233

WHAT DO YOU THINK?

1 What color are stars?

2 Are most stars isolated from other stars, like the Sun?

IN THE NINETEENTH CENTURY, astronomers realized that the Sun is just another star, one of several hundred billion in the Milky Way alone. It would be impossible to study them individually. Fortunately, however, ma... stars have identical properties, so by studying one... learn about millions. Astronomers look for meaning... categories in the seemingly endless variety of space. Th... have identified common properties of stars, includi... masses, chemical compositions, and rotation rates. Th... then developed laws to bring order to what they know... laws that will later help us to trace the lives and deat... of distant stars. The story begins with a fact we mig... easily overlook: Stars are not all the same color.

The Temperature of Stars

10-1 A star's color reveals its surface temperature

1 Even with the naked eye, you can see that stars have di... ferent colors. For example, in the constellation of Ori...

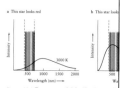

a This star looks red   b This star looks...

Figure 10-1 **Temperature and Color** This diagram sho... the relationship between the color of a star and its surface te... perature. The intensity of light emitted by three hypoth... stars is plotted against wavelength (compare Figure...

Betelgeuse appears red and Rigel appears blue (see Figure 3-30c). Because a star behaves very nearly like a perfect blackbody, its color reveals its surface temperature. As we discussed in Chapter 4, this relationship is described by Wien's law (see Figure 4-2). Accordingly, the intensity of light from a cool star peaks at long wavelengths, and so the star looks red (Figure 10-1a); the intensity of light from a hot star peaks at shorter wavelengths, making it look blue (Figure 10-1c). The maximum intensity of a star of intermediate temperature, such as the Sun, is found near the middle of the visible spectrum (Figure 10-1b).

252

Chapter 11

Figure 11-5 **The Core of the Rosette Nebula** The Rosette Nebula is a large, circular nebula near one end of a sprawling giant molecular cloud in the constellation of Monoceros. Radiation from young, hot stars has blown gas away from the center of this nebula. Some of this gas has become clumped in dark globules that appear silhouetted against the glowing background gases.

greater pressure. When cooled, the balloon shrinks, because the pressure decreases and the walls are more effective in compressing the gas. Similarly, when part of an interstellar cloud cools, its gravitational attraction allows the cloud to shrink. When a region of a cloud is sufficiently cold and dense, the gravitational attraction completely overwhelms the pressure and pulls the gas together

to form a new star. This collapse is called the **Jeans instability** after the British physicist James Jeans, who in 1902 calculated the conditions for it to occur.

Infrared observations show compact regions called **dense cores** inside many interstellar clouds. Their temperatures, around 10 K, are so low that the dense cores are destined to form stars. Often a giant molecular cloud will

---

EYES ON . . . Nebulae

Binoculars and the naked eye are enough to let you "get your hands dirty" exploring star dust. Distant nebulae—clusters of stars and glowing gases—are among the most impressive objects in the night sky.

During the winter the Great Nebula of Orion can be observed even with the naked eye. In all likelihood, you've seen it dozens of times without knowing it. To locate the Great Nebula, find the constellation Orion in the night sky (using, for example, the star charts at the end of the book). Locate Orion's belt. Due south of the belt are three stars in a row making up Orion's sword. Examine the sword very carefully with your naked eye. Do any of the stars in it look at all odd? Now look at them through a pair of binoculars. Which one is different from the others? That one is the Great Nebula of Orion, not a

star at all! How does what you see compare with Figure 11-10?

The North American Nebula and the Pelican Nebula in Cygnus are best spotted in the fall. Pick a dark, moonless night, and use binoculars rather than a telescope. Higher magnification reveals too small a region of the sky for you to see the entirety of these vast, dim nebulae. To find them, first locate the bright star Deneb on the tail of Cygnus (using, for example, the star charts at the end of the book). The North American Nebula is located 3° east of Deneb, while the Pelican Nebula is located 2° southeast of it. These are both small angles, so sweep around the sky east of Deneb. If your binoculars are powerful enough, you should be able to see the outlines that give these nebulae their names.

Eyes on. . . . boxes help students do astronomy for themselves. The excitement and clarity of color illustrations show how astronomy is practiced today.

What If... essays at the end of each part show how changes in the universe could have profound effects on Earth. Additional resources include an introduction to astronomy on the Internet.

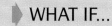

The Earth Were Closer to the Center of the Galaxy?

I<small>T IS NIGHT</small>. The Sun and Moon are down, and stars by the thousands twinkle serenely against the ebony darkness of space. Overhead the soft white of the Milky Way catches your attention and you try to see individual stars in the glowing haze. Most of the 6000 stars visible throughout the year are within 300 ly of the Earth. The rest of the Galaxy's 200 billion or so stars are too dim or too obscured by interstellar gas and dust to be easily observed. What if the solar system were one-third of its present distance from the center of our Galaxy? At that distance we would still be in the realm of the spiral arms, extremely close to the nuclear bulge, and none of the stars we see now would be visible.

Out in the galactic suburbs, where we are today, there is about one star per 300 cubic light-years. The dis-

Earth, would deposit lethal radiation, damage life, and cause mass extinctions. As a result, the direction of evolution would change more frequently.

Today, the closest star, Proxima Centauri, is over 3 ly from Earth. Would there be a danger of colliding with a star if the solar system were closer to the center of the Galaxy? There would certainly be several stars much closer than Proxima Centauri is now. However, stars are so small compared to the vastness of a galaxy that the likelihood of a collision would still be nil. Much more likely would be the passage of a star so close to the Sun that the Earth's orbit would be disturbed. If the Earth's orbit became more elliptical, the change of seasons would be noticeably affected. With the seasonal effect of the Earth's tilt compounded by greater changes in distance between the Earth and Sun, one hemisphere of the Earth would have much more extreme temperatures than it does today, while the other would have less variation. This would affect the evolution of life, of course, and the distribution of life forms on the planet.

Earth-evolved life at our new location would be more likely to encounter sentient beings on other planets. Today, after 75 years of broadcasting radio and television signals, there is a 150-ly-diameter sphere of space centered on the Earth filled with such signals. There are about 6300 stars in that sphere. At our new location there would be 31,500 stars in the same volume and many more stars with life-supporting planets orbiting them. Perhaps one of the reasons that we have been able to develop so long undisturbed is because the solar system exists near the fringe of the Galaxy.

The Milky Way Galaxy · 321

The motion of the gas clouds orbiting near the galactic nucleus can be deduced from the Doppler shifts of infrared spectral lines. In the late 1970s, for example, astronomers discovered that neon emission lines are severely broadened, perhaps from the orbital speed of the gas around the galactic nucleus. Because the spectra of receding and approaching gases are shifted in opposite directions, the spectral lines are smeared, with a range of wavelengths corresponding to a range of line-of-sight velocities. On one side of the galactic nucleus, gas is coming toward us at speeds up to 200 km/s, but on the other side it is rushing away from us at the same speeds.

Something must be holding this high-speed gas in such tight orbits about the galactic nucleus. Using Kepler's third law, astronomers estimate that 10^6 M$_\odot$ is needed to prevent this gas from flying off into interstellar space. The observed broadening of spectral lines suggests that an object with the mass of a million Suns is concentrated at Sagittarius A*. This object must be extremely compact—much smaller than a few light-years across. Many astronomers argue that an object so massive and compact must be a supermassive black hole. As we saw in Chapter 13, extraordinary activity is also occurring in the nuclei of many other galaxies, which indicates the

presence of supermassive black holes at their centers, as well.

However, some astronomers disagree about the presence of a supermassive black hole in the Milky Way. High-resolution infrared views of the galactic center show no indications of the presence of a supermassive black hole. During the coming years, observations from Earth-orbiting satellites as well as from radio telescopes on the ground may elucidate the mysterious core of the Milky Way.

WHAT DID YOU THINK?

1 *Where in the Milky Way is the solar system located?* The solar system is about 28,000 ly from the center of the Galaxy, near the Orion spiral arm.

2 *How fast is the Sun moving in the Milky Way?* The Sun orbits the center of the Milky Way Galaxy at a speed of 828,000 km per hour.

3 *How many stars are in the Milky Way Galaxy?* The Milky Way has about 200 billion stars.

Key Words

dark matter (missing mass)	halo (of a galaxy) nuclear bulge	Sagittarius A spin (of an electron or proton)	spiral arm synchrotron radiation 21-cm radio radiation

---

240 · Chapter 10

both its spectral type and its luminosity class; for example, the Sun is called a G2 V star. This notation supplies a great deal of information about the star, since its spectral type is correlated with its surface temperature. Thus an astronomer knows immediately that a G2 V star is a main sequence star with a luminosity of 1 L$_\odot$ and a surface temperature of 5800 K. Similarly, knowing that Aldebaran is a K5 III star tells an astronomer that it is a red giant with a luminosity of around 500 L$_\odot$ and a surface temperature of about 4000 K (see Figure 10-6).

The first important lesson to learn from the H–R diagram is that there are fundamentally different types of stars. These different kinds of stars represent different stages of stellar evolution. To truly appreciate the H–R diagram we must understand the life cycles of stars: how they are born, what happens as they mature, and how they die. We explore stellar evolution in Chapter 11, but before doing so, we need to know how much mass stars have. You will see that a star's mass is the key to determining its evolutionary history.

Binary Stars and Stellar Mass

10-6 Binary stars provide information about stellar masses

No device on Earth can measure the mass of an isolated star. However, imagine a satellite in orbit about the star.

As we saw in Chapter 2, its orbit must obey Newtonian mechanics. Newton provided the formula we need when he proved Kepler's laws. We can rewrite Kepler's third law as a relation between the masses of a star and its satellite, the satellite's orbital period, and the length of its orbit's semimajor axis:

The sum of the masses, multiplied by the square of the orbital period, gives the cube of the semimajor axis.

By observing the separation between a star and its satellite and how long the satellite takes to complete its orbit, we can calculate the mass of the star.

We cannot send a satellite to orbit a distant star. Fortunately for astronomers, however, many stars have natural companions, and we can use their observed motions just like the orbit of our imagined satellite. About two-thirds of the stars near our solar system are members of star systems in which two or more stars orbit each other; half the objects we see as stars are actually pairs of stars so distant that they appear to us as one. Using telescopes to observe the periods of the orbits and the distances between stars, astronomers can determine stellar masses. To see how this works, look at Toolbox 10-1.

A pair of stars located at nearly the same position in the night sky is called a *double star*, and between 1782 and 1838 William Herschel and his son John catalogued thousands of them. Some double stars are not actually held by each other's gravity; in fact, they are not even

---

AN ASTRONOMER'S TOOLBOX 10-1

Kepler's Third Law and Stellar Masses

The same gravitational force that holds the Earth or a satellite in orbit can also keep a pair of stars in orbit about each other. For *any* such system, the orbits are ellipses, and Kepler's third law becomes

$$M_1 + M_2 = a^3/P^2$$

Here M_1 and M_2 are the two masses (expressed in solar masses), a is the length of the semimajor axis of the ellipse (in astronomical units), and P is the orbital period (in years). Thus, the sum of the masses can be found once the orbital period and semimajor axis are known.

For example, suppose two stars make up a double system, and one star follows an ellipse with a semimajor axis of 4 AU. We find that it takes 2.5 years to complete one orbit. Then the sum of the stars' masses is

$$M_1 + M_2 = (4)^3/(2.5)^2 = 10.2 \text{ M}_\odot$$

The *total* mass of the system is 10.2 M$_\odot$.

As a simple test of what you have just learned, recall from Chapter 2 that we could write Kepler's third law more simply as

$$P^2 = a^3$$

Do you see why? For the Sun and a planet, the combined mass is pretty close to just one solar mass.

Chapters again respond to student misconceptions before listing Key Terms and Key Ideas in full. Abundant end-of-chapter questions now include Observing Projects.

Toolboxes help even poorly prepared students with essential formulas. Full-sentence headings highlight the main ideas.

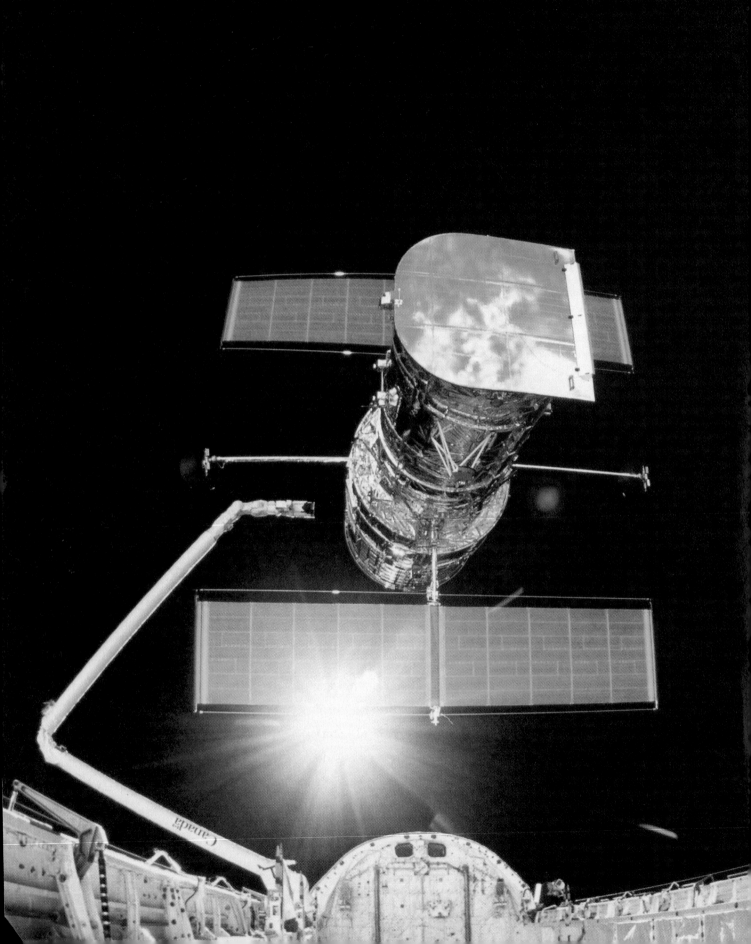

Modern Astronomy

The Hubble Space Telescope Our quest for knowledge about the universe is epitomized by the building of such technological marvels as the Hubble Space Telescope (HST), shown here in 1990 as it was being launched. Since the installation of corrective optics, in 1993, this telescope, now fully functional, is providing new insights into the nature of everything in space.

IN THE PAGES AHEAD

You will learn how astronomers study the universe and why they feel justified in believing the theories that are developed to explain what they see. You will learn what makes a theory "scientific" and how scientific theories are continually tested, modified, and sometimes rejected. You will discover that astronomy today is done as much with pencil, paper, and computers as with telescopes. You will also learn the physical laws essential for understanding what you see in the sky. Part I of the book focuses on the "everyday" aspects of astronomy, from naked-eye observations of the sky to the motion of the planets and the use of telescopes to widen our vistas.

WHAT DO YOU THINK?

1 What is a scientific theory?

2 What do astronomers do?

Look for answers in the numbered paragraphs.

• •

DAYLIGHT HIDES ALL SIGNS of the stars and other wonders shining above the sky's azure curtain. Our thoughts about the universe overhead focus on the Sun, the Moon, and our ever-changing weather. But, ah, the night! No light show, no artist's brush, no poet's words can truly capture the beauty of this breathtaking panorama (Figure I-1).

For thousands of years people have looked up at the sky and found themselves inspired to contemplate the nature of the universe. How was it created? Where did the Earth, Moon, and Sun come from? What are the planets and stars made of? What are our place and role in the cosmic scope of space and time? Astronomers are addressing and answering such questions faster and more accurately now than ever before.

The beauty of the star-filled night sky or the drama of an eclipse alone would make astronomy fascinating. But there are also practical reasons for an interest in the universe. The ancient Greeks knew the connection between the changing height of the noontime Sun and the different patterns of stars in the night throughout the year. This information enabled them to predict the seasons, a useful skill for farming. Many early seafaring cultures were also aware that the positions of the Moon and Sun influence the tides; this knowledge helped these people plan sailing voyages.

What causes these relationships between the Earth and heavenly bodies? Astronomical phenomena were first explained as the result of supernatural forces and divine intervention. The heavens were thought to be populated by demons and heroes, gods and goddesses. Yet, in spite of superstitious beliefs, some people have long realized that the universe is logical and comprehensible. Astronomical cycles, such as the seasons, the reappearance of stars, the tides, and day and night, led many early societies to study the patterns and motions found in the night sky. Improved technology, such as the telescope and the discovery of the fundamental laws of physics, yielded further insights.

Over the past 400 years the scientific method of exploring nature through systematic observations and refined theories has led to an understanding of events and phenomena that our ancestors never could have imagined. In Chapter 1 you will discover the systematic behavior of the stars and other heavenly bodies. In Chapter 2 you will learn about gravitational force and its effects on the planets and moons. In Chapter 3 you will be introduced to the nature of light and to telescopes, astronomers' primary observational tools. In Chapter 4 you will learn more of the physics of light and how it provides astronomers with information about the nature, origin, and evolution of the universe.

Figure I-1 **The Starry Sky** The star-filled sky is a beautiful and inspiring sight. This photograph, taken from northern Mexico, shows Halley's Comet and a portion of the Milky Way (upper right). To get a good view of the heavens, you must be far from any city lights.

Figure I-2 **Galileo's Telescope** When Galileo turned his telescope toward the sky in the early 1600s, he discovered craters on the Moon, spots on the Sun, the phases of Venus, and satellites orbiting Jupiter. These controversial discoveries flew in the face of conventional wisdom and threatened to undermine the teachings of the Church.

I-1 The scientific method demands systematic observations

The **scientific method** uses observations, mathematically based theories, and experimentation to explore physical reality. In essence, the scientific method requires that our ideas about the natural world agree with what we actually observe. Observations lead scientists to create a **scientific theory,** which is an idea or collection of ideas that seems to explain the phenomenon under study. A scientific theory should be consistent with known observations and experiments, because a discrepancy with existing facts implies that the theory is wrong. A scientific theory must also make predictions that can be tested using new observations and experiments. This crucial aspect of the scientific method, **verification,** requires that the theory accurately forecast the results of new observations. Only then do scientists feel confident that they are on firm ground. Thus the scientific method can be summarized in three words: observe, theorize, test.

A theory is considered to be scientific only if it can potentially be disproved. Because the idea that the Sun's gravitational force pulls the planets in elliptical orbits around it can be tested, it is a scientific theory. Because the idea that God created the Earth in six days cannot be tested, it is not a scientific theory.

If a theory is inconsistent with accurate observations, then the theory is either modified, applied only in limited circumstances, or discarded in favor of a more comprehensive explanation of the world around us. For example, Newton's law of gravitation is entirely adequate for describing motion on the Earth around the Sun but is incorrect and useless in the vicinity of a black hole. A scientist must be open-minded, willing to discard even the most cherished ideas if they fail to agree with observation and experiment.

Astronomy is a quantitative science; its discoveries are expressed in terms of numbers and associated units, like 1800 seconds or 8.3×10^{12} kilograms. Throughout this book we use the metric system, the standard for measuring distances in science. To understand astronomy, it is essential that you know this system of measurement. If you are not familiar with scientific notation, the metric system, or units of astronomical length, please study Toolboxes I-1 and I-2.

The power of the scientific method was demonstrated during the late Renaissance, when a few courageous astronomers proposed that the Earth goes around the Sun. Prior to the sixteenth century, most people believed that the Earth was the center of everything. To the untrained mind even today, it appears that all the astronomical objects orbit our planet! But the Earth-centered theory could not fully predict motions of the stars, planets, Moon, and Sun.

In the mid-1500s, the Polish mathematician Nicolaus Copernicus resurrected the idea, unused for nearly fifteen centuries, that the Earth orbits the Sun. He was motivated by an effort to simplify the celestial scheme. This Sun-centered view of the known universe was strengthened by the observations of the Italian astronomer Galileo Galilei, the first person to point a telescope toward the sky (Figure I-2). Among other things, Galileo saw the moons of Jupiter orbit that planet. This discovery flew in the face of the widely held belief that every object in the sky orbits the Earth. It helped fuel the search for the correct relationship between the Earth and the rest of the cosmos, which is a major theme of the first three chapters of this book.

I-2 Astronomers do much more than just make observations

The process of astronomical discovery is much more than looking through a telescope and noting what you see. The research activities of astronomers can roughly be divided into three categories: observing and analyzing observations, theorizing, and computer modeling. Most people think that an astronomer spends his or her time observing the sky, spending long nights directing the most powerful eyes on Earth to reveal the secrets of space. In fact, most astronomers consider themselves very lucky to get even a week's worth of observing time at research observatories each year. Planning their observations and analyzing their data takes up the majority of their time.

Indeed, many astronomers never use telescopes at all. Rather, they create or expand theories to explain what has been seen, or they test theories by using computers to make models that can verify or disprove predictions. These people are more accurately called *astrophysicists*. Astronomical discovery is always exciting and sometimes totally unexpected. Some observations lead astronomers in new directions as they try to reconcile apparent contradictions. For example, using the Hubble Space Telescope, some astronomers recently calculated an age for the universe that is actually younger than the apparent ages for some of the stars in it! Since the stars formed after the universe came into existence, something is obviously amiss.

An underlying theme of this book is that the universe is comprehensible. Physical reality is not a hodgepodge of unrelated things behaving in arbitrary ways. Rather, fundamental **laws of physics** govern the nature and behavior of everything. These laws enable us to explore realms far removed from our Earthly experience. For example, from experiments in a laboratory, scientists can determine the properties of light and the behavior of atoms. They then use this knowledge to discover the life cycles of stars and the structure of the universe.

In turn, the theories developed to explain the universe have led to a much deeper and richer understanding of the Earth. Many people think that astronomy

AN ASTRONOMER'S TOOLBOX I-1

Powers-of-Ten Notation

Astronomy is a subject of extremes. As we examine various environments, we find an astonishing range of conditions, from the incredibly hot, dense centers of stars to the frigid, near-perfect vacuum of interstellar space. To describe such divergent conditions accurately, we need a wide range of both large and small numbers. Astronomers avoid such confusing terms as "a million billion billion" (1,000,000,000,000,000,000,000,000) by using a standard shorthand system: All the cumbersome zeros that accompany a large number are consolidated into one term consisting of 10 followed by an *exponent*, which is written as a superscript and called the **power of ten**. The exponent merely indicates how many zeros you would need to write out the long form of the number. Thus,

$$10^0 = 1$$
$$10^1 = 10$$
$$10^2 = 100$$
$$10^3 = 1000$$
$$10^4 = 10,000$$

and so forth. The exponent tells you how many tens must be multiplied together to give the desired number. For example, ten thousand can be written as 10^4 ("ten to the fourth") because $10^4 = 10 \times 10 \times 10 \times 10 = 10,000$.

With this notation, numbers are written as a figure between 1 and 10 multiplied by the appropriate power of 10. The distance between the Earth and the Sun, for example, can be written as 1.5×10^8 km. Once you get used to it, you will find this notation more convenient than writing "150,000,000 kilometers" or "one hundred and fifty million kilometers."

This shorthand system can also be applied to numbers that are less than 1 by using a minus sign in front of the exponent. A negative exponent tells you that the location of the decimal point is as follows:

$$10^0 = 1$$
$$10^{-1} = 0.1$$
$$10^{-2} = 0.01$$

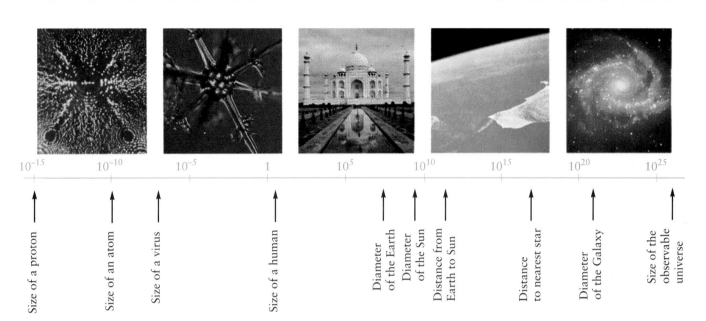

| 10^{-15} | 10^{-10} | 10^{-5} | 1 | 10^{5} | 10^{10} | 10^{15} | 10^{20} | 10^{25} |

Size of a proton

Size of an atom

Size of a virus

Size of a human

Diameter of the Earth

Diameter of the Sun

Distance from Earth to Sun

Distance to nearest star

Diameter of the Galaxy

Size of the observable universe

Examples of Powers-of-Ten Notation The scale gives the sizes of objects in meters, ranging from subatomic particles at the left to the entire observable universe on the right. The photograph at the left shows tungsten atoms, 10^{-10} m in diameter. Second from left is a crystalline skeleton, 10^{-4} m (0.1 mm) in size, of a diatom, a single-celled organism. At the center is the Taj Mahal, which is of a magnitude within reach of our unaided senses. On the right, looking across the Indian Ocean toward the South Pole, we see the curvature of the Earth, 10^{7} m in diameter. At the far right is a galaxy, 10^{21} m in diameter (it would take light, which moves at 3×10^{5} km/s, 100,000 years to traverse that distance).

$$10^{-3} = 0.001$$
$$10^{-4} = 0.0001$$

and so forth. For example, the diameter of a hydrogen atom is 1.1×10^{-8} cm. That is more convenient than saying "0.000000011 centimeter" or "11 billionths of a centimeter."

Using the powers-of-ten notation, one can write large or small numbers like these compactly:

$$3,416,000 = 3.416 \times 10^{6}$$
$$0.000000807 = 8.07 \times 10^{-7}$$

Because powers-of-ten notation bypasses all the awkward zeros, a wide range of circumstances can be numerically described conveniently:

$$one\ thousand = 10^{3}$$
$$one\ million = 10^{6}$$
$$one\ billion = 10^{9}$$
$$one\ trillion = 10^{12}$$

and also

$$one\ thousandth = 10^{-3} = 0.001$$
$$one\ millionth = 10^{-6} = 0.000001$$
$$one\ billionth = 10^{-9} = 0.000000001$$
$$one\ trillionth = 10^{-12} = 0.000000000001$$

The illustration above shows how clearly the powers-of-ten notation expresses the scale of objects, ranging from subatomic particles like the proton to the size of the observable universe.

Astronomical Distances

As we turn toward the stars in Part III, we shall find that some of our traditional units of measure become cumbersome. It is fine to use kilometers to measure the diameters of craters on the Moon or the heights of volcanoes on Mars. But it is as awkward to use kilometers to express distances to planets, stars, or galaxies as it would be to talk about the distance from New York City to San Francisco in millimeters. Astronomers have therefore devised new units of measure.

When discussing distances across the solar system, astronomers use a unit of length called the **astronomical unit** (abbreviated AU), which is the average distance between the Earth and the Sun:

$$1 \text{ AU} = 1.50 \times 10^8 \text{ km} = 93 \text{ million miles}$$

Thus, the distance between the Sun and Jupiter can be conveniently stated as 5.2 AU.

When talking about distances to the stars, astronomers choose between two different units of length. One is the **light-year** (abbreviated ly), which is the distance that light travels in one year:

$$1 \text{ ly} = 9.46 \times 10^{12} \text{ km} \approx 63,000 \text{ AU}$$

One light-year is roughly equal to six trillion miles. Proxima Centauri, the nearest star other than the Sun, is 4.2 ly from Earth.

The second commonly used unit of length is the **parsec** (abbreviated pc). Imagine taking a journey far into space, beyond the orbits of the outer planets. As you look back toward the solar system, the angle between the Sun and the Earth becomes smaller and smaller the farther you travel from the Sun. The distance at which 1 AU makes an angle of $\frac{1}{3600}°$ (called 1 arc second) is defined as 1 parsec, as shown in the drawing. The parsec turns out to be longer than the light-year. Specifically,

$$1 \text{ pc} = 3.09 \times 10^{13} \text{ km} = 3.26 \text{ ly}$$

Thus, the distance to the nearest star can be stated as 1.3 pc as well as 4.2 ly. Whether one uses light-years or parsecs is a matter of personal taste.

For even greater distances, astronomers commonly use *kiloparsecs* and *megaparsecs* (abbreviated kpc and Mpc), in which the prefixes simply mean "thousand" and "million," respectively:

$$1 \text{ kpc} = 10^3 \text{ pc}$$
$$1 \text{ Mpc} = 10^6 \text{ pc}$$

For example, the distance from Earth to the center of our Milky Way Galaxy is about 8.6 kpc, and the rich cluster of galaxies in the direction of the constellation Virgo is 20 Mpc away.

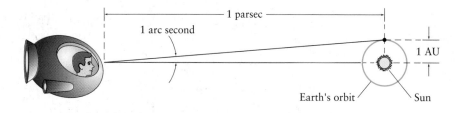

A Parsec The parsec, a unit of length commonly used by astronomers, is equal to 3.26 ly. The parsec is defined as the distance at which 1 AU perpendicular to the observer's line of sight subtends an angle of 1 arc second.

deals only with faraway places and is of no significance to everyday life. This opinion is quite wrong. For example, the seventeenth-century scientist Isaac Newton succeeded in explaining why the planets orbit the Sun. His law of gravitation explains the motions of the planets and also why everything around us is held on the Earth's surface. It explains the time it takes an egg falling off a table to hit the floor and the "hang time" for basketball players under the net. By understanding the law of gravitation, engineers can control the friction between your car's tires and the road or design an airplane wing to lift a jumbo jet. As another example of the utility of astronomical research, helium and other gases were first discovered on the Sun, not on the Earth.

I-3 Telescopes open the door to the universe

How do astronomers gather information? At first, astronomers peered directly through telescopes to observe and analyze visible starlight. By the end of the nineteenth century, however, scientists had begun discovering nonvisible forms of radiation. Along with visible light, these radio waves, infrared rays, ultraviolet rays, x rays, and gamma rays are called **electromagnetic radiation**. During the past half century, astronomers have constructed telescopes that can detect all of these nonvisible forms of electromagnetic radiation emitted by objects in space (Figure I-3). Remarkably, every type of electromagnetic radiation has revealed new information about space.

Whether located on Earth or in orbit above the obscuring effects of the atmosphere, telescopes give us views vastly superior to anything our eyes can see. They are crucial to our understanding of familiar objects like the Sun, and they also give important clues about such exotic objects as neutron stars, pulsars, quasars, and black holes.

Until the 1970s, all our information about the universe came from visible light and other electromagnetic radiations. Today, however, astronomers are also able to detect exotic particles in space such as neutrinos, cosmic rays, and other forms of energy such as gravitational radiation (actual fluctuations of space). Using such nonelectromagnetic radiation, we may soon be able to see directly into the hearts of stars.

What, then, have astronomers seen of the universe? Figure I-4 shows the features we shall explore in this text, including planets, stars, black holes, galaxies, and quasars. Each of these objects is constantly changing; each had an origin and each will have an end. These things, too, we will study.

The dreams of Jules Verne and H. G. Wells pale when compared with current reality. Ours is an age of

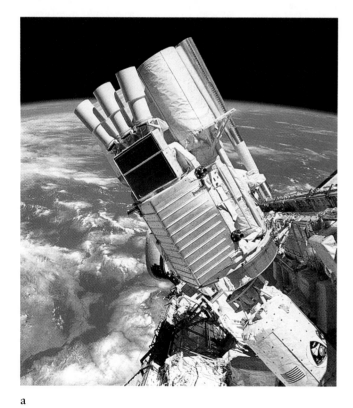

a

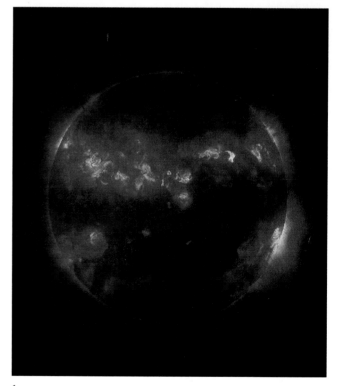

b

Figure I-3 **Telescopes in Space** (a) This battery of telescopes was carried aloft in 1994 by the Space Shuttle. Telescopes such as these give astronomers views of the universe quite different from those obtained by ground-based observatories. (b) An x-ray image of the Sun shows a profoundly different body than that seen in visible light. Such nonvisual images provide astronomers with insights that are essential to a complete understanding of how the universe operates.

Saturn and some of its moons

Our star—the Sun

Veil Nebula—remnant of
an exploded star

Orion Nebula—birthplace
of stars

Galaxy M83—home to
200 billion stars

Quasar 2355+490

Figure I-4 **Inventory of the Universe** This sequence of images shows many of the major types of objects that have been found throughout the universe. Each type will be discussed in the following chapters.

exploration and discovery more profound than any since Columbus and Magellan set sail. We have walked on the Moon and dug into Martian soil. We have discovered active volcanoes and barren ice fields on the satellites of Jupiter. We have visited the shimmering rings of Saturn. Never before has so much been revealed in so short a time. As you proceed through this book, you will come to realize the awesome power of the human mind to reach out, to explore, to observe, and to comprehend. One of the great lessons of modern astronomy is that we can transcend the limitations of our bodies and the brevity of human life.

WHAT DID YOU THINK?

1 *What is a scientific theory?* A theory is an idea or set of ideas explaining something about the natural world. A theory is scientific if it can be objectively tested and potentially disproved.

2 *What do astronomers do?* Most astronomers spend their time analyzing data, building equipment, creating theories, and carrying out computer simulations. A few days each year might be spent looking through telescopes.

Key Words

astronomical unit (AU)	laws of physics	power of ten	scientific theory
electromagnetic	light-year (ly)	scientific method	verification
radiation	parsec (pc)		

EYES ON . . .   Our Sun-Centered Universe

At the end of the twentieth century most of us find it hard to understand why anyone would believe that the Sun, Moon, planets, and stars orbit the Earth. After all, we *know* that the Earth spins on its axis; we *know* that the gravitational force from the Sun holds the planets in orbit, just like the Earth's gravitational force holds the Moon in orbit; and we *know* that the stars in the night sky are all so far away that they are not part of our solar system. These facts have become part of our understanding of the motions of the heavenly bodies. Psychologists call this background information that we use to help explain things a conceptual framework.

A conceptual framework contains all the information we take for granted when viewing and thinking about anything. For example, today when we see the Sun rise, move across the sky, and set, we take for granted that it is the Earth's rotation causing the Sun's apparent motion.

Our ancestors didn't know that the Earth rotates; they didn't know that the (then-mysterious) force that held them to the ground is the same force that attracts the Earth to the Sun and the Moon to the Earth; they didn't know that the Sun is a star, just like the fixed points of light in the sky; and they didn't know any of the other laws of physics related to motion that we take for granted. Their conceptual framework for understanding the motions of the heavenly bodies was strictly based on their senses. That is, they observed motions and drew "obvious" conclusions.

They did not feel the Earth move under their feet, nor did they see any other indication that the Earth is in motion. Thus our forebears sensed nothing to support the belief that the Earth moves. (We now know that one indication is the wind patterns, but they didn't know the cause of the winds either.) The obvious conclusion, if one has a prescientific conceptual framework, is that the Earth does not move.

Today we incorporate the known and tested laws of physics in our understanding of the natural world. Many of these concepts are utterly nonintuitive, and therefore the conceptual frameworks we have are less consistent with common sense than those held in the past. Studying science helps us develop intuition that is consistent with the workings of nature.

Key Ideas

• The universe is comprehensible.

• The scientific method is a procedure for formulating theories. These theories are tested by observation or experimentation in order to build consistent models that accurately describe phenomena in the universe.

• Observations of the heavens have led to discovery of some of the fundamental laws of nature.

Review Questions

The answers to all computational problems, which are preceded by an asterisk, are at the end of the book.

1   Give five examples of nonintuitive ideas about astronomy, like "the Earth rotates," that we now accept as correct.

∗ 2   Convert the following distances into scientific notation:
   (a) Earth's mass: 5,974,000,000,000,000,000,000,000 kg
   (b) Sun's mass:
        1,989,999,000,000,000,000,000,000,000,000 kg
   (c) Sun's radius: 695,990 km
   (d) One year: 31,558,000 s

1 ▶ *Discovering the Night Sky*

IN THIS CHAPTER

You will discover how astronomers divide up the night sky to help them locate objects in it. You will learn how the Earth's spin on its axis causes day and night, how the tilt of the Earth's axis of rotation and the Earth's motion around the Sun combine to create the seasons, and how the Moon's orbit around the Earth creates the phases of the Moon and lunar and solar eclipses. You will also learn how the year is defined and how the calendar was developed.

Circumpolar Star Trails This long exposure is aimed at the north celestial pole and shows the apparent rotation of the sky. The foreground building houses the Canada-France-Hawaii telescope, one of the largest telescopes in the northern hemisphere. Notice the variety of colors in the star streaks. The lines of light on the ground were caused by cars.

WHAT DO YOU THINK?

1 How bright is the North Star, Polaris, compared to other stars?

2 Are constellations just mythic figures in the sky?

3 What causes the seasons?

4 How many zodiac constellations are there?

5 When, if ever, is the Moon visible during the daytime?

....

SOMETIMES WHEN YOU GAZE at the sky on a clear, dark night, there seem to be millions of stars twinkling overhead. In reality, the unaided human eye can detect only about six thousand stars over the entire sky. At any one time, you can see roughly three thousand stars because

only half of the sky is above the horizon. You probably have noticed patterns formed by bright stars and may even be familiar with some common names for these patterns, such as the ladle-shaped Big Dipper and broad-shouldered Orion. Patterns of dimmer stars are harder to identify and often require a vivid imagination to see at all. Recognizable patterns of stars are called **constellations** (Figure 1-1*a*). Many constellations, such as Orion, have names from ancient legends.

Patterns of Stars

1-1 Constellations make locating stars easy

Constellations (from the Latin word meaning "group of stars") can help you orient yourself on Earth. For instance, if you live in the northern hemisphere, you can use the Big Dipper to find the direction north. To do this, first locate the two stars of the bowl farthest from the

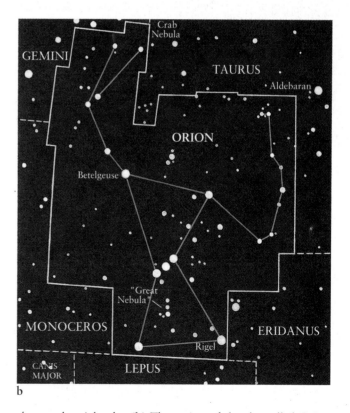

a

b

Figure 1-1 **The Constellation Orion** (a) The pattern of stars called Orion is a prominent winter constellation. From the United States, it is easily seen high above the southern horizon from December through March. You can see in this picture that stars have different colors, something to watch for when you observe the night sky. (b) The region of the sky called Orion and other nearby constellations are depicted on this star map. *All* the stars inside the boundary of Orion are members of that constellation. The celestial sphere is covered by 88 constellations of differing sizes and shapes.

Figure 1-2 **The Big Dipper as a Guide** The Big Dipper is an easily recognized constellation composed of seven bright stars. This star chart shows how the Big Dipper can be used to point out the North Star as well as the brightest stars in three other constellations. Note that the Big Dipper appears upside down in this drawing, but at other times of the night it appears right side up. (Why?) The angular distance from Polaris to Spica is 101°, so a large portion of the sky is shown in this star chart.

Big Dipper's handle, and draw a mental line through these stars leading above the bowl, as shown in Figure 1-2. The first moderately bright star you then encounter is Polaris, also called the North Star because it is located almost directly over the Earth's north pole. Whenever you stand facing Polaris, you are facing north. East is then on your right, south is behind you, and west is on your left.

Once you find your bearings, the well-known constellations make it easy to locate other stars. For example, imagine gripping the handle of the Big Dipper and slamming the pot straight downward. The first group of bright stars you hit is the constellation Leo (the Lion). As Figure 1-2 shows, the brightest star in this group is Regulus, in the "sickle" that traces the lion's mane. As another example, follow the arc of the handle of the Big Dipper away from the dipper's pot. The first bright star you encounter along that arc beyond the handle is Arcturus in Boötes (the Shepherd). Follow the same arc further to the prominent bluish star Spica in Virgo (the Virgin). Spotting these stars is easy if you remember the saying "Arc to Arcturus and speed on to Spica."

During the winter months in the northern hemisphere, you can see some of the brightest stars in the sky.

Many of them are in the vicinity of the "winter triangle," which connects bright stars in the constellations of Orion (the Hunter), Canis Major (the Large Hunting Dog), and Canis Minor (the Small Hunting Dog), as shown in Figure 1-3. The winter triangle is nearly overhead during the middle of winter at midnight. It is easy to find Sirius, the brightest star in the night sky, by locating the belt of Orion and following a line from it to the left (as you face Orion). The first bright star you encounter is Sirius.

The "summer triangle" graces the summer sky. As shown in Figure 1-4, this triangle connects the bright stars Vega in Lyra (the Harp), Deneb in Cygnus (the Swan), and Altair in Aquila (the Eagle). A conspicuous portion of the Milky Way forms a beautiful background for these constellations, which are nearly overhead during the middle of summer at midnight.

Because astronomers require more accuracy in locating dim objects than is possible by just moving from constellation to constellation, they have created a celestial map and applied a coordinate system to it. If a star's coordinates are known, it can be quickly located. For such a map to be useful in finding stars, the stars must be fixed on it like cities are fixed on maps of the Earth.

Figure 1-3 **The Winter Triangle** This star chart shows the eastern sky as it appears during the evening in December. Three of the brightest stars in the sky make up the winter triangle. In addition to the constellations involved in the triangle, Gemini (the Twins), Auriga (the Charioteer), and Taurus (the Bull) are also shown.

Figure 1-4 **The Summer Triangle** This star chart shows the northeastern sky as it appears in the evening in June. In addition to the three constellations involved in the summer triangle, the faint constellations of Sagitta (the Arrow) and Delphinus (the Dolphin) are also shown.

....

EYES ON . . .   The Constellations

Many students take an introductory astronomy course expecting to be taught the constellations, something most teachers just don't have time to cover. You can, however, learn them on your own using two helpful techniques for memorizing the night sky. First, go from easily identified constellations to lesser-known ones. Second, make the connections between different constellations as bizarre as possible. Chances are that you will remember "slam the Big Dipper's bowl downward to hit Leo, the Lion, on the head" much longer than you will remember "the first bright group of stars directly below the Big Dipper is Leo."

An extremely efficient way to learn the constellations is to use the board and card game *Stellar 28*. In addition, astronomy computer programs are available, including *Redshift*, *The Sky*, *Dance of the Planets*, *Deep Space*, *PC-Sky*, *SkyGlobe*, and *Voyager II*.

Another good way to familiarize yourself with the night sky is to use star charts to see which constellations are up each night. You will find a set of star charts from the *Griffith Observer* magazine at the front and back of this book. To use the charts, first select the one that best corresponds to the date and time of your observation. Take the chart outside at night and compare it directly with the sky. Hold the chart vertically and turn it so that the direction you are facing shows at the bottom. Using a

flashlight with a red plastic coating over the light will make it easier to read the chart without constricting your vision. You will find stargazing a surprisingly enjoyable experience.

1-2 The celestial sphere aids in navigating the sky

From our perspective, the stars appear fixed relative to each other and seem to orbit the Earth. We model this Earth-based view of the heavens by pretending that the stars are attached to the inside of an enormous hollow shell, called the **celestial sphere,** centered on the Earth (Figure 1-5). The concept of the celestial sphere harkens back to the days when the Earth was believed to be the center of the universe. The image of the celestial sphere is nevertheless an extremely useful surface on which to make star maps.

The constellations divide the celestial sphere into 88 unequal regions, some (like Ursa Major) very large and

others relatively small. It is often convenient to speak of all the stars in such a region as members of that constellation. When we speak of a constellation, then, we often mean not only a particular pattern of stars but also an entire region of the sky and all the objects in it (see Figure 1-1*b*). Thus, we might speak of "the star Albireo in the constellation Cygnus" much as we would refer to "the Ural Mountains in Russia." (Astronomers sometimes call the traditional star patterns *asterisms* rather than constellations when there is a danger of confusion.)

The stars seem fixed on the celestial sphere only because of their remoteness. In reality, they are all at varying distances from the Earth, and they do move relative to each other. The reason we don't see such motion is because of how far all the stars are from here. You can understand this by imagining a jet plane traveling at one thousand kilometers an hour roaring just overhead. Its motion is unmistakable. But the same plane, moving at the same speed, barely seems to budge when located along the distant horizon. The stars are all over a trillion kilometers away from us. Therefore, although the patterns of stars in the sky do change, their great distances prevent us from seeing changes over a human lifetime.

As shown in Figure 1-5, we can project key geographic features from Earth out into space to establish directions and bearings. If we expand the Earth's equator onto the celestial sphere, we obtain the **celestial equator.** The celestial equator divides the sky into northern and southern hemispheres, just as the Earth's equator divides the Earth into two hemispheres. We can also imagine extending the Earth's north and south poles out into space along the Earth's axis of rotation. Doing so gives us the **north celestial pole** and the **south celestial pole,** also shown in Figure 1-5. With the celestial equator and poles as reference features, astronomers denote the position of an object in the sky in much the same way that latitude and longitude are used to specify a location on Earth.

Just as we need two coordinates to find any location on Earth (i.e., latitude and longitude), so, too, are two coordinates needed to locate any object on the celestial sphere. The equivalent to latitude on Earth is **declination** on the celestial sphere. It is measured north or south of the celestial equator. The equivalent of longitude on Earth is **right ascension** on the celestial sphere, measured around the celestial equator. We will see later in this chapter that the Sun moves in a closed line around the celestial sphere during the course of a year. The celestial equator and the Sun's path intersect at two points.

The equivalent on the celestial sphere of the Earth's prime meridian (from which longitude angles are measured on Earth) is where the Sun crosses the celestial equator moving northward. Angles of right ascension are measured from this point, called the vernal equinox.

In navigating on the celestial sphere, astronomers measure the distance between objects in terms of angles. If you are not familiar with measuring the distance between objects using this method, please read Toolbox 1-1.

1-3 Earth's rotation causes the stars to appear to move

The Earth spins on its axis. Such motion is called **rotation.** The Earth's 24-hour rotation causes the constellations—as well as the Sun, Moon, and planets—to appear to rise in the east, move across the sky, and set in the west. This **diurnal,** or daily, **motion of the stars** is apparent in time exposure photographs such as the picture on the first page of this chapter. *Rotation causes day and night* because it makes the Sun follow a diurnal path across the sky.

People who spend time outdoors at night are familiar with the diurnal motion of the stars. If you lack this knowledge, take a friend and go outside on a warm night to observe this astronomical phenomenon yourself. Soon after dark, find a spot away from bright lights and note the constellations in the sky relative to some prominent landmarks near you on Earth. A few hours later, check again from the same place. You will find that the entire pattern of stars (as well as the Moon, if it is visible) has shifted. New constellations will have risen above the eastern horizon, while other constellations will have disappeared below the western horizon. If you check again just before dawn, you will find, low in the western sky, the stars that were just rising in the east when the night began.

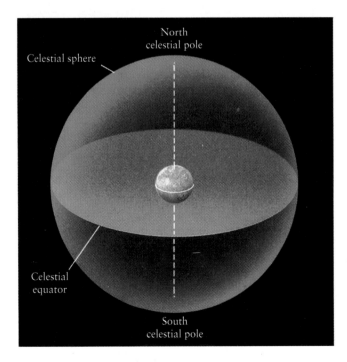

Figure 1-5 **The Celestial Sphere** The celestial sphere is the apparent sphere of the sky. The celestial equator and poles are obtained by projecting the Earth's equator and axis of rotation out into space. The north celestial pole is therefore located directly over the Earth's north pole, while the south celestial pole is directly above the Earth's south pole.

Observational Measurements Using Angles

Astronomers have inherited many useful concepts from antiquity. For example, ancient mathematicians invented angles and a system of angular measure that is still used to denote the positions and apparent sizes of objects in the sky. Angular measure was, and still is, an essential tool for astronomers. For example, to locate stars, we don't need to know their distances from Earth (which are all different and few of which are well-known). All we need to know is the angle from one star to another in the sky, a property that remains fixed over our lifetimes.

An **arc angle,** often just called an **angle,** is the opening between two lines that meet at a point. Angular measure is a method of describing the size of an angle. The basic unit of angular measure is the **degree,** designated by the symbol °. A full circle is divided into 360°. A right angle measures 90°. As shown in the figure on the right, the angle between the two "pointer stars" in the Big Dipper is about 5°.

Astronomers also use angular measure to describe the apparent sizes of celestial objects. For example, imagine looking up at the full Moon. The angle covered by the Moon's diameter is nearly $\frac{1}{2}°$. We therefore say that the **angular diameter,** or **angular size,** of the Moon is $\frac{1}{2}°$. Al-

The Big Dipper The angular distance between the two "pointer stars" at the front of the Big Dipper is about 5°. For comparison, the angular diameter of the Moon is about $\frac{1}{2}°$.

ternatively, astronomers say that the Moon "subtends" an angle of $\frac{1}{2}°$. In this context, *subtend* means "to extend across."

The adult human hand held at arm's length provides a means of estimating angles. For example, your fist covers an angle of 10°, whereas the tip of your finger is

---

Further, the constellations that you see at night gradually change over the course of a year. This shift occurs as the Earth orbits, or *revolves,* around the Sun. **Revolution** is the motion of any astronomical object around another astronomical object. The Earth takes one year, or about $365\frac{1}{4}$ days, to go once around the Sun. As a result of this motion, the darkened, nighttime side of the Earth is gradually turned toward different parts of the heavens. Another result of the Earth's motion around the Sun is that every star rises approximately 4 minutes earlier each night than it did the night before.

We spoke earlier of stars rising in the east and setting in the west. Depending on your latitude, some of the stars and constellations never go below the horizon. Instead, they trace complete circles in the sky. To under-

stand why this happens, imagine that you are standing on the Earth's north pole at night. Looking straight up, you would see Polaris. Because the Earth would be spinning around its axis directly under your feet, all the stars would appear to be moving from left to right in horizontal rings around you except Polaris, which would always remain directly overhead. As seen from the north pole, no stars rise or set (Figure 1-6). They just seem to go around Polaris in horizontal circles. Stars that never go below the horizon are called **circumpolar.** All stars are circumpolar as seen from the north (or south) pole.

Travel now halfway to the equator, where the motion of the sky is very different. For example, in Orono, Maine (latitude 44°45′ N), where this book was written, most of the stars rise at an angle of nearly 45° to the east-

Estimating Angles with the Human Hand Various parts of the adult human hand extended to arm's length can be used to estimate angular distances and sizes in the sky.

about 1° wide. Various segments of your index finger extended to arm's length can be similarly used to estimate angles a few degrees across.

To talk about smaller angles, we subdivide the degree into 60 arc minutes (abbreviated "60 arcmin" or "60'"). An arc minute is further subdivided into 60 arc seconds (abbreviated "60 arcsec" or "60''"). A dime viewed from a distance of one mile has an angular diameter of about 2 arcsec.

From everyday experience, we know that an object looks big when it is nearby but small when it is far away. *The angular size of an object therefore does not necessarily tell you anything about its actual physical size.* For instance, the fact that the Moon's angular diameter is $\frac{1}{2}°$ does not tell you how big the Moon really is. But if you also happen to know the distance to the Moon, then you can calculate the Moon's physical diameter.

ern horizon and set at an angle of 45° to the western horizon. Polaris is 45° up in the sky, not directly overhead as it is at the Pole, but still it never sets. At this latitude only the stars and constellations between Polaris and the land directly below it are circumpolar. Heading farther south, Figure 1-7 shows stars setting at 32°N latitude.

Go now to the equator. There all stars appear to rise straight up in the eastern sky (Figure 1-8) and set straight

◄ Figure 1-6 **Motion of Stars at the North Pole** Because the Earth rotates around the north pole, stars seen from there appear to move in huge, horizontal circles. This is the same effect you would get by standing up in a room and spinning around; everything would appear to move in circles around you.

Figure 1-7 **Rising and Setting of Stars at Middle North Latitudes** Unlike the motion of the stars at the poles (see Figure 1-6), the stars at all other latitudes do change angle above the ground throughout the night. This time-lapse photograph shows stars setting over the Kitt Peak National Observatory located at 32° N. The latitude determines the angle at which the stars rise and set.

Figure 1-8 **Rising of Stars at the Equator** Standing on the equator, you would be perpendicular to the axis around which the Earth rotates. As seen from there, the stars rise straight up in the east and set straight down in the west. This is the same effect you get when driving straight over the crest of a hill; the objects on the other side of the hill appear to move straight upward as you descend.

down in the western sky. Polaris is barely visible on the northern horizon, and none of the constellations are circumpolar. Generally, the farther north you go in the northern hemisphere, the greater the number of constellations that are circumpolar.

Earthly Cycles

The everyday rhythms of the Earth arise from three celestial motions: the Earth's rotation, which causes day and night; the Earth's revolution around the Sun, which creates the year; and the Moon's revolution around the Earth, which creates the lunar phases and causes eclipses.

Contrary to common sense, the seasons are not caused by the change in the Earth's distance from the Sun. The Earth is closest to the Sun on January 3 of each year—the dead of winter in the northern hemisphere!

1-4 The rotation of the Earth determines the length of the day

The Sun's motion provided our earliest reference for time, since the Sun's location determines whether we are awake or asleep and whether it is time for breakfast or dinner. In other words, the Sun determines the length of the **solar day,** our day. However, the Sun is not a perfect timekeeper because the Earth's orbit around the Sun is not circular, as we shall see in Chapter 3. Because the Earth moves more rapidly along its orbit when it is near the Sun than when it is farther away, the speed of the Sun across the sky also varies throughout the year. Using the average time interval between consecutive noontimes corrects this problem, and this average is the basis of our 24-hour day.

Astronomically, noon is defined as the moment when the Sun is highest in the sky. However, at different longitudes the Sun is highest at different times. Thus, astronomical noon in New York City occurs slightly earlier than it does in Philadelphia. Fortunately, the concept of time zones removes this problem. In a **time zone,** everyone agrees to set their clocks alike. Time zones, originally developed for scheduling rail transportation, are based on the time in Greenwich, England (Figure 1-9). With some variation due to geopolitical boundaries, every 15° of longitude begins a new time zone. The time zones for North America are shown in Figure 1-10. Going from one time zone to the next requires you to change the time on your wristwatch by exactly one hour.

Figure 1-9 **The Time Marker at Greenwich, England** Time and longitude are measured with reference to an arbitrary spot on the ground at Greenwich, England. Time there is called Greenwich Mean Time (abbreviated "GMT").

things out, Pope Gregory XIII reformed the calendar in 1582. He began by dropping ten days (October 5, 1582, was proclaimed to be October 15, 1582), which brought the first day of spring back to March 21. Next he modified Caesar's system of leap years.

Caesar had added February 29 to every calendar year that is evenly divisible by four. Thus, for example, 1984, 1988, and 1992 were all leap years with 366 days. But this system produces an error of about three days every four centuries. To solve the problem, Pope Gregory decreed that only the century years evenly divisible by 400 should be leap years. For example, the years 1700, 1800, and 1900 (which would have been leap years under Caesar's system) were not leap years under the improved Gregorian system. But the year 2000—which can be divided evenly by 400—is a leap year.

We use the Gregorian system today. It assumes that the year is 365.2425 mean solar days long, which is very close to the length of the *tropical year,* defined as the time interval from one vernal equinox to the next. In fact, the error is only one day in every 3300 years. That won't cause any problems for a long time. The major impact of our annual journey around the Sun is the unfolding of the seasons. While their influence on our lives is obvious enough, their cause is a fairly subtle combination of effects related to the Earth's orbit.

1-5 The position of the Sun among the stars determines the year

Just as the day originates in the Earth's rotation, the *year* is the unit of time based on the Earth's revolution about the Sun. Unfortunately, nature has not arranged things for our convenience. The year does not divide into exactly 365 whole days. Ancient astronomers realized that the length of a year is approximately $365\frac{1}{4}$ days, and Julius Caesar established the system of leap years to account for this extra quarter of a day. By adding an extra day to the calendar every four years, Caesar hoped to ensure that seasonal astronomical events, such as the beginning of spring, would occur on the same date year after year.

Caesar's system would have been perfect if the year were exactly $365\frac{1}{4}$ days long and the Earth's rotation axis never changed direction. Unfortunately, this is not the case. So, over time, the discrepancy between Caesar's system and actual time accumulated and annual events began to fall on different dates each year. To straighten

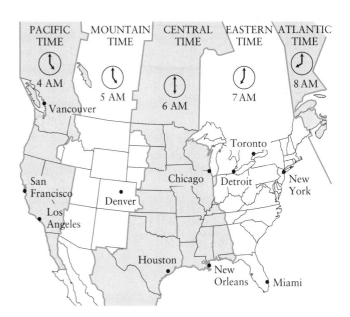

Figure 1-10 **Time Zones in North America** For convenience, the Earth is divided into 24 time zones. There are four time zones across the continental United States, giving a three-hour time difference between New York and California.

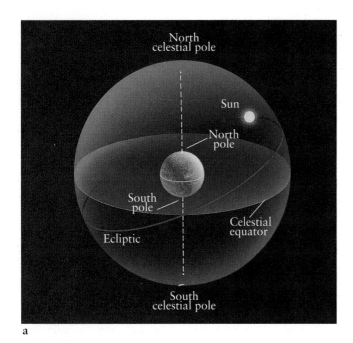

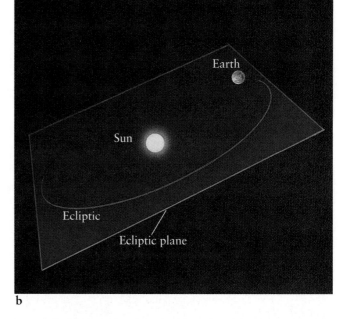

a   b

Figure 1-11 **The Ecliptic** (a) The ecliptic is the apparent annual path of the Sun on the celestial sphere. (b) The ecliptic is

also the plane described by the Earth's path around the Sun.
The planes described by the two ecliptics exactly coincide.

1-6 The seasons are caused by the tilt of the Earth's axis of rotation

The seasonal changes we experience actually result
from the tilt of the Earth's axis of rotation compared to
the plane of our orbit around the Sun. Imagine that you
could see the stars even during the day, so that you could
follow the Sun's apparent motion against the background constellations throughout the year. (The Sun appears to move among the stars, of course, because the
Earth orbits around it.) From day to day the Sun would
trace a straight path along the celestial sphere called the
ecliptic. As you can see in Figure 1-11*a*, the ecliptic
makes a closed circle bisecting the celestial sphere.

The name *ecliptic* has another usage. The Earth orbits the Sun in a flat plane also called the ecliptic. You
can see that the two ecliptics exactly coincide: Imagine
yourself on the Sun watching the Earth move day by day.
The path of the Earth on the celestial sphere as seen from
the Sun is precisely the same as the path of the Sun as
seen from the Earth (Figure 1-11*b*).

The Earth's axis is tilted $23\frac{1}{2}°$ away from a line per-
pendicular to the ecliptic as shown in Figure 1-12. Ex-
cept for tiny changes each year discussed below, the
Earth maintains this tilted orientation as it orbits the
Sun; Polaris is above the north pole throughout the year.
Thus, during part of the year the northern hemisphere is

tilted toward the Sun and the southern hemisphere is
tilted away. Half a year later the situation is reversed.

When the north pole is tilted toward the Sun, the
Sun rises higher in the northern sky than when that pole
is pointing away from the Sun (Figure 1-13). The higher
the Sun rises during the day, the more daylight hours
there are. When it is up longer, the Sun has more time to
send light and heat to our hemisphere each day. Furthermore, when the Sun is higher in the sky, its energy is
more concentrated on the Earth's surface (Figure 1-14).
The temperature at any particular location on Earth depends on the total amount of energy deposited by the
Sun, which is determined by the duration of daylight and
the height of the Sun in the sky.

Because of the tilt of the Earth's axis of rotation, the
ecliptic and the celestial equator are inclined to each
other by $23\frac{1}{2}°$, as shown in Figure 1-15. These two circles
intersect at only two points, which are exactly opposite
each other on the celestial sphere. Both points are called
equinoxes (from the Latin words meaning "equal night")
because when the Sun appears at either point, it is directly over the Earth's equator and there are 12 hours of
daytime and 12 hours of nighttime everywhere on Earth
on that day.

The **vernal equinox** occurs about March 21 when
the Sun crosses the celestial equator heading northward;
the **autumnal equinox** occurs six months later, around

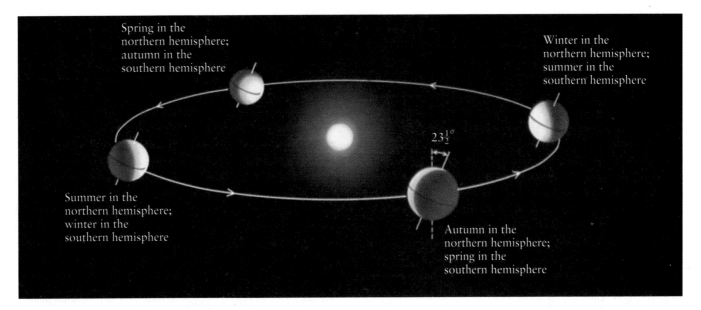

Figure 1-12 **The Tilt of the Earth's Axis** The Earth's axis of rotation is tilted $23\frac{1}{2}°$ away from being perpendicular to the plane of the Earth's orbit. The Earth maintains this orientation (with its north pole aimed at the celestial north pole near the star Polaris) throughout the year as the Earth orbits the Sun. Consequently, the amount of solar illumination and the number of daylight hours at any location on Earth varies in a regular fashion throughout the year.

September 22, with the Sun heading southward. The vernal equinox is the "prime meridian" of the celestial sphere, as discussed above. Between the vernal and autumnal equinoxes lie two other significant locations along the ecliptic. The point on the ecliptic farthest north of the celestial equator is called the **summer solstice.** The Sun reaches this point around June 21 each year. Six months later on December 21 the Sun is farthest south of the celestial equator at the **winter solstice.**

The Sun is highest in the northern sky on the summer solstice. This marks the beginning of summer in the northern hemisphere. As the Sun moves southward, the

Figure 1-13 **The Height of the Sun** The maximum height of the Sun in the sky varies throughout the year because of the $23\frac{1}{2}°$ tilt of the Earth's axis. This photograph shows the height of the Sun in the sky at the same clock time at 10-day intervals throughout the year, as well as its changing east–west location. The streaks show the path of the Sun on three different days.

Figure 1-14 **The Energy Deposited by the Sun** The angle of the Sun above the southern horizon determines how much heat and light strike each square meter of ground. (**a**) During the summer at middle latitudes, a shaft of sunlight illuminates a nearly circular patch of ground at noon. (**b**) During the winter, the same shaft of sunlight at noon strikes the ground at a steeper angle, spreading the same amount of sunlight over a larger oval shape. Because the sunlight's energy is diluted over a larger area, the ground receives less heat during the winter than during the summer.

a

b

amount of daylight decreases. The autumnal equinox marks a midpoint in the amount of heat deposited by the Sun onto the northern hemisphere and is the beginning of the fall, or autumn. When the Sun reaches the winter solstice, it is lowest in the northern sky and is above the horizon for the shortest time of the year. This is the be-

ginning of the winter. Returning northward, the Sun crosses the celestial equator once again on the vernal equinox, the beginning of the spring.

Seasonal changes in the Sun's daily path across the sky are diagrammed in Figure 1-16. On the first day of spring or fall (when the Sun is at one of the equinoxes), the Sun rises directly in the east and sets directly in the west. Daytime and nighttime are of equal duration everywhere on Earth on those days only.

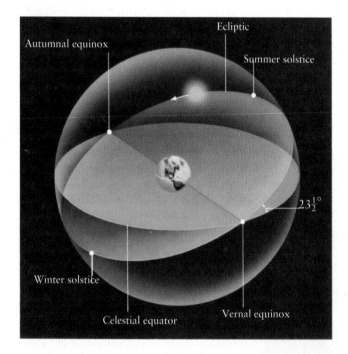

Figure 1-15 **The Seasons Are Coupled to Equinoxes and Solstices** The ecliptic is inclined to the celestial equator by $23\frac{1}{2}°$ because of the tilt of the Earth's axis of rotation. The ecliptic and the celestial equator intersect at two points called the equinoxes. The northernmost point on the ecliptic is called the summer solstice; the southernmost point is called the winter solstice.

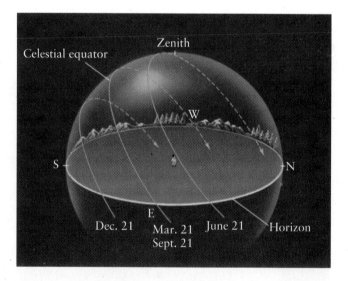

Figure 1-16 **The Sun's Daily Path** This sketch shows the path of the Sun on certain days of the year as seen from middle latitudes in the northern hemisphere. On the first day of spring and the first day of fall, the Sun rises precisely in the east and sets precisely in the west. During summer, the Sun rises in the northeast and sets in the northwest. In the winter, the Sun rises in the southeast and sets in the southwest.

During the summer months, when the northern hemisphere is tilted toward the Sun, the Sun rises in the northeast and sets in the northwest. The Sun spends more than 12 hours above the horizon in the northern hemisphere and passes high in the sky at noontime. At the summer solstice, the Sun is as far north as it gets, giving the greatest number of daylight hours to the northern hemisphere.

During the winter months, when the northern hemisphere is tilted away from the Sun, the Sun rises in the southeast. Daylight lasts for less than 12 hours as the Sun skims low over the southern horizon and sets in the southwest. Night is longest in the northern hemisphere when the Sun is at the winter solstice.

People at different latitudes see the noontime Sun at different angles in the sky each day. The farther from the equator you are, the lower the Sun is in the sky at noon. The farther north you go, the less heat and light are deposited and, therefore, the colder the land is. At latitudes above $66\frac{1}{2}°$ north latitude (and below $66\frac{1}{2}°$ south latitude) there are times of the year when the Sun does not rise at all. Six months later these same regions of the Earth have continuous sunlight for weeks or months, hence the name "Land of the Midnight Sun."

The Sun takes one year to complete a trip around the ecliptic. Since there are about $365\frac{1}{4}$ days in a year and 360° in a circle, the Sun appears to move along the ecliptic at a rate of slightly less than 1° per day. The 13 constellations through which the Sun moves throughout the year are called **zodiac** constellations. (The thirteenth zodiac constellation is Ophiuchus, the Serpent Holder. The Sun is "in" Ophiuchus from December 1 through December 19 of each year.) Table 1-1 lists all the zodiac constellations and the dates the Sun passes through them.

1-7 Precession is a slow, circular motion of the Earth's axis of rotation

As noted above, the position of the Earth's axis of rotation changes somewhat with respect to the celestial sphere (i.e., it "points" in a slightly different direction) each year. This change in orientation is caused by gravitational forces from the Moon and Sun. We shall discuss gravity in greater detail in Chapter 2. For now, it is sufficient to realize that gravity is the universal force of attraction between all matter. The Earth's rotation creates an "equatorial bulge"—our planet is about 43 km (27 mi) fatter around the equator than around the poles.

TABLE 1-1

The Sun's Passage Through Constellations of the Zodiac

Constellation	Dates
Pisces	March 13–April 20
Aries	April 20–May 13
Taurus	May 13–June 21
Gemini	June 21–July 20
Cancer	July 20–August 11
Leo	August 11–September 18
Virgo	September 18–November 1
Libra	November 1–November 22
Scorpius	November 22–December 1
Ophiuchus	December 1–December 19
Sagittarius	December 19–January 19
Capricorn	January 19–February 18
Aquarius	February 18–March 13

Because of the Earth's tilted axis of rotation, the Sun and Moon are not usually located over the Earth's equator. As a result, their gravitational attraction on the Earth tries to force the equatorial bulge to be as close to them as possible. The Earth does not respond to these forces from the Sun and Moon by straightening up. Instead, it changes the direction in which its axis of rotation points on the celestial sphere—a motion called **precession**. This is exactly the same behavior that is exhibited by a spinning top (Figure 1-17). If the top were not spinning, gravity would pull it over on its side. But because it is spinning, the combined actions of gravity and rotation cause the top's axis of rotation to precess in a circular path. As with the toy top, the combined actions of gravity and rotation cause the Earth's axis to trace out a circle in the sky while remaining tilted about $23\frac{1}{2}°$ away from the perpendicular.

The Earth's rate of precession is slow compared to human time scales. It takes 26,000 years for the north celestial pole to trace out a complete circle around the sky, as shown in Figure 1-18. (Of course, the south celestial pole executes a similar circle in the southern sky.) At the present time, the Earth's axis of rotation points within 1° of the star Polaris. In 3000 BC, it was pointing near the star Thuban in the constellation of Draco (the Dragon). In AD 14,000, the pole star will be Vega in Lyra (the Harp).

As the Earth's axis of rotation precesses, its equatorial plane also moves. Because the Earth's equatorial

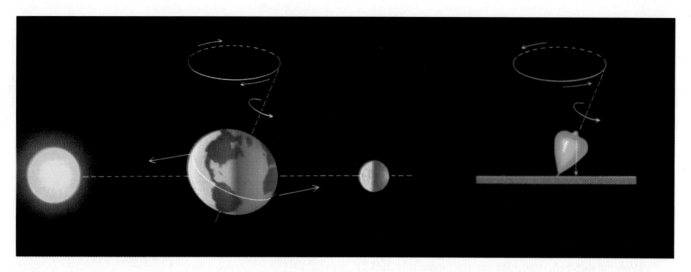

Figure 1-17 Precession The gravitational pulls of the Moon and the Sun on the Earth's equatorial bulge cause the Earth to precess. As the Earth precesses, its axis of rotation slowly traces out a circle in the sky. The situation is analogous to that of a spinning top. As the top spins, the Earth's gravitational pull causes the top's axis of rotation to move in a circle.

plane defines the location of the celestial equator in the sky, the celestial equator also precesses. Because the intersections of the celestial equator and the ecliptic define the equinoxes, these key locations in the sky also shift slowly from year to year. This entire phenomenon is often called the **precession of the equinoxes**. This change

Figure 1-18 The Path of the North Celestial Pole As the Earth precesses, the north celestial pole slowly traces out a circle among the northern constellations. At the present time, the north celestial pole is near the moderately bright star Polaris, which serves as the pole star.

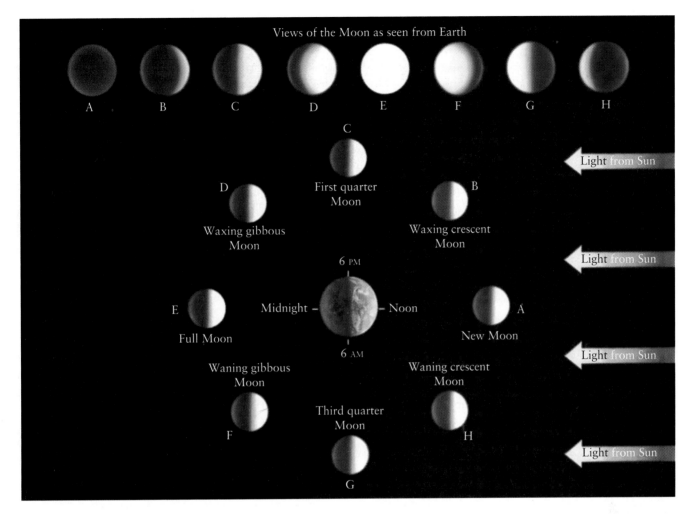

Figure 1-19 The Phases of the Moon This diagram shows the Moon at eight locations on its orbit (as viewed from above the Earth's north pole) along with the resulting lunar phases. Light from the Sun illuminates one-half of the Moon while the other half is dark. It takes $29\frac{1}{2}$ days for the Moon to go through all its phases.

was discovered by the great Greek astronomer Hipparchus in the second century BC. Today the vernal equinox is located in the constellation Pisces (the Fishes). Two thousand years ago, it was located in Aries (the Ram). Around the year AD 2600, the vernal equinox will move into Aquarius (the Water Bearer).

1-8 The phases of the Moon originally determined the month

As the Moon orbits the Earth we see different **lunar phases.** The Moon's phase depends on how much of its sunlit hemisphere is exposed to our Earth-based view. When the Moon is closest to the Sun in the sky, its dark hemisphere faces the Earth. This phase, during which the Moon is at most a tiny crescent, is called *new Moon* (Figure 1-19).

During the seven days following the new phase, more of the Moon's illuminated hemisphere becomes exposed to our view, resulting in a phase called the *waxing crescent Moon.* At *first quarter Moon,* we see half of the illuminated hemisphere and half of the dark hemisphere. "Quarter Moon" refers to how far in its cycle the Moon has gone rather than what fraction of the Moon appears lit by sunlight.

During the next week still more of the illuminated hemisphere can be seen from Earth, giving us the phase called a *waxing gibbous Moon.* When the Moon arrives on the opposite side of the Earth from the Sun, we see almost all of the fully illuminated hemisphere, which is called the *full Moon.*

Figure 1-20 **The Moon's Appearance** As the Moon orbits the Earth, we see varying amounts of the Moon's illuminated hemisphere.

| Waxing crescent (age: 4 days) | First quarter (age: 7 days) | Waxing gibbous (age: 10 days) |

Over the following two weeks, we see less and less of the illuminated hemisphere as the Moon continues along its orbit. This movement produces the phases called *waning gibbous Moon, third quarter Moon,* and *waning crescent Moon.* The Moon completes a full cycle of phases in $29\frac{1}{2}$ days.

Figure 1-20 shows the Moon at various positions in its orbit. Remember that the bright side of the Moon is on the right (west) side of the waxing Moon, while the bright side is on the left (east) side of the waning Moon. Also, when looking at the Moon through a telescope, the best place to see details is where the shadows are longest. This occurs at the boundary between the bright and dark regions, called the **terminator.**

Figure 1-19 also shows local time around the globe, from noon, when the Sun is highest in the sky, to mid-night, when it is on the opposite side of the Earth. These time markings can be used to correlate the phase and position of the Moon with the time of day. For example, at first quarter the Moon is 90° east of the Sun in the sky; hence moonrise occurs approximately at noon. At full Moon, the Moon is opposite the Sun in the sky and thus moonrise occurs at sunset. Using this information, you can see that the Moon is visible during the daytime for half of its cycle of phases (Figure 1-21) and during the night for the other half.

Since the dawn of civilization, people have needed accurate timekeeping systems. Ancient Egyptians wanted to know when the Nile would flood. Farmers needed to know when to plant crops. Migratory tribes wanted to know when the weather was due to change. Religious leaders had to schedule observances. Astronomers have

Figure 1-21 **The Moon During the Day** The Moon is visible at some time during daylight hours virtually every day. The time it is up depends on its phase.

| Full moon | Waning gibbous | Third quarter | Waning crescent |
| (age: 14 days) | (age: 20 days) | (age: 22 days) | (age: 26 days) |

traditionally been responsible for telling time. Indeed, of the four ways in which time cycles are set, three are astronomical in origin: Time is determined by the positions of the Moon, Sun, or stars or by technological means, such as atomic clocks.

The approximately four weeks that the Moon takes to complete one cycle of its phases inspired our ancestors to invent the concept of a month. Astronomers find it useful to define two types of months, depending on whether the Moon's motion is measured relative to the stars or to the Sun. Neither type corresponds exactly to the months of our usual calendar.

The **sidereal month** is the time it takes the Moon to complete one full orbit of the Earth, measured with respect to the stars (Figure 1-22). This true orbital period is equal to 27.3 days. The **synodic month**, or **lunar month**, is the time it takes the Moon to complete one $29\frac{1}{2}$-day cycle of phases (i.e., from new Moon to new Moon or from full Moon to full Moon) and thus is measured with respect to the Sun (rather than the stars).

The synodic month is longer than the sidereal month because the Earth is orbiting the Sun while the Moon goes through its phases. As Figure 1-22 shows, the Moon must travel *more* than 360° along its orbit to complete a cycle of phases (e.g., from one new Moon to the next), which takes about two days longer than the sidereal month.

Both the sidereal and synodic month vary somewhat, because the gravitational pull of the Sun on the Moon affects the Moon's speed as it orbits the Earth. The sidereal month can vary by as much as 7 hours, while the synodic month can vary by as much as 12 hours.

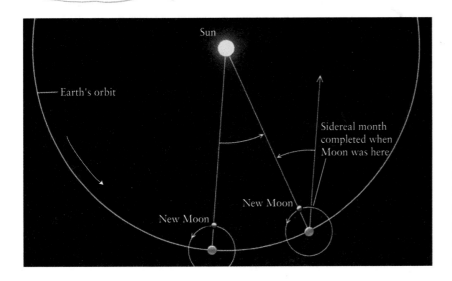

Figure 1-22 **The Sidereal and Synodic Months** The sidereal month is the time it takes the Moon to complete one revolution with respect to the background stars. However, because the Earth is constantly moving in its orbit about the Sun, the Moon must travel through more than 360° to get from one new Moon to the next. Thus the synodic month is slightly longer than the sidereal month.

The terms *sidereal* and *synodic* are also used in discussing the motion of the other bodies in the solar system. Any orbit measured with respect to the distant stars is called sidereal, including orbits of the planets around the Sun as well as orbits of moons around their planets. A synodic period of a planet is the time between consecutive straight alignments between the Sun, Earth, and that planet (during which time the planet also goes through a cycle of phases as seen from the Earth).

Eclipses

1-9 Eclipses occur only when the Moon crosses the ecliptic during the new or full phase

Eclipses are among the most spectacular of nature's phenomena. On a cloud-free night, the brilliant full Moon gradually darkens to a deep red; or broad daylight is transformed into an eerie twilight as the Sun seems to be blotted from the sky. A **lunar eclipse** occurs when the Moon passes through the Earth's shadow. This can happen only when the Sun, Earth, and Moon are in a straight line at full Moon. A **solar eclipse** occurs when the Moon's shadow moves across the Earth's surface. As seen from Earth, the Moon moves in front of the Sun. This can happen only when the Sun, Moon, and Earth are aligned at new Moon.

At first glance, it would seem that eclipses should happen at every new and full Moon, but in fact they occur much less often. Eclipses occur infrequently because the Moon's orbit is tilted 5° out of the ecliptic, as shown in Figure 1-23. Because of this tilt, new Moon and full Moon usually occur when the Moon is either above or below the plane of the Earth's orbit. In such positions, a perfect alignment between the Sun, Moon, and Earth is not possible and an eclipse cannot occur. When the Moon crosses the plane of the ecliptic during its new or full phase, an eclipse takes place. The Moon crosses the ecliptic at what is called the **line of nodes** (Figure 1-24). By calculating the number of times a new Moon takes place on the line of nodes, we find that there are at least two and no more than five solar eclipses each year. Lunar

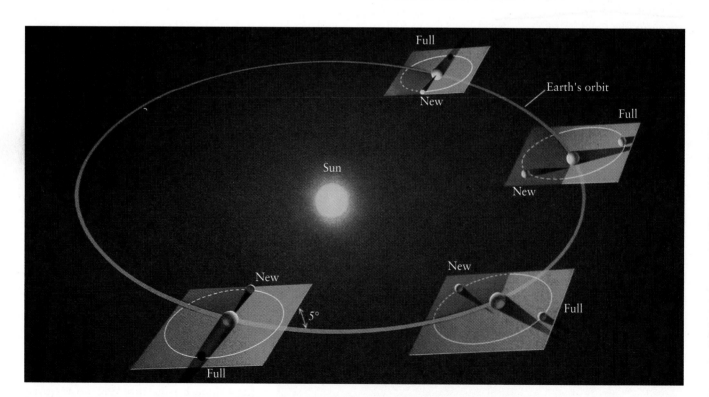

Figure 1-23 **Conditions for Eclipses** A solar eclipse occurs only if the Moon is very nearly on the ecliptic at new Moon. A lunar eclipse occurs only if the Moon is very nearly on the ecliptic at full Moon. When new Moon or full Moon phases occur away from the ecliptic, no eclipse is seen because the Moon and the Earth do not pass through each other's shadows.

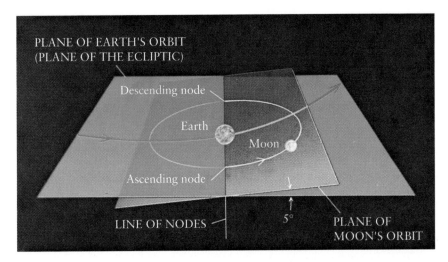

Figure 1-24 **The Line of Nodes** The plane of the Moon's orbit is tilted slightly with respect to the plane of the Earth's orbit. These two planes intersect along a line called the line of nodes.

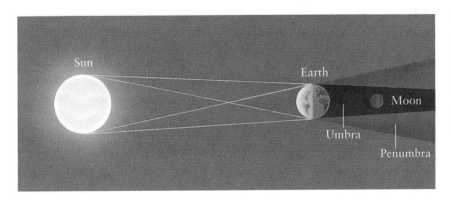

Figure 1-25 **The Geometry of a Lunar Eclipse** People on the nighttime side of the Earth see a lunar eclipse when the Moon moves through the Earth's shadow. The umbra is the darkest part of the shadow. In the penumbra, only part of the Sun is covered by the Earth.

eclipses occur just about as frequently as solar eclipses, but the maximum number of eclipses (both solar and lunar) possible in a year is seven.

1-10 There are three types of lunar eclipse

There are three kinds of lunar eclipse, depending on how the Moon travels through the Earth's shadow. The Earth's shadow has two distinct parts, as diagrammed in Figure 1-25. The **umbra** is the part of the shadow where all direct sunlight is blocked by the Earth; the **penumbra** of the shadow is where the Earth only blocks some of the sunlight. A **penumbral eclipse,** when the Moon passes through only the Earth's penumbra, is easy to miss. The Moon still looks full, just a little dimmer than usual.

When only part of the lunar surface passes through the umbra, a bite seems to be taken out of the Moon and we see a **partial eclipse.** When the Moon travels completely into the umbra, as sketched in Figure 1-26, we see

a **total eclipse** of the Moon. Totality is the time when the entire Moon is in the Earth's umbra. Eclipses with the maximum duration of totality, lasting for up to 1 hour 47 minutes, occur when the Moon is closest to the Earth and travels directly through the center of the umbra.

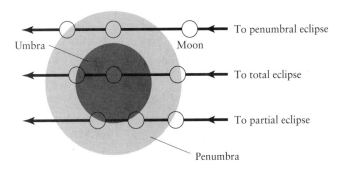

Figure 1-26 **Various Lunar Eclipses** This diagram shows the Earth's umbra and penumbra at the distance of the Moon's orbit. Different kinds of lunar eclipses are seen, depending on the Moon's path through the Earth's shadow.

TABLE 1-2		
Lunar Eclipses, 1996–2000		
Date	Percentage eclipsed (100% = total)	Duration of totality (h:min)
1996 April 4	100	1:26
1996 September 27	100	1:10
1997 March 24	93	——
1997 September 16	100	1:02
1999 July 28	42	——
2000 January 21	100	1:16
2000 July 16	100	1:46

hues during totality, as shown in Figure 1-27. At sunrise and sunset the Sun appears red or orange for the same reason; namely, at those times red and orange light are deflected from the Sun toward you.

1-11 There are also three types of solar eclipse

Because the Sun and the Moon have nearly the same angular diameter as seen from Earth—about $\frac{1}{2}°$—the Moon can just "fit" over the Sun, creating a *total solar eclipse*. You must be within the Moon's umbra in order to see a total solar eclipse.

During those few precious moments, hot gases (called the **solar corona**) surrounding the Sun can be photographed (Figure 1-28).Astronomers can then learn more about the Sun's temperature, chemistry, and atmospheric activity. You can see in Figure 1-29 that only the tip of the Moon's umbra ever reaches the Earth's surface. As the Earth turns, the tip traces an **eclipse path** across it. Only people within this path are treated to the spectacle of a total solar eclipse. Figure 1-30 shows the dark spot produced by the Moon's umbra on the Earth's surface during a total solar eclipse.

Table 1-2 lists all the total and partial lunar eclipses from 1996 through 2000. Penumbral eclipses are not included in this listing.

Even during a total eclipse, the Moon does not completely disappear. A small amount of sunlight passing through the Earth's atmosphere is bent into the Earth's umbra. The light deflected into the umbra is primarily red, and so the darkened Moon glows faintly in reddish

Figure 1-27 **A Total Eclipse of the Moon** This photograph was taken by an amateur astronomer during the lunar eclipse of September 6, 1979. Notice the distinctly reddish color of the Moon.

Figure 1-28 **A Total Eclipse of the Sun** During a total solar eclipse, the Moon completely covers the Sun's disk, and the solar corona can be seen. This halo of hot gases extends for thousands upon thousands of kilometers into space. This photograph was taken by an amateur astronomer during the solar eclipse of July 11, 1991.

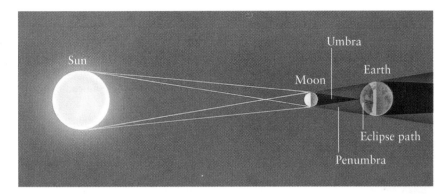

Figure 1-29 **The Geometry of a Total Solar Eclipse** During a total solar eclipse, the tip of the Moon's umbra traces an eclipse path across the Earth's surface. People inside the eclipse path see a total solar eclipse, whereas people inside the penumbra see only a partial eclipse.

The Earth's rotation and the orbital motion of the Moon cause the umbra to race along the eclipse path at speeds in excess of 1700 km/h (1050 mph). Totality never lasts for more than $7\frac{1}{2}$ minutes at any one location on the eclipse path, and it usually lasts for only a few moments.

The Moon's umbra is also surrounded by a penumbra. During a solar eclipse, the Moon's penumbra extends over a large portion of the Earth's surface. When only the penumbra sweeps across the Earth's surface, the Sun is only partly covered by the Moon. This circumstance results in a *partial eclipse of the Sun.*

The Moon's orbit around the Earth is not quite a perfect circle. The distance between the Earth and the Moon, which averages 384,400 km (238,900 mi), varies by a few percent as the Moon goes around the Earth. The width of the eclipse path depends primarily on the Earth–Moon distance during an eclipse. The eclipse path is widest—up to 270 km (170 mi)—when the Moon happens to be at the point in its orbit nearest the Earth. Usually, however, the path is much narrower.

Sometimes the Moon's umbra does not even reach the Earth's surface. If the alignment for a solar eclipse occurs when the Moon is farthest from the Earth, then the Moon's umbra falls short of the Earth and no one sees a total eclipse. From the Earth's surface, the Moon appears too small to cover the Sun completely, and a thin ring of light is seen around the edge of the Moon at mid-eclipse. This type of eclipse is called an **annular eclipse** (Figure 1-31). The length of the Moon's umbra is nearly 5000 km (3100 mi) shorter than the average distance between the Moon and the Earth's surface. Thus, the Moon's shadow often fails to reach the Earth, making annular eclipses more common than total eclipses. Table 1-3 lists all the total, partial, and annular solar eclipses from 1996 through 2000.

A total solar eclipse is a dramatic event. The sky begins to darken, the air temperature falls, and the winds increase as the Moon's umbra races toward you. All nature responds: Birds go to roost, flowers close their petals, and crickets begin to chirp as if evening had arrived. As totality approaches, the landscape is bathed in shimmering bands of light and dark as the last few rays of sunlight peek out from behind the edge of the Moon. Finally the corona blazes forth in a star-studded daytime sky. It is an awesome sight.

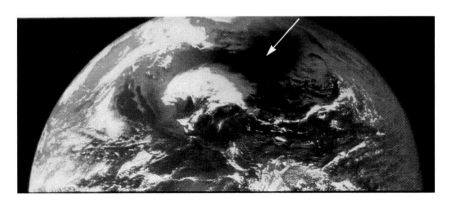

Figure 1-30 **The Moon's Shadow on the Earth** This photograph was taken from an Earth-orbiting satellite during the total eclipse of March 7, 1970. The Moon's umbra appears as a dark spot on the eastern coast of the United States.

Figure 1-31 **An Annular Eclipse of the Sun** This composite of five exposures taken at sunrise in Costa Rica shows the progress of an annular eclipse of the Sun that occurred on December 24, 1974. Note that at mid-eclipse the edge of the Sun is visible around the Moon.

WHAT DID YOU THINK?

1 *How bright is the North Star, Polaris, compared to other stars?* Polaris is a star of medium brightness compared to other stars visible to the naked eye.

2 *Are constellations just mythic figures in the sky?* A constellation is one of two things: It is a pattern of stars (sometimes called an asterism), or it is an entire region of the celestial sphere and all the stars and other objects in it. The entire sky is covered by 88 different-sized constellations.

3 *What causes the seasons?* The tilt of the Earth's rotation axis with respect to the ecliptic causes the seasons. They are not caused by the changing distance from the Earth to the Sun.

4 *How many zodiac constellations are there?* There are 13 zodiac constellations, the "unknown" one being Ophiuchus.

5 *When, if ever, is the Moon visible during the daytime?* The Moon is visible during the day for half of the lunar cycle of phases. Different phases are visible during different times of the day.

TABLE 1-3			
Solar Eclipses, 1996–2000			
Date	Area	Type	Notes
1996 April 17	South Pacific, Antarctic	Partial	88% eclipsed
1996 October 12	North Atlantic, Arctic, Europe, Middle East	Partial	76% eclipsed
1997 March 9	China, Russia, Arctic	Total	Maximum length 2 min 50 s
1997 September 2	Australia, New Zealand, Antarctic	Partial	90% eclipsed
1998 February 26	Pacific, Central America, Atlantic	Total	Maximum length 3 min 56
1998 August 22	Indian Ocean, East Indies, Pacific	Annular	
1999 February 16	Indian Ocean, Australia, Antarctic	Annular	
1999 August 11	Atlantic, Europe, Middle East, India	Total	Maximum length 2 min 23 s
2000 February 5	Antarctic, southern Indian Ocean	Partial	58% eclipsed
2000 July 1	Southern Atlantic	Partial	48% eclipsed
2000 July 31	Northern Europe, Greenland, Canada, western United States	Partial	60% eclipsed
2000 December 25	Canada, United States, Mexico	Partial	72% eclipsed

Key Words

angle	diurnal motion	precession (of the	south celestial pole
angular diameter	eclipse path	Earth)	summer solstice
(angular size)	ecliptic	precession	synodic month (lunar
annular eclipse	equinox	precession of the	month)
arc angle	line of nodes	equinoxes	terminator
autumnal equinox	lunar eclipse	revolution	time zone
celestial equator	lunar phases	right ascension	total eclipse
celestial sphere	north celestial pole	rotation	umbra
circumpolar stars	partial eclipse	sidereal month	vernal equinox
constellation	penumbra	solar corona	winter solstice
declination	penumbral eclipse	solar day	zodiac
degree (°)		solar eclipse	

Key Ideas

• The surface of the celestial sphere is divided into 88 unequal regions called constellations.

• The celestial sphere appears to rotate around the Earth once in each day–night cycle. In fact, of course, it is the Earth's rotation that causes that apparent motion.

• The poles and equator of the celestial sphere are determined by extending the axis of rotation and the equatorial plane of the Earth to the celestial sphere.

• Earth's axis of rotation is tilted at an angle of $23\frac{1}{2}°$ from the perpendicular to the plane of the Earth's orbit. This tilt causes the seasons.

• Equinoxes and solstices are significant points along the Earth's orbit that are determined by the relationship between the Sun's path on the celestial sphere (the ecliptic) and the celestial equator.

• The Earth's axis of rotation slowly changes over thousands of years, a phenomenon called precession. Precession is caused by the gravitational pull of the Sun and Moon on the Earth's equatorial bulge.

• The length of the day is based upon the average motion of the Sun along the celestial equator, which produces the 24-hour day upon which our watches are based.

• The phases of the Moon are caused by the relative positions of the Earth, Moon, and Sun. The Moon completes one cycle of phases in a synodic month, which averages $29\frac{1}{2}$ days.

• The Moon completes one orbit around the Earth with respect to the stars in a sidereal month, which averages 27.3 days.

• The shadow of an object has two parts: the umbra, where the light source is completely blocked, and the penumbra, where the light source is only partially obscured.

• A lunar eclipse occurs when the Moon moves through the Earth's shadow. This phenomenon occurs when the Sun, Earth, and Moon are in alignment.

• A solar eclipse occurs when the Earth passes through the Moon's shadow. This phenomenon occurs when the Sun, Earth, and Moon are in alignment.

• Depending on the relative positions of the Sun, Moon, and Earth, lunar eclipses may be penumbral, partial, or total, and solar eclipses may be annular, partial, or total.

Review Questions

1 How are constellations useful to astronomers?

2 What is the celestial sphere and why is this ancient concept still useful today?

3 What is the celestial equator and how is it related to the Earth's equator? How are the north and south celestial poles related to the Earth's axis of rotation?

4 What is the ecliptic and why is it tilted with respect to the celestial equator?

5 About how many degrees does the Earth move along the ecliptic each day?

6 Using a diagram, explain why the tilt of the Earth's axis relative to the Earth's orbit causes the seasons as the Earth revolves around the Sun.

7 What are the vernal and autumnal equinoxes? What are the summer and winter solstices? How are these four points related to the ecliptic and the celestial equator?

8 What is precession and how does it affect our view of the heavens?

9 How does the daily path of the Sun across the sky change with the seasons?

10 Why is it warmer in the summer than in the winter?

11 Why is it convenient to have the Earth divided into time zones?

12 Why does the Moon exhibit phases?

13 What is the difference between a sidereal month and a synodic month? Which is longer? Why?

14 What is the line of nodes and how is it related to solar and lunar eclipses?

15 What is the difference between the umbra and the penumbra of a shadow?

16 What is a penumbral eclipse of the Moon? Why do you suppose that it is easy to overlook such an eclipse?

17 Which type of eclipse—lunar or solar—do you think most people have seen? Why?

18 How is an annular eclipse of the Sun different from a total eclipse of the Sun? What causes this difference?

19 If changes in the Earth's distance to the Sun caused the seasons, how should seasons in the northern and southern hemispheres be related? Are they?

20 When is the next leap year?

21 At which phase(s) of the Moon does a solar eclipse occur? A lunar eclipse?

22 Is it safe to watch a solar eclipse without eye protection? A lunar eclipse?

Advanced Questions

23 Why can't a person in Australia use the Big Dipper to find the north direction?

24 Are there any stars in the sky that are not members of a constellation?

25 At what location on Earth is the north celestial pole seen on the horizon?

26 Where do you have to be on the Earth in order to see the Sun at the zenith? If you stay at one such location for a full year, on how many days will the Sun pass through the zenith?

27 Where do you have to be on Earth in order to see the south celestial pole directly overhead? What is the maximum possible elevation of the Sun above the horizon at that location? On what date is this maximum elevation achieved?

28 At what point on the horizon does the Sun rise on the vernal equinox?

29 Consult a star map of the southern hemisphere and determine which, if any, bright southern stars could someday become south celestial pole stars.

30 Are there stars in the sky that never set where you live? Are there stars that never rise where you live? Does your answer depend on your location on Earth?

31 Using a diagram, demonstrate that your latitude on Earth is equal to the altitude of the north celestial pole above your northern horizon.

32 Using a star map, determine which bright stars, if any, could someday mark the location of the vernal equinox. Give the approximate years when this should happen.

33 What is the phase of the Moon if it (**a**) rises at 3 AM? (**b**) sets at 9 PM? At what time does (**c**) the full Moon set? (**d**) the first quarter Moon rise?

34 What is the phase of the Moon if, on the first day of spring, the Moon is located at (**a**) the vernal equinox, (**b**) the summer solstice, (**c**) the autumnal equinox, (**d**) the winter solstice?

* **35** How many more sidereal months than synodic months are there in a year? Why?

36 How do we know that the phases of the Moon are not due to the Moon moving in the Earth's shadow?

37 Why do you suppose that paths of total solar eclipses fall more frequently on oceans than on land?

38 Can one ever observe an annular eclipse of the Moon? Why?

39 During a lunar eclipse, does the Moon enter the Earth's shadow from the east or the west? Explain why.

Discussion Questions

40 Examine a list of the 88 constellations. Are there any constellations whose names obviously date from modern times? Where are these constellations located? Why do you suppose they do not have archaic names?

41 Describe what the seasons would be like if the Earth's axis of rotation were tilted (**a**) 0° and (**b**) 90° to its orbital plane.

42 Describe the cycle of lunar phases that would be observed if the Moon moved about the Earth in an orbit perpendicular to the plane of the Earth's orbit. Would solar and lunar eclipses be possible under these circumstances?

43 In his novel *King Solomon's Mines,* author H. Rider Haggard described a total solar eclipse that was seen in both South Africa and the British Isles. Is such an eclipse possible? Why or why not?

44 Describe how a lunar eclipse would look if the Earth had no atmosphere.

45 Examine a listing of total solar eclipses over the next several decades. What are the chances that you might be able to travel to one of the eclipse paths? Do you think you might go through your entire life without ever seeing a total eclipse of the Sun?

Observing Projects

46 On a clear, cloud-free night, use the star charts within the covers of this book to see how many constellations of the zodiac you can identify. Which ones are easy to find? Which are difficult?

47 Examine the star charts that are published monthly in such popular astronomy magazines as *Sky & Telescope* and *Astronomy.* How do they differ from the star charts within the covers of this book? On a clear, cloud-free night, use one of these star charts to locate the celestial equator and the ecliptic. Note the inclination of the Milky Way to the ecliptic and celestial equator. What do your observations tell you about the orientation of the Earth and its orbit around the Sun relative to the rest of the Galaxy?

48 Observe the Moon on each clear night over the course of a month. Note the Moon's location among the constellations and record that location on a star chart that also shows the ecliptic. After a few weeks, your observations will begin to trace the Moon's orbit. Identify the orientation of the line of nodes by marking the points where the Moon's orbit and the ecliptic intersect. On what dates is the Sun near the nodes marked on your star chart? Compare these dates with the dates of the next solar and lunar eclipses.

49 It is quite possible that a lunar eclipse will occur while you are taking this course. If so, look up the precise time it will happen in the current issue of a reference from the U.S. Naval Observatory such as the *Astronomical Almanac* or *Astronomical Phenomena,* You can also consult such magazines as *Sky & Telescope* and *Astronomy,* which generally run articles about eclipses the month before they happen. Make arrangements to observe the next lunar eclipse. Note the times at which the Moon enters and exits the Earth's umbra.

2 ▶ *Gravitation and the Waltz of the Planets*

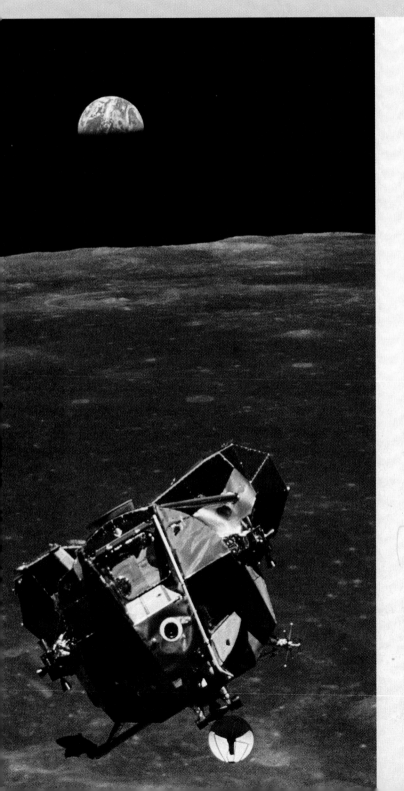

IN THIS CHAPTER

You will see how several brilliant thinkers during the Renaissance led a scientific revolution that dethroned Earth from its location as the center of the universe. Copernicus argued that the planets go around the Sun. A few years later Kepler demonstrated that the planetary orbits are mathematical ovals called ellipses. When Galileo turned his telescope toward the skies and especially toward the moons of Jupiter, the ideas of Copernicus and Kepler became indisputable. You will then see how these discoveries set the stage for Isaac Newton, who formulated the basic laws of physics. You will also learn Newton's mathematical description of the force of gravity, which holds the planets in their orbits about the Sun.

Apollo 11 Leaving the Moon A lunar module returns from the Moon after completing a successful manned mission. This photograph was taken from the spacecraft in which the astronauts returned to Earth. All of the orbital maneuvers to the Moon and back were based on Newtonian mechanics and Newton's law of gravity. These same principles are used by astronomers to understand a wide range of phenomena, from the motions of double stars to the rotation of the Galaxy.

WHAT DO YOU THINK?

1 What is the shape of the Earth's orbit around the Sun?

2 Do the planets orbit the Sun at constant speeds?

NO

THE GROUNDWORK FOR MODERN science was laid around 550 BC when Pythagoras and his Greek followers put forth the idea that natural phenomena could be described by mathematics. About 200 years later, Aristotle asserted that the universe is governed by physical laws. These two concepts found their highest expression when the great scientists of later generations explained planetary motion using mathematics and physics.

Early Greek astronomers were also among the first to leave a written record of their attempts to explain the motion of the five then-known planets: Mercury, Venus, Mars, Jupiter, and Saturn. Most Greeks assumed that the Sun, the Moon, the stars, and the planets revolve about the Earth, and thus their view of the universe is said to have been *geocentric*. A theory about the Earth's place in the universe is called a **cosmogony,** and thus a Greek thinker such as Pythagoras or Aristotle gave credence to a **geocentric cosmogony.** (In contrast, *cosmology* deals with the origin and evolution of the entire universe, without giving the Earth and solar system special status.)

Origins of a Sun-Centered Solar System

The Greeks knew that the positions of the planets slowly shift against the background of "fixed" stars in the constellations. In fact, the word *planet* comes from a Greek term meaning "wanderer." The Greeks observed that planets do not move at uniform rates through the constellations. From night to night they usually move slowly to the left (eastward) relative to the background stars. This eastward movement is called **direct motion.** Occasionally, however, a planet seems to stop and then back up for several weeks or months. This westward movement is called **retrograde motion.**

All of these motions are much slower than the daily movement of the sky caused by the Earth's rotation. Therefore, the planets always rise in the east and set in the west as the stars do. Both direct and retrograde motion are best detected by mapping the nightly position of a planet against the background stars over a long period. An example is shown in Figure 2-1, which shows the path of the planet Mars from October 1996 through August 1997.

Explaining the motions of the five planets in a geocentric universe was one of the main challenges facing the astronomers of antiquity. The effort resulted in an increasingly contrived model, especially in explaining retrograde motion. The mechanical description of the geocentric cosmogony (Toolbox 2-1) was complex, and as

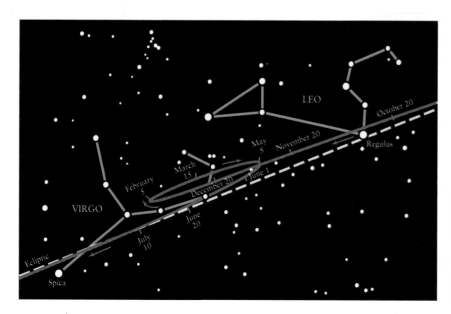

Figure 2-1 **The Path of Mars in 1996–1997** From the fall of 1996 through the summer of 1997, Mars will move across the constellations of Leo and Virgo. From February 4 through April 26, 1997, Mars' motion will be retrograde.

observations improved, the model increasingly failed to fit the data. Because simplicity and accuracy are hallmarks of science, the geocentric cosmogony had to give way to a more elegant model.

The ancient Greek astronomer Aristarchus proposed a more straightforward explanation of planetary motion, namely, that all the planets, including the Earth, revolve about the Sun. The retrograde motion of Mars, for ex-

AN ASTRONOMER'S TOOLBOX 2-1

Geocentric Cosmology

The Greeks developed many theories to account for retrograde motion and the loops that the planets trace out against the background stars. One of the most successful ideas was expounded by the last of the great Greek astronomers, Ptolemy, who lived in Alexandria during the second century AD. The basic concept is sketched in the accompanying figure. Each planet is assumed to move in a small circle called an **epicycle,** the center of which

A Geocentric Explanation of Planetary Motion

Each planet revolves about an epicycle, which in turn revolves about a deferent centered approximately on the Earth. As seen from Earth, the speed of the planet on the epicycle alternately adds to or subtracts from the speed of the epicycle on the deferent, thus producing alternating periods of direct and retrograde motion.

moves in a larger circle called a **deferent,** whose center is offset from the Earth. As viewed from Earth, the epicycle moves eastward along the deferent, and both circles rotate in the same direction (counterclockwise in the figure).

Most of the time the motion of the planet on its epicycle adds to the eastward motion of the epicycle on the deferent. Thus the planet is seen to be in direct (eastward) motion against the background stars throughout most of the year. However, when the planet is on the part of its epicycle nearest the Earth, its motion along the epicycle subtracts from the motion of the epicycle along the deferent. The planet thus appears to slow and then halt its usual eastward movement among the constellations, even seeming to go backward for a few weeks or months. Using this concept of epicycles and deferents, Greek astronomers were able to explain the retrograde loops of the planets.

Using the wealth of astronomical data in the library at Alexandria, including records of planetary positions covering hundreds of years, Ptolemy deduced the sizes of the epicycles and deferents and the rates of revolution needed to produce the recorded paths of the planets. After years of arduous work, Ptolemy assembled his calculations in the *Almagest,* in which the positions and paths of the Sun, Moon, and planets were described with unprecedented accuracy. In fact, the *Almagest* was so successful that it became the astronomer's bible. For over a thousand years Ptolemy's cosmology endured as a useful description of the workings of the heavens.

Eventually, however, things began to go awry. Errors and inaccuracies that were unnoticeable in Ptolemy's day compounded and multiplied over the years, especially errors due to precession. Fifteenth-century astronomers made some cosmetic adjustments to the Ptolemaic system. However, the system became less and less satisfactory as more complicated and arbitrary details were added to keep it consistent with the observed motions of the planets.

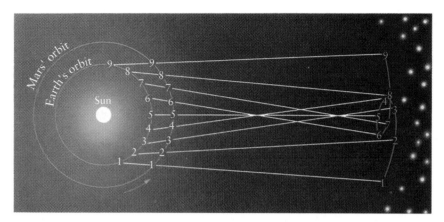

Figure 2-2 **A Heliocentric Explanation of Planetary Motion** The Earth travels around the Sun more rapidly than does Mars. Consequently, as the Earth overtakes and passes this slower-moving planet, Mars appears (from points 4 through 6) to move backward among the background stars for a few months.

ample, occurs when the Earth overtakes and passes it, as shown in Figure 2-2. The occasional retrograde movement of a planet is then merely the result of our changing viewpoint—an idea that is beautifully simple compared to a geocentric system with all its complex planetary motions.

2-1 Nicolaus Copernicus devised the first comprehensive heliocentric cosmology

Seventeen hundred years elapsed before someone had the insight and determination to work out the details of a **heliocentric** (Sun-centered) **cosmogony.** That person was the sixteenth-century Polish lawyer, physician, mathematician, economist, monk, and artist Nicolaus Copernicus.

Copernicus turned his attention to astronomy in the early 1500s. He found that by assuming that the heliocentric cosmogony is correct, he could determine which planets are closer to the Sun than the Earth and which are farther away. Because Mercury and Venus are always observed fairly near the Sun, Copernicus correctly concluded that their orbits must lie inside the Earth's. The other visible planets—Mars, Jupiter, and Saturn—can be seen high in the sky in the middle of the night, when the Sun is far below the horizon. This can occur only if the Earth comes between the Sun and a planet. Copernicus therefore concluded that the orbits of Mars, Jupiter, and Saturn lie outside the Earth's orbit. Three additional planets (Uranus, Neptune, and Pluto), discovered after the telescope was invented, also have orbits outside the Earth's.

The geometrical arrangements between the Earth, another planet, and the Sun are called **configurations.** For example, when Mercury or Venus is between the Earth and the Sun, as in Figure 2-3, we say the planet is in a configuration called an **inferior conjunction;** when a planet is beyond the Sun, it is in a configuration called a **superior conjunction.**

The angle between the Sun and a planet as viewed from the Earth is called the planet's **elongation.** At **greatest elongation,** Mercury or Venus is as far from the Sun in angle as it can be. This is about 28° for Mercury and about 47° for Venus. Since Mercury and Venus can never be farther than their greatest elongations from the Sun, neither of these planets is ever seen very high in the night sky. When either Mercury or Venus rises before the Sun, it is visible in the eastern sky as a very bright "star" and is often called the "morning star." Similarly, when either of these two planets sets after the Sun, it is always low in the western sky and is then called the "evening star." Because they are so bright and sometimes appear to change color due to motion of the Earth's atmosphere, Venus and Mercury are often mistaken for UFOs. (The same motion of the air causes the road in front of your car to shimmer on a hot day.)

Planets whose orbits are larger than Earth's have different configurations. For example, when Mars is located beyond the Sun, as seen from Earth, it is said to be in **conjunction.** When it is opposite the Sun in the sky, the planet is at **opposition.** It is not difficult to determine when a planet happens to be located at one of the key positions in Figure 2-3. For example, when Mars is at opposition, it appears high in the sky at midnight.

It is easy to follow a planet as it moves from one configuration to another. However, these observations

Figure 2-3 **Planetary Configurations** It is useful to specify key points along a planet's orbit, as shown in this diagram. These points identify specific geometric arrangements between the Earth, another planet, and the Sun.

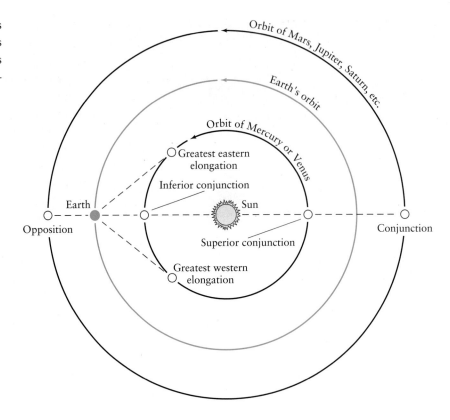

alone do not tell us the planet's actual orbit, because the Earth, from which we make the observations, is also moving. Copernicus was therefore careful to distinguish between two characteristic time intervals, or *periods*, of each planet.

The **sidereal period** is the true orbital period of a planet, the time it takes the planet to complete one orbit of the Sun. The **synodic period** is the time that elapses between two successive identical configurations as seen from the Earth—from one opposition to the next, for example, or from one conjunction to the next. The synodic period of a planet can be determined by observing the planet against the changing background stars, but the sidereal period must be calculated. Copernicus was able to obtain the first six entries shown in Table 2-1 (the others are contemporary results included for completeness).

Having determined the sidereal periods of the planets, Copernicus devised a straightforward geometric method for determining the distances of the planets from the Sun. His answers turned out to be remarkably close to the modern values, as shown in Table 2-2. From these two tables it is apparent that the farther a planet is from the Sun, the longer the planet takes to complete its orbit.

Copernicus explained his heliocentric cosmogony, including supporting observations and calculations, in a

book entitled *De revolutionibus orbium coelestium* ("On the Revolutions of the Celestial Spheres"), which was published in 1543, the year of his death.

It is important to realize that Copernicus had assumed, incorrectly, that the Earth travels around the Sun along a circular path. Indeed, until Kepler determined the correct shape of planetary orbits, the heliocentric

TABLE 2-1

Synodic and Sidereal Periods of the Planets

	Period	
	Synodic	**Sidereal**
Mercury	116 d	88 d
Venus	584 d	225 d
Earth	—	1.0 yr
Mars	780 d	1.9 yr
Jupiter	399 d	11.9 yr
Saturn	378 d	29.5 yr
Uranus	370 d	84.0 yr
Neptune	368 d	164.8 yr
Pluto	367 d	248.5 yr

TABLE 2-2

Average Distances of the Planets from the Sun

	Measurement (AU)	
	By Copernicus	Modern
Mercury	0.38	0.39
Venus	0.72	0.72
Earth	1.00	1.00
Mars	1.52	1.52
Jupiter	5.22	5.20
Saturn	9.07	9.54
Uranus	——	19.19
Neptune	——	30.06
Pluto	——	39.53

theory wasn't even more accurate than the geocentric theory. Copernicus' belief in the heliocentric cosmogony was based on its simplicity compared to the geocentric theory. This idea of using the simplest of several competing theories that describe the same concepts with the same accuracy was formally expressed by William of Occam in the fourteenth century. One of the most basic tenets in science today, this concept is known as **Occam's razor.**

2-2 Tycho Brahe made astronomical observations that disproved ancient ideas about the heavens

In November of 1572, a bright star suddenly appeared in the constellation of Cassiopeia. At first it was even brighter than Venus, but then it began to grow dim. After 18 months, it faded from view.

Modern astronomers recognize this event as a supernova explosion, the violent death of a certain type of star (see Chapter 12). In the sixteenth century, however, the prevailing opinion was quite different. Teachings dating back to Aristotle and Plato argued that the heavens were permanent and unalterable. Consequently, the "new star" of 1572 could not really be a star at all, because the heavens do not change; it must instead be some sort of bright object quite near Earth, perhaps not much farther away than the clouds overhead. A 25-year-old Danish astronomer named Tycho Brahe realized that straightforward observations might reveal the distance to the new star.

It is everyone's common experience that when you walk from one place to another, nearby objects appear to change position against the background of more distant objects. Furthermore, the closer an object is, the more you have to change the angle at which you observe it as you move. This phenomenon, whereby the apparent position of an object changes because of the changing position of the observer, is called **parallax** (Figure 2-4).

Brahe reasoned as follows: If the new star is nearby, then its position should shift against the background stars over the course of a night, because the Earth's rotation changes our viewpoint, as shown in Figure 2-5a. His careful observations failed to disclose any parallax, and so the new star had to be quite far away, farther from Earth than anyone had imagined (Figure 2-5b). Brahe summarized his findings in a small book, *De stella nova* ("On the New Star"), published in 1573.

Tycho Brahe's astronomical records were soon to play an important role in the development of a heliocentric cosmogony. From 1576 to 1597, Brahe made comprehensive observations, measuring planetary positions with an accuracy of 1 arcmin, about as precise as is possible with the naked eye. Upon his death in 1601, most of these invaluable records fell into the hands of his gifted assistant, Johannes Kepler.

Figure 2-4 **Parallax** Nearby objects are viewed at different angles from different places. These objects also appear to be in a different place with respect to more distant objects when viewed at the same time by observers located at different positions. Both effects are called parallax. They are used by astronomers, surveyors, and sailors to determine distances.

Figure 2-5 **The Parallax of a Nearby Object in Space** (a) Tycho Brahe argued that if an object is near the Earth, its position relative to the background stars should change over the course of a night. (b) When Brahe failed to measure such changes for a supernova in 1572, he concluded that the object was far from the Earth.

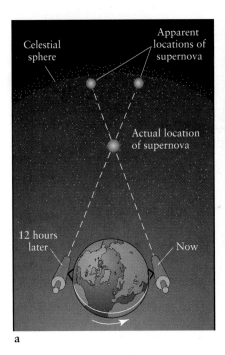

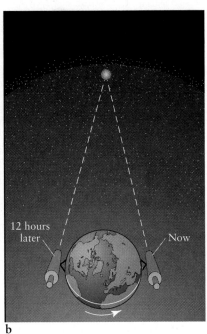

a   b

Kepler's and Newton's Laws

1 Until Johannes Kepler's time, astronomers had assumed that heavenly objects move in circles. For philosophical and aesthetic reasons, circles were considered the most perfect and most harmonious of all geometric shapes. However, assuming circular orbits failed to yield accurate predictions for the positions of the planets. For years Kepler tried in vain to find oval-shaped orbits that would fit Tycho Brahe's observations of the planets against the background of distant stars. Finally he began working with **ellipses**.

2-3 Kepler's laws describe the orbital shapes, changing speeds, and lengths of planetary years

An ellipse can be drawn with a loop of string, two thumbtacks, and a pencil, as shown in Figure 2-6a. Each thumbtack is at a **focus** (plural **foci**). The longest diameter across an ellipse, called the *major axis,* passes through both foci. Half of that distance is called the **semimajor axis,** whose length is usually designated by the letter *a.*

To Kepler's delight, the ellipse turned out to be the curve he had been searching for: Predictions of the locations of planets based on elliptical paths were in very

close agreement with where the planets actually were. He published this discovery in 1609 in a book known today as *New Astronomy.* This important discovery is now considered the first of **Kepler's laws:**

The orbit of a planet about the Sun is an ellipse with the Sun at one focus.

Ellipses have two extremes. The roundest ellipse is a circle. The most elongated ellipse approaches being a straight line. The shape of a planet's orbit around the Sun is described by its **orbital eccentricity,** where orbital eccentricity, designated by the letter *e,* ranges from 0 (circular orbit) to 1 (nearly a straight line). Figure 2-6*b* shows a sequence of ellipses and their associated eccentricities. Observations have since revealed that there is no object at the second focus for each elliptical planetary orbit.

Brahe's observations also showed Kepler that planets do not move at uniform speeds along their orbits. Rather, a planet moves most rapidly when it is nearest **2** the Sun, a point on its orbit called **perihelion.** Conversely, a planet moves most slowly when it is farthest from the Sun, a point called **aphelion.**

After much trial and error, Kepler discovered a way to describe how fast a planet moves anywhere along its orbit. This discovery, also published in *New Astronomy,* is illustrated in Figure 2-7. Suppose that it takes 30 days for a planet to go from point *A* to point *B.* During that time, the line joining the Sun and the planet sweeps out a

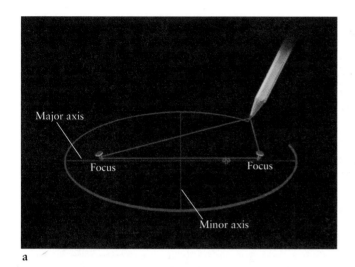

a

$e = 0$ $e = 0.3$ $e = 0.5$

Focus

$e = 0.7$ $e = 0.96$

b

Figure 2-6 **Ellipse** (a) The construction of an ellipse: An ellipse can be drawn with a pencil, a loop of string, and two thumbtacks, as shown in this diagram. If the string is kept taut, the pencil traces out an ellipse. The two thumbtacks are located at the two foci of the ellipse. (b) A series of ellipses with different eccentricities. Eccentricities range between 0 (circle) to just under 1.0 (virtually a straight line).

nearly triangular area. Kepler discovered that the line joining the Sun and the planet sweeps out the same area during any other 30-day interval. In other words, if the planet also takes 30 days to go from point C to point D, then the two shaded segments in Figure 2-7 are equal in area. *Kepler's second law,* also called the **law of equal areas,** can be stated thus:

A line joining a planet and the Sun sweeps out equal areas in equal intervals of time.

Physically this means that each planet's speed decreases as it moves from perihelion to aphelion and increases as it moves from aphelion toward perihelion.

One of Kepler's later discoveries, published in 1619, stands out because of its impact on future developments in astronomy. Now called *Kepler's third law,* it states a relationship between the sidereal period of a planet and the length of the semimajor axis of the planet's orbit:

The square of a planet's sidereal period is proportional to the cube of the length of its orbit's semimajor axis.

If we let P represent the sidereal period and a represent the length of the semimajor axis measured in astronomical units (see Toolbox I-2), we can express Kepler's third law more concisely as

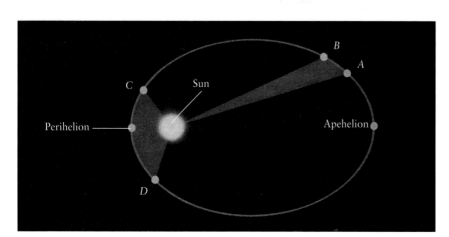

Figure 2-7 **Kepler's First and Second Laws** According to Kepler's first law, every planet travels around the Sun along an elliptical orbit with the Sun at one focus. According to his second law, the line joining the planet and the Sun sweeps out equal areas in equal intervals of time.

TABLE 2-3

A Demonstration of Kepler's Third Law

	Sidereal period P (yr)	Semimajor axis a (AU)	P^2	a^3
Mercury	0.24	0.39	0.06	0.06
Venus	0.61	0.72	0.37	0.37
Earth	1.00	1.00	1.00	1.00
Mars	1.88	1.52	3.53	3.51
Jupiter	11.86	5.20	140.7	140.6
Saturn	29.46	9.54	867.9	868.3
Uranus	84.01	19.19	7,058	7,067
Neptune	164.79	30.06	27,160	27,160
Pluto	248.54	39.53	61,770	61,770

$$P^2 = a^3$$

This law implies that the closer a planet is to the Sun, the more rapidly it orbits and the shorter its year.

The length of the semimajor axis is also the average distance between a planet and the Sun. Using data from Tables 2-1 and 2-2, we can demonstrate Kepler's third law as shown in Table 2-3.

It is testimony to Kepler's genius that his three laws apply to any situation where two objects orbit each other solely under the influence of their mutual gravitational attraction. Thus, Kepler's laws apply not only to planets circling the Sun but also to moons orbiting planets, artificial satellites orbiting the Earth, and two stars revolving about each other in binary star systems. Throughout this book, we shall see that Kepler's three laws have a wide range of practical applications.

2-4 Galileo's discoveries strongly supported a heliocentric cosmogony

While Kepler was making rapid progress in central Europe, an Italian physicist was making equally dramatic observations in southern Europe. Galileo Galilei did not invent the telescope, but he was one of the first people to point the new device toward the sky and publish his observations. He saw things that no one had ever imagined. He saw mountains on the Moon and sunspots on the Sun. He also discovered that Venus exhibits phases (Figure 2-8).

After only a few months of observation, Galileo noticed that the apparent size of Venus as seen through his telescope was related to the planet's phase. Venus appears smallest at gibbous phase and largest at crescent phase, a phenomenon inconsistent with the geocentric

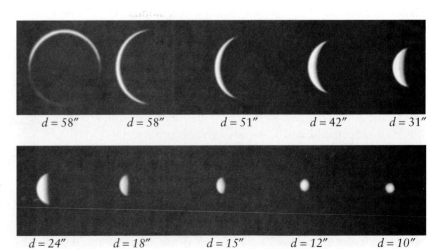

$d = 58''$ $d = 58''$ $d = 51''$ $d = 42''$ $d = 31''$

$d = 24''$ $d = 18''$ $d = 15''$ $d = 12''$ $d = 10''$

Figure 2-8 **The Phases of Venus** This series of photographs shows how the appearance of Venus changes as it moves along its orbit. The number below each view is the angular diameter (d) of the planet in arc seconds.

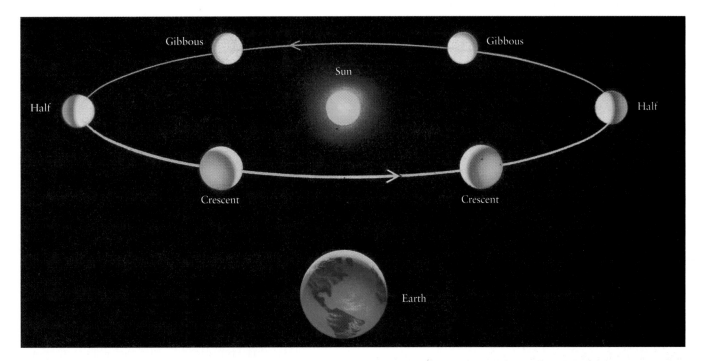

Figure 2-9 **The Changing Appearance of Venus** The phases of Venus are correlated with the planet's angular size and its angular distance from the Sun, as sketched in this diagram. These observations clearly support the idea that Venus orbits the Sun.

but consistent with the heliocentric cosmogony. These observations supported the conclusion that Venus goes around the Sun (Figure 2-9).

In 1610 Galileo also discovered four moons near Jupiter (Figure 2-10). He concluded that they are orbiting Jupiter because they move back and forth from one side of the planet to the other. Confirming observations were made in 1620 (Figure 2-11). Astronomers soon realized that these four moons obey Kepler's third law: The square of a moon's orbital period about Jupiter is proportional to the cube of its average distance from the planet. In honor of their discoverer, these four moons are today called the **Galilean moons** (or **satellites**).

Galileo's telescopic observations constituted the first fundamentally new astronomical data since humans began recording what they saw in the sky. In contradiction to then-prevailing opinions, these discoveries strongly supported a heliocentric view of the universe. Because Galileo's ideas could not be reconciled with certain passages in the Bible or with the writings of Aristotle and Plato, the Roman Catholic church condemned him. He was forced to spend his latter years under house arrest "for vehement suspicion of heresy."

There was a major stumbling block that prevented seventeenth-century thinkers from accepting Kepler's

laws and Galileo's conclusions about the heliocentric cosmogony. At that time the relationships between matter, motion, and forces were not understood. Consequently, people then did not know about the gravitational force of the Sun, which keeps the planets in orbit. They did not know how planets, once they had started in orbit around the Sun, could keep moving. Once anything

Figure 2-10 **Jupiter and Its Largest Moons** This photograph, taken by an amateur astronomer with a small telescope, shows the four Galilean satellites alongside an overexposed image of Jupiter. Each satellite would be bright enough to be seen with the unaided eye were it not overwhelmed by the glare of Jupiter.

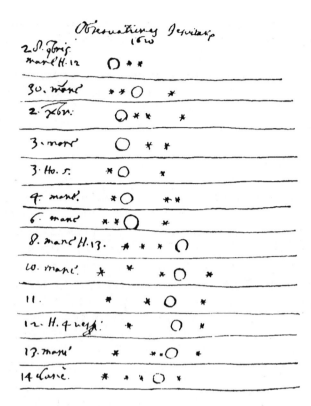

Figure 2-11 **Early Observations of Jupiter's Moons** In 1610 Galileo discovered four "stars" that move back and forth across Jupiter. He concluded that they are four moons that orbit Jupiter much as our Moon orbits the Earth. This drawing shows observations made by Jesuits in 1620.

on Earth is put in motion, it quickly comes to rest. Why didn't the planets orbiting the Sun stop, too?

All those mysteries were soon explained by a brilliant, eccentric scientist, Isaac Newton. Newton was born on Christmas Day, 1642, less than a year after Galileo died. In the decades that followed, Newton revolutionized science more profoundly than any person before him, and in doing so he found physical and mathematical proof of the heliocentric cosmogony.

2-5 Newton formulated three laws that describe fundamental properties of physical reality

Until the mid-seventeenth century, virtually all mathematical astronomy used the same approach. Astronomers from Ptolemy to Kepler worked empirically, or

directly from data and observations. They adjusted their ideas and calculations until they finally came out with the right answers.

Isaac Newton introduced a new approach. He began by making three assumptions about motion that he applied to all forces and bodies. Newton then showed that Kepler's three laws logically follow from these laws of motion and from a formula for the force of gravity that Newton also derived. Using this formula, Newton accurately described the observed orbits of the Moon, comets, and other objects in the solar system. Newton's laws also apply to the motions of all bodies on the Earth.

The first of **Newton's laws of motion** is known as *Newton's first law* or the **law of inertia**:

A body remains at rest or moves in a straight line at a constant speed unless acted upon by an unbalanced outside force.

At first, this law might seem to conflict with your everyday experience. For example, if you shove a chair, it does not continue at a constant speed forever, but rather comes to rest after sliding only a short distance. From Newton's viewpoint, however, an "unbalanced outside force" does indeed act on the moving chair, namely, friction between the chair's legs and the floor. If there were no friction, the chair would continue in a straight path at a constant speed.

Newton's first law tells us that there must be an outside force acting on the planets. If there were no such force acting, the planets would leave their curved orbits and move away from the Sun along straight-line paths at constant speeds. Because this does not happen, Newton concluded that the continuous action of this force confines the planets to their elliptical orbits.

Newton's second assumption formalizes the concept of a force and describes how a force changes the motion of an object. To appreciate these concepts, we must first understand quantities that describe motion: speed, velocity, and acceleration.

Imagine an object in space. Push on the object and it begins to move. At any moment, you can describe the object's motion by specifying both its speed and direction. Speed and direction of motion together constitute the object's **velocity**. If you continue to push on the object, its speed will increase—it will accelerate.

Acceleration is the rate at which velocity changes with time. Since velocity involves both speed and direction, acceleration does not apply only to increases in speed. A slowing down, a speeding up, or a change in direction are all types of acceleration.

Suppose an object revolved about the Sun in a perfectly circular orbit. This body would have acceleration that involved only a change of direction. As this object moved along its orbit, its speed would remain constant, but its direction of motion would be continuously changing. Therefore, it would be continuously accelerating.

Newton's second law says that the acceleration of an object is proportional to the force acting on it. In other words, the harder you push on an object, the greater the resulting acceleration. This law can be succinctly stated as an equation. If a force acts on an object, the object will experience an acceleration such that

Force = mass × acceleration

The **mass** of an object is a measure of the total amount of material in it, which we measure in kilograms. For example, the mass of the Sun is 2×10^{30} kg, the mass of a hydrogen atom is 1.7×10^{-27} kg, and the mass of the author of this book is 83 kg. The Sun, a hydrogen atom, and the author have these same masses regardless of where they happen to be in the universe.

It is important not to confuse the concept of mass with that of weight. **Weight** is the force with which an object presses down on the ground (due to gravity's pull), and force is usually expressed in pounds or newtons. For example, the force with which the author presses down on the ground is 174 pounds.

Note that the author weights 174 pounds only on the Earth. I would weigh less on the Moon and more on Jupiter. Floating in space or orbiting in the Space Shuttle, I would have no weight at all; that is, the same scale that gave my weight as 174 pounds on the Earth would indicate I had no weight if I put it under my feet while aboard the Shuttle. Whenever we describe the properties of planets, stars, or galaxies, we speak of their masses, never of their weights.

Newton's final assumption, called *Newton's third law*, is the famous statement about the forces of action and reaction:

Whenever one body exerts a force on a second body, the second body exerts an equal and opposite force on the first body.

For example, if I weigh 174 pounds, then I press down on the floor with a force of 174 pounds. Newton's third law says that the floor is also pushing up against me with an equal force of 174 pounds. (If it were not, I would fall through the floor.) In the same way, Newton realized that because the Sun is exerting a force on each planet to keep it in orbit, each planet must also be exerting an equal and opposite force on the Sun. As each planet accelerates toward the Sun, the Sun in turn accelerates toward each planet.

Conservation of angular momentum, a fundamental law of physics that affects every object in the cosmos, is a consequence of Newton's laws. Angular momentum is a measure of how much energy is stored in an object due to its rotation or revolution. Angular momentum depends on three things: how fast the body rotates or revolves, how much mass it has, and how spread out that mass is. The greater a body's angular motion or mass, or the more the mass is spread out, the greater its angular momentum. The law of conservation of angular momentum states that an object's angular momentum remains constant unless acted on by an outside force.

Consider twirling ice skaters as examples of rotating objects with a constant mass free of outside forces. When spinning skaters wish to rotate more rapidly, they decrease the spread of their mass distribution by pulling their arms in closer to their bodies. According to the conservation of angular momentum, as the spread of mass decreases, either the amount of mass or the rotation rate must increase. Since the skaters don't change mass, they spin faster. In astronomy, we encounter many instances of the effects of conservation of angular momentum on contracting objects.

We have now reconstructed the essential relationships between matter and motion. Scientific belief in the heliocentric cosmogony still requires a force to hold the planets in orbit around the Sun and the moons in orbit around the planets. That, too, Newton identified.

2-6 Newton's description of gravity accounts for Kepler's laws

Isaac Newton did not invent the idea of gravity. An educated seventeenth-century person had a vague appreciation of the fact that some force pulls things down to the ground. It was Newton, however, who gave us a precise description of the action of gravity. Using his first law, Newton proved mathematically that the force acting on each of the planets is directed toward the Sun. This discovery led him to suspect that the force pulling a falling apple straight down to the ground is the same as the

EYES ON . . .   Early Contributors to Modern Gravitational Theory

Between 1525 and 1725, human understanding of the motion of celestial bodies and the nature of the gravitational force that keeps them in orbit surged forward as never before. Theories related to this subject were developed by Nicolaus Copernicus, Johannes Kepler, Galileo Galilei, and others. Their work established and verified the heliocentric model of the solar system and the role of gravity. Below are brief biographies of these important contributors to our understanding of the heavens.

Nicolaus Copernicus (1473–1543) was born in Toruń, Poland, the youngest of four children. He received a doctorate in canon law from the University of Ferrara in Italy. He also studied medicine at the University of Padua. Shortly after Copernicus returned to Poland in 1506, his uncle secured him an appointment as a canon at the Cathedral of Frauenburg. Income from this position supported him for the rest of his life, freeing him in his mature years to develop a heliocentric theory of the known universe. Copernicus published his work in 1543 under the title *De revolutionibus orbium coelestium*. While the book was still in press, Copernicus became ill. He died just after receiving the first copy.

Tycho Brahe (1546–1601) was born to nobility in the Danish city of Knudstrup, which is now part of Sweden. In 1576 the king offered him the use of the island of Hven and funding to build an astronomical observatory

and an alchemical laboratory. Brahe worked there for 20 years making extremely precise astronomical measurements. He rejected Copernicus' heliocentric theory and the Ptolemaic geocentric system, too, instead devising a halfway theory called the "Tychonic system." According to Brahe's theory, the Earth is stationary, with the Sun and Moon revolving around it, while all the other planets revolve around the Sun. Following Brahe's death in 1601, Kepler inherited the precious observations on which Brahe's fame still rests today.

Johannes Kepler (1571–1630) was educated at the Swabian University at Tübingen in Germany, where he spent three years studying mathematics, philosophy, and theology. In 1596 Kepler published a booklet in which he attempted to mathematically predict the planetary orbits. Although this idea was altogether wrong, its boldness and originality attracted the attention of Tycho Brahe, who had recently moved to Prague. At Brahe's invitation, Kepler joined his staff in 1600. A memorable collaboration between these two great astronomers—the gifted, aging observer and the young, brilliant theoretician—lasted only 22 months before Brahe's death. In those months, Kepler had so distinguished himself that Holy Roman Emperor Rudolf II soon appointed him to succeed Brahe as Imperial Mathematician—a position Kepler held for the rest of his life.

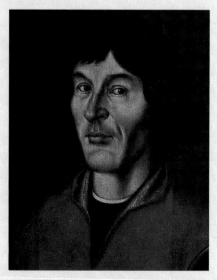

Nicolaus Copernicus

Tycho Brahe

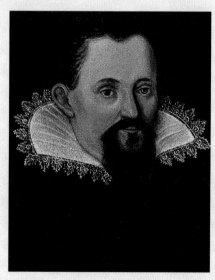

Johannes Kepler

Galileo Galilei (1564–1642) was born in Pisa, Italy. Encouraged by his father to become a doctor, Galileo studied medicine and philosophy at the University of Pisa. He dropped out of school in 1585 for lack of money and abandoned medicine in favor of mathematics. At the age of 25, however, he returned to the University of Pisa as a professor of mathematics. It was about this time that Galileo formulated his famous law of falling bodies: All objects fall with the same acceleration regardless of their weight. Followers of Aristotle, who said that heavier bodies fall faster than lighter ones, bitterly opposed Galileo and forced him to leave the university. In 1592 Galileo became professor of mathematics at the University of Padua, where he taught for the next 18 years. In 1609 he learned about the invention of a telescope by a Dutch optician and proceeded to construct one of his own. When he turned the telescope skyward, he made a host of discoveries that contradicted the teachings of Aristotle and the Roman Catholic church. He summed up his life's work on motion, acceleration, and gravity in the book *Dialogue on the Two New Sciences,* published in 1637.

Isaac Newton (1642–1727) was a quiet boy who seemed to be more interested in making mechanical devices than studying. He constructed sundials, model windmills, a water clock, and a mechanical carriage. By his own admission later in life, he was very inattentive at school. A former teacher who recognized Newton's intellect persuaded his mother to allow him to go to Cambridge University, where he showed no exceptional ability and received a bachelor's degree in 1665. About the time of his graduation, there was an outbreak of the plague which had caused a general exodus from the cities. Newton returned to his home town in the country where he spent two years developing calculus and formulating his thoughts on gravitation. He returned to Cambridge in 1667 and pursued experiments in optics. A year later, he constructed a telescope that used a concave mirror to focus incoming light and magnify images of remote objects. Using a prism, Newton also discovered that white light is actually a mixture of all colors. This discovery was controversial and drew considerable criticism, most of it unfounded. Newton was so sensitive to such criticism that his friends often had to plead with him to publish his most important ideas. His major work on forces and gravitation was the tome *Philosophae naturalis principia mathematica,* which appeared in 1687. In 1704, Newton published his second great treatise, *Opticks,* in which he described his experiments and theories about light and color. Upon his death in 1727, Newton was buried in Westminster Abbey, the first scientist to be so honored.

Galileo Galilei

Isaac Newton

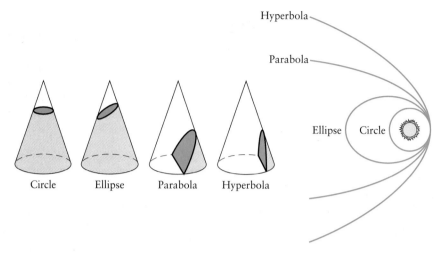

Figure 2-12 **Conic Sections** A conic section is any one of a family of curves obtained by slicing a cone with a plane, as shown in this diagram. The orbit of one body about another can be an ellipse, a parabola, or a hyperbola.

force on the planets that is always aimed straight at the Sun (Toolbox 2-2).

Using his own three laws and Kepler's three laws, Newton succeeded in formulating a general statement describing the nature of the force called **gravity** that keeps the planets in their orbits. Newton's **universal law of gravitation** states:

Two bodies attract each other with a force that is directly proportional to the product of their masses and inversely proportional to the square of the distance between them.

This law means that gravitational force decreases with distance; move twice as far away from a body and you feel only one-quarter of the force from it that you felt before.

Using his law of gravity, Newton found that he could mathematically prove the validity of Kepler's three laws. For example, whereas Kepler discovered by trial and error that $P^2 = a^3$, Newton demonstrated mathematically that this equation follows logically from his law of gravity.

Newton also discovered other types of orbits around the Sun. For example, his equations soon led him to conclude that the orbits of objects passing the Sun could also be **parabolas** and **hyperbolas** (Figure 2-12). For example, comets hurtling toward the Sun from the depths of space often follow parabolic orbits.

Newton's ideas turned out to be applicable in an incredibly wide range of situations. The orbits of the planets and their satellites could now be calculated with unprecedented precision. Using Newton's laws, mathematicians proved that the Earth's axis of rotation must precess because of the gravitational pull of the Moon and the Sun on the Earth's equatorial bulge (recall Figure 1-17).

In addition, Newton's laws and mathematical techniques could be used to predict new phenomena. For example, one of Newton's few friends, Edmund Halley, was intrigued by historical records of a comet that was sighted about every 76 years. Using Newton's methods, Halley worked out the details of the comet's orbit and predicted its return in 1758. It was first sighted on Christmas night of that year, and to this day the comet bears Halley's name (Figure 2-13).

Perhaps the most dramatic confirmation of Newton's ideas was their role in the discovery of the eighth planet from the Sun. The seventh planet, Uranus, had been discovered accidentally by William Herschel in 1781 during

Figure 2-13 **Halley's Comet** Halley's Comet orbits the Sun with an average period of about 76 years. During the twentieth century, the comet passed near the Sun twice—once in 1910 and again in 1986. This photograph shows how the comet looked in 1986. (Australian National University)

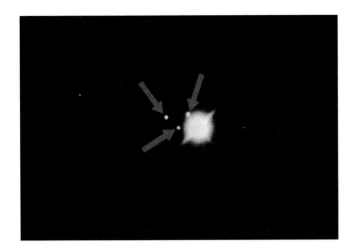

Figure 2-14 **Uranus and Neptune** The discovery of Neptune was a major triumph for Newtonian mechanics. In an effort to explain why Uranus (shown on the left, with three of its moons) was deviating from its predicted orbit, astronomers predicted the existence of Neptune (shown on the right, with one of its moons). Uranus and Neptune are nearly the same size; both have diameters about four times that of Earth.

a telescopic survey of the sky. Fifty years later, however, it was clear that Uranus was not following its predicted orbit. Two mathematicians, John Couch Adams in England and U. J. Leverrier in France, independently calculated that the deviations of Uranus from its orbit could be explained by the gravitational pull of a yet unknown, more distant planet. Each man predicted that the planet would be found at a certain location in the constellation of Aquarius. A telescopic search on September 23, 1846, revealed Neptune less than 1° from its calculated position. Although sighted with a telescope, Neptune was really discovered with pencil and paper (Figure 2-14).

....

AN ASTRONOMER'S TOOLBOX 2-2

Gravitational Force

If two objects have masses m_1 and m_2 and are separated by a distance r, then the gravitational force F between these two masses is

$$F = G\left(\frac{m_1 m_2}{r^2}\right)$$

In this formula, G is a number called the **universal constant of gravitation**, whose value has been determined from laboratory experiments:

$$G = 6.67 \times 10^{-11} \text{ m}^3/\text{kg} \cdot \text{s}^2$$

If m_1 is the mass of the Earth (6.0×10^{24} kg), for example, m_2 is the mass of the Sun (2.0×10^{30} kg), and r is the distance from the Earth to the Sun (1.5×10^{11} m), this equation gives the force from the Sun on the Earth and, equivalently, from the Earth on the Sun:

$$F = 3.6 \times 10^{22} \text{ kg/m} \cdot \text{s}^2$$

This number can then be used in Newton's second law, $F = ma$, to find the acceleration of the Earth due to the Sun. This yields

$$a_{\text{Earth}} = F/m_1 = 6.0 \times 10^{-3} \text{ m/s}^2$$

Newton's third law says that the Earth exerts the same force on the Sun, so the Sun's acceleration due to the Earth's gravitational force is

$$a_{\text{Sun}} = F/m_2 = 1.8 \times 10^{-8} \text{ m/s}^2$$

In other words, the Earth pulls on the Sun, causing the Sun to move toward it. Because of the Sun's greater mass, however, the amount that the Earth accelerates the Sun is over 300,000 times less than how much the Sun accelerates the Earth.

....

Over the years, Newton's ideas were successfully used to predict and explain many phenomena. Even today, as we send astronauts into Earth orbit and probes to the outer planets, Newton's equations are used to calculate the orbits and trajectories of these spacecraft.

It is a testament to Isaac Newton's genius that his three laws were precisely the three basic ideas needed to understand the motions of the planets. Newton brought a new dimension of elegance and sophistication to our understanding of the workings of the universe.

WHAT DID YOU THINK?

1 *What is the shape of the Earth's orbit around the Sun?* All planets have elliptical orbits around the Sun.

2 *Do the planets orbit the Sun at constant speeds?* The speed varies inversely with distance from the Sun. The farther a planet is in its elliptical orbit from the Sun, the slower it moves.

Key Words

acceleration	epicycle	Kepler's laws	retrograde motion
aphelion	focus (of an ellipse;	law of equal areas	semimajor axis
configuration	*plural* foci)	law of inertia	(of an ellipse)
(of a planet)	force	mass	sidereal period
conjunction	Galilean moons	Newton's laws	superior conjunction
conservation of	(satellites)	of motion	synodic period
angular momentum	geocentric cosmogony	Occam's razor	universal constant
cosmogony	gravity	opposition	of gravitation
deferent	greatest elongation	orbital eccentricity	universal law
direct motion	heliocentric cosmogony	parabola	of gravitation
ellipse	hyperbola	parallax	velocity
elongation	inferior conjunction	perihelion	weight

Key Ideas

• Copernicus' heliocentric (Sun-centered) theory simplified the general explanation of planetary motions compared to the geocentric theory.

• In a heliocentric system, the Earth is but one of several planets that orbit the Sun.

• The sidereal period of a planet, which is measured with respect to the stars, is the planet's true orbital period. Its synodic period is measured with respect to the Sun as seen from the moving Earth (for example, from one opposition to the next).

• Ellipses describe the paths of the planets around the Sun much more accurately than do circles. Kepler's three laws give important details about elliptical orbits.

• The invention of the telescope led Galileo to new discoveries, such as the phases of Venus and the moons of Jupiter, that supported a heliocentric view of the universe.

• Newton based his explanation of the universe on three assumptions called the laws of motion. These laws and his universal law of gravitation can be used to deduce Kepler's laws and to describe planetary motions with extreme accuracy.

• The mass of an object is a measure of the amount of matter in the object; its weight is a measure of the force with which the gravity of some other object pulls on the object's mass.

• In general, the path of one astronomical object about another, such as that of a comet about the Sun, is an ellipse, a parabola, or a hyperbola.

Review Questions

1 How did Copernicus explain the retrograde motions of the planets?

2 Which planets can never be seen at opposition? Which planets can never be seen at inferior conjunction?

3 At what configuration (superior conjunction, greatest eastern elongation, etc.) would it be best to observe Mercury or Venus with an Earth-based telescope? At what configuration would it be best to observe Mars, Jupiter, or Saturn? Explain your answers.

4 What is the difference between the synodic and sidereal periods of a planet?

5 What are Kepler's three laws. Why are they important?

6 In what ways did the astronomical observations of Galileo support a heliocentric cosmogony?

7 How did Newton's approach to understanding planetary motions differ from that of his predecessors?

8 What is the difference between mass and weight?

9 Why is the discovery of Neptune a major confirmation of Newton's universal law of gravitation?

Advanced Questions

10 Is it possible for an object in the solar system to have a synodic period of exactly one year? Explain your answer.

11 Explain qualitatively the systematic decrease in the synodic periods of the planets from Mars outward, as shown in Table 2-1.

∗ 12 A line joining the Sun and an asteroid was found to sweep out 5.2 square astronomical units of space in 1994. How much area was swept out in 1995? In five years?

∗ 13 A comet moves in a highly elongated orbit about the Sun with a period of 1000 years. What is the length of the semi-major axis of the comet's orbit? What is the farthest the comet can get from the Sun?

∗ 14 The orbit of a spacecraft about the Sun has a perihelion distance of 0.5 AU and an aphelion distance of 3.5 AU. What is the spacecraft's orbital period?

∗ 15 Suppose that the Earth were moved to a distance of 10 AU from the Sun. How much stronger or weaker would the Sun's gravitational pull be on the Earth?

∗ 16 Look up orbital data for the four largest moons of Jupiter. Demonstrate that these data obey Kepler's third law.

Discussion Questions

17 Which planet would you expect to exhibit the greatest variation in apparent brightness as seen from Earth? Explain your answer.

18 Use two thumbtacks, a loop of string, and a pencil to draw several ellipses. Describe how the shapes of the ellipses vary as you change the distance between the thumbtacks.

Observing Projects

19 It is quite probable that within a few weeks of your reading this chapter one of the planets will be in opposition or at greatest eastern elongation, making it readily visible in the evening sky. Consult a reference book such as the current issue of the *Astronomical Almanac* or the pamphlet entitled *Astronomical Phenomena* (both published by the U.S. government) to select a planet that is at or near such a configuration. At that configuration, would you predict the planet to be moving rapidly or slowly from night to night against the background stars? Verify your predictions by observing the planet once a week for a month, recording your observations on a star chart.

20 If Jupiter happens to be visible in the evening sky, observe the planet with a small telescope on five consecutive clear nights. Record the positions of the four Galilean satellites by making nightly drawings, just as the Jesuit priests did in 1620 (see Figure 2-11). From your drawings, can you tell which moon orbits closest to Jupiter and which orbits farthest? Is there a night when you could see only three of the moons? What do you suppose happened to the fourth moon on that night?

21 If Venus happens to be visible in the evening sky, observe the planet with a small telescope once a week for a month. On each night, make a drawing of the phase that you see. From your drawings can you determine if the planet is nearer or farther from the Earth than the Sun is? Do your drawings show any changes in the shape of the phase from one week to the next? If so, can you deduce if Venus is coming toward us or moving away from us?

3 ▶ *Light and Telescopes*

•••

IN THIS CHAPTER

You will explore the nature of visible light and other types of electromagnetic radiation. Then you will find out how telescopes collect and focus light so that it can be carefully examined. Finally, you will discover how astronomers use all kinds of electromagnetic radiation to observe the stars and other astronomical objects.

The National Optical Astronomy Observatories at Kitt Peak Astronomers prefer to build ground-based observatories on isolated mountaintops far from city lights, where the air is dry, stable, and cloud-free. The Kitt Peak Observatory, 54 mi from Tucson, Arizona, in the Quinlan Mountains at an altitude of 2098 m, is home for more than half a dozen optical telescopes, a telescope to observe the Sun, and a radio telescope.

WHAT DO YOU THINK?

1 What is the main purpose of a telescope?

2 Why do stars twinkle?

· · · · · · · · · · · · · · · · · · · ·

THE TELESCOPE IS THE single most important tool of astronomy. Using a telescope, we see objects in space far more brightly, clearly, and at a greater distance than we can with the naked eye. Telescopes have played a major role in revealing the universe since Galileo saw craters on the Moon for the first time with a telescope four centuries ago. Refracting telescopes, which use large lenses to collect incoming starlight, were popular in the nineteenth century. Modern astronomers strongly prefer reflecting telescopes, which gather light with large concave mirrors. Astronomers attach a variety of equipment to telescopes with which to record and analyze incoming starlight.

The Nature of Light

Light is radiant energy. This fact is apparent to anyone who has felt the warmth of the sunshine on a summer's day. But what exactly is light? How is it produced? What is it made of? How does it move through space? Scientists have struggled with these questions for the past four centuries.

3-1 Early discoveries explained white light and revealed the speed of light

The first major breakthrough in understanding light came from a simple experiment performed by Isaac Newton in the late 1600s. He passed a beam of sunlight through a glass prism, which spread the light out into the colors of the rainbow, as shown in Figure 3-1. This rainbow, called a **spectrum** (*plural* **spectra**), suggested to Newton that white light is actually a mixture of all colors. In Newton's day, most people erroneously believed that the colors were somehow added to the light by the glass. Newton disproved this idea by passing the spectrum through a second prism inverted with respect to the first. Since only white light emerged from this second prism, he had shown that the second prism had reassembled the colors of the rainbow to give back the original beam of sunlight.

Next came the question of how swiftly light moves. Does it travel instantaneously from one place to another, or does it move with a measurable speed? Whatever may be the nature of light, it travels incredibly quickly, far faster than sound. This is why we see lightning before we hear the accompanying thunderclap.

The first evidence for the finite speed of light came in 1675, when Ole Roemer, a Danish astronomer, carefully timed eclipses of Jupiter's moons (Figure 3-2). Roemer discovered that the moment at which a moon enters Jupiter's shadow depends on the distance between the Earth and Jupiter. When the Earth–Jupiter distance is short (when the two planets are on the same side of the

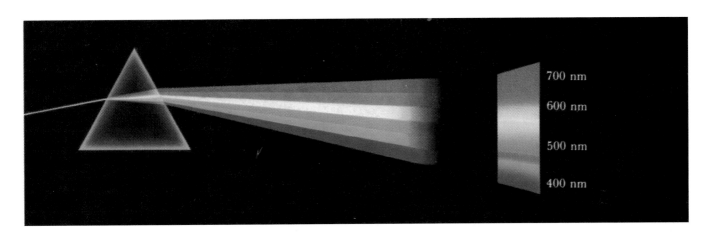

Figure 3-1 **A Prism and a Spectrum** When a beam of white light passes through a glass prism, the light is separated or refracted into a rainbow-colored band called a spectrum. The numbers on the right side of the spectrum indicate wavelengths, as described in the text.

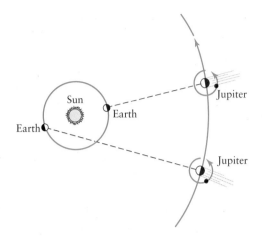

Figure 3-2 Measuring the Speed of Light An eclipse seen at opposition (Earth and Jupiter on the same side of the Sun) appears to occur earlier than an eclipse seen near conjunction (Earth and Jupiter on opposite sides of the Sun). The difference in apparent time of the eclipses is due to the extra time it takes light to travel the additional distance when the planets are near conjunction. The actual time of the eclipse is determined using Kepler's laws.

Sun), eclipses are observed slightly earlier than when Jupiter and the Earth are widely separated.

Roemer correctly interpreted the difference in time as the result of the different lengths of time it takes visible light to travel different distances across space. The

greater the distance to Jupiter, the longer the image of an eclipse takes to reach our eyes. From his timing measurements, Roemer concluded that it takes $16\frac{1}{2}$ minutes for visible light to traverse the diameter of the Earth's orbit (2 AU). Incidentally, Roemer's experiment supported the heliocentric cosmogony but not the geocentric one.

Because no one then knew the length of the astronomical unit, Roemer could not actually compute the speed of light in kilometers per second. The first accurate laboratory experiments to measure the speed of visible light were performed in the mid-1800s. The speed of light in a vacuum, usually designated by the letter c, is 299,792.458 km/s, which we round in everyday conversation to

$$c = 300,000 \text{ km/s} = 186,000 \text{ mi/s}$$

(Standard abbreviations for units of speed, like "km/s" for kilometers per second and "mi/h" for miles per hour, will be used throughout the rest of this book.)

The value c is a fundamental property of the universe. It appears in many equations that describe, among other things, atoms, gravity, electricity, and magnetism. Furthermore, according to Einstein's special theory of relativity (Chapter 13), nothing can travel faster than the speed of light in a vacuum. In addition, visible light traveling through air, water, glass, or any other transparent substance always moves more slowly than it does in a vacuum.

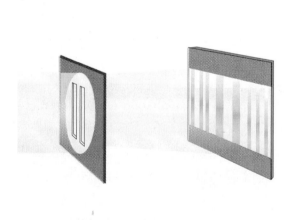

a

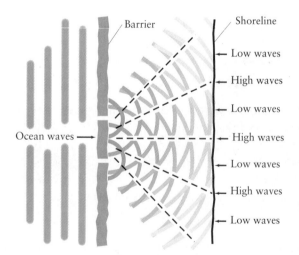

b

Figure 3-3 Electromagnetic Radiation Travels as Waves (a) Thomas Young's interference experiment shows that light of a single color passing through a barrier with two slits creates alternating light and dark patterns on a screen. (b) The intensity of light on the screen is analogous to the height of water

waves that strike the shore after passing through a barrier with two openings. In certain places, ripples from both openings reinforce each other to produce extra high waves. At intermediate locations along the shore line, ripples and troughs cancel each other, producing still water.

3-2 The complex nature of light became apparent only this century

From his many experiments in optics, Newton argued that light is composed of tiny particles of energy. In the mid-1600s, an alternative explanation was proposed by the Dutch scientist Christian Huygens, who suggested that light travels in the form of waves rather than particles.

The English physicist Thomas Young confirmed the wave nature of light in 1801. Young demonstrated that the shadows of objects in light of a single color are not crisp and sharp. Instead, the boundary between illuminated and shaded areas is overlain with patterns of closely spaced dark and light bands. These patterns are similar to the patterns produced by water waves passing through a barrier in the ocean (Figure 3-3). Young showed that a wavelike description of light could explain the results of his experiment, but Newton's particle theory could not. Analyses of similar experiments soon produced overwhelming evidence for the wavelike behavior of light.

Further insight into the wave character of light came from calculations by the Scottish physicist James Clerk Maxwell in the 1860s. Maxwell succeeded in describing all the basic properties of electricity and magnetism in four equations. By combining these equations, Maxwell demonstrated that electrical and magnetic effects should travel through space together in the form of waves. Maxwell's suggestion that these waves exist and are observed as light was soon confirmed by a variety of experiments. Because of its electric and magnetic properties, visible light is called **electromagnetic radiation** (Figure 3-4).

Newton showed that sunlight is actually composed of all the colors of the rainbow. Young and others showed that light travels as waves. What makes the colors of the rainbow distinct from each other? The answer is surprisingly simple: The only difference between different colors is the distance between two successive wave crests in the light wave. This distance is called the **wavelength** of the light, usually designated by the lowercase Greek letter λ (lambda; see Figure 3-4). Maxwell's calculations proved that light waves all travel in empty space at the same speed, regardless of wavelength.

The wavelengths of all colors are extremely small, less than a thousandth of a millimeter. To express these tiny distances conveniently, scientists use a unit of length called the *nanometer* (abbreviated "nm"), where 1 nm = 10^{-9} m. Experiments demonstrated that visible light has wavelengths covering the range from about 400 nm for violet light to about 700 nm for red light. Intermediate colors of the rainbow fall between these wavelengths (see Figure 3-1). The complete spectrum of colors from longest wavelength to shortest is red, orange, yellow, green, blue, and violet.

3-3 Einstein saw that light sometimes behaves as particles, sometimes as waves

In 1905, while scientists were becoming comfortable with the wave nature of light, Albert Einstein proposed that *light is actually composed of discrete bundles of waves that sometimes behave as waves and sometimes as particles!* He based this proposal on observations that shorter wavelengths of light can knock some electrons off the surfaces of metals while longer wavelengths of light cannot. This is called the *photoelectric effect.* Electrons are held onto a metal's surface by electric forces. The fact that some colors can remove the electrons and others cannot implies that the electrons receive different amounts of energy from different colors (or wavelengths) of light. Einstein proposed that light travels in discrete packets, called **photons**, and that photons of different wavelengths have different amounts of energy. He proposed that *the shorter a photon's wavelength, the higher its energy.* Specifically,

$$\text{Photon energy} = \frac{\text{Planck's constant} \times \text{speed of light}}{\text{wavelength}}$$

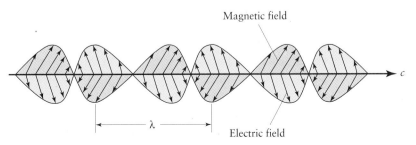

Figure 3-4 Electromagnetic Radiation All forms of electromagnetic radiation (radio waves, infrared radiation, visible light, ultraviolet radiation, x rays, and gamma rays) consist of oscillating electric and magnetic fields that move through empty space at a speed of 3×10^5 km/s. The distance between two successive crests is called the wavelength of the light.

where Planck's constant (denoted *h*), which is named for the German physicist Max Planck, has the value 6.6×10^{-34} kg·m²/s (Toolbox 3-1).

Each photon is a packet of electromagnetic waves of the same wavelength. All photons of the same wavelength are identical, and so every photon of a fixed wavelength carries the same amount of energy as every other photon of that wavelength. Einstein proposed that either the energy delivered by a photon to an electron is enough to rip it off the surface of the metal or it is not. This is exactly what is observed. There is no middle ground, as there would be if light of a fixed color came in waves with different numbers of cycles, each carrying different amounts of energy. Since that time, the concept of photons (wave packets) has been thoroughly proven, and astronomers incorporate this concept in their model of light.

3-4 Light is only one type of electromagnetic radiation

Although visible light has a narrow range of wavelengths, Maxwell's equations placed no restrictions on the wavelengths that electromagnetic radiation can have. It was therefore possible that electromagnetic waves existed with wavelengths both longer and shorter than the 400-to-700-nm range of visible light. Researchers began to look for invisible forms of light, forms to which the cells of the human retina do not respond.

Around 1800 the British astronomer William Herschel discovered **infrared radiation** in an experiment with a prism, when he held a thermometer just beyond the red end of the visible spectrum. The thermometer registered a temperature increase, indicating that it was being exposed to an invisible form of energy. In experiments with electric sparks in 1888, the German physicist Heinrich Hertz first succeeded in producing electromagnetic radiation a few centimeters long in wavelength, now known as **radio waves.** In 1895 Wilhelm Roentgen invented a machine that produces electromagnetic radiation with wavelengths shorter than 10 nm, now called **x rays.** Modern versions of Roentgen's machine are found in medical and dental offices. Over the years radiation has been discovered with many other wavelengths.

We now know that visible light occupies only a tiny fraction of the full range of possible wavelengths, collectively called the **electromagnetic spectrum.** As shown in Figure 3-5, the electromagnetic spectrum stretches from the longest-wavelength radio waves, through infrared ra-

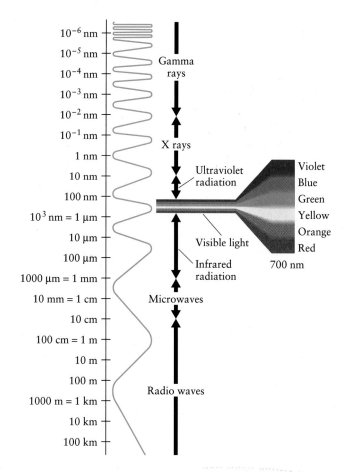

Figure 3-5 The Electromagnetic Spectrum The full array of all types of electromagnetic radiation is called the electromagnetic spectrum. It extends from the shortest-wavelength gamma rays to the longest-wavelength radio waves. Visible light forms only a tiny portion of the full electromagnetic spectrum.

diation, visible light, ultraviolet radiation, and x rays, to the shortest-wavelength photons, gamma rays. On the long-wavelength side of the visible spectrum, infrared radiation covers the range from about 700 nm to 1 mm. Astronomers interested in infrared radiation often express wavelength in *micrometers* or *microns* (abbreviated μ), where 1 μ = 1000 nm = 10^{-6} m. From roughly 1 mm to 10 cm is the range of microwaves, which are sometimes considered to be infrared radiation and sometimes radio waves.

At wavelengths shorter than those of visible light, **ultraviolet (UV) radiation** extends from about 400 nm down to 10 nm. Next are x rays, which consist of wavelengths between about 10 and 0.01 nm, and beyond them are **gamma rays.** It should be noted that these boundaries are approximate and are primarily used as

AN ASTRONOMER'S TOOLBOX 3-1

Photon Energies

The energies of photons with the same wavelength are always the same, implying that these photons are identical. The energy E of any given photon can be calculated from the equation

$$E = hc/\lambda$$

where Planck's constant h is 6.6×10^{-34} kg·m²/s, the speed of light c is 300,000 km/s, and the photon's wavelength is λ. For example, every photon of red light with wavelength 700 nm has an energy of

$$E_{red} = \frac{(6.6 \times 10^{-34} \text{ kg·m}^2/\text{s})(300,000 \text{ km/s})}{700 \text{ nm}}$$

$$E_{red} = 2.8 \times 10^{-19} \text{ J}$$

(A joule, abbreviated "J," is a unit of energy.) Each photon of red light of wavelength 700 nm has an energy of 2.8×10^{-19} J. For comparison, in one second a 25-watt light bulb emits 25 J of energy.

---

convenient divisions in the electromagnetic spectrum, which is actually continuous.

These various types of electromagnetic radiation share many basic properties. For example, they are all photons; they all travel at the speed of light; they all sometimes behave as particles and sometimes as waves. But because of their different wavelengths (and therefore different energies), they interact very differently with matter. Your body is virtually transparent to x rays but not to visible light; your eyes respond to visible light but not to gamma rays; your radio detects radio waves but not ultraviolet light.

The Earth's atmosphere is relatively transparent to both visible light and radio waves, meaning that both pass through freely to reach ground-based telescopes sensitive to these forms of electromagnetic radiation. We say that the atmosphere has *windows* for these parts of the electromagnetic spectrum (Figure 3-6). Infrared has a limited window through the atmosphere. We detect this radiation as heat.

Likewise, the longest-wavelength ultraviolet radiation has a limited window. This radiation causes tanning and sunburns. The Earth's atmosphere is completely opaque to the other types of electromagnetic radiation,

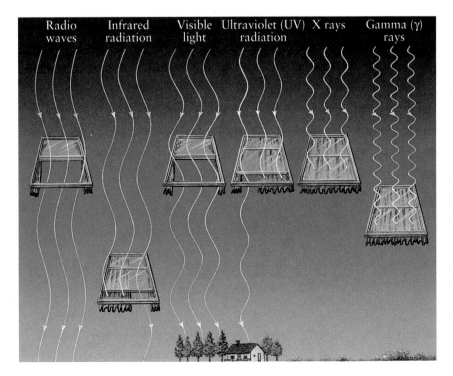

Figure 3-6 **Windows Through the Atmosphere** The atmosphere allows different types of electromagnetic radiation to penetrate into it by varying amounts. Visible light and radio waves reach all the way to the Earth's surface. Some infrared and ultraviolet radiation can reach the ground, too. The other types of radiation are absorbed or reflected by the gases in the air at different characteristic altitudes. While the atmosphere does not have actual "windows," astronomers use the term to characterize the passage of radiation through it.

meaning they do not reach the Earth's surface. (This is a good thing, because x rays, gamma rays, and shorter-wavelength ultraviolet radiation are lethal to living tissue.) Observations of these latter radiations must be performed high in the atmosphere or, ideally, from space.

Detecting electromagnetic radiation is the essence of observational astronomy. Until recently, photons were the only sources of detailed astronomical information that we had. However, in the past three decades, detectors of other energies and particles from space have been developed. These devices—neutrino detectors and gravity wave detectors—are discussed in Chapters 9 and 13, respectively.

Optics and Telescopes

Since the time of Galileo, astronomers have been designing instruments to collect more light than the human eye can collect on its own. As we will see, the ability to collect more light enables us to see things more brightly, in more detail, and at a greater distance than without the technological assistance. There are two basic types of telescopes: those that collect light through lenses and those that collect it from mirrors. We will study each separately, beginning with those using lenses, which were developed first.

3-5 A refracting telescope uses a lens to concentrate incoming light

Although light travels at about 300,000 km/s in a vacuum, it moves more slowly through a dense substance such as glass. The abrupt slowing of light entering a

piece of glass is analogous to the motion of a person walking from a boardwalk onto a sandy beach: Her pace suddenly slows as she steps from the smooth pavement into the sand. And just as the person stepping back onto the boardwalk easily resumes her original pace, light exiting a piece of glass resumes its original speed.

In addition to undergoing an abrupt change in speed, the direction light travels changes as it passes from one transparent medium into another. You see this phenomenon, called **refraction,** every day when looking through windows. Imagine a stream of photons from a star entering a window, as shown in Figure 3-7a. Astronomers call such photon flows light rays. As a light ray goes from the air into the glass, the light ray's direction changes so that it is more perpendicular to the surface of the glass than it was before entering. Once inside the glass, the light ray travels in a straight line. Upon emerging from the other side, the light ray bends once again, resuming its original direction and speed. This is why when we look through windows or car windshields we see exactly what we would see if we were actually on the other side of the glass (with a slight displacement due to the refraction of light in the glass).

To use refraction in designing telescopes, the light rays leaving the lens must go in different directions than the directions in which they were going upon entering the glass. To achieve this effect, lenses have curved surfaces (Figure 3-7b). As a result, not all the light striking the top surface is refracted in the same direction upon entering the glass. Rather, parallel light rays start converging once they enter the glass. As you can see from the figure, this convergence is further enhanced when the light emerges from the glass.

By shaping the lens correctly, the refracting property of glass can be used to converge the light passing

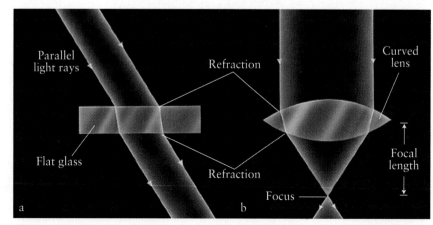

Figure 3-7 **Refraction Through Flat and Curved Glass (a)** A light ray entering a flat piece of glass is bent or refracted to an angle closer to perpendicular to the surface than the angle at which it was originally traveling. As the ray leaves the piece of glass, it is bent away from the perpendicular. **(b)** When the glass is curved to form a convex lens, parallel light rays converge to a focus. The distance from the lens to the focus is the focal length of the lens.

through it at a distance called the **focal length** (Figure 3-7b). If the light source is extremely far away, like the stars, then the incoming light rays are parallel (Figure 3-8 shows why distant light rays are parallel), and they come to a focus at a specific point called the **focus** of the lens. If the object is close enough to be more than just a dot as seen through the telescope, then all the light from it does not converge at the focus but rather in the **focal plane** (Figure 3-9). The Moon and planets are examples of such extended objects.

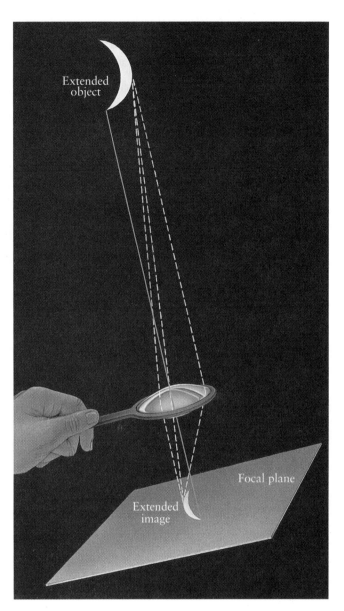

Figure 3-9 **Extended Objects Create a Focal Plane** Light from objects larger than points in the sky does not all converge to the focal point of a lens. Rather, the object creates an image at the focal length in what is called the focal plane.

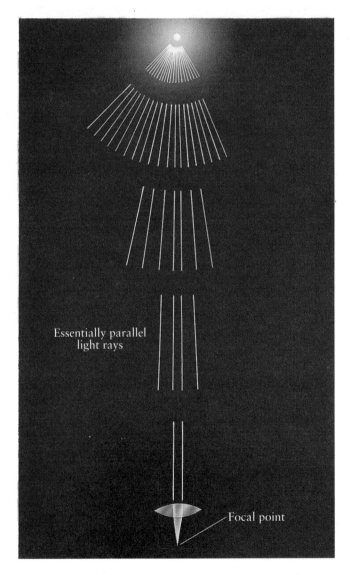

Figure 3-8 **Parallel Light Rays from Distant Objects** As light travels away from any object, the light rays, all moving in straight lines, separate. By the time light has traveled millions of miles, only the light rays moving in parallel tracks are still near each other.

After passing the focus or focal plane, a second lens can be used to restraighten the light rays. By using lenses of the proper shape, the resulting image will be a brighter, and often bigger and clearer, view of the object. Such an arrangement of lenses is called a **refracting telescope**, or **refractor** (Figure 3-10). The large-diameter, long-focal-length lens at the front of the telescope is the **objective lens**. Its purpose is to collect as much light as possible. The smaller, short-focal-length lens at the

Figure 3-10 **Essentials of a Refracting Telescope** A refracting telescope consists of a large, long-focal-length objective lens and a small, short-focal-length eyepiece lens that re-straightens light rays. The eyepiece lens also magnifies the image formed at the focal length of the objective lens.

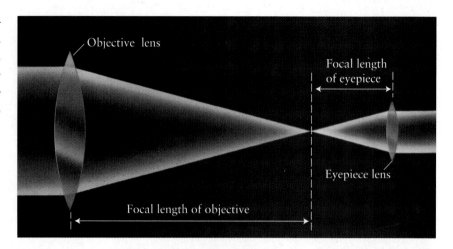

rear of the telescope is the **eyepiece lens**. It is designed to make light rays parallel once again, but the light is now more concentrated than it was before entering the telescope.

3-6 Telescopes brighten, resolve, and magnify

The most important function of a telescope is to provide the astronomer with as many photons as possible from an object. The result of collecting more photons, of course, is to increase each object's *brightness*. Indeed, telescopes are sometimes colloquially called "photon buckets." The more photons available, the more information the astronomer can extract, even if the object still appears as a pinpoint. For example, the relative intensities of different colors emitted by a star provide information about its temperature, chemical composition, age, and motion.

Brightness depends on the total number of photons collected, which in turn depends on the area of the objective lens. A large objective lens intercepts and focuses more starlight than does a small lens (Figure 3-11). A large objective lens can therefore produce brighter images and detect fainter stars than a small objective lens can.

The **light-gathering power** of a telescope is directly related to the area of the telescope's objective lens. Recall that the area and diameter of a circle are related by:

$$\text{Area} = \frac{\pi \times \text{diameter} \times \text{diameter}}{4}$$

where π (pi) is a constant approximately equal to 3.14.

Consequently, a lens with twice the diameter of another lens has 4 times the area of the smaller lens and therefore collects 4 times as much light as the smaller one does. For example, a 36-in.-diameter lens has 4 times the area of an 18-in.-diameter lens. Therefore, the 36-in. telescope has 4 times the light-gathering power of a telescope half its size. The general rule is: *Double the diameter, quadruple the light-gathering power.*

Figure 3-11 **Light-Gathering Power** Because a large lens intercepts more starlight than does a small lens, a large lens produces a brighter image. The same principle applies to telescopes that collect light using a primary mirror rather than an objective lens. Doubling the telescope's diameter quadruples its light-gathering power.

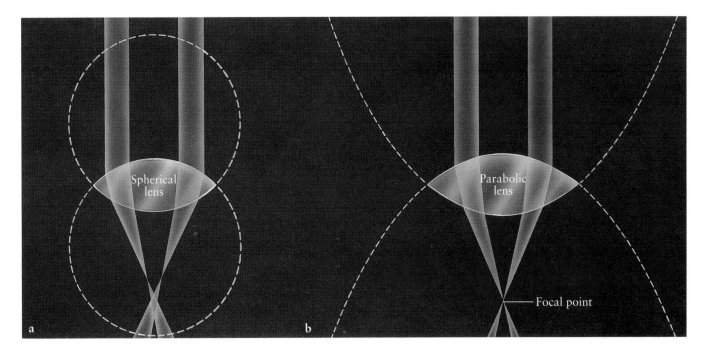

Figure 3-12 **Spherical Aberration and Parabolic Lenses** (a) Different parts of a spherical lens refract light to different focal points. (b) The ideal shape for a lens is a parabola. A par- abolic lens focuses all the light passing through it at the focal plane.

The second most important function of any telescope is to reveal more details of the extended objects under study. A large telescope increases the sharpness of the image and the degree of detail that can be seen. **Angular resolution** measures the clarity of images. Poor angular resolution causes images of galaxies or planets to be fuzzy and blurred. A telescope with good angular resolution produces extended images that are sharp and crisp.

The angular resolution of a telescope is measured as the arc angle between two adjacent stars whose images can just barely be distinguished by that telescope. The smaller the angle, the sharper the image. Large, modern telescopes, like the Keck telescopes in Hawaii, have angular resolutions better than 0.1 arcsec. The general rule is: *Double the diameter, double the detail that can be seen.*

The final function of a telescope—often the least important one—is to make objects appear larger. This property, as you know, is called **magnification**. The magnification of a refracting telescope is equal to the focal length of the objective lens divided by the focal length of the eyepiece lens:

$$\text{Magnification} = \frac{\text{focal length of the objective}}{\text{focal length of the eyepiece}}$$

For example, if the objective of a telescope has a focal length of 100 cm and the eyepiece has a focal length of 0.5 cm, then the magnifying power of the telescope is

$$\text{Magnification} = \frac{100 \text{ cm}}{0.5 \text{ cm}} = 200$$

This property is usually written as 200×.

There is a limit to the magnification that any telescope can have. Try to magnify beyond that limit and the image becomes distorted. The general rule for maximum magnification is: *Double the objective lens's diameter, double the telescope's maximum magnification.*

For convenience in manufacturing, objective lenses for telescopes are ground with spherical surfaces. However, this shape distorts the images of objects seen through the telescope. This distortion and other negative effects greatly limit the use of objective lenses in large research telescopes.

3-7 Refracting telescopes have several severe problems

The spherical surfaces of objective lenses are not ideal for bringing all the light into focus at the same focal length (Figure 3-12*a*). The effect is a slightly smeared image. This effect, called **spherical aberration,** can be overcome by grinding the lenses to have parabolic surfaces (Figure 3-12*b*). However, grinding parabolic lenses is so difficult that for practical purposes objective lenses always have spherical surfaces. The problem of spherical aberration is minimized by making the objective lens so thin that its spherical surfaces nearly coincide with parabolic surfaces. The thinness of the lens, however, requires that it have an extremely long focal length, which is why all research-sized refracting telescopes are many meters long (Figure 3-13).

Second, the amount of refraction that light undergoes varies with color (recall Figure 3-1). Because different colors have different focal lengths, the colors of images are distorted (Figure 3-14), a problem called **chromatic aberration**. This problem is corrected with a *compound lens* composed of two different types of glass with different refractive properties. Because compound lenses in telescopes and cameras bring all the wavelengths into focus at the same focal length, they are called **achromatic lenses** (Figure 3-15).

Sagging of the objective lens is a third problem that became apparent in the last century, as larger and larger objective lenses were ground. A large lens distorts as the telescope changes angle in the sky. This distortion occurs because the lens can only be supported around its edge, which is very thin. The thick center of a lens weighs it

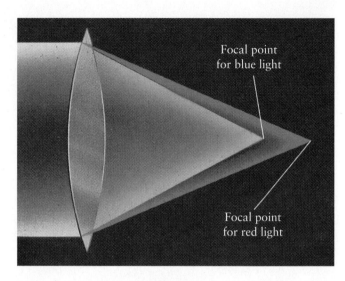

Figure 3-14 **Chromatic Aberration Spreads Colors Out** Light of different colors passing through a lens is refracted by different amounts. This effect is called chromatic aberration. As a result, different-colored objects have different focal lengths, even if the lens is parabolic. Consequently, the images seen through uncorrected lenses are blurred.

Figure 3-13 **A Large Refracting Telescope** This giant refracting telescope, built in the late 1800s, is housed at Yerkes Observatory near Chicago. The objective lens is 102 cm (40 in.) in diameter, and the telescope tube is $19\frac{1}{3}$ m ($63\frac{1}{2}$ ft) long.

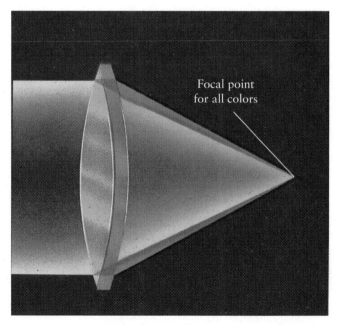

Figure 3-15 **An Achromatic Lens Corrects for Chromatic Aberration** A second lens that refracts colors by different amounts than does the objective lens can bring all colors into focus at the focal length.

down and causes the glass to deform from the desired shape. As a result, starlight comes into focus at different focal lengths, thus distorting the images. Moreover, as the telescope tracks, or follows, a star for a long exposure (often one hour), the angle of the telescope changes and, along with it, the distortion. There is no way of knowing the exact distortion that will result as the angle of the telescope changes.

Unwanted refractions are a fourth problem. Lenses are ground from large, thick disks of glass that are formed by pouring molten glass into a mold. As the liquid glass cools, gas in it becomes trapped, creating air bubbles in the solidified disk. When the lens is then ground into shape, any air bubbles inside it create unwanted and unpredictable extra refractions that blur the images. Part of the expertise of lensmakers is to choose a volume of glass for grinding with as few air bubbles as possible.

A fifth problem is that glass is *opaque* to certain kinds of light. This means that some wavelengths of light cannot pass through the glass. Even visible light is dimmed substantially in passing through the slab of glass at the front of a refractor, and ultraviolet radiation is largely absorbed by the glass lens.

Nineteenth-century master opticians devoted their lives to overcoming these problems, and several magnificent refractors were constructed in the late 1800s. The largest refracting telescope, completed in 1897, is located at the Yerkes Observatory near Chicago (see Figure 3-13) and has an objective lens 102 cm (40 in.) in diameter. The second largest refracting telescope is located at Lick Observatory near San Jose, California. This refractor has an objective lens whose diameter is 91 cm (36 in.). All refractors have extremely long focal lengths. For example, the Yerkes refractor has a focal length of $19\frac{1}{3}$ m (63.5 ft). No major refracting telescopes have been constructed in the twentieth century.

The alternative to using lenses to collect light is to use mirrors. Called **reflecting telescopes** or **reflectors**, such telescopes easily overcome most of the problems inherent in refracting telescopes. As a result, all contemporary research telescopes are reflectors. We turn now to studying how they work.

3-8 Reflecting telescopes use mirrors to concentrate incoming starlight

Prompted by the problem of chromatic aberration, Isaac Newton set about to replace the objective lens with a

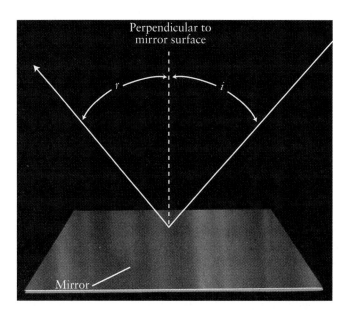

Figure 3-16 **Reflection** The angle at which a beam of light strikes a mirror (the angle of incidence *i*) is always equal to the angle at which the beam is reflected from the mirror (the angle of reflection *r*).

curved mirror that would collect the light. When using a mirror, the light is collected using the principle of reflection rather than refraction. To understand **reflection**, imagine a perpendicular line coming out of a flat mirror's surface at the point where a light ray strikes the mirror, as shown in Figure 3-16. The angle between the arriving light ray and the perpendicular is always equal to the angle between the reflected light ray and the perpendicular, regardless of the light's wavelength.

Using this principle, Isaac Newton realized that a concave mirror will cause parallel light rays to converge to a focus, as shown in Figure 3-17. The distance between the reflecting surface and the focus, where the image of the distant object is formed, is the focal length of the mirror.

In order to view the image, Newton placed a small, flat mirror at a 45° angle in front of the focal point, as sketched in Figure 3-18a. This **secondary mirror** deflects the light rays to one side of the telescope, where the astronomer can place an eyepiece lens through which to view the image. A telescope having this optical design is appropriately called a **Newtonian reflector**. The magnifying power of such a reflecting telescope is calculated in the same way as for a refractor: The focal length of the large or **primary mirror** is divided by the focal length of the eyepiece.

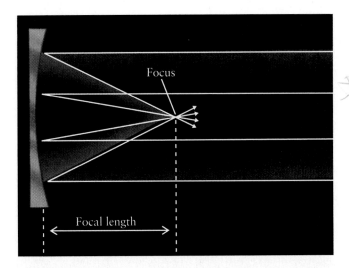

Figure 3-17 **A Concave Mirror** A concave mirror causes parallel light rays to converge to a focus. The distance between the mirror and the focus is the focal length of the mirror.

Newtonian telescopes are very popular with amateur astronomers because they are convenient to use while the observer is standing up. However, they are not used in research observatories because they are too lopsided. If

astronomers attached their often heavy and bulky research equipment onto its side, the telescope would sag and distort the image in unpredictable ways.

There are three basic designs for the reflecting telescopes used in research, all of which employ a concave primary mirror. In the first design, the primary mirror is so large that the astronomer actually sits at the undeflected focal point, directly in front of the primary mirror, and controls the photographic and other equipment directly. This arrangement is called a **prime focus** (Figure 3-18*b*). The astronomer rides throughout the night in the "observing cage," which in the middle of winter is a chilling experience.

The second popular design, called a **Cassegrain focus**, places the focal point at a convenient, accessible location. A hole is drilled directly through the center of the primary mirror, and a convex secondary mirror placed in front of the original focal point is used to reflect the light rays back through the hole (Figure 3-18*c*). Heavy but compact equipment can be bolted to the bottom of the telescope without distorting the telescope frame.

In the third design, a series of mirrors channels the light rays away from the telescope to a remote focal point. Particularly long and bulky optical equipment that cannot be mounted directly onto the telescope is located

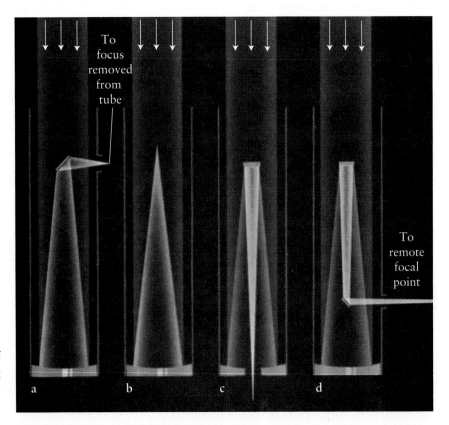

Figure 3-18 **Reflecting Telescopes** Four of the most popular optical designs for reflecting telescopes: (**a**) Newtonian focus, (**b**) prime focus, (**c**) Cassegrain focus, and (**d**) coudé focus.

at the resulting **coudé focus,** named after a French word meaning "bent like an elbow" (Figure 3-18*d*).

Reflecting telescopes have numerous advantages over refracting ones. First, the problem of chromatic aberration that Newton set out to overcome is automatically avoided because, unlike bathroom mirrors, all telescope mirrors have coated top surfaces. Therefore, the light never enters the glass at all! Second, the mirrors, which can weigh tons, don't warp because they can be supported anywhere they need to be from underneath; the light never gets to the bottom of the mirror. Third, air bubbles are much less of a problem since the mirror maker only needs to find a surface, rather than an entire volume, that is free of bubbles. Finally, since the light doesn't enter the glass, the problem of different opacities to different wavelengths never arises, either.

Nonetheless, reflecting telescopes aren't perfect; there are several prices to pay. Two of the most important are blocked light and spherical aberration. Let's consider each problem briefly.

3-9 Reflecting telescopes also have limitations

You have probably noticed that the secondary mirror blocks some of the incoming light. This is, indeed, an unavoidable price that astronomers must pay. Typically, a secondary mirror prevents about 10% of the incoming light from reaching the primary mirror. This problem is addressed by constructing primary mirrors with sufficiently large surface areas to compensate for the loss of light.

You might think that because light is missing from the center of the telescope there is a corresponding central "hole" in the images. However, this problem does not occur because, as discussed above, all astronomical objects are so far away that the light from them arrives as parallel rays. Therefore, light from all parts of each object enters all parts of the telescope (Figure 3-19).

As with refracting telescopes, reflecting telescopes can also suffer from the problem of **spherical aberration.** To make a reflector, an optician grinds and polishes a large slab of glass into the appropriate concave shape. It is much easier to grind a spherical mirror than a parabolic one. However, light entering a spherical telescope mirror at different distances from the mirror's center comes into focus at different focal lengths, as occurs for a spherical lens. Consequently, images taken with such telescopes appear blurry.

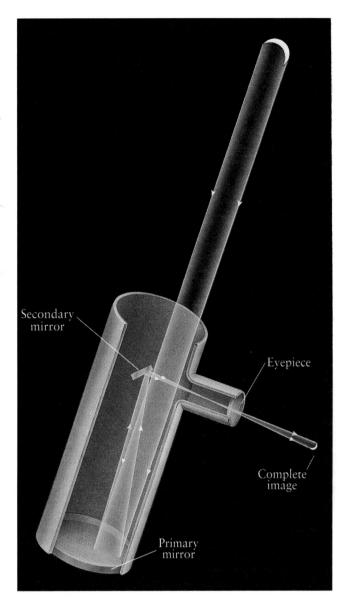

Figure 3-19 **The Secondary Mirror Does Not Create a Hole in the Image** Because the light rays from distant objects are parallel, light from the entire object reflects off all parts of the mirror. Therefore, every part of the object sends photons to the eyepiece. This picture shows the reconstruction of the entire Moon from light passing through just part of this telescope.

Spherical aberration may be overcome at some expense by making the mirror parabolic. An alternative solution is to place a thin correcting lens called a **Schmidt corrector plate** at the top of the telescope. The light coming into the telescope is refracted by the plate just enough to compensate for spherical aberration and to bring all the light into focus at the same focal length.

These correctors have the added benefit of allowing light from a larger angle in the sky to come into focus than would be in focus without the plate. Schmidt corrector plates enable astronomers to take relatively few photographs with moderately high magnification of the entire sky. In other words, the Schmidt corrector plate acts like a wide-angle lens on a camera. However, the plate does not allow for as much magnification as a telescope with a parabolic mirror.

3-10 Earth's atmosphere hinders astronomical research

Besides the problems created by the optics of telescopes, the Earth's atmosphere also introduces difficulties. The problems arise because the Earth's atmosphere is turbulent and filled with impurities. You have probably seen turbulence on a hot day while driving in a car when the road ahead appears to shimmer. This is due to blobs of air, heated by the Earth, moving upward. Light passing through such a blob is refracted because each such air mass has a different density than the air around it; hence it behaves like a lens. Images of objects beyond the blob therefore appear distorted.

The air over our heads is similarly moving, and the starlight passing through it is similarly refracted. Since the air changes rapidly, this refraction makes the star appear to change brightness and position rapidly, an effect we see as **twinkling**. Therefore, when photographed through large telescopes, twinkling smears out the star's image, causing it to look like a disk rather than a pinpoint of light. Astronomers use the expression "seeing" to describe how much twinkling is occurring and, therefore, how smeared out the images from their telescopes are. Next time you visit an observatory, ask an astronomer how the seeing has been lately, but be prepared for a series of expletives; it is rarely ideal.

The angular diameter of a star's smeared-out image, called the **seeing disk,** is a realistic measure of the best possible resolution. The size of the seeing disk varies from one observatory site to another. At Palomar, the seeing disk is roughly 1 arcsec. The best conditions in the world (with a seeing disk of 0.2 arcsec) have been reported at the observatory on the 14,000-ft summit of Mauna Kea, the tallest volcano on the island of Hawaii.

Without the effects of the Earth's atmosphere, stars don't twinkle; as a result, photographs taken from telescopes in space reveal much finer points for stars and more detail for extended objects (Figure 3-20). The Hub-

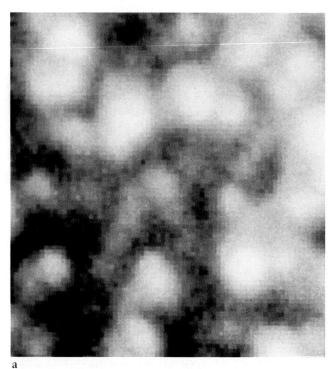

a

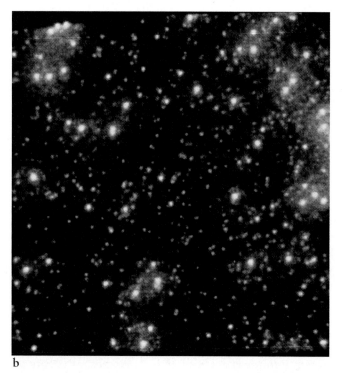

b

Figure 3-20 **Effects of Twinkling** The same star field photographed with (**a**) a ground-based telescope, which is subject to twinkling, and (**b**) the Hubble Space Telescope, which is free from the effects of twinkling.

Figure 3-21 **Light Pollution** These two images of Tucson, Arizona, were taken from the Kitt Peak National Observatory, which is 54 miles away. They show the dramatic growth in groundlight output between 1959 (top) and 1980 (bottom).

Since 1972, light pollution, a problem for many observatories around the world, has been at least partially controlled by a series of local ordinances.

ble Space Telescope, as we shall see often in this book, now achieves magnificent resolution.

Light pollution from cities is another problem for Earth-based telescopes (Figure 3-21). Keep in mind that the larger the primary mirror, the more light gathered and therefore the more information astronomers can obtain. The 200-in. (5-m) telescope at the Palomar Observatory between San Diego and Los Angeles, California, was the first truly great large telescope, providing astronomers with invaluable insights into the universe for decades. However, light pollution from the two cities now fills the night sky, seriously reducing the ability of that telescope to collect light from the stars. The best observing sites in the world are high on mountaintops —above smog, water vapor, and clouds—far from city lights.

3-11 Advanced technology is spawning a new generation of superb telescopes

Until the 1980s, telescopes with primary mirrors of between 3 and 6 m were the largest and most powerful in the world. Now new technologies in mirror building and computer control allow us to construct 8- and 10-m tele-

scopes. Twin 8-m reflectors are being constructed by the Gemini Project, a multinational consortium of astronomers. One Gemini telescope will be located at the summit of Mauna Kea on Hawaii, and the other on Cerro Pachón in Chile. Astronomers hope that the Hawaii telescope will be finished by 1998, with its Chilean twin to follow two years later. There are at least 17 reflectors around the world today with primary mirrors measuring 3 m or more in diameter.

Because the cost of building very large mirrors is enormous, astronomers have recently devised less expensive ways to collect the same light. One approach is to join several smaller mirrors together and aim them at the same focal point. The largest example of this segmented-mirror technique is the recently completed 10-m Keck I telescope on the summit of Mauna Kea in Hawaii. Thirty-six hexagonal mirrors are mounted side by side to collect the same amount of light as a single primary mirror 10 m (400 in.) in diameter, as shown in Figure 3-22. This design is so successful that another telescope, called Keck II, is under construction only a short distance from Keck I on Mauna Kea.

Adaptive optics—perhaps the most promising technique for building huge reflectors—uses a very thin mirror whose precise shape is maintained by computer-

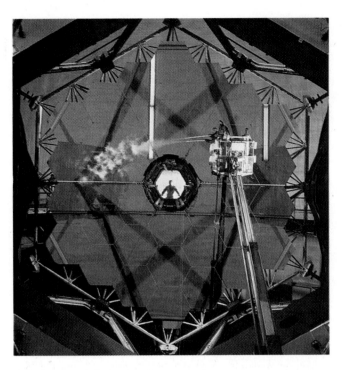

Figure 3-22 **The 10-m Keck Telescope** The world's largest telescope consists of 36 hexagonal mirrors, each measuring 1.8 m (5.9 ft) across. It has the light-gathering, resolving, and magnifying ability of one mirror 10 m in diameter. Here the mirror is being cleaned with carbon dioxide snow.

activated, motorized supports. These supports are coupled to optical sensors that measure the distortion of incoming starlight caused by atmospheric turbulence. A computer rapidly calculates the mirror shape needed to compensate for the distortion, and the motorized supports promptly deform the mirror accordingly. Adaptive optics effectively eliminates atmospheric distortion and produces remarkably sharp images, almost as good as if the telescope were in the vacuum of space.

The most ambitious project to use adaptive optics is the Very Large Telescope (VLT), being built by the European Southern Observatory in Chile. When completed, the VLT will consist of four 8.2-m reflectors, shown in Figure 3-23. The four reflectors can be used individually and thus pointed toward separate targets, or they can observe the same object simultaneously. When used together, the light-gathering power of the four reflectors will equal that of a single 16.4-m (53.8-ft) mirror! The first 8.2-m telescope will be finished in 1996, and all four should be completed by 2000.

The invention of photography during the nineteenth century was a boon for astronomy. By taking a long exposure with a camera mounted at the focus of a telescope, an astronomer could record extremely faint features that could not be seen just by looking through the telescope. The reason is that our eyes clear the images in them several times a second, whereas film adds up the intensity of all the photons affecting its emulsion. Brighter, higher-resolution images therefore reveal details in galaxies, star clusters, and planets that we could not otherwise see.

Astronomers have long realized, however, that a photographic plate is an inefficient detector of light because it depends on a chemical reaction to produce an image. Typically, only 2% of the light striking a photographic plate triggers a reaction in the sensitive material of the plate, called the *photographic emulsion*. Thus, roughly 98% of the light falling onto a photographic plate is wasted.

Technology has changed all that. We now have highly efficient electronic light detectors to replace photographic film. Called **charge-coupled devices (CCDs)**, each one is

Figure 3-23 **The Very Large Telescope** This artist's conception shows four 8.2-m reflectors being constructed by the European Southern Observatory on top of a mountain in Chile. The light of the four reflectors can be combined to give the light-gathering power of a single 16.4-m (53.8-ft) mirror! The project is scheduled to be finished by 2000.

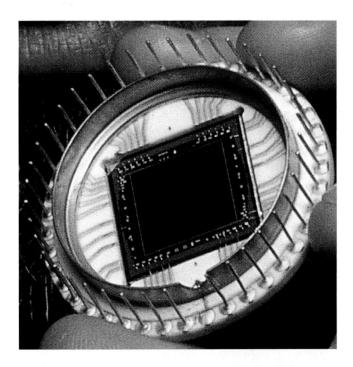

Figure 3-24 **A Charge-Coupled Device (CCD)** This tiny silicon square contains 16,777,216 light-sensitive electric circuits that store images in a one-piece CCD array. Additional circuits transfer the data to a waiting computer.

roughly the size of a large postage stamp (Figure 3-24). A CCD is divided into an array of small, light-sensitive squares called picture elements or, more commonly, **pixels.** For example, one of the latest CCDs has over 16 million pixels arranged in 4096 rows by 4096 columns. When an image from a telescope is focused on the CCD, an electric charge builds up in each pixel in direct proportion to the intensity of the light falling on that pixel. When the exposure is finished, the amount of charge on each pixel is read into a computer. The computer then transfers the image onto ordinary photographic film or a television monitor. CCDs commonly respond to 70% of the light falling on them. Furthermore, their resolution is better than that of film, and they respond more uniformly to light of different colors.

Figure 3-25 shows one photograph and two CCD images of the same region of the sky, all taken with the same telescope. Notice that many details visible in the CCD images are totally absent in the ordinary photograph. In fact, the CCD pictures in Figure 3-25 show some of the faintest stars and galaxies ever recorded. Because of their extraordinary sensitivity and their use in conjunction with computers, CCDs are playing an increasingly important role in astronomy.

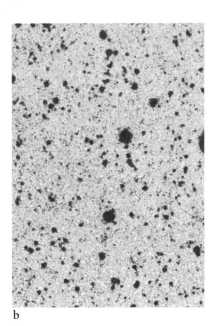

a b c

Figure 3-25 **Ordinary Photography Versus a CCD Image** These three views of the same part of the sky, each taken with the same 4-m telescope, compare CCDs to ordinary photographic plates. (a) A negative print (black stars and white sky) of a 45-min exposure on a photographic plate. (b) The sum of fifteen 500-s CCD images. Notice that many faint stars and galaxies virtually invisible in the ordinary photograph can be clearly seen in this CCD image. (c) This color view was produced by combining a series of CCD images taken through colored filters. The total exposure time was 6 h.

··

EYES ON . . .   Your First Telescope

Because the stars move continuously across the sky, it is essential to buy a telescope that can follow them reliably and easily. The tracking motion of the telescope is determined by the "mount" that it is connected to. The standard mount that enables you to track the stars by moving the telescope around one axis is called an *equatorial* mount. Unless you are prepared to motorize and computerize the mount yourself, you should avoid alt-azimuth mounts, which make tracking a strictly hit-or-miss proposition. You should be sure that the mount is sturdy enough to hold the telescope you choose, but light and compact enough so that you can transport the telescope.

You will also want to get a few different *eyepieces* (three is a good number to start with), so that you can look at large areas of the sky under low magnification and details of small areas under high magnification. As discussed in the text, the magnification of a telescope depends on the focal lengths of the objective lens and the eyepiece. The objective lens or primary mirror is fixed in the telescope. However, all telescopes come with removable eyepieces. Change the eyepiece with another of different focal length, and you change the telescope's magnification.

It is also essential to have a *finder scope*, which is a small, low-magnification, very-wide-field telescope attached directly onto your main telescope. If the finder and your main scope are well aligned, you can quickly zero in on an object. The finder scope should have crosshairs to help you locate the object accurately.

Finally, you will need a flashlight with a red gel coating. Red light does not cause the pupils of your eyes to contract, and so they will remain dilated (wide open) while you use the red flashlight to inspect your equipment and your star charts. Later we will describe other worthwhile features for your telescope.

Equatorial Telescope Mount Equatorial mounts are designed to rotate parallel to, but in the opposite direction from, the Earth's spin. This motion of the telescope allows it to stay fixed on a star as the Earth turns. A simple motor attached to the mount can keep an object in view for hours.

Radio Astronomy— and Beyond

Until the 1930s, all information that astronomers gathered from the universe was based on visible light. With the discovery of nonvisible electromagnetic radiation, scientists began to wonder if objects in the universe might also emit radio waves, infrared and ultraviolet radiation, and perhaps even x rays or gamma rays. Little did anyone realize back then the enormous range of objects and activities in the universe that give off one or more of these radiations without giving off any detectable visible light!

3-12 A radio telescope uses a large concave dish to reflect radio waves to a focus

The first evidence of nonvisible radiation from outer space came from the work of a young radio engineer, Karl Jansky, working at Bell Telephone Laboratories. Using long antennas, Jansky was investigating the sources of radio static that affect short-wavelength radiotelephone communication. In 1932, he realized that a certain kind of radio noise is strongest when the constellation Sagittarius is high in the sky. The center of our Galaxy is located in the direction of Sagittarius, and

Figure 3-26 **A Radio Telescope** The dish of this radio telescope is 45.2 m (148 ft) in diameter. It is one of several large instruments at the National Radio Astronomy Observatory near Green Bank, West Virginia.

Jansky concluded that he was detecting radio waves from elsewhere in the Galaxy.

At first, astronomers were not enthusiastic about detecting radio noise from space, in part because of the poor angular resolution of early radio telescopes. The angular resolution of any telescope worsens as the wavelength increases. In other words, the longer the wavelength, the fuzzier the picture. Because radio radiation has very long wavelengths, astronomers then thought that radio telescopes could only produce blurry, indistinct views.

The standard **radio telescope** today has a large, reflecting, concave dish (Figure 3-26). A small antenna tuned to the desired wavelength is located at the focus. The incoming signal is relayed to amplifiers and recording instruments, which are typically located in a room at the base of the telescope's pier. Very large radio telescopes create sharper radio images because, as with optical telescopes, the bigger the dish, the better the angular resolution. For this reason, most modern radio telescopes have dishes more than 30 m in diameter. Nevertheless, even the largest radio dish in existence (305-m diameter in Arecibo, Puerto Rico) cannot come close to the resolution of the best optical telescopes. For example, a 6-m optical telescope has 2000 times better resolution than a 6-m radio telescope detecting radio waves of 1-mm wavelength.

A very clever technique was devised to circumvent this limitation and produce high-resolution radio images. Unlike ordinary light, radio signals can be carried over electrical wires, which means that two radio telescopes separated by many kilometers can be hooked together. This technique is called **interferometry,** because the incoming radio signals are made to "interfere," or blend together, so that the combined signal is sharp and clear. The result is impressive: The effective angular resolution is equivalent to that of one gigantic dish with a diameter equal to the distance between the two telescopes.

Interferometry, exploited for the first time in the late 1940s, gave astronomers their first detailed views of radio objects in the sky. More recently, radio telescopes separated by thousands of kilometers have been linked together to produce images that are much sharper and crisper than even those from optical telescopes. This technique is called **very-long-baseline interferometry (VLBI).** The best angular resolution would be obtained by two telescopes on opposite sides of the Earth. In that case, features as small as 0.00001 arcsec could be distinguished at radio wavelengths—100,000 times better than the sharpest pictures from ordinary optical telescopes.

One of the finest systems of radio telescopes began operating in 1980 in the desert near Socorro, New Mexico. Called the Very Large Array (VLA), it consists of 27 concave dishes, each 26 m (85 ft) in diameter. The 27 telescopes are positioned along the three arms of a gigantic Y covering an area 27 km (17 mi) in diameter. Working together they can create radio images with 0.1-arcsec resolution. Only a portion of the VLA is shown in Figure 3-27. This system produces radio views of the sky with

Figure 3-27 **The Very Large Array (VLA)** The 27 radio telescopes of the VLA system are arranged along the arms of a **Y** in central New Mexico. The north arm of the array is 19 km long; the southwest and southeast arms are each 21 km long.

a

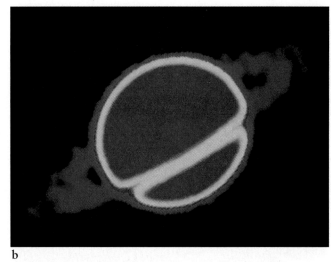

b

Figure 3-28 **Optical and Radio Views of Saturn** (a) This picture was taken by a camera on board a spacecraft as it approached Saturn. Sunlight reflected from the planet's cloudtops and rings is responsible for this view. (b) This false-color picture, taken by the VLA, shows radio emission from Saturn at a wavelength of 2 cm.

resolution comparable to that of the very best optical telescopes.

To make radio images more comprehensible, radio astronomers often use "false color" or gray scales to display their radio views of astronomical objects. An example of the use of false colors is shown in Figure 3-28. The most intense radio emission is shown in red, the least intense in blue. Intermediate colors of the rainbow represent intermediate levels of radio intensity. Black indicates that there is no detectable radio radiation. Astronomers working at other nonvisible-wavelength ranges also frequently use false-color techniques to display views obtained from their instruments.

The recently completed Very Long Baseline Array (VLBA) consists of ten 25-m radio telescopes located across the United States from Hawaii to New Hampshire. With a maximum baseline of 8000 km, VLBA has a resolving power of a milli–arc second. It has already started making major contributions to astronomy, such as the discovery in 1994 of a black hole in a nearby galaxy (discussed in Chapter 13).

3-13 Observations at other wavelengths are revealing sights previously invisible

As the success of radio astronomy began to mount, astronomers started exploring the possibility of making observations at other nonvisible wavelengths. Because water vapor is the main absorber of infrared radiation from space, locating infrared observatories at sites of low humidity can overcome much of the atmosphere's hindrance. For example, the summit of Mauna Kea on Hawaii is exceptionally dry (most of the moisture in the air is below the height of Mauna Kea), and infrared observations are the primary function of NASA's 3-m infrared telescope there.

The best way of avoiding water vapor is to place a telescope in orbit around the Earth. In 1983 the Infrared Astronomical Satellite (IRAS) was launched into a 900-km-high polar orbit (Figure 3-29). During its ten-month mission, this 60-cm reflector revealed the full richness and variety of the infrared sky. For the first time, astronomers saw dust bands in our galaxy, dust disks around nearby stars, and distant galaxies that emit most of their radiation at infrared wavelengths. All of these features are invisible to optical telescopes. Altogether, IRAS located over 200,000 infrared sources in the sky.

In 1994 the European Space Agency (ESA) launched another infrared telescope that will spend several years making observations at infrared wavelengths that cannot penetrate the Earth's atmosphere.

The best observations of ultraviolet radiation are also made from space. During the early 1970s, both Apollo and Skylab astronauts used small telescopes above the Earth's atmosphere to give us some of our first views of the ultraviolet sky. Small rockets have also been used to place ultraviolet cameras briefly above the

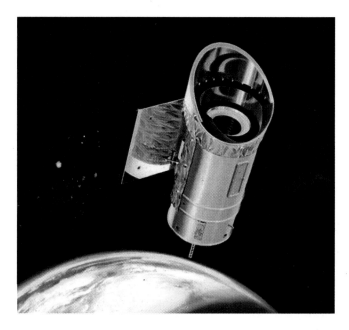

Figure 3-29 **The Infrared Astronomical Satellite (IRAS)** This satellite's small reflecting telescope gave astronomers their first in-depth look at the infrared sky. Launched in 1983, IRAS contributed to many research topics, from asteroids to the large-scale distribution of matter in the universe.

Earth's atmosphere. A typical ultraviolet view is shown in Figure 3-30, along with a corresponding infrared view from IRAS, a view in visible light, and a star chart. The Space Shuttle had a highly successful mission using ultraviolet telescopes in 1995 (Figure 3-31).

Some of the finest ultraviolet astronomy has been accomplished by the International Ultraviolet Explorer (IUE), which was launched in 1978 and still functions today. The satellite is built around a Cassegrain telescope with a 45-cm (18-in.) mirror and a total focal length of 6.74 m (22 ft). Observations cover the ultraviolet range from 116 to 320 nm. In 1992 the Extreme Ultraviolet Explorer (EUVE) was launched. It is sensitive to photons with wavelengths between 7 and 76 nm, at the short end of the ultraviolet spectrum. As with infrared observations, ultraviolet images reveal sights previously invisible and often unexpected.

Because neither x rays nor gamma rays penetrate the Earth's atmosphere, observations at these extremely short wavelengths must be done from space. Astronomers got their first look at the x-ray sky during brief rocket flights in the late 1940s. Several small satellites launched during the early 1970s viewed the entire x-ray

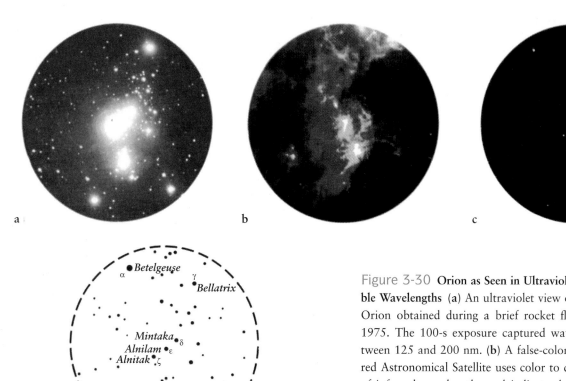

Figure 3-30 **Orion as Seen in Ultraviolet, Infrared, and Visible Wavelengths** (a) An ultraviolet view of the constellation of Orion obtained during a brief rocket flight on December 5, 1975. The 100-s exposure captured wavelengths ranging between 125 and 200 nm. (b) A false-color view from the Infrared Astronomical Satellite uses color to display specific ranges of infrared wavelengths: red indicates long-wavelength radiation; green, intermediate-wavelength radiation; and blue, short-wavelength radiation. For comparison, (c) an ordinary optical photograph and (d) a star chart are included.

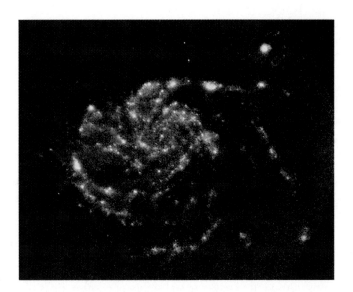

Figure 3-31 **The Ultraviolet Sky** This image of the spiral galaxy M101 in the constellation Ursa Major was photographed by an Astro-2 telescope aboard the Space Shuttle. It shows hot stars and ultraviolet emission from interstellar gas and dust. Such ultraviolet images provide details of the spiral structure of galaxies, which, among other things in space, we cannot see through visible light telescopes.

and gamma-ray sky, revealing hundreds of previously unknown short-wavelength sources, including several black hole candidates (see Chapter 13).

To date thousands of x-ray sources have been discovered all across the sky. Among these are stars, vast clouds of intergalactic gas, jets of gas emitted by galaxies, black holes, and quasars.

The electromagnetic radiation with the shortest wavelengths and the most energy are gamma rays. In 1991 the Compton Gamma-Ray Observatory was carried aloft by the Space Shuttle. Named in honor of Arthur Holly Compton, an American physicist who made important discoveries about gamma rays, this orbiting observatory carries four instruments that are performing a variety of observations, giving us tantalizing views of the gamma-ray sky. However, its resolution is only about 5′.

3-14 The Hubble Space Telescope is crystal clear at last

For decades astronomers have dreamed of having a major observatory in space. Although satellites like IRAS and IUE gave us excellent views at selected wavelengths,

astronomers were enthusiastic about the prospect of one large telescope that could be operated over a wide range of wavelengths—from the infrared through the visible range and far out into the ultraviolet. This was the mission of the Hubble Space Telescope (HST), which was carried aloft by the Space Shuttle in 1990 (Figure 3-32).

Soon after HST was placed in orbit, astronomers discovered that the telescope's 2.4-m primary mirror did not have the proper curvature, which caused star images to be surrounded by a hazy glow. During a repair mission in December 1993, astronauts installed corrective optics that eliminated the problem. Now HST has a resolution of better than 0.1″. Figure 3-33 compares HST pictures of a bright star in a star cluster taken before and after the repair mission.

Already the observations taken by HST have staggered the astronomy community. It has dished up new results related to the planets, planetary systems forming around other stars, the distances to other galaxies, black holes, quasars, the lower-than-expected numbers of very dim stars, and the age of the universe, among many other things. In the coming years we can look forward to a wealth of crystal clear images of distant galaxies and nebulae coming from HST.

Figure 3-32 **The Hubble Space Telescope (HST)** This photograph of HST hovering above the Space Shuttle's cargo bay was taken in 1993, at completion of the first servicing mission. During its 15-year lifetime, HST will study the heavens at wavelengths from the infrared through the ultraviolet.

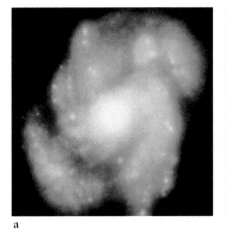
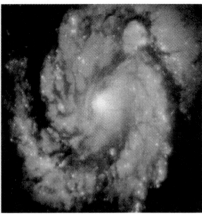

a   b

Figure 3-33 **A Comparison of HST Images** These pictures of a galactic nucleus compare HST images taken before and after the repair mission in December 1993. (a) Although atmospheric blurring is absent, HST's optical defect produced a "skirt" around the nucleus. (b) After astronauts installed corrective optics, HST took this sharp, clear picture of the galaxy.

The advantages and benefits of all these Earth-orbiting observatories cannot be overemphasized. We are no longer limited to the narrow ranges of wavelengths that manage to leak through our shimmering, hazy atmosphere. Until this century astronomers had been like the blind person trying to describe the elephant. Our ancestors couldn't see the big picture of the universe. We are beginning to, and our understanding is increasing dramatically. In this book we describe just a few of the most important observations that have been made this century and the theories that have arisen to explain them.

WHAT DID YOU THINK?

1. *What is the main purpose of a telescope?* A telescope is designed to collect as much light as possible. It also improves resolution and magnifies images.

2. *Why do stars twinkle?* Rapid changes in the density of the Earth's atmosphere cause passing starlight to change direction, making the star appear to twinkle.

Key Words

achromatic lens
adaptive optics
angular resolution
Cassegrain focus
charge-coupled device (CCD)
chromatic aberration
coudé focus
electromagnetic radiation
electromagnetic spectrum
eyepiece lens

focal length
focal plane
focus (of a lens or concave mirror)
gamma ray
infrared radiation
interferometry
light-gathering power
magnification
Newtonian reflector
objective lens
photon
pixel

primary mirror
prime focus
radio telescope
radio wave
reflecting telescope (reflector)
reflection
refracting telescope (refractor)
refraction
Schmidt corrector plate
secondary mirror
seeing disk

spectrum (*plural* spectra)
spherical aberration
twinkling
ultraviolet (UV) radiation
very-long-baseline interferometry (VLBI)
wavelength
x ray

Key Ideas

• Electromagnetic radiation consists of vibrating electric and magnetic fields that carry energy through space at the speed of light (300,000 km/s).

• Visible light, radio waves, infrared and ultraviolet radiation, x rays, and gamma rays are all forms of electromagnetic radiation.

Visible light occupies only a small portion of the electromagnetic spectrum.

• The wavelength of visible light is associated with its color; wavelengths of visible light range from about 400 nm for violet light to 700 nm for red light.

Infrared radiation and radio waves have wavelengths longer than those of visible light; ultraviolet radiation, x rays, and gamma rays have wavelengths that are shorter.

• Refracting telescopes, or refractors, produce images by bending light rays as they pass through glass lenses.

Glass impurity, opacity to certain wavelengths, and structural difficulties make it inadvisable to build extremely large refractors.

• Reflecting telescopes, or reflectors, produce images by reflecting light rays from concave mirrors to a focal point or focal plane.

Reflectors are not subject to many of the problems that limit the usefulness of refractors.

Telescopes that employ advanced technologies, such as adaptive optics, produce extremely sharp images.

• Charge-coupled devices (CCDs) are often used at a telescope's focus to record faint images.

• Radio telescopes have large reflecting antennas (dishes) that are used to focus radio waves.

Very sharp radio images are produced with arrays of radio telescopes linked together in a technique called interferometry.

• The Earth's atmosphere is transparent primarily to visible light and radio waves arriving from space, but it absorbs much of the radiation at other wavelengths.

For observations at other wavelengths, astronomers depend upon telescopes carried above the atmosphere by rockets and satellites.

Satellite-based observatories are giving us a wealth of new information about the universe and permit coordinated observation of the sky at all wavelengths.

Review Questions

1 Describe refraction and reflection. Why do these processes enable astronomers to build telescopes?

2 Give everyday examples of refraction and reflection.

3 With the aid of a diagram, describe a refracting telescope.

*4 How much more light does a 3-m-diameter telescope collect than a 1-m-diameter telescope?

5 With the aid of a diagram, describe a reflecting telescope. Describe four different ways in which an astronomer can access the light collected by reflecting telescopes.

6 Explain some of the advantages of reflecting telescopes over refracting telescopes.

7 What are the three major functions of a telescope?

8 What is meant by the angular resolution of a telescope?

9 What limits the angular resolution of the 5-m telescope at Palomar?

10 Why will very large telescopes of the future make use of multiple mirrors or ultrathin mirrors?

11 What is meant by adaptive optics? What problem does adaptive optics overcome?

12 Compare an optical reflecting telescope and a radio telescope. What do they have in common? How are they different?

13 Why can radio astronomers observe at any time during the day, whereas optical astronomers are mostly limited to observing at night?

14 Why must astronomers use satellites and Earth-orbiting observatories to study the heavens at x-ray or gamma-ray wavelengths?

15 Why did Roemer's observations of the eclipses of Jupiter's moons support the heliocentric, but not the geocentric, cosmogony?

Advanced Questions

16 Advertisements for telescopes frequently extol magnifying ability. Is this a good criterion for evaluating telescopes? Explain your answer.

* **17** The observing cage in which an astronomer sits at the prime focus of the 5-m telescope on Palomar Mountain is about 1 m in diameter. What fraction of the incoming starlight is blocked by the cage? (*Hint:* The area of a circle of radius r is πr^2, where $\pi = 3.14$).

* **18** Compare the light-gathering power of the Palomar 5-m telescope with that of the fully dark-adapted human eye, which has a pupil diameter of about 5 mm.

19 Show by means of a diagram why the image formed by a simple refracting telescope is "upside down."

* **20** Suppose your Newtonian reflector has a mirror with a diameter of 20 cm and a focal length of 2 m. What magnification do you get with eyepieces whose focal lengths are **(a)** 9 mm, **(b)** 20 mm, and **(c)** 55 mm?

21 Why do you think no major observatory has a Newtonian reflector as its primary instrument, whereas Newtonian reflectors are extremely popular among amateur astronomers?

22 Several telescope manufacturers build telescopes having a design called "Schmidt-Cassegrain." Consult advertisements in such magazines as *Sky & Telescope* and *Astronomy* to see what these telescopes look like and what they cost. Why do you suppose they are very popular among amateur astronomers?

Discussion Questions

23 Discuss the advantages and disadvantages of using a small telescope in Earth orbit versus a large telescope on a mountaintop.

24 If you were in charge of selecting a site for a new observatory, which factors would you consider important?

Observing Projects

25 Obtain a telescope along with several eyepieces of various focal lengths. If you can determine the telescope's focal length, calculate the magnifying powers of the eyepieces. Focus the telescope on some familiar object during the daytime, such as a distant lamppost or tree. **DO NOT FOCUS ON THE SUN! Looking directly at the Sun can cause blindness.** Describe the image you see through the telescope. Is it upside down? How does the image move as you slowly and gently shift the telescope left and right, up and down? Examine the eyepieces, noting their focal lengths. By changing the eyepieces, examine the distant object under different magnifications. How does the field of view and the quality of the image change as you go from low power to high power?

26 On a clear night, view the Moon, a planet, and a star through a telescope using eyepieces of various focal lengths. (Consult such magazines as *Sky & Telescope* or *Astronomy* to determine the phase of the Moon and the locations of the planets.) How does the image seem to change as you view with increasing magnification? Does it degrade at any point?

27 Many towns and cities have amateur astronomy clubs. If you are so inclined, attend a "star party" hosted by your local club. People who bring their telescopes to such gatherings are delighted to show you their instruments and take you on a telescopic tour of the heavens. Such an experience can lead to a very enjoyable, lifelong hobby.

4 ▶ The Origin and Nature of Light

....

IN THIS CHAPTER

You will examine the origins of electromagnetic radiation. You will discover that stars with different surface temperatures emit electromagnetic radiation with different wavelengths and intensities. Because different chemicals also emit different wavelengths, astronomers can determine the chemical compositions of stars and interstellar clouds by studying electromagnetic radiation. Finally, you will learn how to tell whether an object in space is moving toward or away from us.

Stellar Spectra Starlight passing through a prism or diffraction grating spreads into its component colors. From such spectra we can learn an incredible amount about stars, including their masses, surface temperatures, diameters, chemical compositions, and motions toward or away from us. This image, called a spectrogram, shows the spectra of stars in the Hyades star cluster. Note the distinct differences between the intensities of the various colors emitted by different stars.

1 How hot is a "red hot" object compared to objects glowing with other colors?

2 What color is the Sun?

....

SEEING OBJECTS IN SPACE is one thing, understanding their physical properties—such as their temperature, chemical composition, mass, size, and motion—is something else altogether. Since virtually all of our knowledge about space comes from electromagnetic radiation, we now need to delve deeper into the nature of light. By understanding how light and matter interact, we will be able to glean an amazing amount of information about the heavenly bodies. We begin this study of light by exploring a property of matter we are quite familiar with: temperature.

Blackbody Radiation

4-1 Peak color shifts to shorter wavelengths as an object is heated

Imagine that you have taken hold of an iron rod in a completely darkened room. You cannot see the rod, but you can feel its warmth in your hands. The rod is emitting infrared radiation (heat), but too few visible photons for your eyes to see. Now imagine that there is also a propane torch in the room. You light it and stick the iron rod in the flame for several seconds, until it begins to glow (Figure 4-1*a*).

The rod's first visible color is red—it glows "red hot." Heated a little more, the rod appears orange and brighter (Figure 4-1*b*). Heated more, it appears yellow

and brighter still. After still more heating the rod appears white-hot and brighter yet (Figure 4-1*c*). If it didn't melt, further heating would make the rod appear blue and even brighter. This "thought experiment" shows how the light from any object changes with its temperature:

1 As an object heats up, it gets brighter because it emits more electromagnetic radiation.

2 The color or dominant wavelength of the emitted radiation changes with temperature. A cool object emits most of its energy at long wavelengths, such as infrared or red. A hotter object emits most of its energy at shorter wavelengths, such as blue, violet, or even ultraviolet.

Although the iron rod appears to emit a single color depending on its temperature, it actually gives off all wavelengths of electromagnetic radiation. So do stars, rocks, animals, and most other things in nature. However, a single wavelength of emitted radiation, denoted λ_{max}, dominates. When the object is cool, like a rock or animal, λ_{max} is a radio or infrared wavelength. When it is hot enough, λ_{max} is in the range of visible light, giving a hot object its characteristic color. In fact, sufficiently hot stars even have λ_{max} in the ultraviolet part of the electromagnetic spectrum.

Iron bars and stars are good approximations to an important class of objects that physicists call **blackbodies**. A blackbody absorbs all the electromagnetic radiation that strikes it; none of the incident radiation is reflected or scattered off its surface. The incoming radiation heats up the blackbody, which then re-emits the energy it has absorbed, *but not with the same intensities at each wavelength as it received*. Rather, the amount of each wavelength emitted by the blackbody is determined solely by the object's temperature. Three temperature profiles of the intensity of blackbody radiation at different wavelengths, called **blackbody curves**, are shown in Figure 4-2.

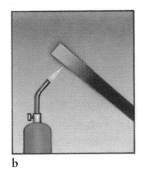
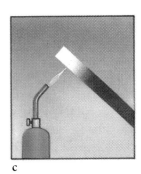

a   b   c

Figure 4-1 **Heating a Bar of Iron** This sequence of drawings shows the changing appearance of a bar of iron as it is heated. As the temperature increases, the amount of energy radiated by the bar increases. The color of the bar also changes because, as the temperature goes up, the dominant wavelength of light emitted by the bar decreases.

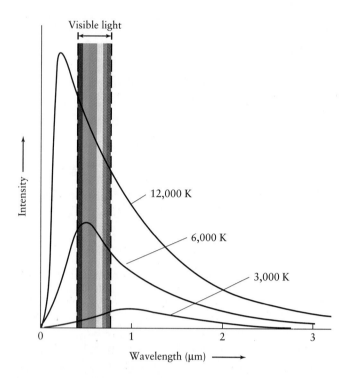

Figure 4-2 **Blackbody Curves** Three representative blackbody curves are shown here. Each curve shows the intensity of radiation at every wavelength emitted by a blackbody at a particular temperature. Note that hotter objects emit more of every wavelength and that the peak of emitted radiation changes with temperature. On this graph, wavelength is measured in micrometers (μm), where 1 μm = 1000 nm. Note the range of wavelengths of visible light.

Stars produce their own electromagnetic radiation, rather than just absorbing and reradiating light from an outside source, as our iron rod did. Even so, they behave as blackbodies and the self-generated radiation they emit closely follows the same blackbody curves. We can therefore measure a star's surface temperature from its color.

In 1879 the Austrian physicist Josef Stefan observed that *an object emits energy at a rate proportional to the fourth power of its temperature measured in kelvins.* (Temperatures throughout this book will be expressed in kelvins. If you are not familiar with this temperature scale, please review "Temperatures and Temperature Scales" at the back of this book.) In other words, if you double the temperature of an object (for example, from 500 to 1000 K), the energy emitted from each square meter of the object's surface each second increases by a factor of $2^4 = 16$. If you triple the temperature (for instance, from 500 to 1500 K), the rate at which energy is emitted increases by a factor of $3^4 = 81$. Stefan's experi-

mental results were put on firm theoretical ground five years later by Ludwig Boltzmann. The intensity–temperature relationship for blackbodies is named the **Stefan–Boltzmann law** in their honor.

4-2 The color of electromagnetic radiation reveals an object's temperature

The mathematical relationship between color peak and temperature was discovered in 1893 by the German physicist Wilhelm Wien. Consistent with the iron rod experiment described above, he found that *the dominant wavelength of radiation emitted by a blackbody is inversely proportional to its temperature.* In other words, the hotter an object, the shorter its λ_{max}, and vice versa. This relationship is today called **Wien's law.**

Wien's law is very useful in computing the surface temperature of a star because the star's size and brightness do not have to be known; all we need to know is the dominant wavelength of the star's electromagnetic radia-

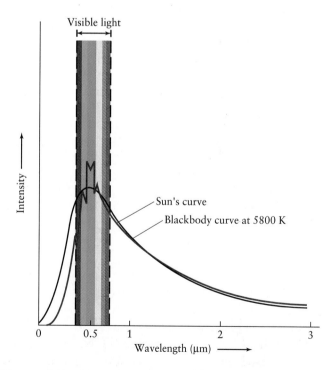

Figure 4-3 **The Sun as a Blackbody** This graph compares the intensity of sunlight over a wide range of wavelengths with the intensity of radiation from a blackbody at a temperature of 5800 K. Measurements of the Sun's intensity were made above the Earth's atmosphere. The Sun mimics a blackbody remarkably well.

··

AN ASTRONOMER'S TOOLBOX 4-1

The Radiation Laws

The energy emitted from each square meter of an object's surface each second is called the **energy flux** (*F*). In this context, flux means "rate of flow," and thus *F* is a measure of how much energy is flowing out of the object. The energy flux also measures the brightness of an object. Using this concept, we can write the Stefan–Boltzmann law as

$$F = \sigma T^4$$

where *T* is the temperature of the object in kelvins and σ (lowercase Greek sigma) is a number called the Stefan–Boltzmann constant:

$$\sigma = 5.67 \times 10^{-8} \text{ J/m}^2 \cdot \text{K}^4 \cdot \text{s}$$

The Stefan–Boltzmann law has several applications. For example, suppose you have two identical pieces of iron, one at room temperature (300 K) and the other at the temperature of the Sun's surface (5800 K). You can use the Stefan–Boltzmann law to determine how much brighter the hotter object is:

$$\frac{F_{\text{hotter}}}{F_{\text{colder}}} = \frac{\sigma \times 5800^4}{\sigma \times 300^4} \approx 140,000$$

Wien's law can be stated as a simple equation. If λ_{max}, the wavelength of maximum flux, is measured in meters, then

$$\lambda_{\text{max}} = \frac{2.9 \times 10^{-3}}{T}$$

where *T* is the temperature of the blackbody measured in kelvins. For example, the Sun emits energy over a wide range of wavelengths, but the maximum intensity of sunlight is at a wavelength of about 500 nm, or 5×10^{-7} m. From Wien's law, we can calculate the Sun's surface temperature as

$$T = \frac{2.9 \times 10^{-3}}{5 \times 10^{-7}} = 5800 \text{ K}$$

··

tion. If you would like to see how to use the two radiation laws to make predictions, see Toolbox 4-1.

The Stefan–Boltzmann law and Wien's law describe only two basic properties of *blackbody radiation*, the electromagnetic radiation emitted by a hypothetical blackbody. Because blackbodies give off all wavelengths of electromagnetic radiation, a more complete picture is given by blackbody curves, such as those already shown in Figure 4-2. These curves show the intensities of electromagnetic radiation emitted at different wavelengths by a blackbody with a given temperature.

Figure 4-3 shows how the intensity of sunlight varies with wavelength. The blackbody curve for a temperature of 5800 K is also plotted in Figure 4-3. Note how closely the observed intensity curve for the Sun matches the blackbody curve. Because the observed intensity curves for most stars and the idealized blackbody curves are so closely correlated, the laws of blackbody radiation can be applied to starlight. Note that the peak of the intensity curve for the Sun is at a wavelength of about 0.5 μm (500 nm), which is in the blue-green part of the visible spectrum. Contrary to our experience and intuition, the Sun emits more turquoise (i.e., a mixture of blue and green) light than any other color.

By the end of the nineteenth century, physicists realized that they had reached an impasse in understanding electromagnetic radiation. While intensities and peak wavelengths were understood, all attempts to explain the characteristic shapes of blackbody curves had failed. A breakthrough finally came in 1900, when the German physicist Max Planck derived a mathematical formula for the blackbody curves. He assumed that electromagnetic radiation is emitted in separate packets of energy. Light—a wave—behaved like a beam of particles. It was a remarkable result verified in 1905 by Albert Einstein, who called these particlelike packets **photons.**

Our thought experiment can now be explained in terms of photons. Recall from Chapter 3 that the energy carried by a photon of light is inversely proportional to its wavelength. In other words, long-wavelength photons, such as radio waves, carry little energy. Short-wavelength photons, like x rays and gamma rays, carry

TABLE 4-1

Some Properties of Electromagnetic Radiation

	Wavelength (cm)	Photon energy (eV)	Blackbody temperature (K)
Radio	$>10^{-1}$	$<10^{-5}$	<0.03
Microwave*	10^{-1} to 10^{-4}	10^{-5} to 10^{-2}	0.03 to 30
Infrared	10^{-4} to 7×10^{-7}	0.01 to 2	30 to 4100
Visible	7×10^{-7} to 4×10^{-7}	2 to 3	4100 to 7300
Ultraviolet	4×10^{-7} to 10^{-9}	3 to 10^3	7300 to 3×10^6
X ray	10^{-9} to 10^{-11}	10^3 to 10^5	3×10^6 to 3×10^8
Gamma ray	$<10^{-11}$	$>10^5$	$>3 \times 10^8$

Note: The symbol > means "greater than;" the symbol < means "less than."
*Microwaves, listed here separately, are often classified as radio waves or infrared radiation.

much more energy. The relationship between the energy of a photon and its wavelength is called **Planck's law** (you may want to refer back to Toolbox 3-1).

Together, Planck's law and Wien's law relate the temperature of an object to the energy of the photons it emits (Table 4-1). A cool object emits primarily long-wavelength photons that carry little energy, while a hot object gives off mostly short-wavelength photons that carry much more energy. In later chapters, we will find these ideas invaluable for understanding how stars of various temperatures interact with gas and dust in space.

Discovering Spectra

In 1814 the German master optician Joseph von Fraunhofer repeated Newton's classic experiment of shining a beam of sunlight through a prism (recall Figure 3-1), but he magnified the resulting rainbow-colored spectrum. Fraunhofer discovered that the solar spectrum contains hundreds of fine dark lines, which became known as **spectral lines.** Fraunhofer counted over 600 such lines, and today we know more than 30,000 of them. Hundreds of spectral lines are visible in the photograph of the Sun's spectrum shown in Figure 4-4.

Half a century later, chemists discovered that they could produce spectral lines in the laboratory. Around 1857 the German chemist Robert Bunsen invented a special gas burner that produces a clean, colorless flame. Certain chemicals are easy to identify by the distinctive colors emitted when bits of the chemical are sprinkled

into the flame of this type of burner, known as a Bunsen burner.

Bunsen's colleague, Prussian-born physicist Gustav Kirchhoff, suggested that light from the colored flames could best be studied by passing it through a prism (Figure 4-5). The chemists promptly discovered that the spectrum from a flame consists of a pattern of thin, bright spectral lines against a dark background. They next found that *each chemical element produces its own characteristic pattern of spectral lines.* Thus was born in 1859 the technique of **spectral analysis,** the identification of chemical substances by their spectral lines.

4-3 Each chemical element produces its own unique set of spectral lines

A chemical **element** is a fundamental substance because it cannot be broken down into more basic chemicals. By the mid-1800s, chemists had already identified such familiar elements as hydrogen, oxygen, carbon, iron, gold, and silver. Spectral analysis promptly led to the discovery of additional elements, many of which are quite rare.

After Bunsen and Kirchhoff had recorded the prominent spectral lines of all the known elements, they began to discover other spectral lines in mineral samples. In 1860, for instance, they found a new line in the blue portion of the spectrum of mineral water. After chemically isolating the previously unknown element responsible for the line, they named it cesium (from the Latin *caesius,* "gray-blue"). The next year a new spectral line in the red

Figure 4-4 **The Sun's Spectrum** Numerous spectral lines are seen in this photograph of the Sun's spectrum. The spectrum is spread so much that it had to be cut into convenient segments to fit on this page.

portion of the spectrum of a mineral sample led to the discovery of the element rubidium (from *rubidus*, "red").

During a solar eclipse in 1868, astronomers found a new spectral line in the light coming from the upper atmosphere of the Sun while the main body of the Sun was hidden by the Moon. This line was attributed to a new

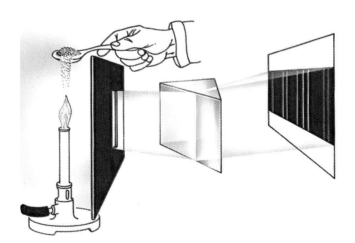

Figure 4-5 **The Kirchhoff–Bunsen Experiment** In the mid-1850s, Kirchhoff and Bunsen discovered that when a chemical substance is heated and vaporized, the resulting spectrum exhibits a series of bright spectral lines. In addition, they found that each chemical element produces its own characteristic pattern of spectral lines.

element that was named helium (from the Greek *helios*, "Sun"). Helium was not actually discovered on the Earth until 1895, when it was identified in gases obtained from a uranium mineral.

A list of the chemical elements is most conveniently displayed in the form of a **periodic table** (Figure 4-6). Each element has a unique atomic number (described later in this chapter), and the elements are arranged in the periodic table by this number. With a few exceptions, this sequence also corresponds to increasing average mass of the atoms of the elements. Thus hydrogen (the symbol H) with atomic number 1 is the lightest element. Iron (Fe) has atomic number 26 and is a moderately heavy element.

All the elements in a single vertical column of the periodic table have similar chemical properties. For example, the elements in the far right column are all gases at the Earth's surface, and they rarely react chemically with other elements.

In addition to the 92 naturally occurring elements, Figure 4-6 lists the artificially produced elements. All the human-made elements are heavier than uranium (U), and all are highly radioactive, which means that they spontaneously decay into lighter elements shortly after being created in the laboratory.

Because each chemical element produces its own unique pattern of spectral lines, scientists can determine

Figure 4-6 **Periodic Table of the Elements** The periodic table is a convenient listing of the elements arranged according to their masses and chemical properties.

1 H																	2 He
3 Li	4 Be											5 B	6 C	7 N	8 O	9 F	10 Ne
11 Na	12 Mg											13 Al	14 Si	15 P	16 S	17 Cl	18 A
19 K	20 Ca	21 Sc	22 Ti	23 V	24 Cr	25 Mn	26 Fe	27 Co	28 Ni	29 Cu	30 Zn	31 Ga	32 Ge	33 As	34 Se	35 Br	36 Kr
37 Rb	38 Sr	39 Y	40 Zr	41 Nb	42 Mo	43 Tc	44 Ru	45 Rh	46 Pd	47 Ag	48 Cd	49 In	50 Sn	51 Sb	52 Te	53 I	54 Xe
55 Cs	56 Ba	57 La	72 Hf	73 Ta	74 W	75 Re	76 Os	77 Ir	78 Pt	79 Au	80 Hg	81 Ti	82 Pb	83 Bi	84 Po	85 At	86 Rn
87 Fr	88 Ra	89 Ac	104 Rf	105 Ha	106	107	108	109	110	111							
				58 Ce	59 Pr	60 Nd	61 Pm	62 Sm	63 Eu	64 Gd	65 Tb	66 Dy	67 Ho	68 Er	69 Tm	70 Yb	71 Lu
				90 Th	91 Pa	92 U	93 Np	94 Pu	95 Am	96 Cm	97 Bk	98 Cf	99 Es	100 Fm	101 Md	102 No	103 Lr

the chemical composition of a remote astronomical object by identifying the spectral lines in its spectrum. For example, Figure 4-7 shows a portion of the Sun's spectrum along with the spectrum of an iron sample taken here on Earth. No other chemical can mimic iron's particular pattern of spectral lines at these wavelengths. It is

EYES ON . . . The Color of the Sun

Different people perceive the Sun to have different colors. To many it appears white, to others yellow. Still others, who notice it at sunset, believe it to be orange or even red. Yet we have seen that the Sun actually gives off all colors. Moreover, the peak in the Sun's spectrum falls between blue and green. Why doesn't the Sun appear turquoise? Several factors affect our perception of its color.

Before reaching our eyes, visible sunlight passes through the Earth's atmosphere. Certain wavelengths are absorbed and re-emitted by the molecules in the air, a process called *scattering.* Violet light is scattered most strongly, followed in decreasing order by blue, green, yellow, orange, and red. That means that more violet, blue, and green photons are scattered by the Earth's atmosphere than are yellow photons. The intense scattering of violet, blue, and green has the effect of shifting the peak of the Sun's intensity entering our eyes from blue-green toward yellow.

The perception of a yellow Sun is further enhanced by our eyes themselves. Our eyes do not see all colors equally well. Rather, the light-sensitive cones in our eyes each respond to one of three ranges of colors, which are centered on red, yellow, and blue wavelengths. None of the cones are especially sensitive to blue-green photons. By adding together the color intensities detected by the three types of cones, our brains *re-create* color. After combining all the light it can, the eye is most sensitive to the yellow-green part of the spectrum. We see blue and orange less well, and violet and red most poorly. Although the eye sees yellow and green light about equally well, the dominant color from the Sun is yellow because the air scatters green light and our eyes are relatively insensitive to blue-green. Therefore, a casual glance at the Sun leaves the impression of a yellow object.

A longer look is extremely dangerous. **Don't try it!** Hypothetically, such a glance would leave the impression of a white Sun. The Sun's light is so intense that it would saturate the color-sensitive cones in our eyes; our brains interpret such saturation of the cones as the color white.

At sunrise and sunset we see an orange or red Sun. This occurs because close to the horizon the Sun's violet, blue, green, and even yellow photons are strongly scattered by the thick layer of atmosphere through which they travel, leaving the Sun looking redder and redder as it sets.

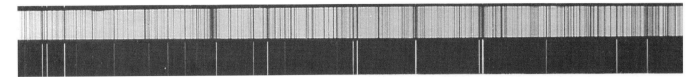

Figure 4-7 **Iron in the Sun's Atmosphere** The upper spectrum is a portion of the Sun's spectrum from 420 to 430 nm. Numerous dark spectral lines are visible. The lower spectrum is a corresponding portion of the spectrum of vaporized iron.

Several bright spectral lines can be seen against the black background. The fact that the iron lines coincide with some of the solar lines proves that there is some iron (albeit a very tiny amount) in the Sun's atmosphere.

iron's own distinctive "fingerprint." Since the spectral lines of iron also appear in the Sun's spectrum, we can safely conclude that the Sun's atmosphere contains some vaporized iron.

Bunsen and Kirchhoff collaborated in designing and constructing the first **spectroscope.** This device consists of a prism and several lenses that magnify the spectrum so that it can be closely examined. After photography was invented, scientists preferred to produce a permanent photographic record of spectra. A device for photographing a spectrum is called a **spectrograph.** This latter instrument is among the astronomer's most important tools.

In its basic form, a spectrograph consists of a slit, two lenses, and a prism arranged to focus the spectrum of an astronomical object onto a small photographic plate, as shown in Figure 4-8. This optical device typically mounts at the focal point of a telescope and the image of the object to be examined is focused on the slit. After the spectrum of a star or galaxy has been photographed, the exposed part of the photographic plate is covered and light from a known source, usually an electric spark that vaporizes a small amount of iron, is focused on the slit. This "comparison spectrum" is photographed next to the spectrum of the object, as in Figure 4-7. Because the wavelengths of the spectral lines in the comparison spectrum are already known from laboratory experiments, these lines can be used to identify and measure the wavelengths of the lines in the spectrum of the star or galaxy under study.

There are drawbacks to this type of spectrograph. A prism does not spread colors evenly: Blue and violet portions of the spectrum are spread out more than the red portion. In addition, because the blue and violet wavelengths must pass through more glass than the red wavelengths (see Figure 3-1), light is absorbed unevenly across the spectrum. Indeed, a glass prism is opaque to ultraviolet wavelengths.

A better device for breaking starlight into the colors of the rainbow is a **grating,** a piece of glass on which thousands of closely spaced lines are cut. Some of the finest gratings have as many as 10,000 lines per centimeter. The spacing of the lines must be very regular. Light rays reflected from different parts of the grating interfere with each other to produce a spectrum. Figure 4-9 shows the design of a modern grating spectrograph.

In recent years the television and electronics industries have produced a variety of light-sensitive devices that are significantly better than photographic film for recording spectra. For instance, observatories now record spectra with charge-coupled devices (CCDs; see Figure 3-24). CCDs produce a graph that plots light intensity against wavelength. Dark lines in the rainbow-colored spectrum appear as depressions or valleys on the

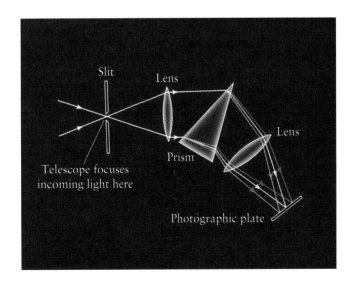

Figure 4-8 **A Prism Spectrograph** This optical device uses a prism to break up the light from an object into a spectrum. The lenses focus that spectrum on a photographic plate.

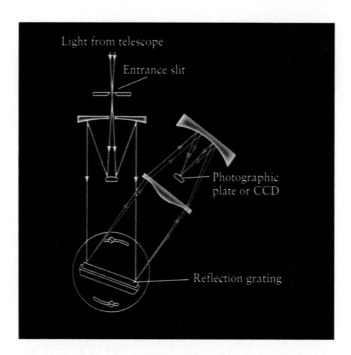

Figure 4-9 **A Grating Spectrograph** This optical device uses a grating to break up the light from an object into a spectrum. An arrangement of lenses and mirrors focuses that spectrum onto a photographic plate or CCD.

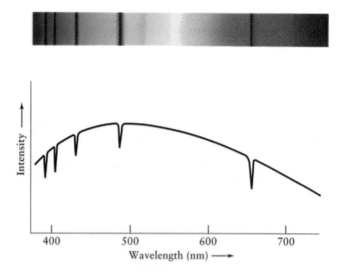

Figure 4-10 **Spectrum of Hydrogen Gas** When photographic film is placed at the focus of a spectrograph, a rainbow-colored spectrum is obtained. When a CCD is placed at the focus, a graph of intensity versus wavelength is produced by computer-controlled equipment attached to the CCD. Note that the dark spectral lines appear as dips in the intensity-versus-wavelength curve.

CCD-produced graph, while bright lines in the spectrum appear as peaks. For example, Figure 4-10 shows both a picture and a plot of a spectrum for hydrogen in which five dark spectral lines appear.

4-4 The brightness of spectral lines depends on conditions in the spectrum's source

During his pioneering experiments with spectra, Kirchhoff sometimes saw dark spectral lines, called *absorption lines,* among the colors of the rainbow. In other experiments, he saw bright spectral lines, called *emission lines,* against an otherwise dark background. By the early 1860s, he had discovered the conditions under which these different types of spectra are observed. His description is summarized today as **Kirchhoff's laws:**

Law 1 A hot object or a hot, dense gas produces a **continuous spectrum**—a complete rainbow of colors without any spectral lines. (This is a blackbody spectrum.)

Law 2 A hot, rarefied gas produces an **emission line spectrum**—a series of bright spectral lines against a dark background.

Law 3 A cool gas in front of a continuous source of light produces an **absorption line spectrum**— a series of dark spectral lines among the colors of the rainbow.

Figure 4-11 shows how absorption and emission lines are formed. Absorption lines are seen if the background is hotter than the gas. Emission lines are seen if the background is cooler. Note that the bright lines in the emission spectrum of a particular gas occur at exactly the same wavelengths as the dark lines in the absorption spectrum of that gas.

Consider, for example, the spectrum of the Sun. We know that the Sun's surface emits a continuous, blackbody spectrum, but here on Earth many absorption lines are seen in it (see Figure 4-4). Kirchhoff's third law explains why. There must be a cooler gas between the surface of the Sun and the Earth. In fact, there are two: the Sun's atmosphere and the Earth's atmosphere. (How can we exclude the Earth's atmosphere as the source of the iron absorption lines in Figure 4-7?) With these observations in hand, we turn now to exploring why the different spectra occur.

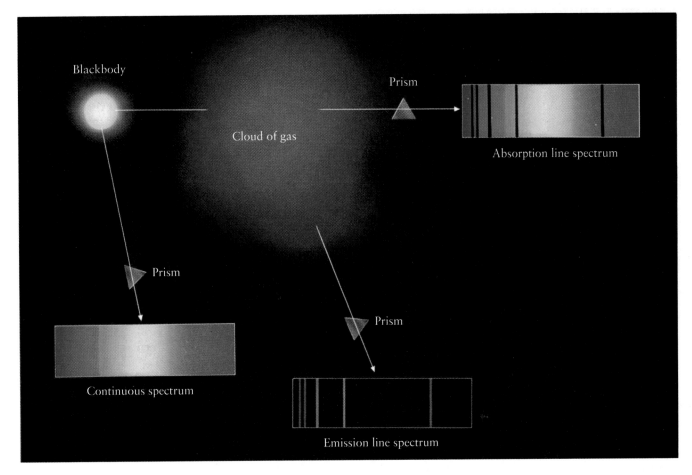

Figure 4-11 Continuous, Absorption Line, and Emission Line Spectra This schematic diagram summarizes how different types of spectra are produced. A hot, glowing object emits a continuous spectrum. If this source of light is viewed through a cool gas, dark absorption lines appear in the resulting spectrum. When the same gas is viewed against a cold, dark background, its spectrum consistes of bright emission lines.

Atoms and Spectra

An atom is the smallest particle of a chemical element that still has the properties of that element. At the time of Kirchhoff's discoveries, scientists knew that all matter is composed of atoms, but they did not know their structures. Scientists could see that atoms of a gas somehow extract light of very specific wavelengths from white light that passes through the gas, leaving dark absorption lines, and they could even guess that the atoms then radiate light of precisely the same wavelengths—the bright emission lines (see Figure 4-7). But traditional theories of electromagnetism could not explain this phenomenon. The answer came early in the twentieth century with the development of quantum mechanics and nuclear physics.

4-5 An atom consists of a small, dense nucleus surrounded by electrons

The first important clue about the internal structure of atoms came in 1910 from Ernest Rutherford, a gifted chemist and physicist from New Zealand. Rutherford and his colleagues at the University of Manchester in England were investigating the recently discovered phenomenon of radioactivity (the natural, spontaneous emission of subatomic particles by some atoms). Certain

Figure 4-12 Rutherford's Model of the Atom Electrons orbit the atom's nucleus, which contains most of the atom's mass. The nucleus contains two types of particles: protons and neutrons. Because the nucleus and the electrons are so small compared to the distance between them, atoms are mostly empty space. The rapid movements of the electrons make the nuclei appear to be surrounded by clouds.

radioactive elements, such as uranium and radium, were known to emit massive particles with considerable speed. Surely a beam of these particles could penetrate a thin sheet of gold. Rutherford and his associates found that almost all the particles passed through the gold sheet with little or no deflection. To the surprise of the experimenters, however, an occasional particle bounced right back. It had struck something very dense, indeed.

Rutherford was quick to realize the implications of this experiment. Most of the mass of an atom is concentrated in a compact, massive lump of matter that occupies only a small part of the atom's volume. Most of the radioactive particles pass freely through the nearly empty space that makes up most of the atom, but a few particles happen to strike the dense mass at the center and rebound.

Rutherford proposed a new model for the structure of an atom. According to this model, a massive, positively charged **nucleus** at the center of the atom is orbited by tiny, negatively charged *electrons* (Figure 4-12). Rutherford concluded that at least 99.98% of the mass of an atom is concentrated in the nucleus, whose diameter is only about one-ten-thousandth the diameter of the atom.

Further research revealed that the nucleus of an atom contains two types of particles: *protons* and *neutrons*. A proton has almost the same mass as a neutron, and each is about 2000 times more massive than an electron. A proton has a positive electric charge and a neutron has no charge. Because like charges repel each other, protons do not naturally stay bound together; they try to move as far away from each other as possible. Neutrons help keep the protons bound together. Conversely, opposite charges attract, and it is this attraction that keeps electrons in orbit around the nucleus, held there by the positively charged protons.

The number of protons in an atom's nucleus determines what element that atom is. Each element is assigned an **atomic number** that equals the number of protons it contains. Of the naturally occurring elements, a hydrogen nucleus always has 1 proton, a helium nucleus always has 2, and so forth, up to uranium with 92 protons in its nucleus.

In contrast, the number of neutrons in the nuclei of atoms of the same element may vary. For example, oxygen, the eighth element in the periodic table, with an atomic number of 8, always has eight protons, but it may have eight, nine, or ten neutrons. These three slightly different kinds of oxygen are called **isotopes**. The isotope of oxygen with eight neutrons is by far the most abundant variety.

Normally, the number of electrons orbiting an atom is equal to the number of protons in the nucleus, thus making the atom electrically neutral. Astronomers denote neutral atoms by writing the atomic symbol followed by the Roman numeral I. For example, neutral hydrogen is written H I; neutral iron is Fe I.

When an atom contains a different number of electrons than protons, the atom is called an **ion**. The process of creating an ion is called **ionization**. Ions are denoted by the atomic symbol followed by a roman numeral that is one greater than the number of missing electrons. Ionized hydrogen (missing its one electron) is denoted H II, while ionized iron with seven electrons missing is denoted Fe VIII. One way to create ions from neutral atoms is *photoionization*, in which photons of sufficiently high energy literally rip an electron completely out of orbit and into the space between atoms. Ionization occurs in stars, as seen by the fact that many of the spectral lines for stars correspond to ionized atoms.

Atoms can also share electrons and, by doing so, remain bound together. Such groups are called **molecules**. They are the essential building blocks of all complex structures, including life.

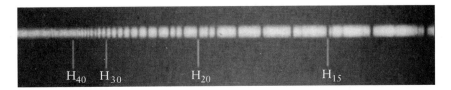

Figure 4-13 Balmer Lines in the Spectrum of a Star This portion of the spectrum of a star called HD 193182 shows nearly two dozen Balmer lines. HD are the initials of Henry Draper, who cataloged this and many other stars. The series converges at 364.56 nm, just to the left of H_{40}. This star's spectrum also contains the first 12 Balmer lines (H_α through H_{12}), but they are not visible in this particular spectrogram.

4-6 Spectral lines occur when an electron jumps from one energy level to another

The challenge of reconciling Rutherford's atomic model with the observations of spectral analysis was undertaken by the young Danish physicist Niels Bohr, who joined Rutherford's group at Manchester in 1911. Bohr began by trying to understand the structure of hydrogen, the simplest and lightest of the elements. A hydrogen atom consists of a single electron and a single proton. Hydrogen has a simple spectrum consisting of a pattern of lines that begins at 656.28 nm and ends at 364.56 nm. The first spectral line is called H_α, the second H_β, the third H_γ, and so forth, ending with H_∞ at 364.56 nm. (The first few lines of the series are identified by Greek-letter subscripts; the remainder are identified by numerical subscripts.) The closer you get to 364.56 nm, the more spectral lines you see.

This spectral pattern had been described mathematically in 1885 by Johann Jakob Balmer, an elderly German schoolteacher. By trial and error, Balmer discovered a formula for calculating the wavelengths of the hydrogen lines. Because of his discovery, the spectral lines of hydrogen at visible wavelengths are called *Balmer lines*, and the entire pattern from H_α to H_∞ is called the *Balmer series*. The spectrum of the star shown in Figure 4-13 exhibits more than two dozen Balmer lines, from H_{13} through H_{40}.

Bohr's goal was to mathematically derive Balmer's formula from basic laws of physics. He began by assuming that the electron in a hydrogen atom moves around the nucleus only in certain specific orbits. As shown in Figure 4-14, it is customary to label these orbits $n = 1$, $n = 2$, $n = 3$, and so on. They are called the *Bohr orbits*.

Bohr argued that for an electron to jump, or to make a **transition**, from one orbit to another, the hydrogen atom must gain or lose a specific amount of energy. *An electron jump from an inner orbit to an outer orbit requires energy; an electron jump from an outer orbit to an inner one releases energy. The energy gained or released by the atom when the electron changes orbits is the difference in energy between these two orbits.*

According to Planck and Einstein, the packet of energy gained or released is a photon whose energy is inversely proportional to its wavelength. Using these ideas, Bohr mathematically derived the formula that Balmer had discovered by trial and error. Furthermore, Bohr's discovery elucidated the meaning of the Balmer series: All the Balmer lines are produced by electron transitions between the second Bohr orbit ($n = 2$) and higher orbits ($n = 3$, 4, 5, and so forth). As an example, Figure 4-15 shows the electron transition that gives rise to the H_α spectral line, which has a wavelength of 656.28 nm.

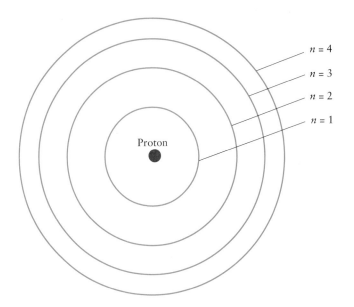

Figure 4-14 Bohr Model of the Hydrogen Atom According to Bohr's model of the atom, an electron circles the nucleus only in allowed orbits $n = 1$, 2, 3, and so on. The first four Bohr orbits are shown here.

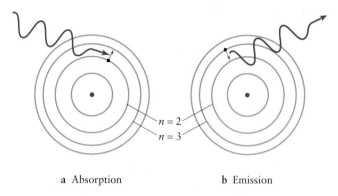

a Absorption   **b** Emission

Figure 4-15 **The Absorption and Emission of an H$_\alpha$ Photon**
This schematic diagram, drawn according to the Bohr model of
the atom, shows what happens when a hydrogen atom absorbs
or emits a photon whose wavelength is 656.28 nm. (a) The
photon is absorbed by the atom as the electron jumps from orbit $n = 2$ up to orbit $n = 3$. (b) The photon is emitted by the
atom as the electron falls from orbit $n = 3$ down to orbit $n = 2$.

An extremely useful way of displaying the structure
of an atom is with an energy level diagram, such as that
shown in Figure 4-16 for hydrogen. The lowest energy
level, called the **ground state,** corresponds to the $n = 1$
Bohr orbit. An electron can jump from the ground state
up to the $n = 2$ level only if the atom absorbs a Lyman-
alpha photon of wavelength 121.6 nm. The energy of a
photon is usually expressed in electron volts (abbreviated "eV"). The Lyman-alpha photon has an energy of
10.19 eV, and so the energy level $n = 2$ is shown on the
diagram as having an energy 10.19 eV above the energy
of the ground state, which is usually assigned a value
of 0 eV. Similarly, the $n = 3$ level is 12.07 eV above
the ground state, and so forth up to the $n = \infty$ level at
13.6 eV. If the atom absorbs a photon having an energy
greater than 13.6 eV, an electron from the ground state
will be knocked completely out of the atom. This is the
process of photoionization mentioned above.

In addition to giving the wavelengths of the Balmer
series, Bohr's formula correctly predicts the wavelengths
of other series of spectral lines that occur at nonvisible
wavelengths. For example, electron transitions between
the lowest Bohr orbit ($n = 1$) and all higher orbits also
produce spectral lines. These transitions create the *Lyman
series,* which is entirely in the ultraviolet wavelengths. At
infrared wavelengths is the *Paschen series,* which arises
out of transitions to and from the third Bohr orbit ($n =
3$). Additional series exist at still longer wavelengths.

Bohr's ideas also help explain Kirchhoff's laws. Each
spectral line corresponds to one specific transition between the orbits of the electrons of a particular element.
An absorption line is created when an electron jumps
from an inner orbit to an outer orbit, extracting the required photon from an outside source of energy such as
the continuous spectrum of a hot, glowing object. An
emission line is produced when an electron falls down to
a lower orbit and gives up a photon.

Today's view of the atom owes much to the Bohr
model, but it is enhanced in certain ways. The modern
theory of atoms is called **quantum mechanics,** a branch
of physics dealing with photons and subatomic particles
that was first developed during the 1920s. As a result of
this work, physicists have moved away from the concept
that electrons are solid particles with planet-like orbits
about the nucleus. Instead, electrons are now known to
have both wave and particle properties; they are said to
occupy certain allowed **energy levels** in the atom.

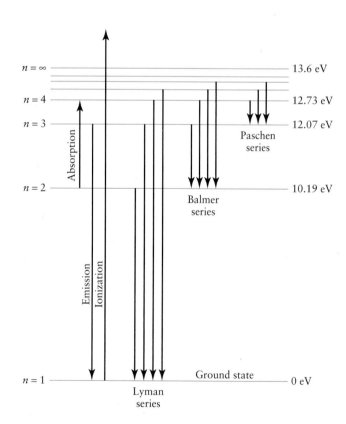

Figure 4-16 **Energy Level Diagram of Hydrogen** The structure of the hydrogen atom is conveniently displayed in a diagram showing the energy levels above the ground state. A variety of electron jumps, or transitions, are shown, including
those that produce some of the most prominent lines in the
hydrogen spectrum.

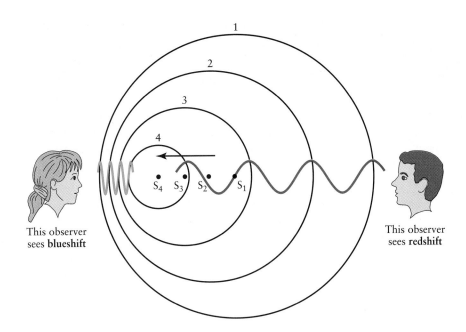

Figure 4-17 **The Doppler Effect** This diagram shows why wavelength is affected by motion between the light source and an observer. A source of light is moving toward the left. The four circles (numbered 1 through 4) indicate the location of light waves that were emitted by the moving source when it was at points S_1 through S_4, respectively. Note that the waves are compressed in front of the source but stretched out behind it. Consequently, wavelengths appear shortened (blueshifted) if the source is moving toward the observer and lengthened (redshifted) if the source is moving away from the observer. Motion perpendicular to an observer's line of sight does not affect wavelength.

4-7 A spectral line is shifted by the relative motion between the source and the observer

Christian Doppler, a professor of mathematics in Prague, pointed out in 1842 that wavelength is affected by motion. As shown for the observer on the left in Figure 4-17, the wavelengths of electromagnetic radiation from an approaching source are compressed. The circles represent waves emitted from consecutive wave peaks as the source moves along. Because each successive wave is emitted from a position slightly closer to you, you see a shorter wavelength than you would if the source were stationary. All the spectral lines in the spectrum of an approaching source are therefore shifted toward the short-wavelength (blue) end of the spectrum. This phenomenon is called a **blueshift.**

Conversely, electromagnetic radiation waves from a receding source are stretched out. You see a longer wavelength than you would if the source were stationary. All the spectral lines in the spectrum of a receding source are shifted toward the longer-wavelength (red) end of the spectrum, producing a **redshift.** In either case, the effect of relative motion on wavelength is called the **Doppler effect** (or **shift**). Toolbox 4-2 gives a numerical example.

A speed determined from the Doppler shift is called the **radial velocity,** because the motion is parallel to our line of sight, or along the "radius" drawn from Earth to the star. Of course, the star may well have motion perpendicular to our line of sight, across the celestial sphere. This **proper motion** does not affect the perceived wave-

length and cannot be determined by Doppler shift. Proper motion is determined by measuring a star's motion relative to background stars (Figure 4-18). The star with the greatest proper motion as seen from Earth is Barnard's star (Figure 4-19).

The Doppler effect is a powerful tool for tracking the motions of objects in space. For instance, careful spectroscopic studies reveal the speed of hot gases on the Sun's surface as they rise and fall. Similarly, Doppler shift measurements of stars in double star systems give crucial data about the speeds of the stars along their orbits. By

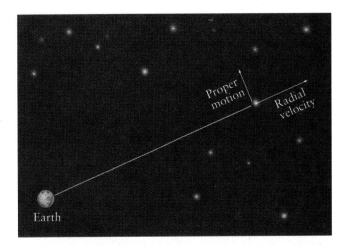

Figure 4-18 **Radial and Proper Motions of a Star** The motion of a star toward or away from the Earth is called radial motion. This motion creates a Doppler shift in the star's spectrum. The motion of the star across the sky, called proper motion, does not affect the star's spectrum.

August 24, 1894

May 30, 1916

Figure 4-19 **Barnard's Star** These two photographs, taken 22 years apart, show the proper motion of Barnard's star in the constellation of Ophiuchus. In addition to having the largest known proper motion (10.3″ per year), Barnard's star is one of the stars nearest to Earth.

measuring the redshifts of distant galaxies, we can determine the rate at which the entire universe is expanding. In later chapters we shall refer to the Doppler effect whenever we need to convert an observed wavelength shift to a speed.

With the work of people like Planck, Einstein, Rutherford, Bohr, and Doppler, the interchange between astronomy and physics came full circle. Modern physics was born when Newton set out to understand the motions of the planets. Two and a half centuries later, physicists in their laboratories discovered the basic properties of electromagnetic radiation and the structures of atoms. As we shall see in later chapters, the fruits of their labors have important implications for astronomy.

AN ASTRONOMER'S TOOLBOX 4-2

The Doppler Shift

Suppose that λ_0 is the wavelength of a spectral line from a stationary source. This is the wavelength that a reference book would list or a laboratory experiment would yield. If the source is moving, this particular spectral line is shifted to a different wavelength λ. The size of the wavelength shift is usually written as $\Delta\lambda$, where $\Delta\lambda = \lambda - \lambda_0$. Thus $\Delta\lambda$ is the difference between the wavelength that you actually observe in the spectrum of a star or galaxy and the wavelength listed in reference books.

Christian Doppler proved that the wavelength shift is governed by the simple equation

$$\frac{\Delta\lambda}{\lambda} = \frac{v}{c}$$

where v is the speed of the source measured along the line of sight between the source and the observer. As usual, c is the speed of light (3×10^8 m/s).

For example, the spectral lines of hydrogen appear in the spectrum of the bright star Vega as shown in the figure. The prominent hydrogen line H_α has a normal wavelength of 656.285 nm, but in Vega's spectrum this line is located at 656.255 nm. The wavelength shift is −0.030 nm, and so the star is approaching us with a speed of −14 km/s. The minus sign indicates that the star is moving toward us.

The Spectrum of Vega Several Balmer lines are seen in this photograph of the spectrum of Vega, the brightest star in Lyra (the Harp). All the spectral lines are shifted very slightly toward the blue side of the spectrum, indicating that Vega is approaching us.

WHAT DID YOU THINK?

1 *How hot is a "red hot" object compared to objects glowing with other colors?* Of all objects that glow from heat stored or generated inside them, those that glow red are the coolest.

2 *What color is the Sun?* The Sun emits all colors (wavelengths of electromagnetic radiation). The colors it emits most intensely are in the blue-green part of the spectrum.

Key Words

absorption line spectrum	emission line spectrum	molecule	spectral analysis
atomic number	energy flux (in an atom)	nucleus (of an atom)	spectral lines
blackbody	energy level	periodic table	spectrograph
blackbody curve	grating	photon	spectroscope
blueshift	ground state	Planck's law	Stefan–Boltzmann law
continuous spectrum	ion	proper motion	transition
Doppler effect (shift)	ionization	quantum mechanics	Wien's law
element	isotope	radial velocity	
	Kirchhoff's laws	redshift	

Key Ideas

• A blackbody is a hypothetical object that is a perfect absorber of electromagnetic radiation at all wavelengths. Because a blackbody does not reflect any electromagnetic radiation from outside sources, the radiation that it does emit depends only on its temperature. Stars closely approximate blackbodies.

The Stefan–Boltzmann law relates the temperature of a blackbody to the rate at which it radiates energy.

Wien's law states that the dominant wavelength of radiation emitted by a blackbody is inversely proportional to its temperature.

The intensities of radiation emitted at various wavelengths by a blackbody at a given temperature are shown by a blackbody curve.

• Electromagnetic radiation has particlelike properties; these particles, called photons, are packets of electromagnetic waves.

• Planck's law says that the energy of a photon is inversely proportional to its wavelength.

• Spectroscopy—the study of spectra—provides important information about the chemical composition of remote astronomical objects.

Kirchhoff's three laws of spectral analysis describe the conditions under which absorption lines, emission lines, and a continuous spectrum can be observed.

Spectral lines serve as distinctive "fingerprints" for the chemical elements and chemical compounds comprising a light source.

• An atom consists of a small, dense nucleus (composed of protons and neutrons) surrounded by electrons that occupy only certain energy levels.

The spectral lines of a particular element correspond to the various electron transitions between allowed energy levels in the atoms of the element.

When an electron shifts from one energy level to another, a photon of the appropriate energy (and hence a specific wavelength) is absorbed or emitted by the atom.

The spectrum of hydrogen at visible wavelengths consists of the Balmer series, which arises from electron transitions between the second energy level of the hydrogen atom and higher levels.

• When an atom loses one or more electrons it is said to be ionized. One way for this to occur is for the atom to absorb an energetic photon.

• The spectral lines of an approaching light source are shifted toward short wavelengths (a blueshift); the spectral lines of a receding light source are shifted toward long wavelengths (a redshift).

The equation describing the Doppler effect states that the size of a wavelength shift is proportional to the radial velocity between the light source and the observer.

Review Questions

1 What is a blackbody? What does it mean to say that a star behaves almost like a blackbody? If stars behave like blackbodies, why are they not black?

2 What is the Stefan–Boltzmann law? Why do you suppose that astronomers are interested in it?

3 What is Wien's law? How could you use it to determine the temperature of a star's surface?

4 Using Wien's law and the Stefan–Boltzmann law, explain the changes in color and brightness that are observed as the temperature of a hot, glowing object increases.

5 Describe the experimental evidence that supported the Bohr model of the atom.

6 Explain how the spectrum of hydrogen is related to the structure of the hydrogen atom.

7 Why do different elements have different patterns of lines in their spectra?

8 What is the Doppler effect, and why is it important to astronomers?

9 Explain why the Doppler effect tells us only about the motion directly along the line of sight between a light source and an observer.

Advanced Questions

* **10** Approximately how many times around the world could a beam of light travel in one second?

* **11** The bright star Regulus in the constellation of Leo (the Lion) has a surface temperature of 12,200 K. Approximately what is the dominant wavelength (λ_{max}) of the light it emits?

* **12** The bright star Procyon in the constellation of Canis Minor, the Little Dog, emits the greatest intensity of radiation at a wavelength (λ_{max}) of 445 nm. Approximately what is the surface temperature of the star?

* **13** Imagine a star twice the diameter of the Sun but with the same surface temperature. At what wavelength would that star emit most of its radiation? How many times brighter than the Sun would that star be?

* **14** The wavelength of H_β in the spectrum of the star Megrez in the Big Dipper is 486.112 nm. Laboratory measurements demonstrate that the normal wavelength of this spectral line is 486.133 nm. Is the star coming toward us or moving away from us? At what speed?

* **15** In the spectrum of the bright star Rigel, H_α has a wavelength of 656.331 nm. Is the star coming toward us or moving away from us? How fast?

* **16** Imagine driving down a street toward a traffic light. How fast would you have to go so that the red light would appear green?

Discussion Questions

17 Compare the technique of identifying chemicals by their spectral line patterns with that of identifying people by their fingerprints.

18 Suppose you look up at the night sky and observe some of the brightest stars with your naked eye. Is there any way of telling which stars are hot and which are cool? Explain.

Observing Projects

19 As soon as weather conditions permit, observe a rainbow. Confirm that the colors are in the same order as shown in the front figure of this chapter.

20 Turn on an electric stove or toaster oven. Watch as the color and brightness of the heating element changes. Confirm that red is the color of the coolest glowing objects, and that as an object heats up it gets brighter and its color shifts from red to orange to yellow.

21 Obtain a glass prism or a diffraction grating, available from science museum stores, catalogues, or your college physics departments. Look through the prism or grating at various light sources, such as an ordinary incandescent light, a neon sign, and a mercury vapor street lamp. **DO NOT LOOK AT THE SUN! Looking directly at the Sun causes blindness.** Do you have any trouble seeing spectra? What do you have to do to see a spectrum? Describe the differences in the spectra of the various light sources you observed.

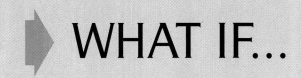
Our eyes are sensitive to less than a trillionth of 1% of the electromagnetic spectrum—what we call "visible light." Yet this minuscule resource provides us with an awesome amount of information about the universe. We interpret visible-light photons as the six colors of the rainbow—red, orange, yellow, green, blue, and violet. These colors combine to form all the others that help make the world so attractive, from the beauty of a sunset to the wonder of the stars.

But the Sun actually gives off photons of all wavelengths, not just the wavelengths of visible light. So, what would we see if our eyes had evolved to see another part of the spectrum? You learned in Chapter 3 how astronomers have developed telescopes that can "see" the entire spectrum.

Gamma rays, x rays, and most ultraviolet radiation do not pass through the Earth's atmosphere. Consequently, the world would look dark, indeed, if our eyes were sensitive to these wavelengths. Radio waves, in contrast, easily pass through our atmosphere. However, in order to see the same detail from radio waves that we now see from visible-light photons, our eyes would have to be 10,000 times larger. Part of this increase is due to the need for larger sensors in our eyes than the existing rods and cones. Another reason is that the Sun emits less than one-billionth as many radio wave photons as visible-light photons. Certainly having eyes the size of baseball fields would have substantial drawbacks!

Finally, what about infrared radiation? Not all incoming infrared photons get through our atmosphere. Some are reflected back into space, while most are absorbed by water and carbon dioxide molecules in the air. However, short-wavelength ("near") infrared radiation passes easily through air. Depending on wavelength, the Sun emits between one-half and one-ten-billionth as many infrared photons as optical photons. Fortunately, most of these are in the near infrared.

To see infrared photons, our eyes would need to be five to ten times larger. Some snakes have evolved infrared vision, and portable infrared "night vision" cameras are available, as many of us saw during the Gulf War. Since everything that emits heat emits infrared photons, infrared sight would be useful for seeing objects on Earth. And not everything we saw would be due just to reflected sunlight—hotter objects would be intrinsically brighter than cooler ones. Such eyes would also be useful for observing changes in a person's emotional state. A person who is angry or excited often has more blood near the skin, and so gives off more infrared (heat) than normal. Conversely, someone who is scared has less blood near the skin, and so emits less heat.

The night sky would be a spectacular sight through infrared-sensitive eyes. Gas clouds in the Milky Way absorb visible light, thus preventing the light of distant stars from getting to the Earth. However, because most infrared radiation passes through these clouds, we would be able to see distant stars that we cannot see today. On the other hand, the white glow of the Milky Way we see today, which is caused by the scattering of starlight by clouds, would be dimmer, since the gas clouds do not scatter infrared light as much as they do visible light.

Our concept of stars would be different than it is today, too. Many stars, especially young, hot ones, are surrounded by cocoons of gas and dust that emit infrared radiation. This dust is heated by the nearby stars. Instead of appearing as pinpoints, many stars would appear to be surrounded by wild strokes of color—we would have an Impressionist sky.

FOUNDATIONS

The Solar System

The Planets This montage of photographs taken by various spacecraft shows nine planets. The images are *not* reproduced to the same scale. At the top (from left to right) are Mercury, Venus, Earth, and Mars along with our Moon. At the bottom (from right to left) are Jupiter, Saturn, Uranus, Neptune, and Pluto.

....

IN THE PAGES AHEAD

In Chapters 5 through 9, you will examine the origin of the solar system and survey its contents. You will discover that all the planets have distinct "personalities" as a result of their differences in size, chemical composition, rotation rate, atmosphere, and surface features. You will also find that the moons throughout the solar system are similarly varied. The solar system also contains debris—asteroids, meteoroids, and comets—left over from its birth. This material provides astronomers with important clues about the formation of the solar system.

WHAT DO YOU THINK?

1   How many stars are there in the solar system?

2   How long has the Earth existed?

3   What planet is farthest from the Sun?

4   What shapes do moons have?

· · · · · · · · · · · · · · · ·

FOR AS LONG AS people have looked up at the heavens, they have wondered about the Sun, the Moon, and the visible planets. After the telescope was invented, scientists discovered countless other bodies—more planets and their moons, asteroids, meteoroids, and comets—in orbit about our star, the Sun. The Sun and all the bodies that orbit it make up our **solar system.**

Did the solar system form all at once, or did it come together by serendipity? And what created the varied building blocks of rock, metal, ice, and gas? Within the past few decades, telescopes and space probes, along with the principles of modern science, have finally provided answers to these age-old questions. Our new wealth of information is giving us a rapidly growing understanding of our nearest neighbors in space.

II-1 The solar system formed from a cloud of gas and dust

To find out where the solar system came from, we look for clues in its pieces. Especially valuable is the interplanetary debris we know as asteroids, meteoroids, and comets. Some of these bodies are believed to be essentially unchanged remnants from the formation of the solar system. Other clues to the origin of our solar system have recently been found in distant stars, where planetary systems like our own are developing.

Today we picture the solar system as forming from a fragment of a vast cloud of interstellar gas and dust. Called the **solar nebula,** this cloud fragment condensed under the influence of its own gravitational force into the planets, the Sun, and other bodies 4.6 billion years ago. To learn how this condensation occurred, we must first know what elements were in the nebula.

Astronomers have determined that hydrogen and helium are by far the most abundant elements in the universe. Together these elements account for 98% of the mass of all the material in existence—all the other elements combined account for only 2%. However, the Earth's mass contains less than 0.15% hydrogen and helium. Somehow the early solar system was enriched with the heavy elements we see today—oxygen, silicon, aluminum, iron, carbon, calcium, and others. Although these elements are rare in the universe as a whole, they are commonplace on Earth.

There is a good reason for the overwhelming abundance of hydrogen and helium throughout the universe. Most astronomers believe that the universe began roughly 15 billion years ago with a violent event called the Big Bang. Only the lightest elements—hydrogen, helium, and tiny amounts of lithium and beryllium—emerged when the cosmos was formed. Galaxies and stars condensed out of this matter hundreds of millions of years later. It was in the stars that the other elements were created.

Stars shine by generating energy in their cores. In this process, called *fusion,* hydrogen and helium change into heavier elements, such as carbon, silicon, aluminum, and iron. Indeed, all of the heavier elements existing in the universe today were manufactured in the stars. We shall explore details of the creation of this heavier matter in Chapters 11 and 12.

Near the ends of their lives, most stars cast matter out into space to form clouds of interstellar gas and dust. This process is often a comparatively gentle one in which a star's outer layers are gradually expelled (Figure II-1). However, some stars end their lives with spectacular detonations called *supernovae,* which blow the stars apart. Either way, space between stars becomes filled with gas containing mostly hydrogen and helium but now enriched with heavy elements created in the stars. New stars and planetary systems then form out of this enriched interstellar gas (Figure II-2). We are literally made of star dust.

Initially it was very cold (well below the freezing point of water) inside the solar nebula. Ice and ice-coated dust grains composed of heavy elements were scattered abundantly across this vast volume, which had a diameter of at least 100 AU and a total mass roughly two or three times the mass of the Sun. Hydrogen and helium gas, accompanied by this ice and dust, fell toward the center of the solar nebula under the influence of their mutual gravitational attraction. As a result, density and pressure at the center of the nebula began to increase, producing a concentration of matter called the **protosun.** As the nebula continued to contract, atoms at its center collided with one another with increasing speed and frequency.

Figure II-1 **A Star Losing Mass** This star is shedding material. The nebulosity around the star is caused by starlight reflected from dust grains that condensed from the material cast off by the star.

Such collisions created heat, causing the temperature deep inside the solar nebula to soar.

The planets and other bodies orbiting the Sun formed because the solar nebula had a slight amount of rotation. Without rotation, everything in the solar nebula would have fallen straight into the protosun, leaving nothing behind to form the planets. Mathematical studies show that the combined effects of gravity and rotation transform even an irregular cloud fragment (which the solar

Figure II-2 **A Dusty Region of Star Formation** These young stars are still surrounded by much of the gas and dust from which they formed. The bluish, wispy nebulosity is caused by starlight reflecting off abundant interstellar dust grains. These grains are made of heavy elements (such as carbon, silicon, and iron) produced by earlier generations of stars.

nebula probably was) into a rotating flattened disk with a warm center and cold edges, as shown in Figure II-3. The transformation of the solar nebula into a flattened disk explains why the orbits of the planets are all nearly in the same plane. Astronomers have found disks of material surrounding other stars, like the one shown in Figure II-4. Planets are probably still forming today in these disks of debris left over from the births of stars.

Temperatures around the newly created protosun soon began to climb. The rising temperatures vaporized all the common icy substances in the inner regions of the solar nebula and pushed light gases like hydrogen and helium outward. Only the heavier elements remained.

The formation of the four inner planets was dominated by the fusing together of these small, rocky particles. Initially neighboring dust grains and pebbles in the solar nebula collided and stuck together. Then, over a period of a few million years, these accumulations of dust and pebbles coalesced into larger objects called **planetesimals,** with diameters of about 100 km.

During the next stage, the planetesimals collided and coalesced into still larger objects called **protoplanets** because of their mutual gravitational attraction. This accumulation of material is called **accretion.** Recent computer simulations based on Newtonian mechanics have shown that accretion continues for roughly 100 million years and typically leads to the formation of about half a dozen planets (Figure II-5). Agreement between these simulations and most characteristics of our inner solar system is very striking.

Figure II-3 The Birth of the Solar System This sequence of drawings shows six stages in the formation of the solar system. (a) A slowly rotating cloud of interstellar gas and dust begins to contract because of its own gravity. (b) A central condensation, the protosun, forms as the cloud flattens and rotates faster. (c) A flattened disk of gas and dust surrounds the protosun, which has begun to shine. (d) The Sun removes the gas from the inner regions, leaving dust and larger debris. (e) The planets have established dominance in their regions of the solar system. (f) The solar system as it appears today.

While the inner regions of the solar system were heating up, temperatures in the outer regions of the solar nebula remained quite cool. Ice and ice-coated dust, along with hydrogen and helium gas, were able to survive in these cooler regions. The outer planets also probably formed from the accretion of planetesimals. For each of the large outer planets—Jupiter, Saturn, Uranus, and Neptune—a rocky core served as a "seed," which for about a million years accumulated a coating of additional rock and gas. When the masses of the rocky core and the gas-rich envelope became equal, the gas began to accumulate at a runaway pace. Thereafter the envelope pulled in all the gas in its vicinity, creating a huge hydrogen-rich shell surrounding an Earth-sized core of rocky material.

As a result of these origins, the outer planets are made primarily of low-density substances, including abundant quantities of hydrogen, helium, methane, ammonia, and water. Many of the moons of the outer planets are partially composed of these substances, too.

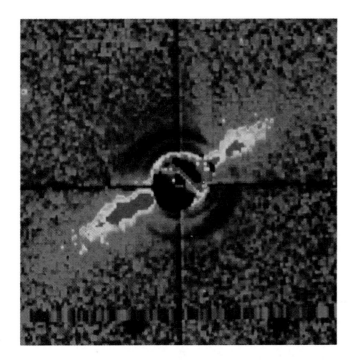

Figure II-4 **A Circumstellar Disk of Matter** This computer-enhanced photograph shows a disk of material orbiting the star called β-Pictoris. This star is located behind a small circular mask at the center of the picture. The mask was needed to block the star's light, which otherwise would have overwhelmed light from the disk. The disk, seen nearly edge-on, is believed to be very young, possibly no more than a few hundred million years old.

During the millions of years that the planets were forming, the temperature and pressure at the center of the contracting protosun continued to climb. Finally, the center of the protosun became hot enough to ignite thermonuclear reactions (fusing hydrogen into helium in its core), and the Sun was born. Sunlike stars take approximately 100 million years to form from a nebula, which means that the Sun must have become a full-fledged star at roughly the same time the accretion of the inner protoplanets was complete. According to radioactive dating of the oldest debris ever discovered from space, the solar system came into the form we know today about 4.6 billion years ago.

II-2 Collisions dominated the early solar system

Astronomers have recently come to appreciate that spectacular collisions helped shape the solar system. Our Moon's surface bears the scars of numerous impacts, most of which date from the final stages of the formation of the solar system (Figure II-6). (We capitalize the word *moon* only when we are referring to the Earth's Moon.) These scars, called **craters**, are also found on those planets and moons that do not have appreciable atmospheres or geological activity that would otherwise erase these features. Indeed, the Moon's airless environment has

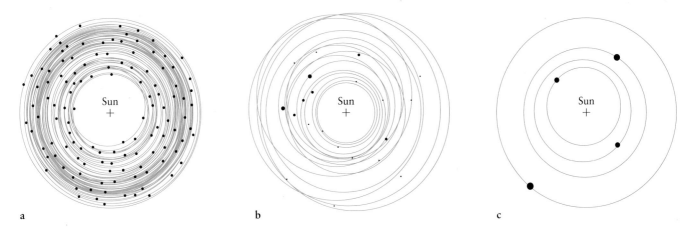

a   b   c

Figure II-5 **Accretion of the Inner Planets** These three drawings show the results of a computer simulation of the formation of the inner planets. (a) The simulation begins with 100 planetesimals. (b) After 30 million years, these planetesimals have coalesced into 22 protoplanets. (c) This final view is for an elapsed time of 441 million years, but the formation of the inner planets is essentially complete after only 150 million years.

◄ Figure II-6 **Our Moon** This photograph, taken by astronauts in 1972, shows thousands of craters, most of them produced by impacts of rocky debris left over from the formation of the solar system. Age-dating of lunar rocks brought back by the astronauts indicates that the Moon is about 4.5 billion years old. Most of the lunar craters were formed when the Moon was less than a billion years old.

preserved important information about the early history of the solar system.

Radioactive dating of Moon rocks brought back by the Apollo astronauts indicates that the rate of impacts declined dramatically about 3.5 billion years ago. Since that time, impact cratering has proceeded at a very low rate. Thus most of the craters on the Moon and planets were formed during the first billion years of the solar system's history as the young planets swept up rocky debris left over from the solar nebula.

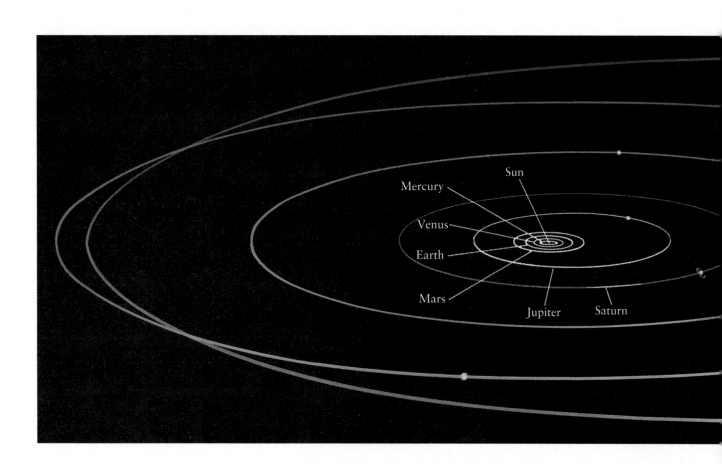

The early violence in the solar system has never completely ceased. Indeed, collisions in the solar system still occur today. But with the decrease in impact activity about 3.5 billion years ago, the chemical composition of the planets evolved more gradually and predictably. Inevitably, biological evolution took place on one planet ideally situated to provide liquid surface water as a solvent in which life could develop.

II-3 There are nine known planets

There are nine known planets: Mercury, Venus, Earth, Mars, Jupiter, Saturn, Uranus, Neptune, and Pluto. As Figure II-7 shows, the orbits of the four inner planets—Mercury, Venus, Earth, and Mars—are crowded close to the Sun. In contrast, the orbits of the four large planets—Jupiter, Saturn, Uranus, and Neptune—are widely spaced at greater distances from the Sun. Pluto is usually at the far fringes of the part of the solar system inhabited by planets.

Kepler's laws (see Chapter 2) showed us that all the planets have elliptical orbits. Even so, most of the planets' orbits are nearly circular. The exceptions are Mercury and Pluto, whose orbits are noticeably elliptical. In fact, Pluto's orbit sometimes takes it nearer the Sun than its neighbor, Neptune. In 1979 Pluto passed inside Neptune's orbit. Neptune will be the most distant planet from the Sun until 1999, when Pluto's elliptical orbit once again takes it farther out than Neptune (Figure II-8).

All the planetary orbits except Pluto's lie in planes close to the ecliptic. Pluto's orbit has a conspicuous tilt (called an *orbital inclination*) compared to the orbits of the other planets. Table II-1 lists some orbital characteristics of the nine planets.

The most obvious physical difference between the inner planets and Jupiter, Saturn, Uranus, and Neptune is in size. The smallest of the four outer planets, Neptune,

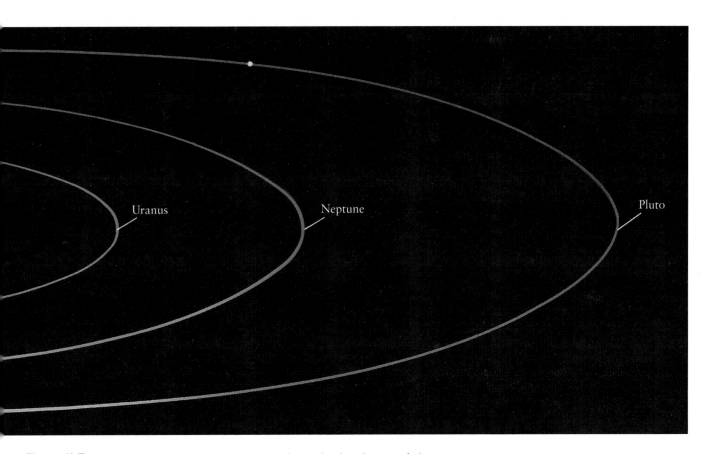

Figure II-7 **The Solar System** This scale drawing shows the distribution of planetary orbits around the Sun. The four inner planets are crowded close to the Sun; the five outer planets orbit at much greater distances from the Sun.

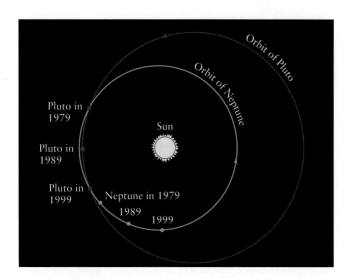

Figure II-8 **Pluto's Orbit Is Highly Elliptical** Pluto has the most elliptical orbit of any planet. Indeed, it spends part of the time closer to the Sun than Neptune. It last moved inside Neptune's orbit in 1979. It continues there today and will move back outside in 1999.

is nearly 4 times larger than the Earth, the largest of the four inner planets. First place among the giant planets goes to Jupiter, whose diameter is about 11 times bigger than Earth's. Pluto is smaller than the inner planets, despite being the outermost planet. It is even slightly smaller than our Moon. Figure II-9 shows the Sun and the planets drawn to the same scale. The diameters of the planets are given in Table II-2.

Mass is another characteristic that distinguishes the inner and outer planets. The four inner planets have low masses compared to the giant planets. Again, first place goes to Jupiter, whose mass is 318 times greater than Earth's (see Table II-2).

Size and mass can be combined in a very useful way to provide information about the chemical composition of a planet (or any other object, for that matter). Matter composed of heavy elements, like iron or lead, has more particles compressed into a given volume than does matter composed of light elements, like hydrogen, helium, or carbon. The measure of an object's compactness is its **average density,** given by the equation

$$\text{Average density} = \frac{\text{mass}}{\text{volume}}$$

The chemical composition (kinds of elements) of any object therefore determines how much space the object occupies. For example, a kilogram of iron takes up less space (is denser) than a kilogram of water, even though both have the same mass.

Average density is expressed in kilograms per cubic meter. The numbers become meaningful only when compared to something familiar, such as water, which has an average density of 1000 kg/m^3. The four inner planets have high average densities compared to water (see Table II-2). The average density of the Earth is 5520 kg/m^3. Since the density of typical rock is only about 3000 kg/m^3, the Earth must contain a large amount of material inside it that is denser than surface rock.

In sharp contrast, the outer planets have quite low densities. For example, Saturn has an average density less than that of water. We derive information about densities from the masses and volumes of the planets: Kepler's laws yield the mass of each planet from the periods of their moons' orbits, while their measured diameters yield their volumes. Their low densities support the belief that the giant outer planets are composed primarily of such light elements as hydrogen and helium.

Pluto is again an oddity. Although it is smaller than any of the inner planets, its average density is between those of the Earthlike and the giant outer planets. Pluto is probably composed of a mixture of rock and ice because its average density is between that for rock and ice.

The planets fall roughly into four groups. The four inner planets form the first group, followed by Jupiter and Saturn as the second, Uranus and Neptune as the third, and finally Pluto as the fourth. Pluto may be only the first of several more distant planets yet to be discov-

TABLE II-1			
Orbital Characteristics of the Planets			
	Average distance from Sun		**Orbital period**
	(AU)	(10^6 km)	(yr)
Mercury	0.39	58	0.24
Venus	0.72	108	0.62
Earth	1.00	150	1.00
Mars	1.52	228	1.88
Jupiter	5.20	778	11.86
Saturn	9.54	1427	29.46
Uranus	19.19	2871	84.01
Neptune	30.06	4497	164.79
Pluto	39.53	5914	248.54

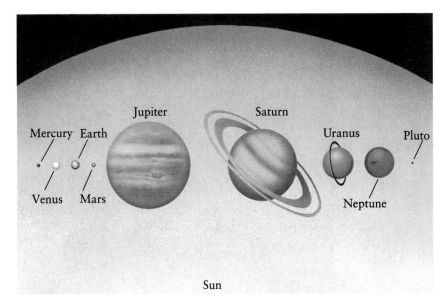

Figure II-9 **The Sun and the Planets** This drawing shows the nine planets in front of the disk of the Sun, with all ten bodies drawn to the same scale. The four planets that have orbits nearest the Sun (Mercury, Venus, Earth, and Mars) are small and made of rock and metal. The next four planets (Jupiter, Saturn, Uranus, and Neptune) are large and composed primarily of light elements such as hydrogen and helium.

ered. Indeed, at least 48 small solar system bodies beyond Pluto have been identified, but few details about them are known.

From their average densities, we can draw conclusions about the composition of the planets. The four inner ones, composed primarily of rock and metal, are called **terrestrial planets** because they resemble the Earth (in Latin, *terra*). Mountains, craters, canyons, and volcanoes are common on their hard, rocky surfaces. Jupiter and Saturn are composed primarily of hydrogen and helium, Uranus and Neptune have extremely large quantities of water, and Pluto is roughly an even mix of rock and ice. Vast, swirling cloud formations dominate the appearance of Jupiter, Saturn, Uranus, and Neptune. The montage of photographs at the beginning of Part II shows the distinctive appearances of the different planets.

The different surfaces (or upper cloud layers) of the planets reflect different amounts of light back out into space. The amount of light returning directly to space is measured by each body's **albedo.** An object that reflects no light has an albedo of 0.0; for example, powdered charcoal has an albedo of nearly 0.0. An object that reflects all the light incident upon it (like a high-quality mirror) has an albedo of 1.0. The number for an albedo multiplied by 100 gives the percentage of light directly reflected by that body.

TABLE II-2

Physical Characteristics of the Planets

	Diameter		Mass		Average density (kg/m³)
	(km)	(Earth = 1)	(kg)	(Earth = 1)	
Mercury	4,878	0.38	3.3×10^{23}	0.06	5430
Venus	12,100	0.95	4.9×10^{24}	0.81	5250
Earth	12,756	1.00	6.0×10^{24}	1.00	5520
Mars	6,786	0.53	6.4×10^{23}	0.11	3950
Jupiter	142,984	11.21	1.9×10^{27}	317.94	1330
Saturn	120,536	9.45	5.7×10^{26}	95.18	690
Uranus	51,118	4.01	8.7×10^{25}	14.53	1290
Neptune	49,528	3.88	1.0×10^{26}	17.14	1640
Pluto	2,300	0.18	1.3×10^{22}	0.002	2030

Besides the Earth, Mars, Jupiter, Saturn, Uranus, Neptune, and Pluto all have moons (sometimes called natural satellites). There are at least 65 moons in the solar system and more are still being discovered. Unlike our Moon, most moons are irregularly shaped, more like potatoes than spheres. We will discover that there is as much variety in moons as there is in planets.

II-4 There is still minor debris left over from the formation of the solar system

In addition to the nine planets and their moons, many other objects orbit the Sun. Between the orbits of Mars and Jupiter are believed to be over one hundred thousand such bodies, called **asteroids,** most of them smaller than a kilometer across. This region is called the **asteroid belt.** Asteroids are composed primarily of metal and rock. The largest asteroid, Ceres, has a diameter of about 900 km. The next largest, Pallas and Vesta, are each about 500 km in diameter. Still smaller ones are increasingly numerous. There are thousands of kilometer-sized asteroids. A close-up picture of the asteroid Gaspra, taken by a spacecraft on its way to Jupiter, is shown in Figure II-10. In addition to the asteroids between Mars and Jupiter, other asteroids orbit in highly elliptical orbits that take them across the paths of some of the planets.

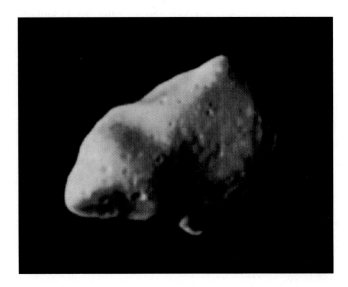

Figure II-10 **An Asteroid** This picture of the asteroid Gaspra was taken in 1991 by the Galileo spacecraft on its way toward Jupiter. The asteroid measures $12 \times 20 \times 11$ km. Several thousand similar chunks of rock orbit the Sun between the orbits of Mars and Jupiter.

Figure II-11 **A Comet** The solid part of a comet is a chunk of ice roughly 10 km in diameter. When a comet passes near the Sun, solar radiation vaporizes some of the comet's ices, and the resulting gases form a tail millions of kilometers long. This photograph shows a comet that was seen in January 1974.

Pieces of rocky and metallic debris smaller than asteroids are called **meteoroids.** Typical meteoroids are boulder-sized or smaller. These apparently exist throughout the disk of the solar system, although most are in the asteroid belt. Most meteoroids are probably small fragments of asteroids that broke off when larger bodies collided.

Quite far from the Sun, well beyond the orbit of Pluto, are hundreds of billions of chunks of ice called **comets.** Some comets orbit in the same plane as the planets, but others are distributed in a sphere around the Sun. Many comets have highly elongated orbits that occasionally bring them close to the Sun. When this happens, the Sun's radiation vaporizes some of the comet's ices, thereby producing long, flowing tails (Figure II-11).

We begin our detailed exploration of the solar system by examining the two bodies we know best: the Earth and its Moon. By understanding them, we will be better able to make some sense out of the remarkably alien worlds we encounter thereafter. Indeed, we will discover that the properties of some planets, moons, and other objects in the solar system are so different from the Earth and its Moon as to make direct comparison almost meaningless.

WHAT DID YOU THINK?

1 *How many stars are there in the solar system?* The solar system is the Sun and everything orbiting it. Therefore, there is only one star in it.

2 *How long has the Earth existed?* The Earth formed along with the rest of the solar system 4.6 billion years ago.

3 *What planet is farthest from the Sun?* Neptune is farthest from the Sun today. Pluto's highly elliptical orbit brought it closer to the Sun than Neptune in 1979. Pluto will be farther from the Sun than Neptune again in 1999.

4 *What shapes do moons have?* While some moons are spherical, most look roughly like potatoes.

Key Words

accretion	average density	planetesimal	solar system
albedo	comet	protoplanet	terrestrial planet
asteroid	crater	protosun	
asteroid belt	meteoroid	solar nebula	

Key Ideas

• The four inner planets of the solar system share many characteristics and are distinctly different from the four giant outer planets and from Pluto.

The four inner (terrestrial) planets are relatively small, have high average densities, and are composed primarily of rock and metal.

Jupiter and Saturn have large diameters and low densities and are composed primarily of hydrogen and helium.

Uranus and Neptune have large quantities of water as well as much hydrogen and helium.

Pluto, the smallest of the nine planets, is of intermediate density.

• Hydrogen and helium, the two lightest elements, were formed shortly after the creation of the universe. The heavier elements were produced much later by stars and cast into space when the stars died.

By mass, 98% of the matter in the universe is hydrogen and helium.

• The solar system formed 4.6 billion years ago from a swirling, disk-shaped cloud of gas, ice, and dust called the solar nebula.

The four inner planets formed through the accretion of dust particles into planetesimals and then into larger protoplanets.

The four large outer planets probably formed through the runaway accretion of gas onto rocky protoplanetary cores.

• The Sun formed at the center of the solar nebula. After about 100 million years, the temperature at the protosun's center was high enough to ignite thermonuclear reactions.

For nearly a billion years after the Sun formed, impacts of asteroidlike objects on the young planets dominated the early history of the solar system.

Review Questions

1 Why—at the present time—is Pluto closer to the Sun than Neptune?

2 Why do astronomers believe that the solar nebula was rotating?

Advanced Questions

3 What two properties of a planet must be known in order for its average density to be determined? How are these properties determined?

4 How can Neptune have more mass but a smaller diameter than Uranus? (*Hint:* See Table II-2.)

5 ▶ The Earth and Its Moon

IN THIS CHAPTER

You will see what makes the Earth such an ideal environment for life and what makes the Moon so forbidding. You will discover that Earth's atmosphere has evolved dramatically, that its insides are churning, and that its surface is constantly in motion. Its magnetic field helps protect the surface from high-energy particles from space. In comparison, the Moon is a dead, airless, unchanging world, whose surface has been pounded by countless impacts from space debris. Just as the Earth has two major types of surface features, its continents and ocean bottoms, the Moon has rugged highlands and flat maria. But unlike the Earth, the Moon has so far been found to be devoid of water.

The Earth and the Moon The Earth is a dynamic planet whose surface is mostly covered with water and whose nitrogen–oxygen atmosphere supports life. In contrast, the Moon is a barren, desolate world. Because the Moon has no atmosphere, lunar rocks have not been subjected to weathering by wind. They thus preserve information about the early history of the solar system. The Earth's diameter is about four times the Moon's. Both worlds are shown to scale.

WHAT DO YOU THINK?

1. Will the ozone layer, which is now being depleted, naturally replenish itself?

2. Does the Moon have a dark side, where it is forever night?

3. Does the Moon rotate, and if so, how fast?

4. What causes the ocean tides?

5. When does the spring tide occur?

• • • • • • • • • • • • • • • •

SUPPOSE YOU HAD COME to the solar system from a distant star system looking for inhabitable worlds. Your spacecraft approaches the inner solar system on the opposite side of the Sun from the Earth. You encounter Mars, with its thin, chilled, unbreathable atmosphere and barren desert landscapes. Not very promising. The next planet you encounter, Venus, is enshrouded with corrosive clouds hiding a menacingly hot surface. Swinging around the Sun you finally spy Earth, located between these two relatively forbidding planets: one too cold, the other too hot. Ever-changing white clouds pirouette above the darker browns and blues of Earth's continents and oceans. Is this a world that your funding agency back home can use to justify more astronomical research?

Earth: A Dynamic, Vital World

The Earth is a geologically active world. Earthquakes shake many regions of it. Volcanoes pour huge quantities of molten rock from inside the Earth onto the surface, and gas from within vents into the atmosphere. Some mountains are still rising, while others are wearing away. Water flow erodes topsoil and carves river valleys. Rain and snow help rid the atmosphere of dust particles. And life teems virtually everywhere, making the Earth unique in the solar system. Table 5-1 lists Earth's important physical and orbital properties.

An interstellar traveler would find Earth an appealing place to investigate because it contains large quantities of liquid water. Water, the universal solvent in which life formed, covers nearly 71% of the Earth's surface (Figure 5-1). A space traveler might first judge the Earth's shifting clouds and its vast bodies of water from

TABLE 5-1	
Earth's Vital Statistics	
Statistic	**Measurement**
Mass	6.0×10^{24} kg (1 $M_\oplus$)
Radius	6378 km = 1 $R_\oplus$
Average density	5520 kg/m^3
Orbital eccentricity	0.017
Sidereal period of revolution (year)	365.26 d (1.00 yr)
Average distance from Sun	149.6×10^6 km (1 AU)
Sidereal rotation period	0.997 d
Solar rotation period (day)	1.00 d
Albedo (average)	0.37

Note: $\oplus$ refers to the Earth.

Figure 5-1 **A View of Earth's Surface** An oasis in the forbidding cold of space, the Earth is a world of unsurpassed beauty and variety. Ever-changing cloud patterns drift through its skies. Two-thirds of its surface is covered with oceans. This liquid water, in combination with a huge variety of chemicals from its lands, led to the formation and evolution of life over most of the planet's surface.

the sunlight they reflect. The Earth has an average albedo of 0.37, meaning it reflects about 37% of the sunlight it receives back into space. (Why does this number change daily and seasonally?)

5-1 The Earth's atmosphere has evolved over billions of years

The Earth's atmosphere is unique among the planets. The air we breathe is predominantly a 4-to-1 mixture of nitrogen to oxygen, two gases that are found only in small amounts on the other planets.

We now know that our atmosphere was not always so amenable to life. Indeed, we are enveloped in the third atmosphere that the Earth has had. The first one, composed of trace remnants of hydrogen and helium left over from the formation of the solar system, didn't last very long. These gases are too light to stay near the Earth (consider what happens to helium balloons). Heated by sunlight, these gases quickly evaporated into space.

The second atmosphere came from inside the Earth and was composed primarily of carbon dioxide along with some nitrogen. In fact, there was roughly 100 times as much gas in the air of this second atmosphere as there is today. The oceans absorbed about half of this atmosphere. (Water holds a lot of carbon dioxide, as you can see by the amount of carbon dioxide fizz in a can of soda.) Then, as life evolved in the oceans, much of that dissolved carbon dioxide was transformed into the shells of many creatures and then into rock as the animals died and their carbon-rich shells sank to the ocean bottoms. Shells piled up and compressed the shells underneath them into rock, such as limestone.

Early plant life in the oceans and then on the Earth's landmasses removed the remaining carbon dioxide in the air by converting it into oxygen plus the nutrients that the plants needed to survive. The most efficient such conversion mechanism is photosynthesis, which today helps maintain the balance between carbon dioxide and oxygen in the air.

The first oxygen that returned to the air from the oceans and from land plants did not stay there for long. Oxygen is wildly reactive. Early oxygen combined with many elements on the Earth's surface, notably iron, to form new compounds. (We will see similar iron compounds on Mars.) Eventually, after all the minerals that could combine with oxygen had done so, the atmosphere began to fill with this gas. With the carbon dioxide gone,

the air became the nitrogen–oxygen mixture that we breathe today.

Because it has mass, the air is pulled down toward the Earth by gravity, thereby creating a pressure on the surface. Pressure is merely a force acting over some area. We formally define pressure by the following formula:

$$\text{Pressure} = \frac{\text{force}}{\text{area}}$$

The average atmospheric pressure at sea level is 14.7 pounds per square inch, or 1 atmosphere (abbreviated "atm"). This pressure is due to the weight of the air pushing down. The atmospheric pressure decreases smoothly with increasing altitude, falling by roughly half with every 5.5 km. Seventy-five percent of the mass of the atmosphere lies within 11 km (roughly 7 mi, or 36,000 ft) of the Earth's surface.

All Earth's weather—clouds, rain, sleet, and snow—occurs in this lowest region of the atmosphere, called the **troposphere**. Commercial jets generally fly at the top of

Figure 5-2 **Comparing the Continents** Africa, Europe, Greenland, and North and South America fit together as though they were once joined. This fit is especially convincing if the edges of the continental shelves (shown in blue) are used rather than today's shoreline. Alfred Wegener used this fit to support his theory of continental drift.

the troposphere to minimize buffeting and jostling from the air. If the Earth were the size of a typical classroom globe, the troposphere would only be as thick as a sheet of paper wrapped around its surface.

5-2 Plate tectonics produce major changes to the Earth's surface

Landmasses protrude through Earth's oceans. These are regions of the Earth's outermost layer, or **crust,** and are composed of relatively low-density rock that literally floats on denser material below it. By studying the various kinds of rocks at a particular location, geologists can deduce the history of that site. For instance, whether an area was once covered by an ancient sea or was flooded by lava from volcanoes is readily apparent from the kinds of rocks that are present.

Beneath the Earth's surface waters, the crust has distinct segments, indicating that the continents are separate bodies. We all believe, or at least hope, that the land under our feet will remain secure and unchanged. But the segmented appearance of the Earth reveals that the planet is undergoing global changes. Earthquakes and volcanoes hint at activity hidden within the Earth's crust.

Anyone who carefully examines the shapes of Earth's continents (Figure 5-2) might conclude that landmasses move. Eastern South America, for example, looks like it would fit snugly against western Africa. The German meteorologist Alfred Wegener first noticed the remarkable fit between landmasses on either side of the Atlantic Ocean. He was inspired to advocate *continental drift*—that the continents on either side of the Atlantic Ocean have drifted apart.

Originally, Wegener argued in 1924, there had been a single gigantic supercontinent called Pangaea (meaning "all lands"), which began to break up and drift apart some 200 million years ago. Initially most geologists ridiculed Wegener's ideas. Although it was generally accepted that the continents do "float" on denser rock beneath them, few geologists could accept that entire continents move across the Earth's face at speeds as great as several centimeters per year. Wegener and other "continental drifters" could not explain what forces could be shoving the massive continents around.

Then, in the mid-1950s, long underwater mountain ranges, such as the Mid-Atlantic Ridge stretching all the way from Iceland to Antarctica (Figure 5-3), were discovered. Careful examination revealed that molten rock from the Earth's interior is being forced upward there,

Figure 5-3 **The Mid-Atlantic Ridge** This artist's rendition shows the floor of the North Atlantic Ocean. The unusual mountain range in the middle of the ocean floor is called the Mid-Atlantic Ridge. It is caused by lava seeping up from the Earth's interior along a rift that extends from Iceland to Antarctica.

a   200 million years ago

b   180 million years ago

Figure 5-4 **The Supercontinent Pangaea** (a) The continents of today are pieces of what was once a bigger, united body called Pangaea. (b) Geologists now argue that Pangaea must have first split into two smaller supercontinents, which they call Laurasia and Gondwanaland. Gondwanaland later split into Africa and South America, with Laurasia dividing to become North America and Eurasia. (c) These bodies then separated into the continents of today.

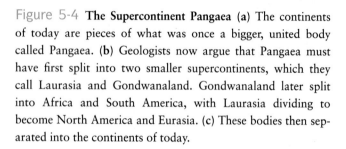

c   Today

pushing the ocean floor apart on either side of the ridge. This **seafloor spreading** is in fact pushing South America and Africa apart at a speed of roughly 3 cm per year.

Seafloor spreading provided just the physical mechanism that had been missing from the theory of continental drift. It established Wegener's theory of crustal motion, which came to be known as **plate tectonics.** The Earth's surface is made up of about a dozen major tectonic plates that move relative to each other. In recent years geologists have uncovered evidence that points to a whole succession of supercontinents that once broke apart and then reassembled. Pangaea is only the most recent supercontinent in this cycle, which repeats on average every 500 million years (Figure 5-4).

The boundaries between plates are the sites of some of the most impressive geological activity on our planet. Earthquakes tend to occur at the boundaries of the Earth's crustal plates, where the plates are colliding, separating, or sliding against each other. The boundaries between tectonic plates stand out clearly when the epicenters of earthquakes are plotted on a map (Figure 5-5). The San Andreas Fault, running along the West Coast of the United States, is an example of a fault where two plates are rubbing against each other.

Great mountain ranges, such as those along the western coasts of North and South America, are thrust up by ongoing collisions between plates. The Himalayan Mountains are still forming as the Indian subcontinent rams into Asia. Where old crust is pushed back down into the Earth, we find deep trenches, such as the ones off the coasts of Japan and Chile. Figures 5-6 and 5-7 show two well-known geographic features that resulted from separating and colliding tectonic plates, respectively.

But what powers all this activity? Some enormous energy source has been moving continents for billions of years. To answer that question, we need to understand something about the Earth's interior.

5-3 Earth's interior consists of a mantle and an iron-rich core

Geologists have calculated that the Earth was entirely molten soon after its formation, about 4.6 billion years ago. The violent impact of space debris, along with energy released by the breakup of radioactive elements, heated and melted the young Earth. The same process of

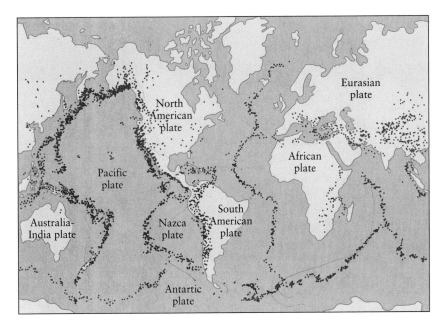

Figure 5-5 **The Earth's Major Plates** The boundaries of Earth's major plates are the scenes of violent seismic and geologic activity. Most earthquakes occur where plates separate or collide. Plate boundaries are therefore easily identified by plotting the locations of earthquakes on a map.

atoms breaking apart, called *nuclear fission*, creates heat used to generate electricity in nuclear power plants.

Iron and the other dense elements sank toward the center of the young, molten Earth, just like a rock sinking in a pond. At the same time, less dense materials were forced upward toward the surface. The process, called **planetary differentiation**, produced a layered structure within the Earth: a very dense central **core** surrounded by a **mantle** of less dense minerals, which in turn was surrounded by a thin crust of relatively light

Figure 5-6 **The Separation of Two Plates** The plates that carry Egypt and Saudi Arabia are moving apart, leaving the trench that contains the Red Sea. This view, taken by astronauts in 1966, shows the northern Red Sea on the right. Egypt is at the bottom, Saudi Arabia is at the upper right, and the Sinai Peninsula dominates the center.

Figure 5-7 **The Collision of Two Plates** The plates that carry India and China are colliding. As a result, the Himalayas have been thrust upward. In this photograph, taken by astronauts in 1968, India is on the left and Tibet on the right; Mount Everest is one of the snow-covered peaks near the center.

Figure 5-8 **Cutaway of the Earth** The early Earth was probably a homogeneous mixture of elements with no continents or oceans. (a) While molten, iron sank to the center and light material floated upward to form a crust. (b) As a result, Earth has a dense iron core and a crust of light rock, with a mantle of intermediate density between them.

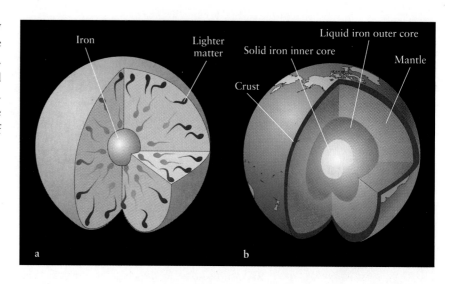

minerals (Figure 5-8). That is why most of the rocks you find on the ground are composed of lower-density elements like silicon and aluminum. The iron, gold, lead, and other very dense elements found on the Earth today actually had to return to the surface through volcanoes and other lava flows. In spite of this return of heavier elements to the surface, the average density of crustal rocks is 3000 kg/m³, considerably less than the average density of the Earth as a whole (5520 kg/m³). Thus the Earth's interior must be composed of substances much denser than the material comprising the crust.

Much of the heat generated when the Earth was young still remains trapped inside the planet. The temperature of the Earth's interior rises steadily from about 290 K on the surface to nearly 5000 K at the center. In addition to temperature, pressure also increases with increasing depth below the Earth's surface. The deeper you go, the more mass there is above your head to press down on you. These factors shape the Earth's layers.

You might think that temperatures of 5000 K would melt virtually any substance. The mantle, in particular, is composed largely of minerals rich in iron and magnesium, both of which have melting points of only slightly over 1250 K at the Earth's surface. However, the melting point of rock also depends on the pressure to which it is subjected: The higher the pressure, the higher the melting point. The high pressures within the Earth preserve a solid mantle down to a depth of about 2900 km.

Geologists have deduced basic properties of the Earth's interior by studying the response of the planet to earthquakes. Earthquakes produce a variety of **seismic waves,** vibrations that travel through the Earth either as ripples like ocean waves or by compressing matter like

sound waves. Geologists use sensitive **seismographs** to detect and record these vibratory motions. The varying density and composition of the Earth's interior affects the direction of seismic waves traveling through the Earth. By studying the deflection of these waves, geologists have been able to determine properties of the Earth's interior.

By 1906 analysis of earthquake recordings led to the discovery that the Earth has a molten iron core with a diameter of about 7000 km (4300 mi). For comparison, the overall diameter of our planet is 12,756 km (7926 mi). More careful measurements in the 1930s revealed that inside the molten core is a solid core with a diameter of about 2500 km (1550 mi). The cores are composed of roughly 80% iron and 20% lighter elements, such as silicon. The interior of our planet therefore has a curious structure: a liquid region sandwiched between a solid inner core and a solid mantle (see Figure 5-8b).

The tremendous heat and molten rock inside the Earth provide the mechanism for plate tectonics. The energy to drive the plate motion is the heat inside the planet. That heat, from the formation of the planet and from the decay of radioactive elements, affects the Earth's surface today. This heat transport is accomplished by **convection,** a process in which lighter, hotter material rises upward, while denser, cooler material sinks downward. You witness convection every time you see simmering soup. Heated from below, blobs of hot soup move upward. Once this soup reaches the surface, it gives off the heat it carries, cools, becomes denser, and sinks back down.

Heated from below, parts of the Earth's upper mantle are hot enough to have an oozing, plastic flow (like

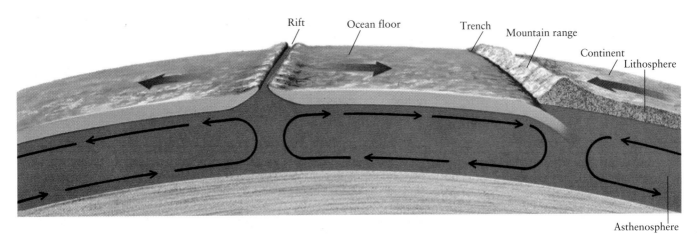

Rift Ocean floor Trench Mountain range Continent Lithosphere Asthenosphere

Figure 5-9 **The Mechanism of Plate Tectonics** Convection currents in the Earth's interior are responsible for pushing around rigid plates on the crust. New crust forms in oceanic rifts, where lava oozes upward between separating plates. Mountain ranges and deep oceanic trenches are formed where plates collide.

stretching Silly Putty). As sketched in Figure 5-9, molten rock, called *magma*, seeps upward along cracks or rifts in the ocean floor, which are found where plates are separating. Cool crustal rock sinks back down into the Earth where plates collide. The continents simply ride on top of the plates as they are pushed around by the convection currents beneath them.

Apparently supercontinents like Pangaea sow the seeds for their own destruction by blocking the flow of heat from the Earth's interior. As soon as a supercontinent forms, temperatures beneath it rise. As heat accumulates, the supercontinent domes upward and cracks. Overheated molten rock wells up to fill the resulting rifts, which continue to widen as pieces of the fragmenting supercontinent move apart.

The Earth's molten, iron-rich interior has another important effect on the Earth's exterior, one that has fewer signs than tectonic plate motion. It is the creation of a magnetic field that extends through the Earth's surface and out into space. As we will now see, that magnetic field results from the convection of molten metal inside the Earth combined with the Earth's rotation.

5-4 The Earth's magnetic field shields us from the solar wind

If you have ever used a compass, you have seen the effect of the Earth's magnetic field. This field is quite similar to the field that surrounds a bar magnet (Figure 5-10a).

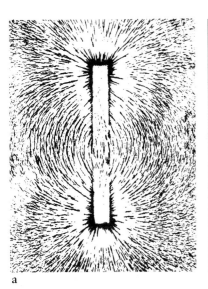

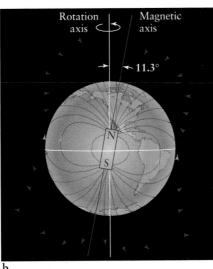

Rotation axis Magnetic axis 11.3° N S

a   b

Figure 5-10 **The Earth's Magnetic Field** (a) The magnetic field of a bar magnet is revealed by the alignment of iron filings on paper. (b) Generated in the Earth's molten, metallic core, the Earth's magnetic field extends far into space. Note that the field is not aligned with the Earth's rotation axis. We will see similar misalignment, often much more exaggerated, in other planets.

Magnetic fields are created by electrical charges in motion. Thus the motion of your compass is caused by electrical charges moving inside the Earth. The wires in your house also carry moving charges called electric currents.

Many geologists believe that convection of molten iron in the outer core combined with the Earth's rotation creates electric currents, which in turn create the Earth's magnetic field. The details of this so-called **dynamo theory** of Earth's magnetic field are still being developed.

Magnetic fields always form complete loops (see Figure 5-10). The magnetic fields near the Earth's south rotation pole loop tens of thousands of kilometers into space and then return near the Earth's north rotation pole. The places where the magnetic fields pierce the Earth's surface are called the north and south magnetic poles. The magnetic and rotation poles of the Earth are only 11.3° apart; other planets have much larger angles between their magnetic and rotation poles. Figure 5-11 is a scale drawing of the magnetic fields around the Earth, which comprise what is called the Earth's **magnetosphere.**

Our planet's magnetic field protects the Earth's surface from bombardment by energetic particles from space. Most of these particles originate in the Sun (see Chapter 9) and are called the **solar wind.** The solar wind is an erratic flow of charged particles away from the Sun's upper atmosphere. Near the Earth, the particles in

the solar wind move at speeds of roughly 400 km/s, or about a million miles per hour.

Magnetic fields can change the direction of moving charged particles. The solar wind particles heading toward the Earth are deflected by our planet's magnetic field. The field is strong enough to trap these charged particles in two huge, doughnut-shaped rings called the **Van Allen radiation belts,** which surround the Earth (see Figure 5-11). These belts, discovered in 1958 during the flight of America's first successful Earth-orbiting satellite, were named after physicist James Van Allen, who insisted that the satellite carry a Geiger counter to detect charged particles.

Sometimes the Van Allen belts overload with particles. The particles then leak through the magnetic fields at their weakest points and cascade down into the Earth's upper atmosphere, usually in a ring-shaped pattern. As these high-speed, charged particles collide with gases in the upper atmosphere, the gases fluoresce (give off light) like the gases in a fluorescent light. The result is a beautiful, shimmering display called the **northern lights (aurora borealis)** or the **southern lights (aurora australis),** depending on the hemisphere in which the phenomenon is observed (Figure 5-12).

Occasionally a violent event on the Sun's surface called a **solar flare** sends a burst of protons and electrons

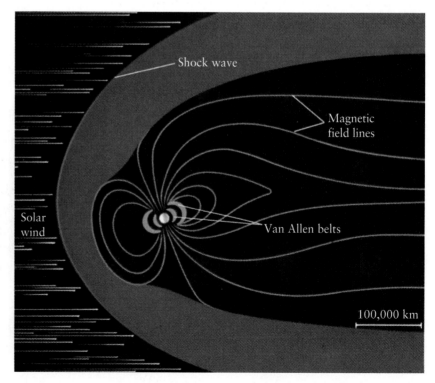

Figure 5-11 Earth's Magnetosphere Earth's magnetic field carves out a cavity in space that excludes particles from the Sun, called the solar wind. Most of the particles of the solar wind are deflected around the Earth by the fields in a turbulent region colored purple in this drawing. Because of the strength of the Earth's magnetic field, our planet can trap charged particles in two huge, doughnut-shaped rings called the Van Allen belts.

Figure 5-12 **The Northern Lights (Aurora Borealis)** A deluge of charged particles from the Sun can overload the Van Allen belts and cascade toward the Earth, producing aurorae that can be seen over a wide range of latitudes. Aurorae typically occur 100 to 400 km above the Earth's surface.

toward the Earth. The resulting auroral display can be exceptionally bright (see Figure 5-12) and can often be seen over a wide range of latitudes and longitudes. Such events also disturb radio transmissions and can damage communications satellites and electrical transmission lines.

It is remarkable that the Earth's magnetosphere, dominated as it is by vast radiation belts completely encircling the Earth, was unknown until a few decades ago. Such discoveries remind us of how much remains to be learned even about our own planet.

5-5 The ozone layer had to form before life could move onto dry land

Atmospheric temperature decreases upward, reaching a minimum of 218 K (–67°F) at the top of the troposphere. Above this level is the region called the **stratosphere,** which extends from 11 to 50 km (roughly 7 to 31 mi) above the Earth's surface. This is the realm of the **ozone layer.** Ozone molecules (three oxygen atoms, O_3, bound together) in the stratosphere efficiently absorb solar ultraviolet rays, thereby heating the air in this layer while preventing most of this lethal radiation from reaching the Earth's surface. The temperature therefore increases upward through the stratosphere to about 273 K (32°F) at its top (Figure 5–13).

There is not much ozone in the ozone layer. If all of it was compressed to the density of the air we breathe,

it would be a layer only a few millimeters thick. It is enough, however, to protect us from the Sun's ultraviolet radiation. Because the Earth's early atmosphere lacked ozone, ultraviolet radiation penetrated to the planet's surface, where it provided much of the energy necessary for life to evolve in the oceans. After life developed, however, the formation of the ozone layer by the same ultraviolet radiation was essential in cutting the amount of

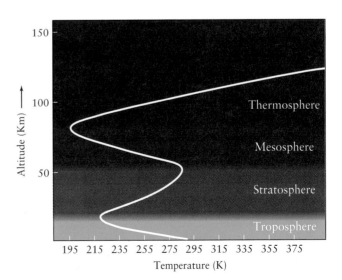

Figure 5-13 **Temperature Profile of Earth's Atmosphere** The atmospheric temperature varies with altitude because of the way sunlight interacts with various atoms at different heights.

ultraviolet rays that reached the surface, thereby allowing life to leave the oceans and survive on land.

The ozone layer is still essential today, because the
energy it absorbs would otherwise cause skin cancer and
various eye diseases. In sufficiently high doses, ultraviolet
radiation even damages the human immune system. The
ozone layer is being depleted today by human-made
chemicals, such as chlorofluorocarbons (CFCs). Fortunately, however, sunlight creates ozone; if the depletion
rate drops, the ozone layer will naturally replenish itself.

Above the stratosphere lies the **mesosphere.** Atmospheric temperature again declines as one moves up
through the mesosphere, reaching a minimum of about
200 K (–103°F) at an altitude of about 80 km (50 mi), as
shown in Figure 5–13. This minimum marks the bottom
of the **thermosphere,** above which the Sun's ultraviolet
light ionizes atoms, producing charged particles that reflect radio waves. You can tune in distant AM radio stations because their transmissions bounce off this region.
On the other hand, FM radio stations use much higher
frequencies that pass through this layer without bouncing back. That is why you must be quite near an FM station in order to receive it.

The Moon and Tides

Earth's only natural satellite provides one of the most
dramatic sights in the nighttime sky (Figure 5-14). The
Moon is so large and so nearby that some of its surface
features are readily visible to the naked eye. Without a
telescope, you can easily see dark gray and light gray areas that cover vast expanses of the Moon. Although people have long known the Moon's effects on the tides, its
surface began to be understood only with Galileo. Most
people back then believed that the Moon had a smooth
face. Galileo's telescope revealed mountains towering
above its barren surface. Table 5-2 lists the Moon's important properties.

5-6 The Moon's surface is covered with craters, plains, and mountains

Perhaps the most familiar and characteristic features on
the Moon are its craters (Figure 5-15). With an Earth-
based telescope some 30,000 craters are visible, with diameters ranging from 1 km to more than 100 km. Following a tradition established in the seventeenth century,
the most prominent craters are named after philoso-

Figure 5-14 **The Moon** Our Moon is one of seven large
satellites in the solar system. The Moon's diameter of 3476 km
(2160 mi) is slightly less than the distance from New York to
San Francisco. This photograph is a composite of first quarter
and last quarter views, in which long shadows enhance the surface features.

TABLE 5-2

The Moon's Vital Statistics

Statistic	Measurement
Mass	0.012 $M_\oplus$ (7.35 × 10²² kg)
Radius	0.272 $R_\oplus$ (3476 km or 2160 mi)
Average density	3340 kg/m³
Average distance from Earth	384,600 km (238,900 mi)
Orbital eccentricity*	0.055
Sidereal period of revolution	27.3 d
Synodic period of revolution†	29.5 d
Sidereal rotation period	27.3 d
Albedo (average)	0.07

*Around the Earth.
†Cycle of lunar phases.

Figure 5-15 **Details of a Lunar Crater** This photograph, taken from lunar orbit by astronauts in 1969, shows a typical view of the Moon's heavily cratered far side. The large crater near the middle of the picture is approximately 80 km (50 mi) in diameter. Note the crater's central peak and the numerous tiny craters that pockmark the lunar surface.

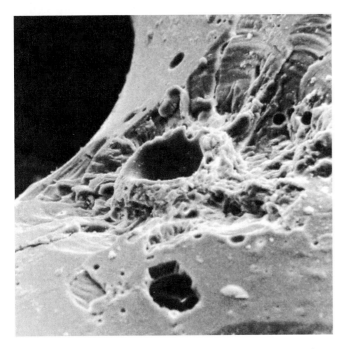

Figure 5-16 **A Microscopic Lunar Crater** This extreme close-up photograph shows tiny microcraters less than 1 mm across on a piece of moon rock.

phers, mathematicians, and scientists, such as Plato, Aristotle, Pythagoras, Copernicus, and Kepler. Close-up photographs from lunar orbit have revealed millions of craters too small to be seen with Earth-based telescopes. Indeed, extreme close-up photos of the Moon's surface reveal countless microscopic craters (Figure 5-16).

Earth, by way of comparison, has only 200 known impact craters. Many other craters on Earth have been drawn inside our planet by plate tectonic motions, while others have been worn away by the effects of wind and weather. Many craters were prevented from forming by the vaporization of space rocks in the Earth's atmosphere. We will discuss Earth's craters further in Chapter 8.

Virtually all lunar craters, both large and small, are the result of bombardment by meteoritic material. Nearly all of these craters are circular, indicating that they were not merely gouged out by high-speed rocks. Instead, the violent collisions with the Moon's surface vaporized the rapidly moving debris, and the resulting explosions of hot gas produced the round craters observed today.

Meteoritic impact causes material from the crater site to be ejected onto the surrounding surface. This pulverized rock is called an **ejecta blanket.** You can see some of the more recent ejecta blankets, which are lighter colored than the older ones (see Figure 5-14). The blanket darkens with time as its surface roughens from impact by particles flowing out of the Sun. The lightness of its ejecta is one clue astronomers use to determine how long ago a crater formed.

Craters larger than about 20 km also form central peaks (see Figure 5-15). These occur because the impact compresses the crater floor so much that afterwards the crater rebounds and pushes the peak upward. As the peak goes up, the crater walls collapse and form terraces.

Besides the craters, the most obvious characteristic of the Moon visible from Earth is that its surface is various shades of gray. Most prominent are the large, dark gray plains called **maria** (pronounced MAR-ee-uh). The singular form of this term, **mare** (pronounced MAR-ay), means "sea" in Latin and was introduced in the seventeenth century when observers using early telescopes thought these features were large bodies of water. We know now that there is no liquid water on our satellite. Nevertheless, we have kept their fanciful names, such as Mare Tranquillitatis (Sea of Tranquillity), Mare Nubium (Sea of Clouds), Mare Nectaris (Sea of Nectar), and Mare Serenitatis (Sea of Serenity).

The largest of the maria is Mare Imbrium (Sea of Showers). It is roughly circular and measures 1100 km (700 mi) in diameter. Although the maria seem quite

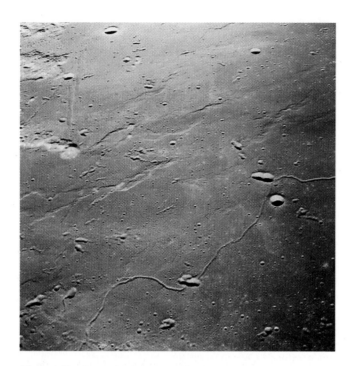

Figure 5-17 **Details of a Lunar Mare** Close-up views of the lunar surface reveal numerous tiny craters and cracks on the maria. Astronauts in lunar orbit took this photograph of Mare Tranquillitatis (Sea of Tranquility) in 1969 while searching for potential landing sites for the first manned landing.

Figure 5-18 **Comparing the Near and Far Sides of the Moon** On the upper right side of this photograph are three maria visible from the Earth. In this view, taken by the crew of Apollo 16, the far side of the Moon is more uniformly and more heavily cratered than the side facing the Earth.

smooth in telescopic views from the Earth, close-up photographs from lunar orbit reveal that they contain small craters and occasional cracks called **rilles** (Figure 5-17). One of the surprises stemming from the early days of lunar exploration is that there is only one small mare on the Moon's far side (Figure 5-18). Except for that lone mare, craters cover the entire side of the Moon that faces away from the Earth.

Detailed measurements by astronauts in lunar orbit demonstrated that the maria on the Moon's Earth-facing side are 2 to 5 km below the average lunar elevation. In contrast, the cratered terrain on the lunar far side is typically 4 to 5 km above the average lunar elevation. The flat, low-lying, dark gray maria cover only 17% of the lunar surface. The remaining 83% is composed of light gray, heavily cratered, mountainous regions called **highlands** (Figure 5-19).

Lunar mountain ranges that rim the vast maria basins suggest ancient impacts far more violent than those that produced typical craters. One such mountain range can be seen around the edges of Mare Imbrium (see Figure 5-19). Although they take their names from

Figure 5-19 **Mare Imbrium and the Surrounding Highlands** Mare Imbrium, the largest of the 14 dark plains that dominate the Earth-facing side of the Moon, is ringed by lighter-colored highlands strewn with craters and several high mountains.

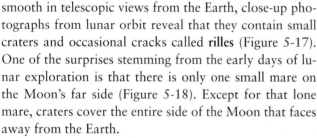

famous terrestrial ranges such as the Alps, the Apennines, and the Carpathians, the lunar mountains were not formed in the same way as mountains on Earth. Analyses of Moon rocks and observations from lunar orbit imply that violent impacts of huge asteroids thrust up the lunar mountain ranges.

The ideal place to learn about the Moon is, of course, on its surface. There were six manned lunar landings. The first two, Apollo 11 and Apollo 12, set down in maria. Human voyages into space were briefly set back when Apollo 13 experienced a nearly fatal explosion en route to the Moon. Fortunately, a titanic effort by the astronauts and ground personnel brought it safely home. Apollo 14 through Apollo 17 took on progressively more challenging lunar terrain. The program culminated when Apollo 17 landed amid rugged mountains just east of Mare Serenitatis (Figure 5-20).

Astronauts discovered that the lunar surface is covered with a layer of fine powder and rock fragments. This layer, ranging in thickness from 1 to 20 m, is called the **regolith** (from the Greek meaning "blanket of stone"; Figure 5-21). It formed as a result of 4.5 billion years of relentless meteoritic bombardment. We don't refer to this layer as soil because the term *soil* suggests the presence of decayed biological matter, which is not found on the Moon's surface. The rough, powdered regolith absorbs most of the light incident on it. Therefore, the Moon's albedo is only 0.07, meaning that it reflects only 7% of the light striking it.

A major goal of the astronauts of the 1960s and 1970s was to study the geology of the Moon and help determine its history. The astronauts brought back a total of 382 kg (842 lb) of lunar rocks, which have provided important information about the early history of the Moon. Lunar rock from the maria is solidified lava, which tells us that the Moon's surface was once molten. Indeed, these Moon rocks are composed mostly of the same minerals that are found in volcanic rocks on Hawaii or Iceland. The rock of these low-lying lunar plains is called **mare basalt** (Figure 5-22).

In contrast to the dark maria, the lunar highlands are covered with a light-colored rock called **anorthosite** (Figure 5-23). On Earth, anorthositic rock is found only in very old mountain ranges, such as the Adirondacks in the eastern United States. Compared to the mare basalts, which have more of the heavier elements like iron, manganese, and titanium, anorthosite is rich in calcium and aluminum. Anorthosite is therefore less dense than basalt. Many highland rocks brought back to Earth are

Figure 5-20 **An Apollo Astronaut on the Moon** Apollo 17 astronaut Harrison Schmitt entered the Taurus-Littow Valley on the Moon. Here an enormous boulder had once slid down a mountain, fracturing on the way. This final Apollo mission landed in the most rugged terrain of any Apollo flight.

Figure 5-21 **The Regolith** The Moon's surface is a powder created by billions of years of bombardment by space debris. Called the regolith, the surface material sticks together like wet sand, as seen in this photograph of an Apollo 11 astronaut's bootprint.

Figure 5-22 **Mare Basalt** This 1.53-kg (3.38-lb) specimen of mare basalt was brought back by Apollo 15 astronauts in 1971. Small holes that cover about a third of its surface suggest that gas must have dissolved under pressure in the lava from which this rock solidified. When the lava reached the airless lunar surface, bubbles formed as the pressure dropped. Some of the bubbles were frozen in place as the rock cooled.

Figure 5-24 **Impact Breccias** These rocks are created from shattered debris fused together under high temperature and pressure. Such conditions prevailed immediately following impacts on the Moon's surface.

impact breccias (Figure 5-24), which are composites of different rock fused together as a result of meteorite impacts.

By carefully measuring trace amounts of radioactive elements, geologists determined that lunar rocks on the mare are between 3.1 and 3.8 billion years old, while typical anorthositic specimens from the highlands are be-

tween 4.0 and 4.3 billion years old. All these extremely ancient specimens are probably samples of the Moon's original crust (Figure 5-25).

Although lunar rocks bear a strong resemblance to terrestrial rocks, every terrestrial rock contains some water. No evidence has yet been found that water ever existed on the Moon. Thus it is not surprising that the astronauts have found no traces of life there.

Life could not survive for long on the Moon anyway, for it lacks air. Atmospheres are held around a planet or moon by the body's gravitational force. Our Moon has so little gravitational attraction (about one-sixth that of the Earth) that it is unable to hold any of the gases that make up the Earth's atmosphere. The Earth, in contrast, has enough mass to hold down such gases as nitrogen, oxygen, carbon dioxide, and argon.

Figure 5-23 **Anorthosite** The lunar highlands are covered with an ancient type of rock called anorthosite, which is believed to be the material of the original lunar crust. This sample, called the "Genesis rock" by the Apollo 15 astronauts who picked it up, has an age of approximately 4.1 billion years.

5-7 The Moon probably formed from debris cast into space when a huge planetesimal struck the young Earth

Before the Apollo program, there were three different theories about the origin of the Moon. The **fission theory** holds that the Moon was pulled out from a rapidly rotating proto-Earth. This theory does not explain why the

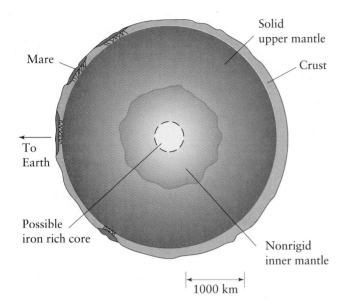

Figure 5-25 **The Moon's Interior** Like the Earth, the Moon probably has a crust, a mantle, and a core. The lunar crust has an average thickness of about 60 km on the Earth-facing side but about 100 km on the far side. The crust and solid upper mantle extend to about 800 km, where the nonrigid inner mantle begins. If the Moon has a core, it is probably 400 km in radius or less. Although the main features of the Moon's interior are analogous to those of the Earth, the proportions are quite different. The information here is based on analysis of data from seismographs left on the Moon by astronauts.

Moon has relatively little iron and other dense elements compared to the Earth, or why the Moon has so little water. The **capture theory** posits that the Moon was formed elsewhere in the solar system and then drawn into orbit about the Earth by gravitational forces. Since bodies formed in different places have different chemistries, this theory does not explain the similar chemistry between the Moon and the Earth's surface. The **cocreation theory** proposes that the Earth and the Moon were formed near each other at the same time. However, this theory also fails to explain why the Moon lacks dense elements compared to the Earth.

A fourth theory, first proposed in the 1980s, is called the **collision–ejection theory.** It proposes that the Earth was struck by a Mars-sized object within the first 100 million years after our planet formed. This collision literally splashed debris from the Earth's surface into orbit. A computer simulation of this cataclysm is shown in Figure 5-26. This matter became a short-lived disk orbiting the Earth that eventually condensed into the Moon, just like planetesimals condensed to form the Earth.

The collision–ejection theory is in agreement with many of the known facts about the Moon. For example, rock vaporized by the impact would have been depleted of volatile (easily evaporated) elements and water, leaving the parched Moon rocks we now know. Furthermore, most of the debris from the collision would lie near the plane of the ecliptic, as long as the orbit of the impacting asteroid had also been in that plane. Indeed, the impact of an object large enough to create the Moon could also have tipped the Earth's axis of rotation and so inaugurated the seasons.

The surface of the newborn Moon probably stayed molten for many years, both from heat released during the impact of rock fragments falling onto the young satellite and from the decay of short-lived radioactive isotopes. As the Moon gradually cooled, low-density lava floating on the Moon's surface began to solidify into the anorthositic crust that exists today in the highlands. By correlating the ages of the Moon rocks with the density of craters at the sites where they were collected, geologists have found that the rate of impacts on the Moon changed over the ages. The ancient, heavily cratered lunar highlands are evidence of intense bombardment that dominated the Moon's early history. For nearly a billion years, rocky debris left over from the formation of the planets rained down upon the Moon's young surface. As Figure 5-27 shows, this barrage extended from 4.6 billion years ago, when the Moon's surface solidified, until about 3.8 billion years ago.

The frequency of impacts gradually tapered off as meteoroids and planetesimals were swept up by the newly formed planets. Recorded among the final scars at the end of this crater-making era are the impacts of more than a dozen objects, each measuring at least 100 km across. As these huge rocks pounded the young Moon, they may have blasted gaping craters in what would later become the maria. Meanwhile, heat from the decay of long-lived radioactive elements like uranium and thorium began to re-melt the inside of the Moon. Then, from 3.8 to 3.1 billion years ago, great floods of molten rock gushed up from the lunar interior through the weak spots in the crust created by the large impacts. Lava filled the impact basins and created the maria we see today.

Relatively little has happened on the Moon since those ancient times. A few fresh large craters have been formed (see Figure 5-14), but the astronauts visited a world that has remained largely unchanged for over three billion years. At that time, the biological history of our own planet was just beginning, as the first organisms began populating the Earth's oceans.

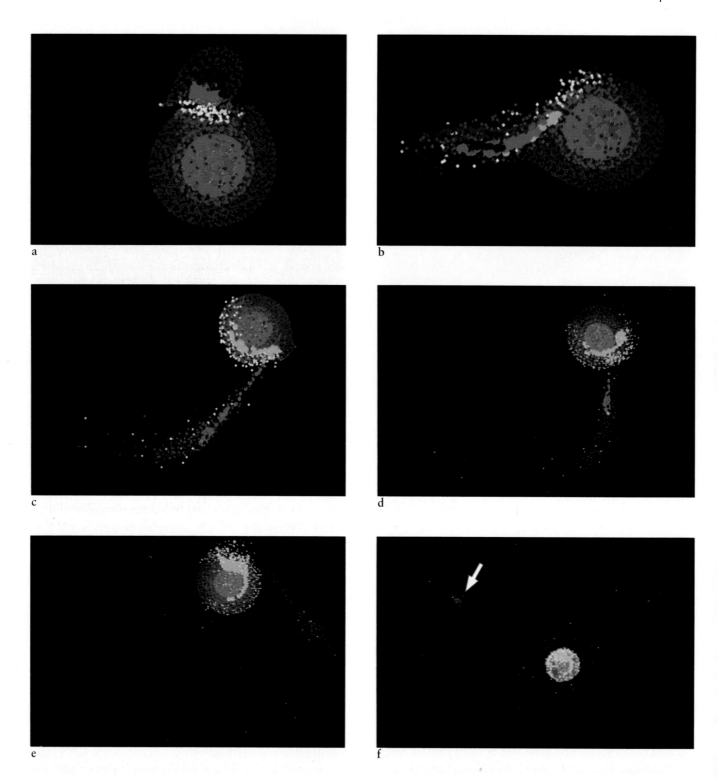

Figure 5-26 **The Moon's Creation** This supercomputer simulation shows the creation of the Moon from material ejected by the impact of a large asteroid on the Earth. To follow the ejected material as it moves away from the Earth, successive views show increasingly larger volumes of space. Blue and green indicate iron from the cores of the Earth and the asteroid; red, orange, and yellow indicate rocky mantle material. In this simulation, the impact ejects both mantle and core material, but most of the iron falls back onto the Earth. The surviving ejected rocky matter coalesces to form the Moon (indicated by arrow) during its first orbit of the Earth.

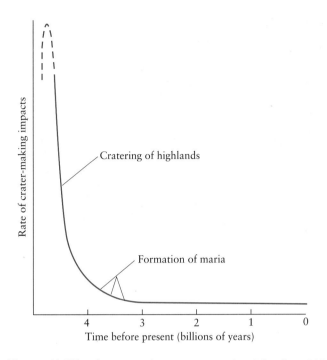

Figure 5-27 **The Rate of Crater Formation** The first 800 million years of the Moon's history was dominated by frequent crater-making impacts. Near the end of this intense bombardment, several large impacts gouged out the mare basins. For the past 3.8 billion years, the impact rate has been quite low.

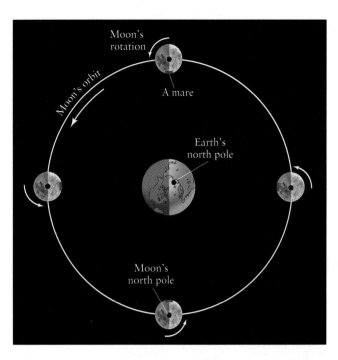

Figure 5-28 **Synchronous Rotation of the Moon** This figure shows the motion of the Moon around the Earth as seen from above the Earth's north polar region. In order for the Moon to keep the same side facing the Earth as it orbits our planet, the Moon must rotate on its axis at precisely the same rate that it revolves around the Earth.

5-8 Gravitational forces keep one side of the Moon always facing the Earth and produce the tides

If you observe the Moon from Earth over many nights, you always see the same maria and highlands. Even as the Moon goes through its phases, the same surface features face the Earth (see Figure 1-20). Only from spacecraft have we seen the "far side."

The far side of the Moon is often, and incorrectly, called the "dark side" of the Moon. Whenever we see less than a full Moon, some of the sunlight is falling on the far side. Throughout each cycle of lunar phases all parts of the Moon get equal amounts of sunlight.

Although the same side of the Moon always faces the Earth, the Moon does rotate. In fact, it must have **synchronous rotation:** It rotates on its axis at the same rate that it orbits the Earth (Figure 5-28). To see why, hold your arm out with your palm facing you. Your palm represents the side of the Moon facing the Earth. Standing still, swing your arm horizontally always keep-

ing the palm facing you. The only way you can do this is if your hand turns (rotates) at exactly the same rate that it swings (revolves). To understand the origin of the Moon's synchronous rotation, we must investigate the gravitational forces acting between the Earth and the Moon.

Figure 5-29a shows the gravitational force from the Moon at several locations on the Earth. Recall from Chapter 2 that gravitational force decreases with distance. Therefore, the side of the Earth closest to the Moon feels a greater gravitational pull than the Earth's center does. Similarly, the Earth's far hemisphere feels less attraction from the Moon than the Earth's center does. **Tidal forces** are these differences in the gravitational pull at different places.

As sketched in Figure 5-29b, tidal forces deform the oceans, causing them to rise at some places around the equator and to settle elsewhere. There are two areas of high tide on the Earth at any given time, one on the side closest to the Moon, the other on the opposite side of the Earth. Low tides occur where the Moon is on the horizon. The high tide on the far side of the Earth from the

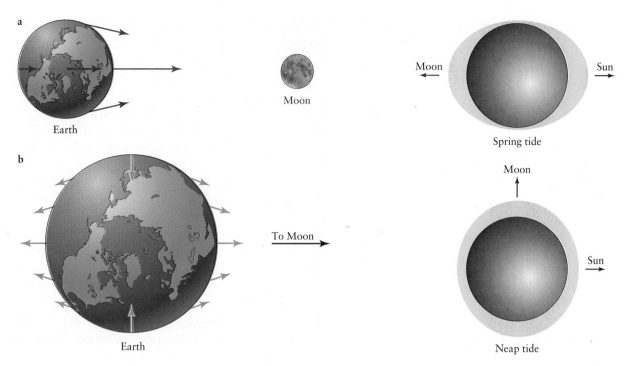

Figure 5-29 **Tidal Forces** The Moon induces tidal forces on the Earth because the strength of the Moon's gravity varies over the Earth's surface. (**a**) Red arrows indicate the strength and direction of the Moon's gravitational pull at selected points on the Earth. (**b**) Blue arrows indicate the strength and direction of the tidal forces acting on the Earth. At any location, the tidal force equals the Moon's gravitational pull at that point minus the gravitational pull of the Moon at the center of the Earth.

Figure 5-30 **Tides on the Earth** The gravitational forces of the Moon and Sun deform the oceans. The greatest deformation (spring tides) occurs when the Sun, Earth, and Moon are aligned with either the Sun and Moon on the same or opposite sides of the Earth. The least deformation (neap tides) occurs when the Sun, Earth, and Moon form a right angle.

Moon occurs because the water there is "left behind" as the Moon pulls the solid center of the Earth toward it.

The Sun also distorts the shape of the oceans, but only half as much as the Moon, because the Sun is nearly 400 times farther away. When the Sun, Moon, and Earth are aligned, the Sun and Moon pull in the same direction and the tides are greatest. These **spring tides** (from the German, meaning to "spring up") occur at every new and full Moon. At first quarter and last quarter, the Sun and Moon form a right angle with the Earth. The forces they exert compete with each other, and so the tidal distortion is the least pronounced. The smaller tidal shifts on these days are called **neap tides** (Figure 5-30).

The Moon has synchronous rotation because of the Earth's tidal force acting on it. When the Moon was young, it had huge tides of molten rock up to 60 ft high. The Earth's tidal force acted to keep the Moon's high

tide directly between the Moon's center and the center of the Earth. As a result, the Moon's rotation rate became the same as its revolution rate. When it solidified, the Moon was locked into this synchronous rotation.

5-9 The Moon is moving away from the Earth

A century ago Sir George Darwin (son of Charles Darwin) proposed that the Moon is moving away from the Earth. To test this theory, the Apollo 12 astronauts placed on the Moon a set of reflectors, similar to the orange and red ones found on cars, to help measure the Moon's distance. Pulses of laser light were fired at the Moon from the Earth (Figure 5-31), and the time it took the light to reach the Moon, bounce off the reflector ar-

Figure 5-31 **Lunar Ranging** A pulse of laser light is fired through this McDonald Observatory telescope at the Moon; the light is then reflected back by the corner reflectors placed on the Moon by Apollo astronauts. From the time it takes the light to reach the Moon and reflect back to Earth, astronomers can determine the distance to the Moon. The long exposure required to show the laser beam in this photograph causes the Moon to be overexposed.

ray, and return to the Earth was carefully recorded. Knowing the speed of light and the time of flight for the light, astronomers can calculate the distance to the Moon with an error of less than a meter. From such measurements over time, astronomers have established that the Moon is spiraling away from the Earth at a rate of roughly 4 cm per year.

This means, of course, that the Moon used to be closer to the Earth. Although the initial distance that the Moon was when it coalesced is not known, its present rate of recession tells us that it was at least half its present distance; more accurate calculations suggest one-tenth. At this distance, the tides on the Earth were a thousand times higher than they are today. (What effects would those tides have had on the geology of the young Earth?)

Where does the Moon get the energy necessary to spiral away from the Earth? The answer, of course, is from the Earth itself. Tidal interactions between the Earth and the Moon gradually enlarge the Moon's orbit. Gravitational pull from the high ocean tide nearest the Moon acts like a lever that pulls the Moon ahead in its orbit, forcing it to spiral outward. Because of gravitational forces, the oceans actually push on the continents opposite to the Earth's direction of rotation. The result is that *the tides are slowing the Earth's rotation.* The rotational energy lost by the Earth is gained by the Moon. Geologists calculate that when the Earth first formed, the day was only between five and six hours long!

Will the Earth's rotation ever slow down enough so that it will be in synchronous rotation with respect to the Moon? Yes, but billions of years before that happens, the Sun is going to self-destruct, so it doesn't really matter!

Many questions and mysteries still remain. The six American and three Soviet lunar landings have barely scratched the Moon's surface, bringing back samples from only nine locations. We still know very little about the Moon's far side and its poles. Could there be ice at the Moon's poles? Is the Moon's interior molten? Does the Moon really have an iron-rich core? How old are the youngest lunar rocks? Did lava flows occur over western Oceanus Procellarum only two billion years ago, as crater densities there suggest? Is the Moon really geologically dead, or does it just look dead, simply because our examination of the lunar surface has been so cursory? Such questions can be answered only by returning to the Moon.

WHAT DID YOU THINK?

1 *Will the ozone layer, which is now being depleted, naturally replenish itself?* Yes. The ozone is created from normal oxygen molecules by the Sun's ultraviolet radiation.

2 *Does the Moon have a dark side, where it is forever night?* Half of the Moon is always dark, but that half is continually changing as the Moon orbits the Earth.

3 *Does the Moon rotate, and if so, how fast?* The Moon rotates at the same rate that it revolves around the Earth. If the Moon didn't rotate, then as it revolved we would see different sides of it, which we don't.

4 *What causes the ocean tides?* The tides are created by gravitational forces, primarily from the Moon and Sun.

5 *When does the spring tide occur?* Spring tide occurs during each full and new Moon.

Key Words

anorthosite
capture theory
cocreation theory
collision–ejection
   theory
convection
core
crust
dynamo theory
ejecta blanket
fission theory

highlands
impact breccia
magnetosphere
mantle
mare (*plural* maria)
mare basalt
mesosphere
neap tide
northern lights
   (aurora borealis;
   plural aurorae)

ozone layer
planetary
   differentiation
plate tectonics
regolith
rille
seafloor spreading
seismic waves
seismograph
solar flare
solar wind

southern lights
   (aurora australis)
spring tide
stratosphere
synchronous rotation
thermosphere
tidal force
troposphere
Van Allen radiation
   belts

Key Ideas

• The outermost layer, or crust, of the Earth offers clues to the history of our planet.

• The Earth's surface is divided into huge plates that move over the upper mantle.

   Movements of these plates, a process called plate tectonics, are caused by the upwelling of molten material along cracks in the ocean floor that produce seafloor spreading.

   Plate tectonics is responsible for most of the major features of the Earth's surface, including mountain ranges, volcanoes, and the shapes of the continents and oceans.

• Study of seismic waves (vibrations produced by earthquakes) shows that the Earth has a small, solid inner core surrounded by a liquid outer core; the outer core is surrounded by the dense mantle, which in turn is surrounded by the thin, low-density crust.

   The Earth's inner and outer cores are composed of almost pure iron with some nickel mixed in. The mantle is composed of iron-rich minerals.

• The Earth's atmosphere is four-fifths nitrogen and one-fifth oxygen. This abundance of oxygen is due to the biological activities of life forms on the planet.

   The Earth's atmosphere is divided into layers called the troposphere, stratosphere, mesosphere, and thermosphere; ozone molecules in the stratosphere absorb ultraviolet light rays.

• The Earth's magnetic field produces a magnetosphere that surrounds the planet and blocks the solar wind from hitting the atmosphere.

   Some charged particles from the solar wind are trapped in two huge, doughnut-shaped rings called the Van Allen radiation belts. A deluge of particles from a solar flare can initiate an auroral display.

• The Moon has light-colored, heavily cratered highlands and dark-colored, smooth-surfaced maria.

• Gravitational interactions between the Earth and the Moon produce tides in the oceans of the Earth and keep the Moon in synchronous rotation.

- Many of the lunar rock samples are solidified lava formed largely of minerals also found in Earth rocks.

> Anorthositic rock in the lunar highlands was formed between 4.0 and 4.3 billion years ago, whereas the mare basalts solidified between 3.1 and 3.8 billion years ago.

> The Moon's surface has undergone very little change over the past 3 billion years.

> Impacts have been the only significant "weathering" agent on the Moon; the Moon's regolith ("soil" layer) was formed by meteoritic action.

Lunar rocks contain no water and are depleted in volatile elements.

- The collision–ejection theory of the Moon's origin holds that the young Earth was struck by a huge asteroid, and debris from this collision coalesced to form the Moon.

- The Moon was molten in its early stages, and the anorthositic crust solidified from low-density magma that floated to the lunar surface. The mare basins were created later by the impact of planetesimals and were then filled with lava from the lunar interior.

Review Questions

1 Why is the Earth's surface not riddled with craters as the Moon is?

2 Describe the process of plate tectonics. Give specific examples of geographic features created by plate tectonics.

3 How do we know about the Earth's interior, given that the deepest wells and mines go down only a few kilometers?

4 Describe the interior structure of the Earth.

5 Why is the center of the Earth not molten?

6 Describe the structure of the Earth's atmosphere.

7 Describe the Earth's magnetosphere.

8 What are the Van Allen belts?

9 What kind of features can you see on the Moon with a small telescope?

10 On the basis of lunar rocks brought back by the astronauts, explain why the maria are dark-colored, but the lunar highlands are light-colored.

11 Why are there so few craters on the maria?

12 Briefly describe the main differences and similarities between Moon rocks and Earth rocks.

13 How do we know that the maria were formed after the lunar highlands?

14 What is a tidal force? How do tidal forces produce tides in the Earth's oceans?

15 What is the difference between spring tides and neap tides?

16 Why do most scientists favor the collision–ejection theory for the Moon's formation?

Advanced Questions

17 Explain how the outward flow of energy from the Earth's interior drives the process of plate tectonics.

18 Why do some geologists believe that Pangaea was the most recent in a succession of supercontinents?

19 Why do you suppose that active volcanoes, such as Mount St. Helens, are usually located in mountain ranges that border on subduction zones?

20 Why do you suppose more lunar detail is visible through a telescope when the Moon is near quarter phase than when it is at full phase?

21 In *The Tragedy of Pudd'nhead Wilson*, Samuel Clemens (Mark Twain) wrote, "Everyone is a Moon, and has a dark side which he never shows to anybody." What, if anything, is wrong with his astronomy?

22 Why do you suppose temperature variations between day and night on the Moon are much more severe than they are on the Earth?

23 Why are Moon rocks so much older than Earth rocks, even though both worlds formed at nearly the same time?

24 Some people who supported the fission theory proposed that the Pacific Ocean basin is the scar left when the Moon pulled away from the Earth. Explain why this idea is wrong.

25 Apollo astronauts left seismometers on the Moon that radioed seismic data back to Earth. The data showed that moonquakes occur more frequently when the Moon is at perigee (closest to the Earth) than at other locations along its orbit. Give an explanation for this finding.

26 Why do you suppose that no Apollo mission landed on the far side of the Moon?

27 How might studying albedo help astronomers locate inhabitable worlds orbiting other stars?

Discussion Questions

28 If the Earth did not have a magnetic field, do you think auroras would be more common or less common than they are today?

29 Cycles of supercontinents imply profound environmental changes that repeat every 500 million years. What sort of global changes might accompany the formation and breakup of a supercontinent? How might these cycles affect the evolution of life?

30 Comment on the idea that without the presence of the Moon in our sky, astronomy would have developed far more slowly.

31 Compare the advantages and disadvantages of exploring the Moon with astronauts as opposed to using mobile, un-manned instrument packages.

32 When was the last earthquake near your hometown? How far is your hometown from a plate boundary? What kinds of topography (e.g., mountains, plains, seashore) dominate the geography of your hometown area? Do the topography and the frequency of earthquakes seem to be consistent with your hometown's proximity to a plate boundary?

Observing Projects

33 Use a telescope to observe the Moon. Compare the texture of the lunar surface you see on the maria with that of the lunar highlands. How does the visibility of details vary with distance from the terminator (the boundary between day and night on the Moon)?

34 If you live near the ocean, observe the tides to see how the times of high and low tides are correlated with the position of the Moon in the sky.

6 ▶ The Other Inner Planets

Earth

Mercury

Mars

Venus

IN THIS CHAPTER

You will explore the other three Earthlike, or terrestrial, planets: Mercury, Venus, and Mars. Mercury is a Sun-scorched planet with a heavily cratered, Moonlike surface and an iron, Earthlike core. Venus is perpetually shrouded in thick, poisonous clouds made of sulfurous compounds. It is mostly covered by gently rolling hills but does have two upraised, continentlike features. Finally, you will journey to Mars—a small world that has inspired much speculation about extraterrestrial life.

The Inner Planets The four inner planets are shown to the same scale. Venus and Earth are nearly the same size. Mars' diameter is about half Earth's diameter; Mercury's diameter is about a third that of Earth's. The picture of Mercury was assembled from many photographs taken at different times by a spacecraft coasting past the planet.

WHAT DO YOU THINK?

1. Is the temperature on Mercury, the closest planet to the Sun, higher than the temperature on Earth?

2. What is the composition of the clouds surrounding Venus?

3. Does Mars have surface liquid water today?

4. Is life known to exist on Mars today?

FAR FROM THE SUN lie planets so alien as to make comparisons with the Earth virtually meaningless. Closer to home, however, a space traveler might feel like an ordinary tourist. The craters on Mercury call to mind the Moon, but what a violent history they reveal! The clouds of Venus look almost comforting, which belies their true chemical composition. A feature on Mars recalls the Grand Canyon, but how can it exist on such an arid world? As we begin our tour of the solar system, it will help to keep asking: They look like things back home, but are they really the same?

Mercury

When it is visible, Mercury is among the brightest objects in the sky. It is best seen when it is farthest in angle from the Sun, that is, at its greatest elongation. When Mercury rises after the Sun, it is seen only for a short

TABLE 6-1

Mercury's Vital Statistics

Statistic	Measurement
Mass	3.30×10^{23} kg ($0.055\ M_{\oplus}$)
Radius	2439 km ($0.382\ R_{\oplus}$)
Average density	5430 kg/m^3
Orbital eccentricity	0.206
Sidereal period of revolution (year)	88 Earth days
Average distance from the Sun	5.79×10^7 km (0.387 AU)
Sidereal rotation period	58.7 Earth days
Solar rotation period (day)	176 Earth days
Albedo (average)	0.06

time after sunset, hovering low over the western horizon as an "evening star." Alternatively, when it rises before the Sun, Mercury heralds the ascending Sun in the brightening eastern sky as a "morning star." Table 6-1 lists Mercury's essential properties.

6-1 Photographs from a spacecraft reveal Mercury's lunarlike surface

Although the telescopic photographs in Figure 6-1 are two of the finest Earth-based views of Mercury, they reveal almost nothing about the planet's surface. It was only in 1974, when Mariner 10 coasted past the planet,

Figure 6-1 **Earth-Based Views of Mercury** These two views are among the finest photographs of Mercury ever produced with an Earth-based telescope. Hazy markings are faintly visible on the tiny planet.

Figure 6-2 **Mercury and Our Moon** Mercury (left) and our Moon (right) are shown here to the same scale. Mercury's diameter is 4878 km and the Moon's is 3476 km. For comparison, the distance from New York to Los Angeles is 3944 km (2451 mi). Both the Moon and Mercury have heavily cratered surfaces and virtually no atmospheres. Daytime temperatures at the equator on Mercury reach 700 K (800°F), hot enough to melt lead or tin.

that astronomers got their first glimpse of Mercury's inhospitable but familiar-looking surface (Figure 6-2). Mariner 10 photographed only half of Mercury's surface, and we still do not know what the other half looks like. Figure 6-3 shows a typical close-up view.

Astronomers conclude that most of the craters on both Mercury and the Moon were produced by impacts in the 800 million years after the solar system formed. As we saw in the previous chapter, the strongest evidence for this belief comes from analyzing and dating Moon rocks. Debris remaining after the planets were formed pounded these young worlds, gouging out most of the craters we see today. Like the Moon, Mercury has an exceptionally low albedo of 0.06 (6% reflection). (Why, then, is it so bright?)

Although first impressions of Mercury evoke a lunar landscape, closer scrutiny reveals decidedly nonlunar characteristics. Lunar craters (see Figure 6-2) are densely packed, with one overlapping the next. In sharp contrast, Mercury's surface has broad plains separating many craters, but no sign of the large, localized, nearly craterless maria seen on the Moon.

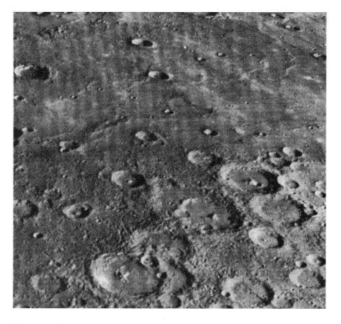

Figure 6-3 **Mercury's Craters and Plains** This view of Mercury's northern hemisphere was taken by Mariner 10 as it sped past the planet in 1974. Numerous craters and broad plains cover an area 480 km (300 mi) wide.

As we learned in Chapter 5, the lunar maria were produced by extensive lava flows that occurred between 3.1 and 3.8 billion years ago. Ancient lava flows also probably formed the Mercurian plains. As large meteoroids punctured the planet's thin, newly formed crust, lava welled up from the molten interior to flood low-lying areas. The number of craters that pit Mercury's plains suggest that the plains formed near the end of the era of heavy bombardment, just over 3.8 billion years ago. Mercury's plains are therefore older than most of the lunar maria, leaving more time for cratering to occur and eradicate marialike features.

The most impressive feature discovered by Mariner 10 was a huge circular region called the Caloris Basin

Figure 6-5 **Unusual, Hilly Terrain** What look like tiny, fine-grained wrinkles on this picture are actually closely spaced hills, part of a jumbled terrain that covers nearly 500,000 km² on the opposite side from the Caloris Basin. The large, smooth-floored crater near the center of this photograph has a diameter of 170 km (106 mi).

Figure 6-4 **The Caloris Basin** Mariner 10 sent back this view of a huge impact basin on Mercury's equator. Only about half of the Caloris Basin appears because it happened to lie on the terminator when the spacecraft sped past the planet. Although the center of the impact basin is hidden in the shadows (just beyond the left side of the picture), several semicircular rings of mountains reveal its extent. The outer rim of the basin is defined by a ring of mountains up to 2 km (6500 ft) high. The diameter of the basin is 1300 km (810 mi).

(Figure 6-4), which measures 1300 km (810 mi) in diameter and lies along the *terminator* (the border between day and night) in the figure. It is surrounded by a 2-km-high ring of mountains, beyond which are relatively smooth plains. Like the lunar maria, the Caloris Basin was probably gouged out by the impact of a large meteoroid that penetrated the planet's crust. Because relatively few craters pockmark the lava flows that filled the basin, the Caloris impact must have occurred toward the end of the crater-making period.

The Caloris impact was a tumultuous event that shook the entire planet. Indeed, the collision even seems to have affected the side of Mercury directly opposite the Caloris Basin: That area has a jumbled, hilly region covering nearly half a million square kilometers, about twice the size of Wyoming. Figure 6-5 is a wide-angle view of this area. The hills, which appear as tiny wrinkles that

Figure 6-6 **Scarps on Mercury** A long, meandering cliff, indicated by the arrow, is seen near the horizon in this view of Mercury's northern hemisphere. The cliff, called a scarp by geologists, extends southward for several hundred kilometers. This photograph covers a square area measuring roughly 550 km (340 mi) on a side.

cover most of the photograph, are about 5 to 10 km wide and between 100 and 1800 m high. Geologists believe that seismic waves from the Caloris impact became focused, like light through a lens, as they passed through Mercury. As this concentrated seismic energy reached the far surface of the planet, jumbled hills were pushed up.

Mariner 10 also revealed gently rolling plains and numerous, long cliffs, called **scarps**, meandering across Mercury's surface (Figure 6-6). These and the lack of recent volcanic activity suggest that the planet's interior is solid to a significant depth. Otherwise, lava would have leaked out. The scarps are believed to have formed as the planet cooled. Virtually everything that cools contracts (water is a notable exception). Therefore, as Mercury's mantle and molten iron core cooled and contracted, its surface moved inward. But being solid, Mercury's crust wrinkled as it contracted, thereby forming the scarps.

Mercury's mass is quite low, only 5.5% that of the Earth's. As on the Moon, the force of gravity on Mercury is too weak to hold a permanent atmosphere. Some gases have been detected around Mercury, but only because they are continually being replenished. Scientists believe that hydrogen and helium gas there come from the Sun, while sodium and potassium gas escape from rocks inside the planet (a process called *outgassing*, which also occurs on the Earth).

6-2 Mercury has an iron core and a magnetic field like Earth

Mercury's average density of 5430 kg/m^3 is quite similar to Earth's (5520 kg/m^3). As we saw in Chapter 5, typical rocks on Earth's surface have a density of only about 3000 kg/m^3 because they are composed primarily of lightweight elements. The high average densities of both Mercury and our planet are caused by their iron cores.

Because Mercury is less dense than the Earth, you might conclude that Mercury has a lower percentage of iron, a heavy element, than our planet does. In fact, it is the Earth's greater mass pressing in that makes it denser; Mercury is actually the most iron-rich planet in the solar system. Figure 6-7 shows a scale drawing of its interior, where an iron core fills 42% of the planet's volume. Surrounding the core is a 600-km-thick rocky mantle. For comparison, the Earth's iron core occupies only 17% of our planet's volume. Whether Mercury's core is at all molten is not yet known.

Events early in Mercury's life must somehow account for its high iron content. We know that the inner regions of the primordial solar nebula were incredibly hot. Perhaps only iron-rich minerals able to withstand the heat formed solids. According to another theory, an especially intense outflow of particles from the young Sun stripped Mercury of its low-density mantle shortly after the Sun formed. A third possibility is that, during the final stages of planet formation, Mercury was struck by a large planetesimal. Computer simulations show that this cataclysmic collision would have ejected much of the lighter mantle (Figure 6-8).

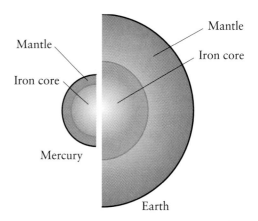

Figure 6-7 **The Interiors of Mercury and Earth** Mercury is the most iron-rich planet in the solar system. Its iron core occupies an exceptionally large fraction of its interior.

Figure 6-8 **The Stripping of Mercury's Mantle** To account for Mercury's high iron content, one theory proposes that a collision with a planet-sized object stripped Mercury of most of its rocky mantle. These four images show a computer simulation of a head-on collision between proto-Mercury and a body one-sixth its mass. Both worlds are shattered by the impact, which vaporizes much of their rocky mantles. Mercury eventually re-forms from the iron-rich debris left behind.

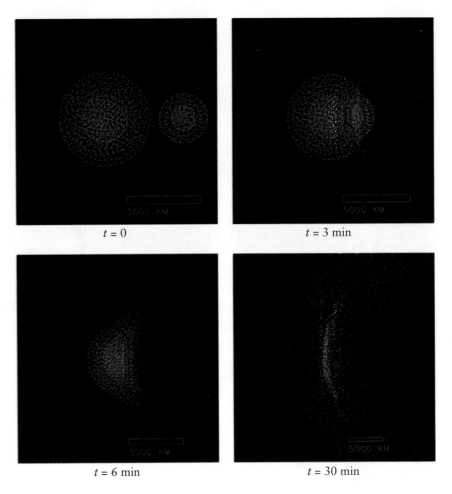

As we saw in Chapter 5, electric currents flowing in a planet's liquid iron core create a planetwide magnetic field. Why, then, did Mariner 10 find only a weak magnetic field on Mercury? The Earth creates its electric current by rotating once a day. Mercury rotates 59 times more slowly (see Table 6-1), hardly fast enough to generate a magnetic field. A solar day on Mercury (noontime to noontime) is actually 176 Earth days—twice as long as a year there! Its magnetic field is still a mystery.

Because of Mercury's slow rotation and minimal atmosphere, the differences in temperature between day and night are far more noticeable there than on the Earth. At noontime with the Sun directly overhead on Mercury, the surface temperature is 700 K (800°F). At the terminator, where day meets night, the temperature is about 425 K (305°F), and on the night side, the temperature falls as low as 100 K (−280°F)! This range of temperature, 600 K (1080°F), on Mercury contrasts tremendously with a typical change in temperature between day and night of 11 K (20°F) on Earth.

Venus

Venus and Earth have almost the same mass, the same diameter, and the same average density. Indeed, if Venus were located at the same distance from the Sun as is the Earth, then it, too, might well have evolved life. However, Venus is 30% closer to the Sun than the Earth, and this one difference between the two planets leads to a host of others, making Venus quite untenable for life as we know it.

6-3 The surface of Venus is completely hidden beneath permanent cloud cover

At nearly twice the distance from the Sun as Mercury, Venus is often easy to view without interference from the Sun's glare. At its greatest elongation, Venus is seen high above the western horizon after sunset, where it is called the "evening star," or high in the eastern sky before sunrise, where it is called the "morning star." Table 6-2 lists Venus' essential properties.

Venus is so easy to identify because it is often one of the brightest objects in the night sky. Only the Sun and the Moon outshine Venus at its greatest brilliance. Venus is often mistaken for a UFO because low on the horizon its bright light is strongly refracted by the Earth's atmosphere, making it appear to change color and position rapidly.

Unlike Mercury, Venus is intrinsically bright (see Table 6-2) because it has a very highly reflective layer exposed to space. This layer is not the planet's surface but a permanent cloud cover.

Earth-based telescopic views cannot penetrate this thick, unbroken layer of clouds. Until 1962, we did not even know how fast Venus rotates. In the 1960s, however, both the United States and the former Soviet Union began sending probes there. The Americans sent fragile, lightweight spacecraft into orbit near Venus (Figure 6-9). The Soviets, who had more powerful rockets, sent massive vehicles directly into the Venusian atmosphere.

Building spacecraft that could survive the descent proved to be more frustrating than anyone had expected. Finally, in 1970, a Soviet probe managed to transmit data for a few seconds directly from the Venusian surface. Soviet missions during the 1970s measured a surface temperature of 750 K (900°F) and a pressure of 98 atm. This is the same force you would feel if you were swimming 0.93 km (3000 ft) underwater.

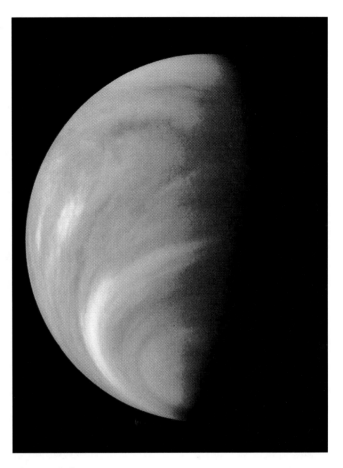

Figure 6-9 **Venus** Venus' thick cloud cover efficiently traps heat from the Sun, resulting in a surface temperature even hotter than that on Mercury. Unlike Earth's clouds, which are made of water droplets, Venus' clouds are very dry and contain droplets of concentrated sulfuric acid. This photograph was taken in 1979 by an American spacecraft in orbit about Venus.

TABLE 6-2	
Venus' Vital Statistics	
Statistic	**Measurement**
Mass	4.87×10^{24} kg (0.815 $M_\oplus$)
Radius	6051 km (0.949 $R_\oplus$)
Average density	5250 kg/m^3
Orbital eccentricity	0.007
Sidereal period of revolution (year)	224.7 Earth days
Average distance from the Sun	1.08×10^8 km (0.723 AU)
Sidereal rotation period	243 days (retrograde)
Solar rotation period (day)	116.8 Earth days
Albedo (average)	0.76

Soviet spacecraft also discovered that Venus' clouds are confined to a 20-km-thick layer located 50 to 70 km above the planet's surface. Below the clouds is a 20-km-thick layer of haze. Beneath this haze the Venusian atmosphere is remarkably clear all the way down to the surface. Standing on the surface of Venus, you would experience a perpetually cloudy day.

Unlike the clouds on Earth, which appear white from above, the cloudtops of Venus appear yellowish or yellow-orange to the human eye. These colors are typical of sulfur and its compounds. Indeed, spacecraft found substantial amounts of sulfur dust in Venus' upper atmosphere and sulfur dioxide and hydrogen sulfide at lower elevations. The clouds are composed of droplets of concentrated sulfuric acid! Because of the tremendous

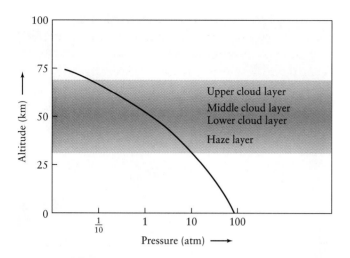

Figure 6-10 **Pressure in the Venusian Atmosphere** The pressure at the Venusian surface is a crushing 98 atm (1440 lb/in.²). Above the surface, atmospheric pressure decreases smoothly with increasing altitude.

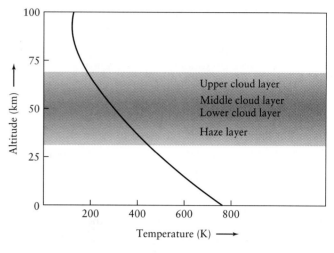

Figure 6-11 **Temperature in the Venusian Atmosphere** The temperature in Venus' atmosphere increases smoothly from a minimum of about 173 K (−150°F) at an altitude of 100 km to a maximum of nearly 750 K (900°F) on the ground.

atmospheric pressure on Venus, the droplets do not fall as a rain; instead, they remain suspended in thick mist.

Compelling evidence suggests that these sulfurous compounds come from active volcanoes. All of the major volcanic chemicals spewed by Earth's volcanoes have been detected in Venus' atmosphere. Because many of these substances are very short-lived, they must be constantly replenished by new eruptions. In fact, the abundance of sulfur compounds on Venus does vary, suggesting varied volcanic activity. Finally, spacecraft have also picked up radio bursts thought to be strokes of lightning, as seen in the plumes of erupting volcanoes on Earth.

The results of the early probing of Venus are summarized in Figures 6-10 and 6-11. Both pressure and temperature decrease smoothly with increasing altitude. From the changes in temperature and pressure above a planet's surface, we can understand the structure of its atmosphere. As shown in Figure 6-11, Venus' surface temperature is very high, higher even than Mercury's.

6-4 The greenhouse effect heats Venus' surface

At first, no one could believe reports that the surface temperature on Venus was hotter than the surface temperature on Mercury, which, after all, is closer to the Sun. After some initial skepticism, however, astronomers quickly found a straightforward explanation—the **greenhouse effect.**

Perhaps you have had the experience of parking your car in the sunshine on a warm summer day. You roll up the windows, lock the doors, and go on an errand. After a few hours, you return to discover that the interior of your automobile has become stiflingly hot, typically 20 K warmer than the outside air temperature.

What happened to make your car so warm? First, sunlight entered your car through the windows. This radiation was absorbed by the dashboard and the upholstery, raising their temperatures. Because they become warm, your dashboard and upholstery emit heat, to which your car windows are opaque. This energy is therefore trapped inside your car and absorbed by the air and interior surfaces. As more sunlight comes through the windows and is trapped, the temperature continues to rise. The same trapping of sunlight warms an actual greenhouse (Figure 6-12).

Carbon dioxide is responsible for a similar warming of Venus' atmosphere and, to a lesser extent, the Earth's. Like your car windows, carbon dioxide is transparent to visible light but opaque to infrared radiation. Warmed by sunlight, the Earth emits infrared radiation, some of which is absorbed by the small fraction of the air that is carbon dioxide, causing the air temperature to rise. The small amount of carbon dioxide in the Earth's atmosphere produces a comparatively gentle greenhouse effect.

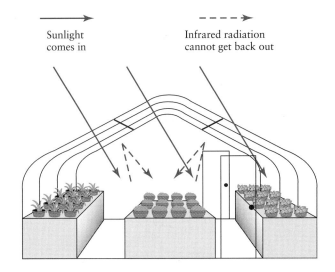

Sunlight
comes in

Infrared radiation
cannot get back out

Figure 6-12 **The Greenhouse Effect** Incoming sunlight easily penetrates the windows of a greenhouse and is absorbed by objects inside. These objects reradiate energy at infrared wavelengths, to which the glass windows are opaque. The trapped infrared radiation causes the temperature inside the greenhouse to rise.

In contrast to Earth, 96% of Venus' thick atmosphere is carbon dioxide; the remaining 4% is mostly nitrogen. Although most of the visible sunlight striking the Venusian cloudtops is reflected back into space, enough light reaches the Venusian surface to heat it. The warmed surface in turn emits infrared radiation, which cannot escape through Venus' carbon-dioxide-rich atmosphere. This trapped radiation produces the high temperatures found on Venus.

Without the greenhouse effect, the surface of Venus would have a noontime temperature of 465 K. Because of it, that temperature is actually a sweltering 750 K. Furthermore, the thick atmosphere keeps the night side of Venus at nearly the same temperature, unlike the night side of Mercury, where the temperature drops precipitously. This high temperature prevents liquid water from existing on Venus; its surface is bone dry.

6-5 Venus is covered with gently rolling hills, two "continents," and numerous volcanoes

Soviet spacecraft that landed on Venus have provided us with close-up images of the planet's arid surface. Figure 6-13 is a view of the Venus' regolith, taken in 1981. Russian scientists believe that this region was covered with a thin layer of lava that fractured upon cooling to create the rounded, interlocking shapes seen in the photograph. Indeed, measurements by several spacecraft indicate that Venusian soil is quite similar to lava rocks called basalt, which are common on Earth and the Moon.

By far the best images of the Venusian surface came from the highly successful Magellan spacecraft that arrived at Venus in 1990. In orbit about the planet, Magellan sent radar signals through the clouds surrounding Venus. It mapped Venus using a radar altimeter that bounced microwaves off the ground directly below the spacecraft. By measuring the time delay of the radar echo, scientists determined the heights and depths of

a

b

Figure 6-13 **The Venusian Surface** (a) This color photograph, taken by a Soviet spacecraft, shows rocks that appear orange because of the thick, sulfur-rich atmosphere. (b) By comparing the apparent color of the spacecraft to the color it was known to be, computers can correct for the sulfurous light. The actual color of the rocks is gray. In this view, the rocky plates covering the ground may be fractured segments of a thin layer of lava.

Figure 6-14 **A Venusian Landscape** Most of Venus is covered with plains and gently rolling hills produced by numerous lava flows. The crater in the foreground is 48 km (30 mi) in diameter. The volcano near the horizon is 3 km (1.9 mi) high. Note the cracks in the Venusian surface toward the right side of the image.

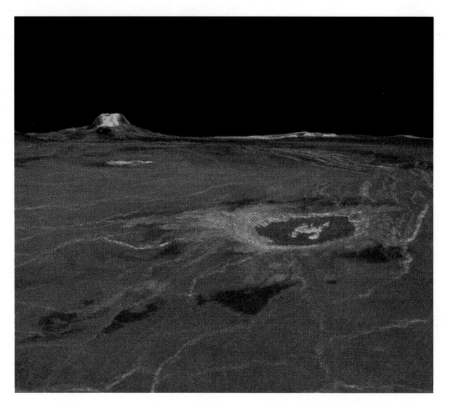

Venus' hills and valleys. As a result, astronomers have been able to construct a three-dimensional map of the planet. The resolution of this map is about 100 m—you could see a football stadium on Venus if there were any (none were detected).

Venus is remarkably flat compared to the Earth. Over 80% of Venus' surface is covered with volcanic plains and gently rolling hills created by numerous lava flows. Figure 6-14 is a computer-generated view showing a typical Venusian landscape.

Magellan's data revealed two large highlands, or "continents," rising well above the generally level surface of the planet (Figure 6–15). The continent in the northern hemisphere is Ishtar Terra, named after the Babylonian goddess of love. Ishtar Terra, approximately the same size as Australia, is dominated by a high plateau ringed by towering mountains. The highest mountain is Maxwell Montes, whose summit rises to an altitude of 11 km above the average surface. For comparison, Mount Everest on Earth rises 9 km above sea level.

Figure 6-15 **A Map of Venus** This false-color map of Venus was produced by the Magellan spacecraft. The Venusian equator extends across the middle of the map. Color indicates elevation: Red for highest, blue for lowest. Grey areas were not mapped. The elevated region in the north is Ishtar Terra, dominated by Maxwell Montes, the planet's highest mountain. The large, scorpion-shaped feature extending along the equator is Aphrodite Terra, a continentlike highland that contains several spectacular volcanoes.

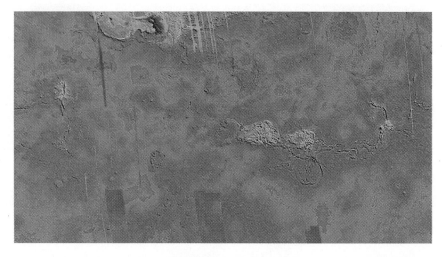

Figure 6-16 **A "Global" View of Venus** A computer was used to map numerous Magellan images onto a simulated globe. Color is used to enhance small-scale structures. Extensive lava flows and lava plains cover about 80% of Venus' relatively flat surface.

ing the lack of extensive cratering is that the entire planet's surface melts! This would occur if the crust is very thick compared to the Earth's crust. A crust 300 km thick would insulate the interior so much that, heated by radioactive elements in it, the mantle could become hot enough to episodically melt the crust. It appears that, every three or four hundred million years, the entire surface of Venus liquefies until the pent-up heat escapes. Then a new solid crust forms and the process begins anew.

The sulfur content of the air and the traces of active volcanoes on the surface are strong clues that, like the Earth, Venus has a molten interior. Because the average density of Venus is similar to that of Earth, its core is predominantly iron. Currents in the molten iron should generate a magnetic field. However, none of the spacecraft sent there has detected one. This is plausible only if Venus rotates exceptionally slowly. In fact, the planet takes 116.8 Earth days to get from one sunrise to the next (that is, if you could see the Sun from Venus' surface). Unlike the Earth, Venus also rotates in a direction opposite to its revolution around the Sun. We say that Venus has **retrograde rotation**. In other words, sunrise

The largest Venusian continent, Aphrodite Terra (named after the Greek goddess called Venus by the Romans) is a vast belt of highlands just south of the equator. Aphrodite is 16,000 km (10,000 mi) in length and 2000 km (1200 mi) wide, giving it an area about one-half that of Africa. The global view of Venus in Figure 6-16 shows that most of Aphrodite is covered by vast networks of faults and fractures.

Fewer than a thousand impact craters have been observed on Venus (Figure 6-17), compared to the hundreds of thousands seen on the Moon and Mercury. Today, of course, Venus' thick atmosphere vaporizes much of the debris that would otherwise create craters. Because its atmosphere is thicker than Earth's, Venus clears such space junk falling toward it more efficiently than Earth does. However, there should have been a period shortly after Venus formed and before its atmosphere developed (from gases escaping from the planet's interior) when impacts were common.

The low number and incredibly random distribution of craters on Venus leads to the belief that the planet's surface is periodically erased and replaced. Because geologists have found no large-scale tectonic plate motion on Venus to refresh the surface, the current theory explain-

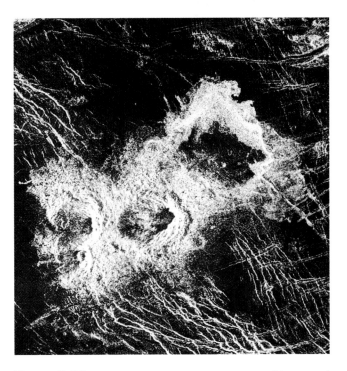

Figure 6-17 **Craters on Venus** Impact craters on Venus tend to occur in clusters, which suggests that they are formed from large, single pieces of infalling debris that are broken up in the atmosphere. Shown here is triple-impact Crater Stein.

on Venus occurs in the west. Venus' rotation axis is tilted more than 177°, compared to the Earth's $23\frac{1}{2}°$ tilt. Because Venus' axis is within 3° of being perpendicular to the plane of its orbit around the Sun, the planet has no seasons. Although we don't know the cause of Venus' retrograde rotation, one likely explanation is that a major impact altered the rotation early in the planet's life.

6-6 Magellan's last gasp reveals irregularities in Venus' upper atmosphere

Magellan finished mapping the surface of Venus in October 1994. Instead of keeping it in orbit, NASA astronomers decided to give it one last mission. They fired braking rockets on the spacecraft, sending it spiraling down toward the planet's surface. By doing so, it was hoped that Magellan would transmit useful data about Venus' dense atmosphere before the spacecraft vaporized. On October 12, 1994, Magellan stopped transmitting, but before it did so, it sent back tantalizing information.

Magellan indicated that the density of Venus' atmosphere is lower at some altitudes than was expected, and higher at other altitudes. This information will be useful in helping astronomers plan future missions to the cloud-shrouded planet.

Mars

Mars is the only planet whose surface features can be seen through Earth-based telescopes. Its distinctive rust-colored hue makes it stand out in the night sky (Figure 6-18). When Mars is near opposition, even telescopes for home use reveal its seasonal changes. Dark markings on the Martian surface can be seen to vary, and prominent polar caps shrink noticeably during the spring and summer months (Figure 6-19). Table 6-3 lists Mars' essential properties.

6-7 Earth-based observations originally suggested that Mars might harbor extraterrestrial life

The Dutch physicist Christian Huygens made the first reliable observations of Mars in 1659. Using a telescope of his own design, Huygens identified a prominent, dark surface feature that re-emerged roughly every 24 hours, suggesting a rate of rotation very much like the Earth's.

TABLE 6-3

Mars' Vital Statistics

Statistic	Measurement
Mass	6.42×10^{23} kg (0.107 M$_\oplus$)
Radius	3393 km (0.532 R$_\oplus$)
Average density	3950 kg/m^3
Orbital eccentricity	0.093
Sidereal period of revolution (year)	687 Earth days (1.88 Earth years)
Average distance from the Sun	2.28×10^8 km (1.52 AU)
Sidereal rotation period	24 h 37 min 22 s
Solar rotation period (day)	24 h 39 min
Albedo (average)	0.16

Huygens' observations soon led to speculation about life on Mars, because the planet seemed so similar to Earth.

In 1877 Giovanni Virginio Schiaparelli, an Italian astronomer, reported seeing 40 lines crisscrossing the Martian surface. He called these dark features *canali,* an Italian term meaning "water channels." It was soon mistranslated into English as *canals,* implying the existence on Mars of intelligent creatures capable of substantial

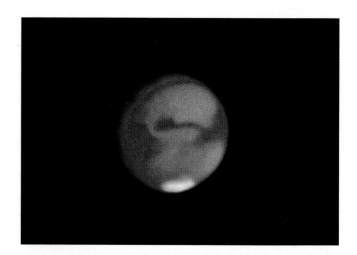

Figure 6-18 **Mars Viewed from the Earth** This high-quality Earth-based photograph of Mars was taken in 1971 when the Earth was about as near to Mars as it can be. This photograph portrays what you might typically see through a moderate-sized telescope under excellent observing conditions. Notice the prominent southern polar cap.

Figure 6-19 **Changing Seasons on Mars** During the Martian winter, the temperature drops so low that carbon dioxide freezes out of the Martian atmosphere. A thin coating of carbon dioxide frost covers a broad region around Mars' south pole. During the summer, the range of this south polar carbon dioxide cap decreases dramatically.

engineering feats. This speculation led Percival Lowell, who came from a wealthy Boston family, to finance a major new observatory near Flagstaff, Arizona. By the end of the nineteenth century, Lowell had allegedly observed 160 Martian canals.

It soon became fashionable to speculate that the Martian canals formed an enormous planetwide irrigation network to transport water from melting polar caps to vegetation near the equator. (The seasonal changes on Mars' dark surface markings can be mistaken for vegetation.) In view of the planet's reddish, desertlike appearance, Mars was thought to be a dying planet whose inhabitants must go to great lengths to irrigate their farmlands. No doubt the Martians would readily abandon their arid ancestral homeland and invade the Earth for its abundant resources. Hundreds of science fiction stories and dozens of monster movies owe their existence to the *canali* of Schiaparelli.

6-8 Space probes to Mars found craters, volcanoes, and canyons—but no canals

Three American spacecraft journeyed past Mars in the 1960s and sent back pictures that clearly showed numerous flat-bottomed craters (Figure 6-20), volcanoes, and canyons, but not one canal of a size consistent with those allegedly seen from Earth. Schiaparelli's *canali* had been optical illusions.

The partially eroded, flat bottoms of the Martian craters probably result from dust storms that frequently rage across the planet's surface. Over the past two centuries, astronomers have seen faint surface markings disappear under a reddish-orange haze as thin Martian winds have stirred up finely powdered surface dust. Some storms actually obscure the entire planet. Over the ages, deposits of dust have filled in the crater bottoms, but three billion years of sporadic storms have not wiped out the craters; the Martian atmosphere is too thin to

Figure 6-20 **Martian Craters** Numerous flat-bottomed craters are seen in this mosaic of images taken by a spacecraft orbiting Mars in 1980. The large, heavily cratered circular area in the upper half of this view is known as Arabia. Carbon dioxide snow covers the floors of craters near the Martian south pole, toward the lower right in this picture.

EYES ON . . .   Desirable Features in a Telescope

Recall from Chapter 3 that objects appear brighter and are magnified more in larger-diameter telescopes than in smaller ones. You should buy the largest-diameter telescope you can afford. Active amateurs normally choose telescopes with diameters of at least four inches. However, keep in mind that the bigger the telescope, the heavier it is and the harder it is to transport.

If you buy a telescope whose eyepiece is positioned so that you have to crawl under it or climb a ladder to see through it, then you will find the experience less satisfying than if you can observe comfortably. If the eyepiece is not easily accessible, you can buy a *diagonal mirror* (a right-angle mirror that goes between the telescope body and the eyepiece) to help correct the problem.

If you plan to take photographs with a CCD camera or other instrument on your telescope or to show the cosmos to large groups of people, you will need a *tracking motor* to enable the telescope to automatically follow

the stars. Tracking motors require their own (often heavy) batteries, a car nearby so they can be connected to the cigarette lighter, or a 110-V outlet. The motor should also be able to run at high speed so you can slew (turn) the telescope rapidly from one object to the next. For photography you will also need a *camera adapter*.

If you want to watch the Sun, it is essential that you buy a good *Sun filter*. Viewing the Sun is very exciting, especially when you can see sunspots. **Never, ever look at the Sun directly either through a telescope or with your naked eye. Doing so for even a second can lead to partial or total blindness!**

If you want to see especially large areas of the sky, you will want to purchase a telescope with a *Schmidt corrector plate*. The most common wide-field telescopes are of the Schmidt-Cassegrain design. These telescopes have a Schmidt corrector plate on top and a Cassegrain mirror on the bottom. Their eyepieces are located underneath.

---

carry enough of this extremely fine-grained powder to eradicate them completely.

In 1971 the Mariner 9 spacecraft went into orbit around Mars and began sending back detailed, close-up pictures that showed enormous volcanoes and vast canyons (Figure 6-21). The largest Martian volcano, Olympus Mons, covers an area as big as the state of Missouri and rises 24 km (15 mi) above the surrounding

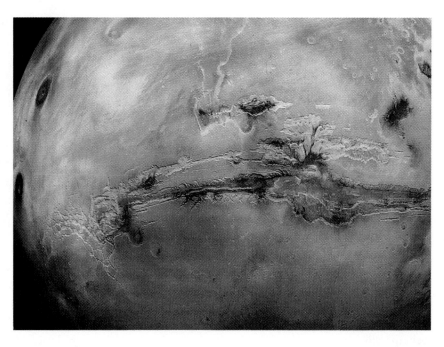

Figure 6-21 Canyons and Volcanoes on Mars This high-altitude photograph shows a variety of the features on Mars, including broad, towering volcanoes (left) on the highland called Tharsis Bulge, impact craters (upper right), and vast, windswept plains. An enormous canyon system called Valles Marineris crosses horizontally just below the center of the image.

plains—nearly three times the height of Mount Everest. The highest volcano on Earth, Mauna Loa in the Hawaiian Islands, has a summit only 8 km above the ocean floor.

The Hawaiian Islands are only the most recent additions to a long chain of extinct volcanoes. They resulted from **hot-spot volcanism,** a process by which molten rock rises to the surface from a fixed hot region far below. The Pacific tectonic plate is slowly moving northwest at a rate of several centimeters per year. As a result, new volcanoes are created above the hot spot while older ones move off and become extinct, eventually disappearing beneath the ocean.

Mars, however, does not appear to have any plate tectonic activity, and so one hot spot can keep pumping lava upward through the same vent for millions of years. One result is Olympus Mons—a single giant volcano rather than a long chain of smaller ones. The volcano's summit has collapsed to form a volcanic crater, called a **caldera,** large enough to contain the state of Rhode Island (Figure 6-22).

For reasons we do not yet understand, most of the volcanoes on Mars are in the northern hemisphere, but most of the impact craters are in the southern hemisphere. Between the two hemispheres, Mariner 9 discovered a

Figure 6-23 **A Segment of Valles Marineris** This mosaic of Viking orbiter photographs shows a segment of Valles Marineris. The canyon is about 100 km (60 mi) wide in this region. The canyon floor has two major levels. The northern (upper) canyon floor is 8 km (5 mi) beneath the surrounding plateau, whereas the southern canyon floor is only 5 km (3 mi) below the plateau.

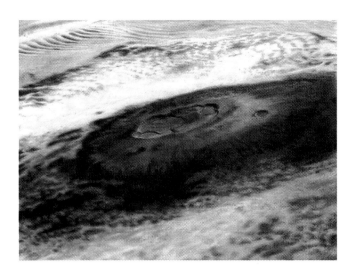

Figure 6-22 **The Olympus Caldera** This view of the summit of Olympus Mons is based on a mosaic of six pictures taken by one of the Viking orbiters. The caldera consists of overlapping, volcanic craters and measures roughly 70 km across. The volcano is wreathed in mid-morning clouds brought upslope by cool air currents. The cloudtops are about 8 km below the volcano's peak.

vast canyon running roughly parallel to the Martian equator (Figure 6-23). In honor of the Mariner spacecraft that revealed so much of the Martian surface, this enormous chasm has been designated Valles Marineris.

Valles Marineris stretches over 3000 km, beginning with heavily fractured terrain in the west and ending with ancient cratered terrain in the east. If this canyon were located on Earth, it would stretch from New York to Los Angeles. Unlike the Grand Canyon, which it superficially resembles, Valles Marineris was not formed by water erosion. Some geologists suspect Valles Marineris is an ancient fracture in the Martian crust somewhat similar to those caused by plate tectonics on Earth. The Martian

Figure 6-24 **Canyon Fog** Delicate fog created by the scant water vapor in Mars' atmosphere fill canyons of Labyrinthus Noctis at sunrise and sunset.

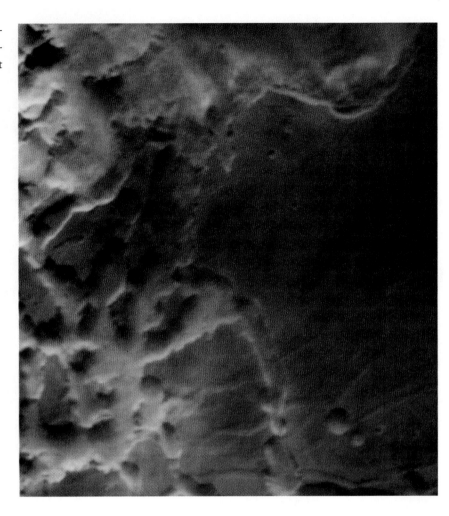

canyons are up to 6 km (4 mi) deep and 190 km (120 mi) wide (Figure 6-24).

6-9 Surface features indicate that water once flowed on Mars

Despite disproving the theory that Mars has broad canals, Mars-orbiting spacecraft did reveal many features that look like dried-up riverbeds (Figure 6-25). These include intricate branched patterns and delicate channels meandering among flat-bottomed craters. Rivers on Earth invariably follow similarly winding courses (Figure 6-26).

Direct evidence that water once flowed on Mars comes from so-called *SNC meteorites* found on Earth. These space rocks are believed to have been pieces of Mars that were ejected into space during especially powerful impacts on that planet's surface. They contain water-soaked clay (Figure 6-27).

The Martian riverbeds were totally unexpected, because liquid water cannot exist today on Mars. Water is liquid over a limited range of temperature and pressure: At low temperatures, water becomes ice; at high temperatures, it becomes steam. The atmospheric pressure also affects the state of water. If the pressure is very low, molecules easily escape from the liquid's surface, causing the water to vaporize. Because the pressure of Mars' atmosphere is only 1% that of the Earth's atmosphere, any liquid water on Mars today would furiously boil and rapidly evaporate into the thin Martian air.

Where did the water on Mars come from and go to? Some may have been frozen in the Martian polar caps, although their seasonal variations are created by the freezing and evaporating of dry ice (carbon dioxide). Other frozen water probably lies under the Martian surface. The temperature at Mars' poles, typically 160 K (−170°F), keeps the water there permanently frozen, like the permafrost found in northern Asia and on Antarctica. In the distant past heat from meteoritic impacts or

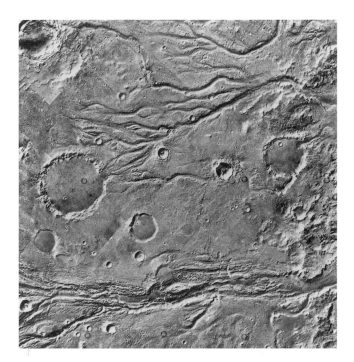

Figure 6-25 **Ancient River Channels on Mars** The existence of braided riverbeds on Mars suggests that the planet's atmosphere was thicker and its climate more Earthlike long ago, when liquid water flowed across its surface. Any water exposed to the sparse Martian atmosphere today would rapidly boil away or freeze solid.

Figure 6-26 **River Flow on Earth** Typical of how rivers wind as they wander downhill are the Ohio river (upper left), Allegheny river (upper right), and Monongahela river (bottom). This Landsat 5 photograph shows the greater Pittsburgh, Pennsylvania, area where the rivers converge. Contrary to intuition, the Monongahela flows northward to meet the Allegheny. Together they create the Ohio river, which flows first northward, and then turns south.

from volcanic activity occasionally melted this layer of permafrost on Mars. The ground then collapsed, and millions of tons of rock pushed the water to the surface. These flash floods could account for the riverbeds we see there today.

6-10 Mars' air is thin and often filled with dust

As on Venus, some 95% of Mars' thin atmosphere is composed of carbon dioxide. The remaining 5% consists of nitrogen, argon, and some traces of oxygen. The planet's gravitational force is just strong enough to hold all its gases, but too weak to prevent most water vapor from evaporating into space. Consequently, the concentration of water vapor in Mars' atmosphere is 30 times lower than the concentration above the Earth. If all the water vapor could somehow be squeezed out of the Martian atmosphere, it would not fill even one of the five Great Lakes.

Figure 6-27 **A Piece of Mars on Earth** Believed to have been knocked off Mars by an impact, this SNC meteorite was recovered in Antarctica. It shows strong evidence of having been exposed to liquid water on Mars, perhaps for hundreds of years.

Figure 6-28 **Mars' Rust-Colored Sky** Taken by Viking 1, this photograph shows the Chryse Planitia (the "Golden Plains"). At the top of the image, the pink color of the Martian sky is evident. Compare the color of the sky in this photograph and in Figure 6-29.

Unlike our blue sky, Mars' atmosphere is often pastel red, sometimes turning shades of pink and russet (Figure 6-28). All these colors are due to fine dust, blown from the planet's desertlike surface during powerful windstorms. The dust is iron-oxide—familiar here on Earth as rust. Mars' sky changes color because the amount of dust in the air varies with the season. During the winter, carbon dioxide ice adheres to the dust particles and drags them to the ground. This helps clear the air and make it a lighter color. In the summer months, the carbon dioxide is not frozen and the dust blown by surface winds remains aloft longer.

The sky color also changes over periods of many years. In 1995 the amount of dust was observed to have dropped dramatically compared to that observed in the 1970s. The reason for the long-term change is still under investigation.

Mars experiences Earthlike seasons because of a striking coincidence, first noted in the late 1700s by the famous German-born English astronomer William Herschel: Just as the Earth's equatorial plane is tilted $23\frac{1}{2}°$ from the plane of its orbit, Mars' equator makes an angle of about 25° with its orbit. However, the Martian seasons last nearly twice as long as Earth's, because Mars takes nearly two Earth years to orbit the Sun.

6-11 The Viking landers sent back close-up views of a barren Martian surface

The discovery in the 1960s that water once existed on the surface of Mars rekindled speculation about Martian life. Although it was clear that Mars has neither civilizations nor fields of plants, microbial life forms still seemed possible. Searching for Martian microbes was one of the main objectives of the ambitious and highly successful Viking missions.

The two Viking spacecraft were launched during the summer of 1975. Each spacecraft consisted of two modules: an orbiter and a lander. Almost a year later, both Viking landers set down on rocky plains north of the Martian equator. Figure 6-29 shows a panoramic view from the Viking 2 lander.

The landers confirmed the long-held suspicion that the red color of the planet is due to large quantities of iron in its soil. Despite the high iron content of its crust, Mars has a lower average density (3950 kg/m³) than the other terrestrial planets (more than 5000 kg/m³ for Mercury, Venus, and Earth). Mars must therefore contain a lower percentage of iron than these other planets. Furthermore, spacecraft detected no measurable magnetic field around Mars. Perhaps Mars' iron is uniformly distributed throughout the body of the planet rather than being concentrated in a dense core.

Each Viking lander was able to dig into the Martian regolith and retrieve rock samples for analysis (Figure 6-30). Bits of the regolith were observed to cling to a magnet mounted on the scoop, indicating that the regolith contains iron (consistent with the assertion above that Mars is rich in iron oxides). Indeed, further analysis showed the rocks at both sites to be rich in iron, silicon, and sulfur. The Martian regolith can best be described as an iron-rich clay.

Each Viking lander also carried a compact biological laboratory designed to test for microorganisms in the Martian soil. Three biological experiments were conducted, each based on the idea that living things alter their environment: They eat, they breathe, and they give off waste products. In each experiment, a sample of the Martian regolith was placed in a closed container, with or without a nutrient substance. The container was then examined for any changes in its contents.

The first data returned by the Viking biological experiments caused great excitement: In almost every case, rapid and extensive changes were detected inside the sealed containers. However, further analysis showed that

Figure 6-29 **A Panorama of the Viking 2 Site** This mosaic of three views shows about 180° of the landing site, a barren, rocky plain. The flat, featureless horizon is about 3 km (2 mi) away.

these changes were due solely to nonbiological chemical processes. Apparently the Martian regolith is rich in chemicals that effervesce (fizz) when moistened. A large amount of oxygen is apparently tied up in the regolith in the form of unstable chemicals called peroxides and superoxides, which break down in the presence of water to release oxygen gas.

The chemical reactivity of the Martian regolith probably comes from ultraviolet radiation that beats down on the planet's surface. Ultraviolet photons easily break apart molecules of carbon dioxide (CO_2) and water vapor (H_2O) by knocking off oxygen atoms, which then become loosely attached to chemicals in the regolith. Ultraviolet photons also produce ozone (O_3) and hydrogen peroxide (H_2O_2), which become incorporated in the regolith. In all these cases, the loosely attached oxygen atoms make the regolith extremely reactive.

Here on Earth, hydrogen peroxide is commonly used as an antiseptic. When you pour this liquid on a wound, it fizzes and froths as the loosely attached oxygen atoms chemically combine with organic material and thereby destroy germs. The Viking landers may have failed to

Figure 6-30 **Digging in the Martian Soil** Viking's mechanical arm with its small scoop protrudes from the right side of this view of the Viking 1 landing site. Several small trenches dug by the scoop in the Martian regolith appear near the left side of the picture.

a

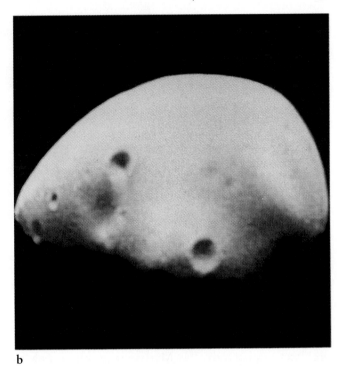

b

Figure 6-31 **Phobos and Deimos** Viking orbiters provided these images of Mars' moons. (a) Phobos, the larger of Mars' two moons, is potato-shaped and measures roughly $28 \times 23 \times 20$ km.

(b) Deimos is less cratered than Phobos and measures roughly $16 \times 12 \times 10$ km.

detect any organic compounds on Mars because the superoxides and peroxides in the Martian regolith make it literally antiseptic.

6-12 Mars' two moons look more like potatoes than spheres

Two tiny moons, Phobos and Deimos, orbit close to Mars' surface. They were not formed like our Moon, by splashing off Mars. Rather, they are captured planetesimals. Phobos (meaning "panic") and Deimos (meaning "fear") are so small that they were not discovered until 1877 by astronomer Asaph Hall. Potato-shaped Phobos is the inner and larger of the two (Figure 6-31a). It rises in the west and gallops across the sky in only five and a half hours, as seen from Mars' equator. Phobos is heavily cratered. Football-shaped Deimos is less cratered than Phobos (Figure 6-31b). As seen from Mars, Deimos rises in the east and takes about three Earth days to creep from one horizon to the other. Both moons are in synchronous rotation as they orbit the red planet.

WHAT DID YOU THINK?

1 *Is the temperature on Mercury, the closest planet to the Sun, higher than the temperature on Earth?* The temperature on the daytime side of Mercury is much higher than on Earth, but the temperature on the nighttime side of Mercury is much lower than on Earth because Mercury rotates so slowly and has little atmosphere to retain heat.

2 *What is the composition of the clouds surrounding Venus?* The clouds are composed primarily of sulfuric acid.

3 *Does Mars have surface liquid water today?* No, but there are strong indications that it had liquid water in the distant past.

4 *Is life known to exist on Mars today?* No life has yet been discovered on Mars.

Key Words

caldera
greenhouse effect

hot-spot volcanism
retrograde rotation

scarp

Key Ideas

• At its greatest elongations, Mercury can be seen only briefly after sunset or before sunrise.

• The Mariner 10 spacecraft passed near Mercury in the mid-1970s, providing pictures of its surface.

   The Mercurian surface is pocked with craters like those of the Moon, but extensive, smooth plains exist between these craters.

   Long cliffs meander across the surface of Mercury. These scarps probably formed as the planet cooled, solidified, and shrank.

• The impact of a large object long ago formed the huge Caloris Basin on Mercury and shoved up jumbled hills on the opposite side of the planet.

• Mercury has an iron core much like that of the Earth.

• Venus is similar to the Earth in size, mass, and average density, but it is covered by unbroken, highly reflective clouds that conceal its other features from Earth-based observers.

• Venus' clouds consist of droplets of concentrated sulfuric acid along with substantial amounts of yellowish sulfur dust. Active volcanoes on Venus may be a constant source of this sulfurous material.

• The surface pressure on Venus is 98 atm, and the surface temperature is 750 K. Both temperature and pressure decrease as altitude increases.

   Venus' exceptionally high temperature is caused by the greenhouse effect, as the dense carbon dioxide atmosphere traps and retains heat emitted by the planet.

• The surface of Venus is surprisingly flat, mostly covered with gently rolling hills. There are two major "continents" and several large volcanoes.

   The surface of Venus shows evidence of local tectonic activity but not the large-scale motions that play a major role in continually reshaping the Earth's surface.

• Earth-based observers found that the Martian solar day is nearly the same as that on the Earth, that Mars has polar caps that expand and shrink with the seasons, and that the Martian surface undergoes seasonal color changes.

• A few observers reported a network of linear features called canals; these observations led to speculation about life on Mars.

• The Martian surface has numerous flat-bottomed craters, several huge volcanoes, a vast canyon, and dried-up riverbeds —but no canals formed by intelligent life.

   The flash-flood features and dried riverbeds on the Martian surface indicate that water once flowed on Mars.

• Liquid water would quickly boil away in Mars' thin atmosphere, but the planet's polar caps contain frozen water, and a layer of permafrost may exist beneath the regolith.

• The Martian atmosphere is composed mostly of carbon dioxide; the atmospheric pressure is about 1% that of Earth's.

• Chemical reactions in the Martian regolith, together with ultraviolet radiation from the Sun, apparently act to sterilize the Martian surface.

• Mars has two potato-shaped moons, Phobos and Deimos, both of which have synchronous rotation.

Review Questions

1 Why is Mercury so difficult to observe? When is the best time to see the planet?

2 Compare the surfaces of Mercury and our Moon. How are they similar? How are they different?

3 Compare the interiors of Mercury and Earth. How are they similar? How are they different?

4 What kind of tectonic features are found on Mercury? Why are they probably much older than tectonic features on the Earth?

5 Briefly describe at least one theory explaining why Mercury has a large iron core.

6 In early astronomy books, Venus is often referred to as the Earth's twin. What physical properties do the two planets have in common? In what ways are the two planets dissimilar?

7 Why is it hotter on Venus than on Mercury?

8 What is the greenhouse effect, and what role does it play in the atmospheres of Venus and the Earth?

9 What evidence exists for active volcanoes on Venus?

10 Describe the Venusian surface. What kinds of features would you see if you could travel around on the planet?

11 Why do astronomers believe that Venus' surface was not molded by the kind of tectonic activity that shaped the Earth's surface?

12 Why is Mars red?

13 When is the best time to observe Mars from Earth?

14 Compare the cratered regions of Mercury, the Moon, and Mars. Assuming that the craters on all three worlds originally had equally sharp rims, what can you conclude about the environmental histories of these worlds?

15 How would you tell which craters on Mars had been formed by meteoritic impacts and which by volcanic activity?

16 Compare the volcanoes of Venus, Earth, and Mars. Do you think hot-spot volcanism is or was active on all three worlds?

17 What geologic features (or lack thereof) on Mars have convinced scientists that plate tectonics did not significantly shape the Martian surface?

18 What is the current knowledge concerning life on Mars? Do you think Mars might be as barren and sterile as the Moon?

Advanced Questions

19 What evidence do we have that the surface features on Mercury were not formed during recent geologic history?

20 Venus takes 440 days to move from greatest western elongation to greatest eastern elongation, but it needs only 144 days to go from greatest eastern elongation to greatest western elongation. With the aid of a diagram, explain why.

21 As seen from Earth, the brightness of Venus changes as it moves along its orbit. Describe the main factors that determine Venus' variations in brightness as seen from Earth. *(Hint:* See the discussion of Venus in Chapter 2.)

22 How might Venus' cloud cover change if all of Venus' volcanic activity suddenly stopped? How might these changes affect the overall Venusian environment?

23 Compare Venus' continents with Earth's. What do they have in common? How are they different?

24 Explain why Mars has the longest synodic period of all the planets, although its sidereal period is only 687 days.

25 With carbon dioxide accounting for about 95% of the atmospheres of both Mars and Venus, why do you suppose there is little greenhouse effect on Mars today?

26 Could the polar regions of Mars reasonably be expected to harbor life forms, even though the Martian regolith is sterile at the Viking lander sites?

Discussion Questions

27 If you were planning a return mission to Mercury, what features and observations would be of particular interest to you?

28 If you were designing a space vehicle to land on Venus, what special features would you think necessary? In what ways would this mission and landing craft differ from a spacecraft designed for a similar mission to Mercury?

29 Suppose someone told you that the Viking mission failed to detect life on Mars simply because the tests were designed to

detect terrestrial life forms, not Martian life forms. How would you respond?

30 Compare the scientific opportunities for long-term exploration offered by the Moon and Mars. What difficulties would there be in establishing a permanent base or colony on each of these two worlds?

31 Imagine you are an astronaut living at a base on Mars. Describe your day's activities, what you see, the weather, the spacesuit you are wearing, and so on.

Observing Projects

32 Refer to the table below to determine the dates of the next two or three greatest elongations of Mercury. Consult such magazines as *Sky & Telescope* or *Astronomy* to see if any of these greatest elongations is going to be especially favorable for viewing the planet. If so, make plans to be one of those rare individuals who has actually seen the innermost planet of the solar system. Set aside several evenings (or mornings) around the date of the favorable elongation to reduce the chances of being "clouded out." Select an observing site that has a clear, unobstructed view of the horizon where the Sun sets (or rises). Make arrangements to have a telescope at your disposal. Search for the planet on the dates you have selected and make a drawing of its appearance through your telescope.

Greatest Elongations of Mercury

Eastern		Western	
1996	January 2	1996	February 11
	April 23		June 10
	August 21		October 3
	December 15	1997	January 24
1997	April 6		May 22
	August 4		September 16
	November 28	1998	January 6
1998	March 20		May 4
	July 17		August 31
	November 11		December 20
1999	March 3	1999	April 16
	June 28		August 14
	October 24		December 2

33 Refer to the table below to see if Venus is currently near a greatest elongation. If so, view the planet through a telescope. Make a sketch of the planet's appearance. From your sketch, can you determine if Venus is closer to us or further from us than the Sun is?

Greatest Elongations of Venus

Eastern		Western	
1996	April 1	1996	August 20
1997	November 6	1998	March 27
1999	June 11	1999	October 30

34 Observe Venus once a week for a month and make a sketch of the planet's appearance on each occasion. From your sketches, can you determine if Venus is approaching us or moving away from us?

35 Consult such magazines as *Sky & Telescope* or *Astronomy* to determine Mars' location among the constellations. If Mars is suitably placed for observation, arrange to view the planet through a telescope. Draw a picture of what you see. What magnifying power seems to give you the best image? Can you distinguish any surface features? Can you see a polar cap or dark markings? If not, can you offer an explanation for Mars' bland appearance?

7 ▶ *The Outer Planets*

Saturn

Earth

Neptune

Uranus

Pluto

Jupiter

IN THIS CHAPTER

You will discover that Jupiter is an active, vibrant, multicolored world, more massive than all the other planets combined. You will explore Jupiter and its diverse system of moons and then examine Saturn, with its spectacular system of thin, flat rings and numerous moons. More distant planets are very different from Jupiter and Saturn. Turning from Uranus and Neptune, which are similar to each other, you will find that tiny Pluto and its moon Charon are more like a bound pair of planets than a planet–moon system.

The Outer Planets and Earth This montage shows Jupiter, Saturn, Uranus, Neptune, Pluto, and Earth reproduced to the same scale. All the outer planets have very different sizes and compositions from our world. Giants Jupiter and Saturn are composed primarily of hydrogen and helium. Intermediate-sized Uranus and Neptune also have large quantities of water. Tiny Pluto is composed of roughly equal amounts of ice and rock. Note that the cloud features on Saturn are much less distinct than those on Jupiter, while Uranus has an exceptionally bland appearance. Recent discoveries of objects orbiting the Sun beyond Pluto suggest that this little planet may be the archetype of another whole class of planets. The large planets were photographed by the Voyager spacecraft that flew past them in the late 1970s and 1980s. Pluto was imaged by the Hubble space telescope.

WHAT DO YOU THINK?

1 Is Jupiter a "failed star" or almost a star?

2 What is Jupiter's Great Red Spot?

3 Does Jupiter have continents and oceans?

4 Are the rings of Saturn solid ribbons?

. . . .

IT IS EASY TO IMAGINE a trip to Mars, exploring its vast canyons and icy polar regions. Even Venus and Mercury are best understood by carefully comparing them to the Earth and its Moon. The outer five planets are an entirely different matter. Jupiter, Saturn, Uranus, and Neptune are so much larger, are rotating so much faster, and have such different chemical compositions that we shall seem indeed to have left the Earth far behind. Nothing on Earth suggests the swirling red and brown clouds of Jupiter, the everchanging ring system of Saturn, and the golden surface of the moon Io. In its own way, Pluto is also very different from Earth. It is so small that the tidal force of its moon, Charon, was able to set Pluto into synchronous rotation, just as the Earth has set our Moon. We begin with Jupiter, an alien world of unsurpassed splendor.

Jupiter

Even from Earth, Jupiter's appearance sets it apart from the terrestrial worlds (Figure 7-1). As seen from the Voyager 1 and 2 spacecraft, which visited Jupiter in 1979, and from the Hubble Space Telescope today, Jupiter's multicolored bands create a world of breathtaking beauty (Figure 7-2). Table 7-1 lists Jupiter's essential properties.

7-1 Jupiter's rotation helps create colorful, worldwide weather patterns

Jupiter is the largest planet in the solar system. Over 1300 Earths could be packed into its volume. Using the orbital periods of its moons and Kepler's laws, astronomers have determined that Jupiter is 318 times more massive than Earth. This means that Jupiter contains more than twice as much mass as all the other planets combined. Nevertheless, it would have to be 75 times 1

Figure 7-1 **Jupiter as Seen from Earth** Many features of Jupiter's clouds are visible from Earth, including the parallel regions of different colors (belts and zones) and several light and dark spots. The Great Red Spot appears toward the bottom. Nevertheless, resolution is poor compared to pictures taken by the Voyager 1 and 2 spacecraft and the Hubble Space Telescope.

. . . .

TABLE 7-1	
Jupiter's Vital Statistics	
Statistic	**Measurement**
Mass	1.90×10^{27} kg (318 $M_{\oplus}$)
Equatorial radius	71,490 km (11.2 $R_{\oplus}$)
Average density	1330 kg/m^3
Orbital eccentricity	0.048
Sidereal period of revolution (year)	11.86 Earth years
Average distance from Sun	7.78×10^8 km (5.20 AU)
Equatorial rotation period	9 h 50 min 28 s
Internal rotation period	9 h 55 min 30 s
Albedo (average)	0.51

Figure 7-2 **Jupiter as Seen from a Spacecraft** This view was sent back from Voyager 1 in 1979. Features as small as 600 km across can be seen in the turbulent cloudtops of this giant planet. Complex cloud motions surround the Great Red Spot.

Figure 7-3 **Close-up of Jupiter's Atmosphere** The dynamic winds and complex chemical composition of Jupiter's atmosphere create this beautiful and complex pattern.

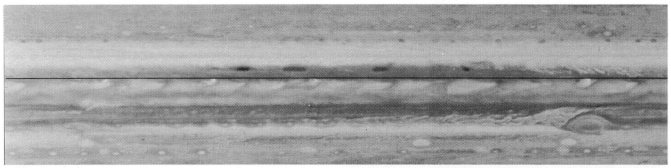

a Voyager 1 view

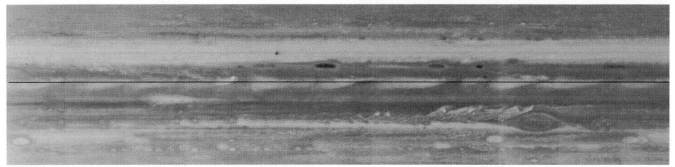

b Voyager 2 view

Figure 7-4 **Jupiter's Northern and Southern Hemispheres** Computer processing shows all of Jupiter's atmosphere from (a) Voyager 1 and (b) Voyager 2. A computer was used to unwrap the planet's surface. Notice the dark ovals in the northern hemisphere and the white ovals in the southern hemisphere. In both views, the banded structure of belts and zones is absent near the poles. Notice that the Great Red Spot moved westward while the white ovals moved eastward during the four months between the two flybys.

more massive still before it could generate its own energy like the Sun and therefore be a star.

What we see of Jupiter are the tops of clouds that permanently cover it (see Figure 7-2). Because the planet rotates once every ten hours—the fastest of any planet—its clouds are in perpetual motion and confined to narrow bands. In contrast, winds on Earth wander over vast ranges of latitude (see Figure 5-1).

Even through a small telescope, you can see on Jupiter dark, reddish bands called **belts** alternating with light-colored bands called **zones.** These relatively permanent features provide a backdrop for ovals and turbulent swirls, which are storms similar to hurricanes on Earth, that last from hours to centuries (Figure 7-3). Computers show us how the cloud features on Jupiter would look if the planet were unwrapped like a piece of paper (Figure 7-4). You can also see ripples, plumes, and light-colored wisps.

One rust-colored oval, called the **Great Red Spot** (Figure 7-5), is so large that it can be seen through a small telescope (at present it is about 26,000 km long by

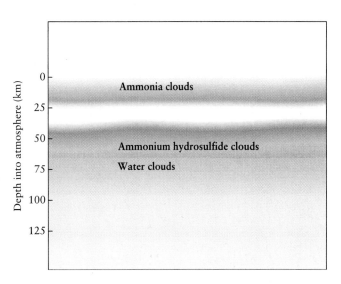

Figure 7-6 **Jupiter's Cloud Layers** This graph depicts Jupiter's three major cloud layers, including the colors that predominate at various depths.

14,000 km wide). Two Earths could fit side by side inside it. Heat welling upward from inside Jupiter has maintained this storm for three centuries. How fortunate we are that Earth's atmosphere sustains storms for only a few weeks.

Jean Domenique Cassini, a French-Italian astronomer, first observed the Great Red Spot in 1665. Twenty-five years later, he noticed that the speeds of Jupiter's clouds vary with latitude, an effect called **differential rotation.** Near the poles, the rotation period of Jupiter's atmosphere is more than five minutes longer than at the equator.

Spectra give more detail of Jupiter's atmosphere. Over 86% of its mass is hydrogen, and 13% is helium. The remainder consists of traces of simple compounds such as methane (CH_4), ammonia (NH_3), and water vapor (H_2O). Of Jupiter's three major cloud layers (Figure 7-6), the uppermost layer is composed of crystals of frozen ammonia. These crystals and frozen water in Jupiter's clouds are white, so what chemicals create the subtle tones of brown, red, and orange? Some scientists think that sulfur, which can assume many different colors, depending on its temperature, plays an important role; others think that phosphorus might be involved, especially in the Great Red Spot.

Astronomers began to determine Jupiter's chemical composition from its average density—only 1330 kg/m^3—which implies that Jupiter itself is composed primarily of lightweight elements such as hydrogen and helium surrounding a relatively small core of

Figure 7-5 **The Great Red Spot** Turbulence surrounding the Great Red Spot is clearly seen in this view. When this picture was taken in 1979, the Great Red Spot measured about 20,000 km long and about 10,000 km wide. It has since grown in size. For comparison, the Earth's diameter is 12,756 km. The prominent white oval south of the Great Red Spot has been observed since 1938.

Figure 7-7 **Why Jupiter Has Belts and Zones** The bright zones and dark belts surrounding Jupiter are created by a combination of the rising and falling of Jupiter's fluid interior and the planet's rapid rotation. The vertical motion is caused by convective heating deep inside Jupiter. This motion is analogous to the behavior of boiling water.

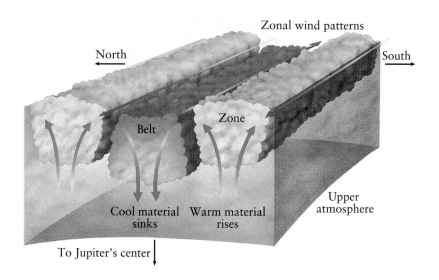

metal. (Recall the discussion of Jupiter's formation in Foundations II.) Because the bulk of Jupiter is hydrogen, we cannot easily distinguish the planet's atmosphere from its "surface." However, at a distance of 150 km below the cloudtops, the pressure (about 10 atm) is enough to liquefy hydrogen. Jupiter's mantle and surface are therefore entirely liquid. Above the planet is the thick atmosphere we observe.

In introducing the solar system, we also noted that a young planet heats up as it coalesces. Radioactive elements continue to heat its interior. On Earth, this heat leaks out to the surface in volcanoes and other vents. Jupiter loses heat everywhere on its surface because, unlike the Earth, it has no landmasses to block the heat loss.

As the heat from within Jupiter warms its liquid mantle, blobs of hydrogen and helium move upward. When these blobs reach the cloudtops, they give off their heat and descend back into the interior. (The same convection drives the motion of the Earth's mantle and its tectonic plates.) Jupiter's rapid, differential rotation then draws the convecting gases into bands around the planet. *The light zones are thus regions of hotter, rising gas, while the dark belts are regions of cooler, descending gas* (Figure 7-7).

7-2 Jupiter's interior has three distinct levels

Because Jupiter is mostly hydrogen and helium, its density is less than one-quarter that of Earth. Yet because of the planet's enormous bulk, the pressure in the liquid hydrogen at a depth of 20,000 km is three million atmospheres, and the temperature there is 11,000 K. Below this depth, the pressure and temperature are high enough to transform hydrogen into **liquid metallic hydrogen.** The electrons in this form of hydrogen are free to roam about, just like electrons in household copper wiring. Electric currents generate a planetary magnetic field that is 19,000 times stronger than the Earth's field.

Despite its preponderance of hydrogen and helium, Jupiter formed around a rocky protoplanet. This core is only 4% of Jupiter's mass, yet it amounts to nearly 13 times the mass of the entire Earth. The tremendous crushing weight of the bulk of Jupiter above the core—equal to the mass of 305 Earths—compresses the terrestrial core down to a sphere only 20,000 km in diameter (Earth's diameter is 12,756 km). The pressure at Jupiter's center is 80 million atmospheres, and the temperature there is about 25,000 K, hotter than the surface of the Sun.

Jupiter's powerful magnetic field is large enough to envelop the orbits of many of its moons. For Earth-based astronomers, the only evidence of this magnetosphere is a faint hiss of radio static, which varies cyclically over a period of nearly 10 hours. However, the two Pioneer and two Voyager spacecraft that journeyed past Jupiter in the 1970s revealed the awesome dimensions of Jupiter's magnetosphere. The volume it engulfs is nearly 30 million kilometers across. In other words, if you could see Jupiter's magnetosphere from the Earth, it would cover an area in the sky 16 times larger than the full Moon.

High-energy particles are trapped in Jupiter's magnetic fields, as in the Van Allen belts around Earth. These

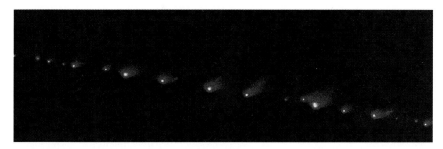

Figure 7-8 **Pieces of Comet Shoemaker-Levy 9 Approaching Jupiter** Comet Shoemaker-Levy 9 was torn apart by Jupiter's gravitational force on July 7, 1992, fracturing into at least 21 pieces. This debris later returned as shown here in May 1994, and struck Jupiter between July 16 and July 22, 1994.

particles can be detected on Earth from radio signals, which change as Jupiter rotates. By timing the radio signals, astronomers determined that the planet rotates once in 9 hours, 55 minutes, and 30 seconds. This matches the circulation of its atmosphere at the poles.

7-3 Pieces of Comet Shoemaker-Levy 9 struck Jupiter

On July 7, 1992, a comet passed so close to Jupiter that the planet's gravitational tidal force ripped it into at least 21 pieces. The debris from this comet was first observed in March 1993 by comet hunters Gene and Carolyn Shoemaker and David Levy. (Because it was the ninth comet they had found together, it was named Shoemaker-Levy 9 in their honor.) Calculations of the comet's orbit showed that the pieces would return to strike Jupiter between July 16 and July 22, 1994 (Figure 7-8).

It is uncommon today for pieces of space debris as large as several kilometers in diameter to collide with planets. Recall from Foundations II that such impacts were common in the early ages of the solar system. However, most of the large bodies destined to strike planets did so billions of years ago. Therefore the discovery that Shoemaker-Levy 9 would hit Jupiter created great excitement in the astronomical community. Seeing how a planet and a comet respond to such an impact would allow astronomers to deduce information about the planet's atmosphere and interior and also about the striking body's properties.

The impacts occurred as predicted, with most of Earth's major telescopes—as well as those on several spacecraft—watching closely (Figures 7-9 and 7-10). At least 20 fragments from Shoemaker-Levy 9 struck Jupiter, of which 15 created detectable impact sites. The impacts resulted in fireballs some 10 km in diameter with temperatures of 7500 K, which is hotter than the surface of the Sun. Impacts were followed by crescent-shaped ejecta containing a variety of chemical compounds, rip-

ples or waves spreading out through Jupiter's clouds, splotches in Jupiter's visible cloud layer that lasted for months, and aurorae.

The observations suggest that the pieces of comet didn't penetrate very far into Jupiter; this fact, in turn, suggests that the pieces were not much larger than a kilometer in diameter. The ejecta from each impact included a dark plume that rose high into Jupiter's atmosphere (see Figure 7-9). The darkness may have been due to carbon from the comet. By watching how rapidly the debris moved, astronomers were able to determine that the winds some 250 km above the cloud layer move at speeds of 3600 km/h. However, the observations led to more questions than answers:

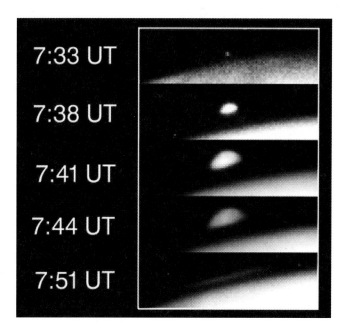

Figure 7-9 **Impact of Comet Shoemaker-Levy 9 Fragment on Jupiter** This sequence of images, sent back from the Galileo spacecraft en route to Jupiter, shows a plume of ejected matter rising and settling back from the July 18, 1994, impact. Fragment G (the seventh to strike and probably the biggest) of Shoemaker-Levy 9 is tracked in "universal time" (UT).

Figure 7-10 **Three Impact Sites of Comet Shoemaker-Levy 9 on Jupiter** Shown here are visible (left) and ultraviolet (right) images of Jupiter taken by the Hubble Space Telescope after three pieces of Comet Shoemaker-Levy 9 struck the planet. The astronomers had expected white remnants (the color of condensing ammonia or water vapor); the darkness of the impact sites may have come from carbon compounds in the comet debris. Note the aurorae in the ultraviolet image. Aurorae and lightning are common on Jupiter.

Where are the water and the carbon- and oxygen-rich compounds astronomers expected to see as a result of the impact, but didn't?

Where did the sulfur compounds, seen on Jupiter for the first time during the impacts, come from?

Why didn't Jupiter vibrate after the impacts as predicted?

Why did the comet break apart in the first place? Was it originally a collection of smaller bodies loosely bound together rather than one object?

What was the comet's chemical composition?

Astronomers are actively analyzing the impact data, hoping to develop more complete models of comets and of Jupiter.

Jupiter's Moons and Rings

Jupiter hosts at least 16 moons, a typical number for the four large planets. Galileo was the first person to observe the four largest moons, in 1610, seen through his meager telescope as pinpoints of light. He called them the "Medicean stars" to attract the attention of a wealthy Florentine patron of the arts and sciences. To Galileo, they provided evidence supporting the then-controversial Copernican cosmology; at that time, many Western theologians asserted that all cosmic bodies orbited the Earth. The fact that the Medicean stars orbited Jupiter therefore raised grave concerns in some circles.

To the modern astronomer, these moons are four extraordinary worlds, different both from the rocky terrestrial planets and from hydrogen-rich Jupiter. Now called the **Galilean satellites,** they are named after the mythical lovers and companions of the Greek god Zeus. From the closest moon outward they are Io, Europa, Ganymede, and Callisto.

The best data and pictures of these four worlds came from the Voyager 1 and Voyager 2 flybys. The two inner Galilean satellites, Io and Europa, are approximately the same size as our Moon. The two outer satellites, Ganymede and Callisto, are comparable in size to Mercury (Figure 7-11). Table 7-2 presents comparative information about these six bodies.

7-4 Io's surface is sculpted by volcanic activity

Sulphury Io is among the most exotic moons in our solar system (Figure 7-12). With a density of 3570 kg/m^3, it is neither terrestrial nor "Jovian" in chemistry. It zooms through its orbit of Jupiter once every 1.8 days. Like our Moon, Io is in synchronous rotation with its planet.

Images from Voyager 1 revealed eight giant erupting volcanoes on Io. These volcanoes are named after gods and goddesses associated with fire in Greek, Norse, Hawaiian, and other mythologies (Figure 7-13). The Voyager cameras also revealed numerous black dots on Io, which apparently are the volcanic vents through which the eruptions occur. Lava flows radiate from many

| Io | Europa | Ganymede | Callisto |

Figure 7-11 **The Galilean Satellites** The four Galilean satellites are shown here to the same scale. Io and Europa have diameters and densities comparable to our Moon and are composed primarily of rocky material. Ganymede and Callisto are roughly as big as Mercury, but their low average densities indicate that each is covered with a thick layer of water and ice.

TABLE 7-2

Vital Statistics of the Galilean Moons, Mercury, and the Moon

	Mean distance from Jupiter (km)	Sidereal period (d)	Diameter (km)	Mass (kg)	Mass (Moon = 1)	Mean density (kg/m³)
Io	421,600	1.77	3630	8.94×10^{22}	1.22	3570
Europa	670,900	3.55	3138	4.80×10^{22}	0.65	2970
Ganymede	1,070,000	7.16	5262	1.48×10^{23}	2.01	1940
Callisto	1,883,000	16.69	4800	1.08×10^{23}	1.47	1860
Mercury	——	——	4878	3.30×10^{23}	4.49	5430
Moon	——	——	3476	7.35×10^{22}	1.00	3340

Figure 7-12 **Io** This close-up view of Io was taken by Voyager 1 in 1979. Notice the extraordinary range of colors from white, yellow, and orange to black. Scientists believe that these brilliant colors result from surface deposits of sulfur ejected from Io's numerous volcanoes.

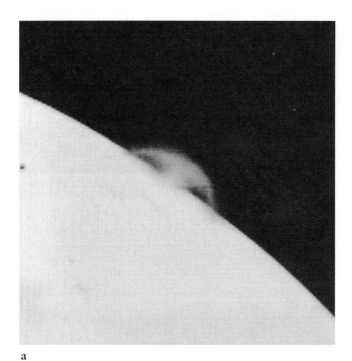

a

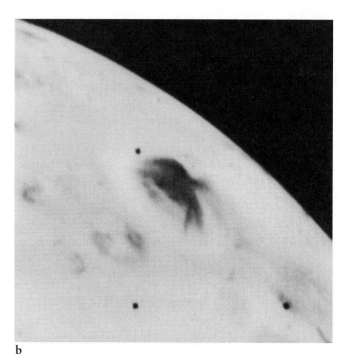

b

Figure 7-13 **Prometheus on Io** These two views from Voyager 1, taken two hours apart, show details of the plume of the volcano called Prometheus rising 100 km above Io's surface. (a) The plume's characteristic umbrella shape is seen silhouetted against the blackness of space. (b) When viewed against the light background of Io's surface, jets of material give the plume a spiderlike appearance.

of these locations, which are typically 10 to 50 km in diameter and cover 5% of Io's surface.

The existence of active volcanoes on Io had been predicted from the forces that the moon is subjected to. As it orbits Jupiter, Io repeatedly passes below one or another of the other Galilean satellites. These moons pull Io farther from Jupiter, changing the tidal forces acting on it. As the distance between Io and Jupiter varies, the resulting tidal stresses alternately squeeze and flex the moon. In turn, this constant tidal stressing heats Io's interior through friction, possibly generating as much energy inside Io as the detonation of 2400 tons of TNT every second. Molten rock eventually makes its way to the satellite's surface. Volcanoes and vents on Io eject roughly one trillion tons of matter each year.

Instruments on board Voyager 1 detected sulfur and sulfur dioxide in the material erupting from Io's volcanoes. Sulfur is normally bright yellow. If heated and suddenly cooled, however, it can assume a range of colors, from orange and red to black, which probably accounts for Io's brilliant colors (see Figure 7-12). Sulfur dioxide (SO_2) is an acrid gas commonly discharged from volcanic vents here on Earth. When eruptions on Io release

this gas into the cold vacuum of space, the gas crystallizes into white flakes. It is likely that the whitish deposits on Io are frozen sulfur dioxide.

7-5 Ice has smoothed the surface of Europa

Voyager 1 did not get near Europa, but Voyager 2 got close enough to capture an excellent view of this intriguing world (Figure 7-14). Europa is a smooth-surfaced body with no mountains and very few craters. It orbits Jupiter every $3\frac{1}{2}$ days and, like Io, is in synchronous rotation. Spectacular streaks and cracks crisscross its surface.

A layer of ice would explain both the smoothness of Europa's surface and the fact that its average density of 2970 kg/m^3 is slightly less than Io's. Some astronomers speculate that Europa has a 100-km-deep layer of slush-covered ice that hides mountain ranges and other topographic features.

Tidal flexing of Europa caused the cracks that cover its surface. Some of the darkest streaks follow paths

Figure 7-14 **Europa** Europa's ice surface is covered by numerous streaks and cracks that give the satellite a fractured appearance. The streaks are typically 20 to 40 km wide. This picture, taken by Voyager 2, reveals surface features as small as 5 km across.

Figure 7-15 **Ganymede** This view from Voyager 2 of Ganymede shows the hemisphere that always faces away from Jupiter. The surface is dominated by a huge, dark, circular region called Galileo Regio, which is the largest remnant of Ganymede's ancient crust.

along which the tidal stresses are strongest. Water gushed up and froze in these stress-induced cracks leaving the streaks we see in the Voyager photographs. Tidal flexing might also supply enough energy to churn the satellite's icy coating. This movement would explain why only a few small impact craters have survived, though Europa's surface is much older than Io's.

In 1995 astronomers discovered an extremely thin atmosphere containing molecular oxygen surrounding Europa. The density of this gas is about 10^{-11} times lower than the air we breathe. The oxygen may come from water molecules broken up on the moon's surface by ultraviolet radiation from the Sun.

7-6 Ganymede is larger than Mercury

Ganymede is the largest satellite in the solar system (Figure 7-15). Its diameter is larger than Mercury's, although its density of 1940 kg/m^3 is much less than Mercury's. Ganymede orbits Jupiter in synchronous rotation once every 7.2 days. Like Europa, Ganymede is believed to have a thick mantle of ice.

Like our Moon, Ganymede has two very different kinds of terrain. Dark, polygon-shaped regions are the oldest surface features, as judged by their numerous craters. Light-colored, heavily grooved terrain is found

between the dark, angular islands. These lighter regions, which are covered with many grooves (Figure 7-16), are much less cratered and therefore younger. Ganymede's

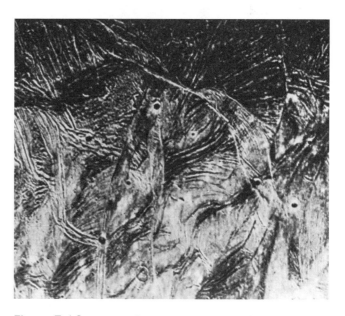

Figure 7-16 **Grooved Terrain on Ganymede** This photograph, taken by Voyager 1, shows an area roughly as large as Pennsylvania. The smallest visible features are about 3 km across. Numerous parallel mountain ridges are spaced 10 to 15 km apart and have heights up to 1 km.

grooved terrain consists of parallel mountain ridges up to 1 km high and spaced 10 to 15 km apart. These features suggest that the process of plate tectonics may have dominated Ganymede's early history. But unlike Europa, where tectonic activity still occurs today, tectonics on Ganymede bogged down three billion years ago as the satellite's crust froze solid.

Unlike most liquids, which shrink upon solidifying, water expands when it freezes. Seeping up through cracks in Ganymede's original crust, water thus forced apart fragments of that crust. This process produced jagged, dark islands of old crust separated by bands of younger, light-colored, heavily grooved ice.

7-7 Callisto wears the scars of a huge asteroid impact

Callisto is Jupiter's outermost Galilean moon. It orbits Jupiter in 16.7 days, and like the other Galilean moons, Callisto's rotation is synchronous. Callisto is 91% as big and 96% as dense as Ganymede. The two moons also share similar surface features. Numerous impact craters are scattered over Callisto's dark, ancient, icy crust (Figure 7-17). There is one obvious difference, however: Unlike Ganymede, Callisto has no younger, grooved terrain. The absence of grooved terrain suggests that tectonic activity never began there: The satellite simply froze too rapidly. It is bitterly cold on Callisto: Voyager instruments measured a noontime temperature of 155 K (−180°F), and the nighttime temperature plunges to 80 K (−315°F).

Callisto carries the cold, hard evidence of what happens when one astronomical body strikes another. Voyager 1 photographed the huge impact basin, Valhalla, on Callisto (see Figure 7-17). An asteroid-sized object produced Valhalla Basin, which is located on Callisto's Jupiter-facing hemisphere. Like throwing a rock into a calm lake, ripples ran out from the impact site along Callisto's surface, cracking the surface and freezing into place for eternity. The largest remnant rings surrounding the impact crater have diameters of 3000 km.

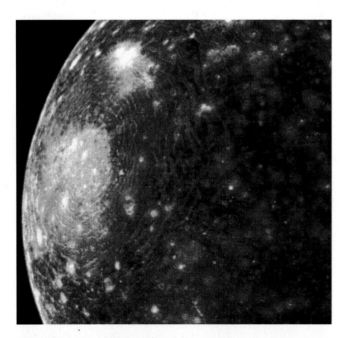

Figure 7-17 **Callisto** Callisto, the outermost Galilean satellite, is almost exactly the same size as Mercury. Numerous craters pockmark Callisto's icy surface. Note the series of faint, concentric rings that cover the left third of the image. These rings outline a huge impact basin called Valhalla that dominates the Jupiter-facing hemisphere of this frozen, geologically inactive world.

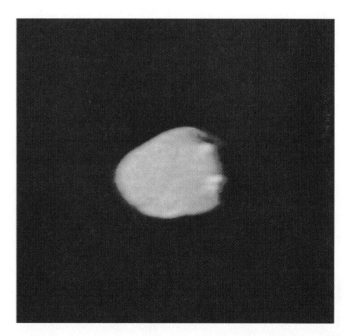

Figure 7-18 **Irregularly Shaped Amalthea** Amalthea is actually more typical of Jupiter's moons than are the Galilean satellites. One of the four innermost satellites, Amalthea measures only 270 km along its longest dimension. This photograph, taken by Voyager 1, has a resolution of 8 km. Amalthea was discovered in 1892 by the American astronomer E. E. Barnard at the Lick Observatory.

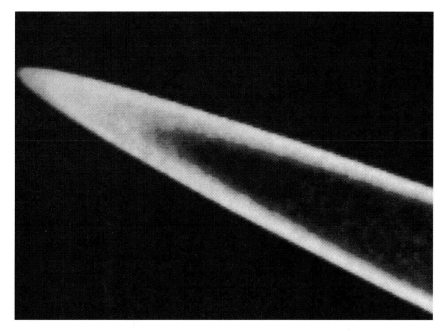

Figure 7-19 **Jupiter's Ring** A portion of Jupiter's faint ring is seen in this photograph from Voyager 2. The ring, which is closer to Jupiter than any of the planet's satellites, is probably composed of tiny rock fragments. The brightest portion of the ring is about 6000 km wide. The outer edge of the ring is sharply defined, but the inner edge is somewhat fuzzy. A tenuous sheet of material extends from the ring's inner edge all the way down to the planet's cloudtops.

7-8 Other debris orbits Jupiter as smaller moons and ringlets

Besides the four Galilean moons, Jupiter has at least 12 other moons and a set of tenuous ringlets. The non-Galilean moons are all irregular in shape and smaller than 150 km in diameter (Figure 7-18). Four of these moons are inside Io's orbit; the other eight are outside. The outer ones appear to be captured asteroids, while the inner ones are probably smaller pieces broken off a larger body.

Voyager's cameras located three ringlets around Jupiter, the brightest of which is seen in Figure 7-19. They consist of very fine dust particles that should be continuously kicked out of orbit by radiation from Jupiter and the Sun. Therefore, these rings are probably being replenished by material from Io.

7-9 The Galileo spacecraft is visiting Jupiter and its moons

The spacecraft Galileo was sent to Jupiter to explore the Jovian system in detail. It arrived at Jupiter in late 1995, when it released a probe into the planet's atmosphere (Figure 7-20). Armed with a heat shield and a parachute, the probe descended through Jupiter's clouds and transmitted back to Earth valuable data about the giant planet's atmosphere. As it descended through the cloud layers, Galileo's probe encountered greater and greater pressure, which eventually crushed it.

After releasing the probe into Jupiter's atmosphere, the spacecraft has continued its journey by visiting Io. Galileo will continue its long-term surveillance of the

Figure 7-20 **Galileo's Atmospheric Probe** Galileo released an atmospheric probe to make direct measurements and chemical analyses as it descended through the Jovian cloud cover.

remaining Galilean satellites, using their gravity to redirect its orbit. The resulting flybys will give scientists the opportunity to study three icy worlds glimpsed only briefly by the Voyagers.

Launched in 1989, Galileo immediately ran into trouble. Its high-gain antenna, which allows it to send large quantities of data rapidly to Earth, failed to deploy correctly. As a result, only the backup, low-gain antenna can send data. Unfortunately, each image that this latter antenna sends takes days to get to Earth. In orbit in the Jovian system, Galileo is storing huge quantities of data on its tape recorders, which will wait for the painfully slow, laborious transmission back to Earth.

Saturn

Saturn, with its ethereal rings, presents the most spectacular visual image of all the planets (Figure 7-21). Giant Saturn has 95 times more mass than the Earth, making it second in mass and size to Jupiter (Table 7-3). Like Jupiter, Saturn has a thick, active atmosphere composed predominantly of hydrogen.

7-10 Saturn's surface and interior are similar to those of Jupiter

Partly obscured by the thick, hazy atmosphere above them, Saturn's clouds lack the colorful contrast visible on Jupiter. Nevertheless, photographs do show faint stripes

	TABLE 7-3	
Saturn's Vital Statistics		
Statistic		**Measurement**
Mass		5.69×10^{26} kg (95.2 $M_\oplus$)
Equatorial radius		60,270 km (9.45 $R_\oplus$)
Average density		690 kg/m^3
Orbital eccentricity		0.056
Sidereal period of revolution (year)		29.5 Earth years
Average distance from Sun		1.43×10^9 km (9.53 AU)
Equatorial rotation period		10 h 13 min 59 s
Internal rotation period		10 h 39 min 25 s
Albedo (average)		0.50

in Saturn's atmosphere similar to Jupiter's belts and zones (Figure 7-22). Changing features there show that Saturn's atmosphere, too, has differential rotation—

Figure 7-21 **Saturn from Voyager 2** Voyager 2 sent back this image when the spacecraft was 34 million kilometers from Saturn. Note that you can see the planet through the gaps in the rings. This creates the illusion that the gaps are empty.

Figure 7-22 **Belts and Zones on Saturn** Voyager 1 took this view of Saturn's cloudtops at a distance of 1.8 million kilometers. Note that there is substantially less contrast between belts and zones on Saturn than there is on Jupiter.

Figure 7-23 **A New Storm on Saturn** An arrowhead-shaped storm was discovered on Saturn in July 1994. Located near Saturn's equator, it stretches 12,700 km across, making its diameter roughly the same as that of the Earth.

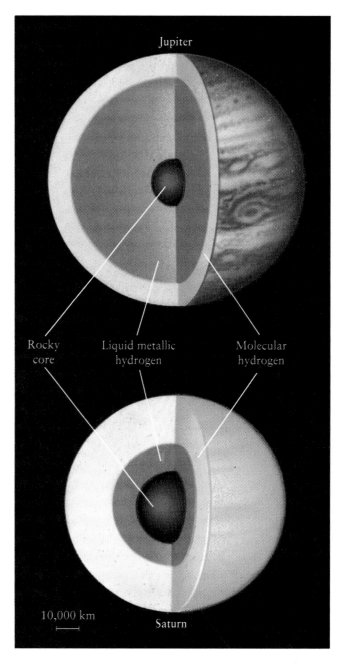

Figure 7-24 **Cutaway of the Interiors of Jupiter and Saturn** The interiors of both Jupiter and Saturn are believed to have three regions: a terrestrial core, a metallic hydrogen shell, and a normal liquid hydrogen mantle. Their atmospheres are thin layers above the normal hydrogen, which boils upward, creating the belts and zones.

ranging from 10 hours 14 minutes at the equator to 10 hours 40 minutes at high latitudes. Although Saturn lacks a long-lived spot like Jupiter's, it does have storms, including a major new one discovered by the Hubble Space Telescope in November 1994 (Figure 7-23).

Astronomers infer that Saturn's interior structure resembles Jupiter's. A layer of molecular hydrogen surrounds a mantle of liquid metallic hydrogen and a solid, terrestrial core (Figure 7-24).

Because of its smaller mass, Saturn's interior is less compressed than Jupiter's. Consequently, Saturn's rocky core is larger, and the pressure is unable to convert as much hydrogen into a liquid metal. Saturn's rocky core is about 32,000 km in diameter, while its layer of liquid metallic hydrogen is 12,000 km thick (see Figure 7-24). There is a wonderful irony in the ancient alchemist's belief that Saturn is a leaden planet. At 690 kg/m³, Saturn is the least dense body in the entire solar system.

7-11 Saturn's spectacular rings are composed of fragments of ice and ice-coated rock

Even when seen through a small telescope (Figure 7-25), Saturn's magnificent rings are among the most spectacular objects in the nighttime sky. Saturn is so far away, however, that our best Earth-mounted telescopes can reveal only the largest features. In 1675 J. D. Cassini discovered one remarkable feature—a dark division in the rings. This 5000-km-wide gap, called the **Cassini division,** separates the dimmer **A ring** from the brighter **B ring,** which lies closer to the planet. By the mid-1800s, astronomers using improved telescopes detected a faint **C ring** just inside the B ring (see Figure 7-25). Cassini's

Figure 7-25 **Saturn from the Earth** This view is one of the best ever produced of Saturn by an Earth-based observatory. Sixteen original color images taken during the same night were combined to make this photograph. Note the prominent Cassini division in the rings and the faint belts and zones in the Saturnian atmosphere.

division exists because the gravitational force from Saturn's moon Mimas conspires with the gravitational force from the planet to keep the region clear of debris. Whenever matter drifts into this region, these two gravitational forces pull it out again, an effect called a **resonance.**

A second gap is visible in the outer portion of the A ring (see Figure 7-25), named the **Encke division** after the German astronomer Johann Franz Encke who allegedly saw it in 1838. (Many astronomers have argued that Encke's report was erroneous because his telescope was inadequate to resolve such a narrow gap.) The first undisputed observation of the 270-km-wide division was made in the late 1880s, with the newly constructed 36-in. refractor at the Lick Observatory in California. Unlike Cassini's division, Encke's division is kept clear because of a small moon within it temporarily called 1981 S13.

Pictures from the Voyager spacecraft show that the rings are razor thin—less than 2 km thick according to recent estimates. This is quite amazing when you consider that the total ring system has a width (from inner edge to outer edge) of over 89,000 km.

Because Saturn's rings are also very bright, the particles that form them must be highly reflective. Astronomers had long suspected that the rings consist of ice and ice-coated rocks. Data from Earth-based observatories and Voyager spacecraft here confirmed this suspi-

cion. The temperature of the rings ranges from 93 K (−290°F) in the sunshine to less than 73 K (−330°F) in Saturn's shadow. Frozen water is in no danger of melting or evaporating at these temperatures.

In order to determine the size of the particles in Saturn's rings, Voyager scientists measured the brightness of the rings from many angles as the spacecraft flew past the planet. They also measured changes in radio signals received as the spacecraft passed behind the rings. The largest particles in Saturn's rings are roughly 10 m across, although snowball-sized particles about 10 cm in diameter are more abundant than larger pieces. The broad rings consist of hundreds upon hundreds of closely spaced thin bands, or **ringlets,** of particles; myriad, dust-sized particles fill the Cassini and Encke divisions. Astronomers still do not know why Saturn's rings are subdivided into these numerous ringlets (Figure 7-26).

The Voyager cameras also sent back the first high-quality pictures of the **F ring,** a thin set of ringlets just beyond the outer edge of the A ring. Two tiny satellites following orbits on either side of the F ring serve to keep the ring intact (Figure 7-27). The outer satellite of the two orbits Saturn at a slower speed than do the ice particles in the ring. As the ring particles pass near it, they experience a tiny, backward gravitational tug, which slows them down, causing them to fall into orbits a bit closer

Figure 7-26 **Numerous Thin Ringlets Comprising Saturn's Rings** Details of Saturn's rings are visible in this photograph sent back by Voyager 2. It shows that the rings are actually composed of thousands of closely spaced ringlets. The colors are exaggerated by computer processing to show the different ringlets more clearly.

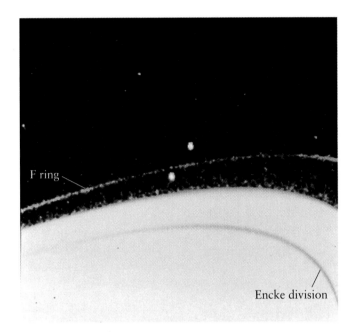

Figure 7-27 **The F Ring and Its Two Shepherds** Two tiny satellites, Prometheus and Pandora, each measuring about 50 km across, orbit Saturn on either side of the F ring. The gravitational effects of these two shepherd satellites confine the particles in the F ring to a band about 100 km wide.

Figure 7-28 **Braided F Ring** This photograph from Voyager 1 shows several strands, each measuring roughly 10 km across, that comprise the F ring. The total width of the F ring is about 100 km.

to Saturn. Meanwhile, the inner satellite orbits the planet faster than the F ring particles. Its gravitational force pulls them forward and nudges the particles into a higher orbit. The combined effect of these two satellites is to focus the icy particles into a well-defined, narrow band about 100 km wide.

Because of their confining influence, these two moons, Prometheus and Pandora, are called **shepherd satellites**. Among the most curious features of the F ring is that the ringlets are sometimes braided and sometimes separate (Figure 7-28).

Saturn's strong magnetic field apparently affects Saturn's rings. The planet's mantle of liquid metallic hydrogen produces a planetwide magnetic field. Saturn's slower rotation and much smaller volume of liquid metallic hydrogen produce a magnetic field only about 3% as strong as that of Jupiter's (but still 570 times stronger than Earth's). Data from spacecraft show that Saturn's magnetosphere contains radiation belts similar to those of Earth. Furthermore, dark **spokes** move around Saturn's rings (Figure 7-29); these are believed to be created by the magnetic field, which lifts charged particles out of the plane in which the rings orbit. Spreading the particles out makes the rings appear darker.

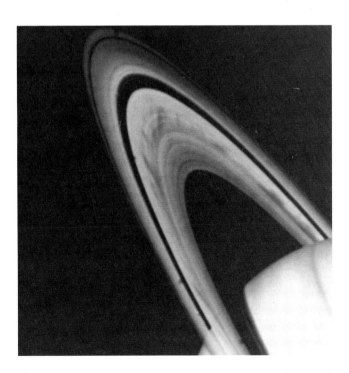

Figure 7-29 **Spokes in Saturn's Rings** Believed to be caused by Saturn's magnetic field temporarily lifting particles out of the ring plane, these dark regions move around the rings like the spokes on a rotating wheel.

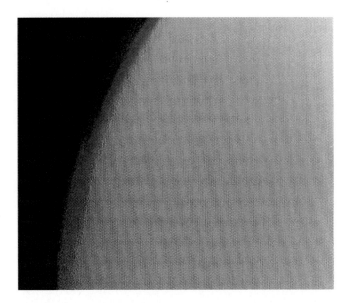

Figure 7-30 **Titan** This view of Titan was taken by Voyager 2. Very few features are visible in the thick, unbroken haze that surrounds this large satellite. The main haze layer is located nearly 300 km above Titan's surface.

7-12 Titan has a thick, opaque atmosphere rich in methane, nitrogen, and hydrocarbons

Saturn has more moons than any other planet. Only 7 of Saturn's 22 known moons are spherical. The rest are oblong, suggesting that they are captured asteroids. Saturn's largest moon, Titan, is second in size only to Ganymede among the moons of the solar system. It is the only moon in the solar system to have an exceptionally dense atmosphere. About ten times more gas lies above each square meter of its surface than lies above each square meter of the Earth.

Christian Huygens discovered Titan in 1655, the same year he proposed that Saturn has rings. By the early 1900s, several scientists had begun to suspect that Titan might have an atmosphere because it is cool enough and massive enough to retain heavy gases.

Because of its atmosphere, Titan was a primary target for the Voyager missions. To everyone's surprise, the Voyagers spent hour after precious hour sending back featureless images (Figure 7-30). Unexpectedly, Titan's thick cloud cover completely blocked any view of its surface. The same dense haze allows little sunlight to penetrate; that moon's surface must be a dark, gloomy place.

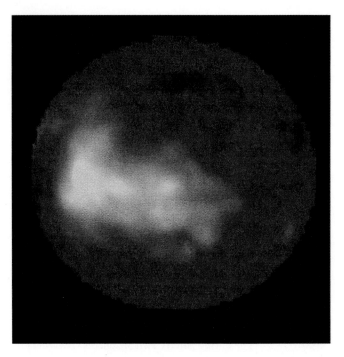

Figure 7-31 **Surface Features on Titan** The Hubble Space Telescope, using an infrared camera, was able to peer through most of Titan's haze (see Figure 7-30). The dark regions in this picture may be oceans of hydrocarbons. If so, they are dark because the hydrocarbons absorb most of the light striking them. The light regions may be vast regions of frozen water and ammonia.

Voyager data did suggest that roughly 90% of Titan's atmosphere is nitrogen. Most of this nitrogen probably came from ammonia (NH_3), which the Sun's ultraviolet radiation breaks down into hydrogen and nitrogen atoms. Because Titan's gravity is too weak to retain hydrogen, it has escaped into space, leaving behind ample nitrogen.

The second most abundant gas on Titan is methane, a major component of natural gas. Sunlight interacting with methane induces chemical reactions that produce a variety of other carbon–hydrogen compounds, or **hydrocarbons**. For example, spacecraft have detected small amounts of ethane (C_2H_6), acetylene (C_2H_2), ethylene (C_2H_4), and propane (C_3H_8) in Titan's atmosphere. Ethane, the most abundant of these compounds, condenses into droplets as it is produced and falls to Titan's surface to form a liquid. Enough ethane may now exist to create rivers, lakes, and even oceans on Titan. Nitrogen combines with these hydrocarbons to produce other compounds. Although one of these compounds, hydrogen cyanide (HCN), is a poison, some of the others are the building blocks of life's organic molecules.

Some molecules can join in long, repeating molecular chains to form substances called **polymers**. Many of the hydrocarbons and carbon–nitrogen compounds in Titan's atmosphere can form such polymers. Droplets of some polymers remain suspended in Titan's atmosphere to form a mist, but the heavier polymer particles settle down onto Titan's surface. As a result, any land that protrudes above Titan's ethane oceans is probably covered with a thick layer of sticky, tarlike goo. The Hubble Space Telescope observed further evidence for oceans on Titan in 1994 (Figure 7-31).

There is little reason to suspect that life exists on Titan, however; its surface temperature of 95 K (–288°F) is prohibitively cold. Nevertheless, a more detailed study of the chemistry of Titan may shed light on the origins of life on Earth.

Uranus

Uranus and its largest moons are so far from the Sun (19.2 AU) that from Earth they appear to be a small cluster of stars. No wonder they seem fixed in the heavens: A Uranian year equals 84 Earth years (Table 7-4). Since its discovery in 1781, Uranus has only gone around the Sun just over $2\frac{1}{2}$ times.

7-13 Uranus is a featureless world

Hopes for seeing detailed surface features on this distant planet rested with Voyager 2, which approached it late in January 1986. Would the planet have distinctive belts and zones? Storm spots? Separate cloud layers? From a visual standpoint, astronomers were disappointed by what they saw. As seen through Voyager 2's eyes, Uranus is the blandest object in the solar system. A uniformly azure blue sphere, it took considerable computer enhancement to discern faint bands of clouds moving parallel to the equator. Eventually smoglike haze was seen over the north pole, but there were no signs of belts and zones or storm spots. The computer-enhanced images did show enough detail to indicate that the atmosphere rotates once every $16\frac{1}{2}$ hours.

Uranus contains $14\frac{1}{2}$ times more mass than the Earth and is 4 times bigger in diameter (Figure 7-32). Its outer layers are composed predominantly of gaseous hydrogen

TABLE 7-4

Uranus' Vital Statistics

Statistic	Measurement
Mass	8.68×10^{25} kg (14.5 M$_\oplus$)
Equatorial radius	25,559 km (4.01 R$_\oplus$)
Average density	1290 kg/m^3
Orbital eccentricity	0.047
Sidereal period of revolution (year)	84.0 Earth years
Average distance from Sun	2.87×10^9 km (19.2 AU)
Equatorial rotation period	16 h 30 min (retrograde)
Interior sidereal rotation period	17 h 14 min (retrograde)
Albedo (average)	0.66

and helium. The temperature in the upper atmosphere of the planet is so low (about 73 K, or –330°F) that the methane and water there condense to form clouds of ice crystals. Because methane freezes at a lower temperature than water, it forms higher clouds over Uranus. Methane efficiently absorbs red light, giving Uranus its blue-green color.

Earth-based observations show that Uranus rotates once every 17 hours 14 minutes on an axis of rotation that lies very nearly in the plane of the planet's orbit. (Consider what life would be like on Earth if its axis of rotation lay on the ecliptic.) With its axis of rotation tilted over by 98°, Uranus is one of only three planets with **retrograde rotation** (Venus and Pluto are the other two). From Voyager photographs planetary scientists have concluded that each of the Uranian moons has probably had at least one shattering impact. A catastrophic collision with an Earth-sized object may also have knocked Uranus on its side, as we see it today.

As Uranus orbits the Sun, its north and south poles alternately point directly toward or directly away from the Sun, producing exaggerated seasons (Figure 7-33). In the summertime near Uranus' north pole, the Sun is almost directly overhead for many Earth years, at which time southern latitudes are subjected to a continuous, frigid winter night. Forty-two Earth years later, the situation is reversed.

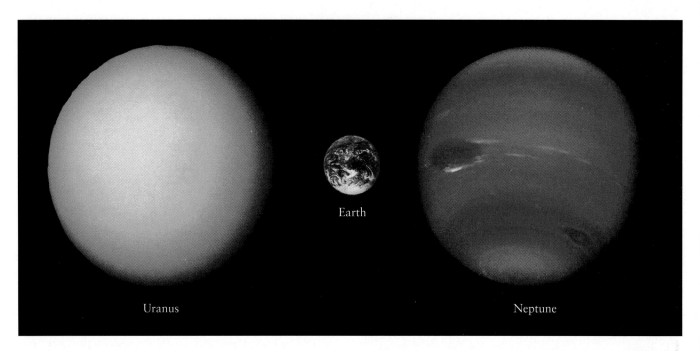

Figure 7-32 Uranus, Earth, and Neptune These images of Uranus, the Earth, and Neptune are to the same scale. Uranus and Neptune are quite similar in mass, size, and chemical composition. Both planets are surrounded by thin, dark rings, quite unlike Saturn's, which are broad and bright. Almost everything we know about the outer two giant planets was gleaned from Voyager 2, which flew past Uranus in 1986 and Neptune in 1989.

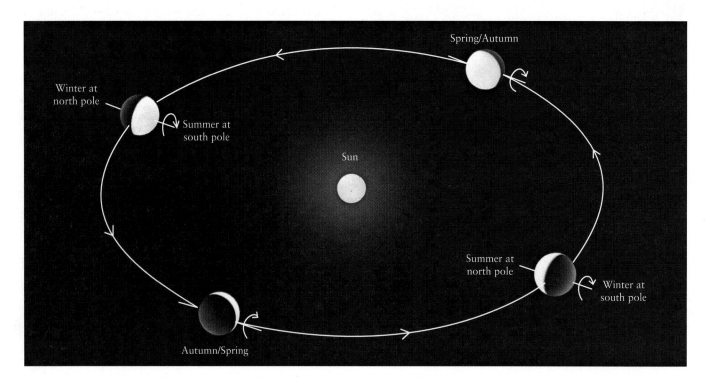

Figure 7-33 Exaggerated Seasons on Uranus Uranus' axis of rotation is tilted so steeply that it lies nearly in the plane of its orbit. Seasonal changes on Uranus are thus greatly exaggerated. For example, during midsummer at Uranus' south pole, the Sun appears nearly overhead for many Earth years, while the planet's northern regions are subjected to a long, continuous winter night. Half an orbit later, the seasons are reversed.

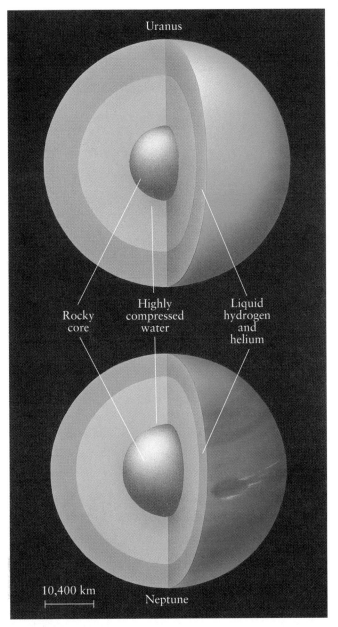

Uranus

Rocky core

Highly compressed water

Liquid hydrogen and helium

10,400 km

Neptune

◀ Figure 7-34 **Cutaway of the Interiors of Uranus and Neptune** The interiors of both Uranus and Neptune are believed to have three regions: a terrestrial core surrounded by a liquid water mantle, which is surrounded in turn by liquid hydrogen and helium. Their hydrogen atmospheres are thin layers at the top of their hydrogen and helium layers.

From its mass and density (1290 kg/m^3), astronomers conclude that Uranus' interior has three layers. The outer 30% of the planet is liquid hydrogen and helium, the next 40% is highly compressed liquid water (with some methane and ammonia), and the inner 30% is a rocky core (Figure 7-34). Indirect evidence for the water layer comes from the apparent deficiency of ammonia on Uranus. This gas dissolves easily in water, which would explain the scarcity of ammonia in the outer layers of the planet.

Voyager 2 passed through the magnetosphere of Uranus, revealing that the planet's magnetic field is 50 times stronger than that of the Earth. That strength is reasonable, but everything else about the magnetic field is extraordinary. It is remarkably tilted—59° from its axis of rotation—and does not even pass through the center of the planet (Figure 7-35).

Because of the large angle between the magnetic field of Uranus and its rotation axis, the magnetosphere of Uranus wobbles considerably as the planet rotates. Such a rapidly changing magnetic field will help us in explaining pulsars, a type of star we will study in Chapter 12.

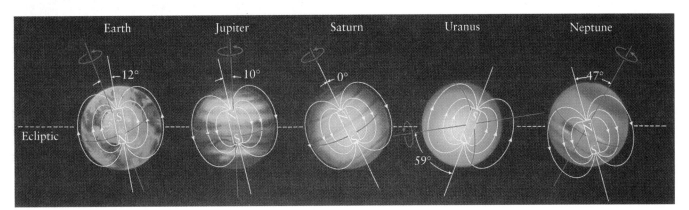

Earth   Jupiter   Saturn   Uranus   Neptune

12°   10°   0°   47°

Ecliptic

59°

Figure 7-35 **The Magnetic Fields of Five Planets** This drawing shows how the magnetic fields of Earth, Jupiter, Saturn, Uranus, and Neptune are tilted relative to their rotation axes. Note that the magnetic fields of Uranus and Neptune are offset from the center of the planets and steeply inclined to their rotation axes. Also note that the magnetic fields of all four outer planets are oriented opposite to that of Earth. A compass on any of these worlds would point southward.

Figure 7-36 **Discovery of the Rings of Uranus** (a) Light from a star is reduced as the rings move in front of it. (b) With sensitive light detectors, astronomers can detect the variation in light intensity. Such dimming led to the discovery of Uranus' rings. When Uranus itself blocks it, of course, the star vanishes completely.

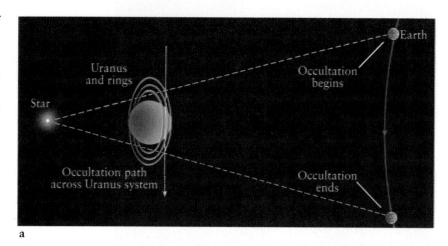

a

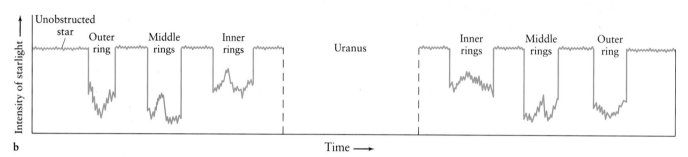

b

7-14 Many satellites and a system of rings revolve around Uranus

Nine of Uranus' thin, dark rings were discovered accidentally in 1977 when Uranus passed in front of a star. The star's light was momentarily blocked, or occulted, by each ring, thereby revealing their existence to astronomers (Figure 7-36). A picture taken while Voyager was in Uranus' shadow revealed many more very thin rings (Figure 7-37).

Uranus' satellites, like its rings, follow the plane of the planet's equator. Five of them, ranging in diameter from 480 to nearly 1600 km, were known before the Voyager mission. However, Voyager's cameras discovered ten additional satellites, each less than 50 km across. Several of these tiny, irregularly shaped moons are shepherd satellites whose gravitational pull confines the particles within the thin rings that circle Uranus.

Of all Uranus' moons, bizarre Miranda, the smallest of Uranus' five main satellites, is the most fascinating. Unusual wrinkled and banded features cover its surface (Figure 7-38). Its highly varied terrain suggests that it was once seriously disturbed. Perhaps a shattering im-

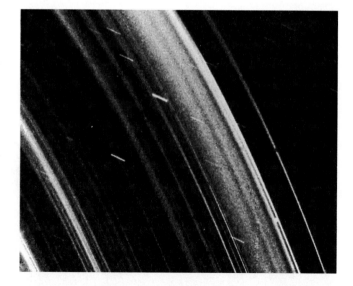

Figure 7-37 **The Rings of Uranus** This view, taken when Voyager 2 was in Uranus' shadow, looks back toward the Sun. Numerous fine dust particles between the main rings gleam in the sunlight. Uranus' rings are much darker than Saturn's, and this long exposure revealed many very thin rings and dust lanes not previously seen. The short streaks are star images blurred because of the spacecraft's motion during the exposure.

Figure 7-38 **Miranda** The patchwork appearance of Miranda suggests that this satellite consists of huge chunks of rock and ice that came back together after an ancient, shattering impact by an asteroid or a neighboring Uranian moon. The curious banded features that cover much of Miranda are parallel valleys and ridges that may have formed as dense, rocky material sank toward the satellite's core.

pact temporarily broke it into several pieces that then recoalesced, or perhaps severe tidal heating, as we saw on Io, stirred its surface.

Miranda's core originally consisted of dense rock, while its outer layers were mostly ice. If a powerful impact did occur, blocks of debris broken off from Miranda drifted back together through mutual gravitational attraction. Recolliding with that moon, they formed a chaotic mix of rock and ice. In this scenario, the landscape we see today on Miranda is the result of huge, dense rocks trying to settle toward the satellite's center, forcing blocks of less dense ice upward toward the surface.

Neptune

Neptune is physically very similar to Uranus (see Figures 7-32 and 7-34). Neptune has 17.1 times the Earth's mass, 3.88 times Earth's diameter, and a density of 1640 kg/m^3 (Table 7-5). Unlike Uranus, however, cloud features can readily be discerned on Neptune. Neptune's whitish, cirruslike clouds consist of methane ice crystals.

The atmosphere of Neptune also has differential rotation, with winds blowing as fast as 1700 km/h—among the fastest in the solar system.

7-15 Neptune was discovered because it had to be there

The discovery of Neptune represents one of those special instances in science of a prediction leading to an expected discovery. In 1781 the great British astronomer William Herschel discovered Uranus. Its position was carefully plotted, and by the 1840s, it was clear that even considering the gravitational effects of all the known bodies in the solar system, Uranus was not following the path predicted by Newton's and Kepler's laws. Either these laws were wrong, or there had to be another, then-undiscovered body in the solar system pulling on Uranus. That planet, Neptune, was located in 1846, within a degree or two of where it had to be to have the observed influence on Uranus.

Over 150 years later, Voyager 2 arrived at Neptune to cap one of NASA's most ambitious and successful space missions. Like Uranus, Neptune presents a tiny, featureless disk that is barely distinguishable from a star when viewed through Earth-based telescopes. Scientists in August 1989 were therefore overjoyed at the detailed, close-up pictures and wealth of data sent back to Earth by Voyager 2 (see Figure 7-32).

At the time Voyager 2 passed it, there was a giant storm in Neptune's atmosphere. Called the **Great Dark Spot**, it was reminiscent of Jupiter's Great Red Spot. The

TABLE 7-5

Neptune's Vital Statistics

Statistic	Measurement
Mass	1.02×10^{26} kg (17.1 M$_\oplus$)
Equatorial radius	24,764 km (3.88 R$_\oplus$)
Average density	1640 kg/m^3
Orbital eccentricity	0.009
Sidereal period of revolution (year)	164.8 Earth years
Average distance from Sun	4.50×10^9 km (30.1 AU)
Internal sidereal rotation period	16 h 7 min
Albedo (average)	0.62

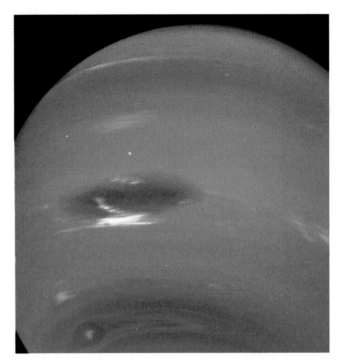

Figure 7-39 **Neptune** This view from Voyager 2 looks down on the southern hemisphere of Neptune. The Great Dark Spot, whose diameter at the time was about the same size as the Earth's diameter, is near the center of this picture. It has since vanished. Note the white, wispy methane clouds. Toward the lower left is a smaller dark spot.

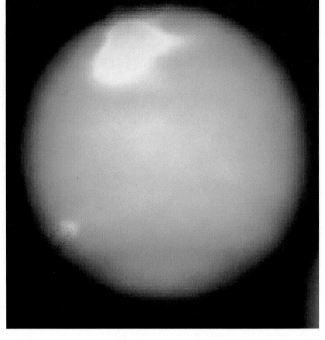

Figure 7-40 **Neptune Changes Its Spots** The Great Dark Spot in Neptune's southern hemisphere vanished in 1994. It was replaced by another dark feature in 1995. This new storm is located in the northern hemisphere.

Great Dark Spot (Figure 7-39) was located at about the same latitude on Neptune and occupied a similar proportion of Neptune's surface as the Great Red Spot does on Jupiter. Although these similarities suggested that similar mechanisms created the spots, the Hubble Space Telescope in 1994 showed that the Great Dark Spot had disappeared. Then, in April 1995, another storm developed in the opposite hemisphere (Figure 7-40).

Neptune's belts and zones are fainter than those on Jupiter. Most prominent is a broad, darkish band at high southern latitudes. Embedded in this band is a smaller dark spot, about a third the size of the Great Dark Spot. Interestingly, white, wispy clouds seem to hover over the smaller dark spot just as they did over the Great Dark Spot.

Neptune's interior is believed to be very similar in composition and structure to that of Uranus, with a rocky core surrounded by ammonia- and methane-laden water (see Figure 7-34). Also as with Uranus, Neptune's magnetic axis, the line connecting its north and south magnetic poles, is tilted from its rotation axis. In this case, the tilt is 47°. Again like Uranus, Neptune's magnetic axis does not pass through the center of the planet. (see Figure 7-35).

7-16 Neptune captured most of its moons

Like Uranus, Neptune is surrounded by a system of thin, dark rings (Figure 7-41). It is so cold at Neptune's distance from the Sun that its ring particles can retain methane ice. Scientists speculate that eons of radiation damage have converted this methane ice into darkish carbon compounds, thus accounting for the low reflectivity of the rings. (The same is true of the rings of Uranus.)

Neptune also has eight known moons. Seven have irregular shapes and highly elliptical orbits which suggests that Neptune captured them. Triton, discovered in 1846, is spherical and was quickly observed to have a

Figure 7-41 **Neptune's Rings** Two main rings are easily seen in this view alongside an overexposed image of Neptune. Careful examination also reveals a faint inner ring. A faint sheet of particles, whose outer edge is located between the two main rings, extends inward toward the planet.

nearly circular, retrograde orbit around Neptune. It is difficult to imagine how a satellite and planet could form together but rotate in opposite directions. Indeed, only a few of the small outer satellites of Jupiter and Saturn have retrograde orbits, and these bodies are probably captured asteroids. Some scientists have therefore suggested that Triton may have been captured three or four billion years ago by Neptune's gravity.

After being captured, Triton most likely began in a highly elliptical orbit. However, with each revolution, the changing distance to Neptune would have stretched and flexed Triton. This activity would have made its or-

bit more circular. It would also have provided enough energy to melt much of the satellite's interior and obliterate Triton's original surface features, including craters. Triton's south polar region is shown in Figure 7-42. Note that very few craters are visible.

Triton does exhibit some surface features seen on other icy worlds, such as long cracks resembling those on Europa and Ganymede. Other features unique to Triton are quite puzzling. For example, near the top of Figure 7-42, you can see a wrinkled terrain that resembles the skin of a cantaloupe. Triton also has a few frozen lakes like the one shown in Figure 7-43. Some scientists have

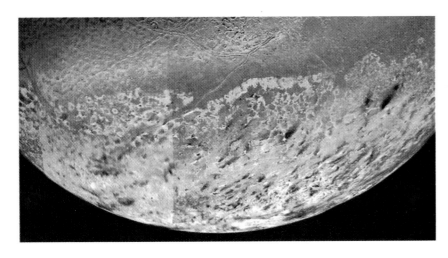

Figure 7-42 **Triton's South Polar Cap** Approximately a dozen high-resolution images were combined to produce this view of Triton's southern hemisphere. The pinkish polar cap is probably made of nitrogen frost. A notable scarcity of craters suggests that Triton's surface was either melted or flooded by icy lava after the era of bombardment that characterized the early history of the solar system.

Figure 7-43 **A Frozen Lake on Triton?** Some scientists believe that this lakelike feature is the caldera of an ice volcano. The flooded basin is about 200 km wide and 400 km long, an area about the size of West Virginia.

speculated that these lakelike features are the calderas of extinct ice volcanoes. A mixture of methane, ammonia, and water, which can have a melting point far below that of pure water, could have formed a kind of cold lava on Triton.

It is unlikely that any lava is flowing on Triton today, however, because the satellite is so very cold. Voyager instruments measured a surface temperature of 36 K (−395°F), making Triton the coldest world we have ever visited. Nevertheless, Voyager cameras did glimpse two towering plumes of gas extending up to 8 km above the satellite's surface. Perhaps these plumes are nitrogen gas escaping through vents or fissures warmed by the feeble summer Sun.

In the same way that our Moon raises tides on Earth, Triton raises tides on Neptune. Whereas the tides on Earth cause our Moon to spiral outward, the tides on Neptune cause Triton (in its retrograde orbit) to spiral inward. Within the next quarter of a billion years, Triton will move close enough to Neptune for the planet to create tides on its moon's solid surface high enough to pull Triton apart. Pieces of Triton will then literally float into space until the entire moon is demolished! By destroying Triton, Neptune will create a new ring system that will be much more substantial than its present one.

Pluto

Clyde W. Tombaugh accidentally discovered Pluto in 1930 (Figure 7-44). He recognized it as a planet because it moved among the background stars from night to night, but he had no idea how strange its orbit is compared to those of the other planets. As we discussed in Foundations II, Pluto's orbit is so elliptical that it is sometimes closer to the Sun than Neptune (as it is today and will be until 1999), although it is usually much farther away. Furthermore, its orbit is tilted with respect to the plane of the ecliptic more than any other planet (see Figure II-7). Table 7-6 lists Pluto's essential data.

Figure 7-44 **Discovery of Pluto** Pluto was discovered in 1930 by searching for a dim, starlike object that slowly moves against the background stars. These two photographs were taken one day apart.

TABLE 7-6

Pluto's Vital Statistics

Statistic	Measurement
Mass	1.30×10^{22} kg (0.002 $M_\oplus$)
Radius	1190 km (0.19 $R_\oplus$)
Average density	1800 kg/m³
Orbital eccentricity	0.249
Sidereal period of revolution (year)	248.6 Earth years
Average distance from Sun	5.91×10^9 km (39.5 AU)
Sidereal rotation period	6.39 Earth days (retrograde)
Albedo (average)	0.50

7-17 Pluto and its moon Charon are about the same size

Pluto was little understood for half a century until astronomers noticed that its image sometimes appears oblong (Figure 7-45). This observation led to the discovery of Pluto's only known moon, Charon, in 1978. From 1985 through 1990, the orbit of Charon was oriented so that Earth-based observers could watch it eclipse Pluto. Astronomers used observations of these eclipses to determine that Pluto is only twice as broad as its satellite: Its diameter is 2380 km, and Charon's is 1190 km.

The Pluto–Charon system is more like a pair of planets in close orbit around each other than a "normal" planet–moon system. The average distance between Charon and Pluto is less than $\frac{1}{20}$ the distance between the Earth and our Moon. Furthermore, Pluto always keeps the same side facing Charon. This is the only case in which a moon and a planet both have synchronous rotation with respect to each other. As seen from the satellite-facing side of Pluto, Charon neither rises nor sets, but instead hovers in the sky, perpetually suspended above the horizon.

The exceptional similarities between Pluto and Charon suggest that this binary system may have formed when Pluto collided with a similar-sized body. Perhaps chunks of matter were stripped from the second body, leaving behind a mass, now called Charon, that was vulnerable to capture by gravity. Alternatively, perhaps gravity captured Charon into orbit during a close encounter between the two worlds.

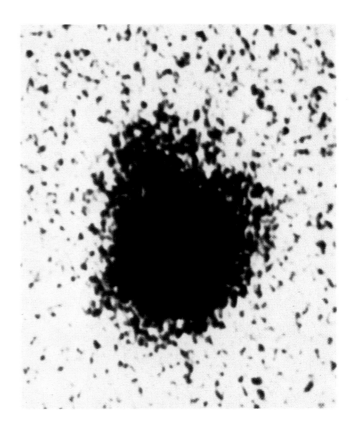

Figure 7-45 **Discovery of Charon** Long ignored as just a defect in the photographic emulsion, the bump on the upper left side of this image of Pluto led astronomer James Christy to discover the moon Charon.

Both of these are unlikely events. For either of these scenarios to be feasible, there must have been many Pluto-like objects in the outer regions of the solar system. One astronomer estimates that there must have been at least a thousand Plutolike bodies in order for a collision or close encounter between two of them to have occurred at least once since the solar system formed 4.6 billion years ago.

The best pictures of Pluto and Charon have come from the Hubble Space Telescope (Figure 7-46). Like Neptune's moon Triton, these worlds are probably composed of nearly equal amounts of rock and ice.

7-18 Our solar system extends beyond Pluto

Ever since Pluto was discovered, astronomers have searched for other planets in our solar system. Because Pluto is different in physical properties from all the other planets, it might be just one example of another whole

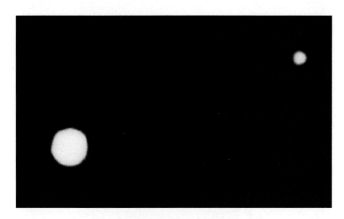

Figure 7-46 **Pluto and Charon** This picture of Pluto with its moon, Charon, was taken by the Hubble Space Telescope. Pluto and Charon are separated by only 19,700 km. The Pluto–Charon system deserves to be called a double planet, because these two objects resemble each other in mass and size more closely than do any other planet–satellite pair in the solar system.

class of small bodies in the outer reaches of the solar system. Neptune's captured moons support the idea of other bodies at the outskirts of the solar system, and astron-

omers have begun to discover them. These distant masses of ice and rock, which we will discuss further in Chapter 8, lie in a region now called the *Kuiper belt*, which is believed to extend out 500 AU from the Sun.

WHAT DID YOU THINK?

1 *Is Jupiter a "failed star" or almost a star?* Jupiter has 75 times too little mass to shine as a star.

2 *What is Jupiter's Great Red Spot?* The Great Red Spot is a long-lived hurricanelike storm.

3 *Does Jupiter have continents and oceans?* Jupiter is completely enveloped in a sea of liquid hydrogen and helium.

4 *Are the rings of Saturn solid ribbons?* Saturn's rings are all composed of thin, closely spaced ringlets consisting of particles of ice and ice-coated rocks. If they were solid ribbons, Saturn's gravitational tidal force would tear them apart.

Key Words

A ring
B ring
belt (on Jupiter)
C ring
Cassini division
differential rotation

Encke division
F ring
Galilean satellite (moon)
Great Dark Spot
Great Red Spot

hydrocarbon
liquid metallic hydrogen
polymer
resonance
retrograde rotation

ringlet
shepherd satellite
spoke
zone (on Jupiter)

Key Ideas

• Jupiter is by far the largest and most massive planet in the solar system.

• Jupiter and Saturn are composed primarily of hydrogen and helium. Both planets have an overall chemical composition very similar to that of the Sun.

• Jupiter and Saturn probably have rocky cores surrounded by a thick layer of liquid metallic hydrogen and an outer layer of ordinary liquid hydrogen.

• The visible features of Jupiter exist in the outermost 100 km of its atmosphere. Saturn has similar features, but they are much fainter. There are three cloud layers in the upper atmospheres of both Jupiter and Saturn. Because Saturn's cloud layers extend to a greater altitude, the colors of the Saturnian atmosphere are somewhat muted.

• The colored ovals visible in the Jovian atmosphere represent gigantic storms, some of which (such as the Great Red Spot) are stable and persist for years.

- Jupiter and Saturn have strong magnetic fields created by electric currents in the metallic hydrogen layer.

- Four large satellites orbit Jupiter. The two inner Galilean moons, Io and Europa, are roughly the same size as our Moon. The two outer Galilean moons, Ganymede and Callisto, are roughly the size of Mercury.

   Io is covered with a colorful layer of sulfur compounds deposited by frequent explosive eruptions from volcanic vents.

   Europa is covered with a smooth layer of frozen water crisscrossed by an intricate pattern of long cracks.

   The heavily cratered surface of Ganymede is composed of frozen water with large polygons of dark, ancient surface separated by regions of heavily grooved, lighter-colored, younger terrain.

   Callisto has a heavily cratered ancient crust of frozen water.

- Saturn is circled by a system of thin, broad rings lying in the plane of the planet's equator.

   Each major ring is composed of a great many narrow ringlets consisting of numerous fragments of ice and ice-coated rock.

   Some of the boundaries between Saturn's rings are produced by shepherd satellites, whose gravitational pull restricts the orbits of the ring fragments.

- Titan, the largest satellite of Saturn, has a dense nitrogen atmosphere that also contains some methane. Sunlight acting on the methane produces hydrocarbons that may cover Triton's surface with a thick sludge.

- Uranus and Neptune are quite similar in appearance, mass, size, and chemical composition.

   Uranus and Neptune each have a rocky core probably surrounded by a dense, watery mantle.

   For both Uranus and Neptune, the axis of the magnetic field is steeply inclined to the axis of rotation.

   Both Uranus and Neptune are surrounded by systems of thin, dark rings.

- Uranus is unique in that its axis of rotation lies nearly in the plane of its orbit, producing greatly exaggerated seasons on the planet.

- Uranus has five moderate-sized satellites, the most bizarre of which is Miranda.

- The largest satellite of Neptune, Triton, is an icy world with a tenuous nitrogen atmosphere.

   Triton moves in a retrograde orbit that suggests it was captured into orbit by Neptune's gravity.

   Triton is spiraling down toward Neptune.

- Pluto, the smallest planet in the solar system, and its satellite, Charon, are icy worlds that may well resemble Triton.

- A population of Plutolike objects may exist far beyond the orbits of Uranus and Neptune.

Review Questions

1   Describe the appearance of Jupiter's atmosphere. Which features are long-lived and which are fleeting?

2   What causes the belts and zones in Jupiter's atmosphere?

3   Why do astronomers believe that Jupiter's core does not create its strong magnetic field?

4   What is liquid metallic hydrogen? Which planets appear to contain this substance?

5   Compare and contrast the surface features of the four Galilean satellites, discussing their geologic activity and their evolution.

6   What is the source of energy that powers Io's volcanoes?

7   Why are numerous impact craters found on Ganymede and Callisto but not on Io or Europa?

8   Describe the structure of Saturn's rings. What are they made of?

9   Why do features in Saturn's atmosphere appear to be much fainter and "washed out" compared to features in Jupiter's atmosphere?

10   Explain how shepherd satellites operate. Is "shepherd satellite" an appropriate term for these objects? Explain.

11   Describe Titan's atmosphere. What effect has sunlight presumably had on Titan's atmosphere?

12   Describe the seasons on Uranus. Why are the Uranian seasons different from those on any other planet?

13   Briefly describe the evidence supporting the idea that Uranus was struck by a large planetlike object several billion years ago.

14   Why are Uranus and Neptune distinctly bluer than Jupiter and Saturn?

15   Compare the ring systems of Saturn and Uranus. Why were Uranus' rings unnoticed until the 1970s?

16   How do the orientations of Uranus' and Neptune's magnetic axes differ from those of the other planets?

17   Suppose you were standing on Pluto. Describe the motions of Charon relative to the horizon. Under what circumstances would you never see Charon?

18   Describe the circumstantial evidence supporting the idea that Pluto is one of a thousand similar icy worlds that once occupied the outer regions of the solar system.

Advanced Questions

19 When this textbook went to press, the Galileo spacecraft had recently arrived at Jupiter. Consult such magazines as *Sky & Telescope* and *Science News* to determine the status of this mission. What data and pictures have the spacecraft sent back?

20 Long before the Voyager flybys, Earth-based astronomers reported that Io appeared brighter than usual for a few hours after emerging from Jupiter's shadow. From what we know about the material ejected from Io's volcanoes, explain this brief brightening of Io.

21 Compare and contrast Valhalla on Callisto with the Caloris Basin on Mercury.

22 Can we infer from naked-eye observations that Saturn is the most distant of the planets visible without a telescope? Explain.

23 As seen by Earth-based observers, the intervals between successive edge-on presentations of Saturn's rings alternate between 13 years 9 months and 15 years 9 months. Why do you think these two intervals are not equal?

24 NASA, the Jet Propulsion Laboratory, and the European Space Agency are currently planning a mission to Saturn that will place the Cassini spacecraft in orbit about the planet. Consult such magazines as *Sky & Telescope* and *Science News* to determine the status of this mission. Has the U.S. Congress approved funds for the mission? What is the estimated launch date? When will Cassini arrive at Saturn?

25 Compare and contrast the internal structures of Jupiter and Saturn with the internal structures of Uranus and Neptune. Can you propose an explanation for the differences between these two pairs of planets?

Discussion Questions

26 Suppose that you were planning a mission to Jupiter employing an airplanelike vehicle that would spend many days (months?) flying through the Jovian clouds. What observations, measurements, and analyses should this aircraft make? What dangers might it encounter, and what design problems would you have to overcome?

27 Speculate on the possibility that Europa, Ganymede, or Callisto might harbor some sort of marine life.

28 Suppose you were planning separate missions to each of Jupiter's Galilean moons. What questions would you want these missions to answer, and what kinds of data would you want your spacecraft to send back? Given the different environments on the four satellites, how would the designs of the four spacecraft differ?

29 NASA and the Jet Propulsion Laboratory have tentative plans to place spacecraft in orbit about Uranus and Neptune early in the twenty-first century. What kinds of data should be collected and what questions would you like to see answered by these missions?

30 Would you expect the surfaces of Pluto and Charon to be heavily cratered? Explain.

Observing Projects

31 Consult such magazines as *Sky & Telescope* and *Astronomy*. If Jupiter is visible in the night sky, make arrangements to view the planet through a telescope. What magnifying power seems to give you the best view? Draw a picture of what you see. Can you see any belts and zones? How many? Can you see the Great Red Spot?

32 Observe Jupiter through a pair of binoculars. Can you see all four Galilean satellites? Make a drawing of what you observe.

33 Observe Jupiter through a telescope on three or four consecutive nights. Make a drawing each night showing the positions of the satellites relative to Jupiter. Record the time and date of each observation. Consult the section called "Satellites of Jupiter" in the *Astronomical Almanac* for the current year to see if you can identify the satellites by name.

34 View Saturn through a small telescope. Make a sketch of what you see. Estimate the angle at which the rings are tilted to your line of sight. Can you see the Cassini division? Can you see any belts or zones in Saturn's clouds? Is there a faint, starlike object near Saturn that might be Titan? What observations could you perform to test whether the starlike object is a Saturnian satellite?

35 Make arrangements to view Uranus and Neptune through a telescope. Throughout the 1990s, both planets are best seen in July and August. To help you find these planets, use the star chart, published each January in *Sky & Telescope*, showing the paths of Uranus and Neptune against the background stars.

8 ▶ Vagabonds of the Solar System

IN THIS CHAPTER

You will learn about asteroids, meteoroids, and comets—the debris orbiting the Sun. Asteroids and meteoroids are pieces of interplanetary rock and metal, while comets have large amounts of frozen ice. You will also survey the space debris that falls through the Earth's atmosphere as meteors and that lands here as meteorites. One such impact 65 million years ago caused the extinction of the dinosaurs—along with two-thirds of all the species on Earth. Even today, life on Earth is threatened by the devastation that could be wrought by a wayward asteroid.

The Head of Comet Halley This photograph shows the bluish head of Comet Halley as it approached the Sun in 1985. Comet Halley orbits the Sun with an average period of 76 years along a highly elliptical path that stretches from just inside the Earth's orbit to slightly beyond the orbit of Neptune. This color photograph was constructed from three black-and-white photographs taken in rapid succession with red, blue, and green filters on the same telescope. Because of the comet's motion during the exposures, the images of background stars are elongated.

1 Were the asteroids a planet that was somehow destroyed?

2 How far apart are the asteroids on average?

3 Why do comets have tails?

4 In which direction does a comet tail point?

5 What is a shooting star?

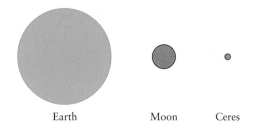

Earth   Moon   Ceres

Figure 8-1 **Comparison of Ceres with the Moon and the Earth** Ceres, the Moon, and the Earth are shown here to scale. Ceres, the largest asteroid, is so small that it is not considered a planet. Since it does not orbit another body (other than the Sun), it is also not a moon.

THE FORMATION OF THE solar system was not a tidy affair. Planetesimals collided by the millions to form planets and large moons. Once formed, these bodies were then struck by innumerable pieces of leftover matter. Dramatic impacts continue today, as vividly shown by the multiple plunges of Comet Shoemaker-Levy 9 into Jupiter. Large interplanetary bodies occasionally pass close to the Earth—closer even than the Moon. Myriad small ones penetrate the atmosphere daily! Some of these objects contain material essential to life. These vagabonds of the solar system are classed as asteroids, meteoroids, and comets.

Asteroids

We know that the solar system formed from a rotating disk of gas and dust. The matter that had too much angular momentum to fall onto the protosun coalesced into planetesimals. Many of these chunks of rock and metal eventually collided, forming the planets and larger moons. Others were captured whole by various planets as small, irregularly shaped moons, like Phobos and Deimos around Mars. However, numerous planetesimals still orbit the Sun today in splendid isolation. These are the **asteroids**, sometimes called **minor planets**.

8-1 Most asteroids orbit the Sun between Mars and Jupiter

The first discovery of asteroids was serendipitous. On New Year's Day 1801, the Sicilian astronomer Giuseppi Piazzi was carefully mapping faint stars in the constella-

tion of Taurus. He noticed a dim, previously uncharted star that shifted its position slightly over the next several nights. Recall that Pluto was discovered the same way (see Figure 7-44). Later that year the orbit of this object was determined to lie between the orbits of Mars and Jupiter. At Piazzi's request, the object was named Ceres (pronounced SEE-reez) after the patron goddess of Sicily. Ceres is spherical but too small to qualify as a full-fledged planet (Figure 8-1). Its diameter is a scant 940 km, only one-quarter the diameter of our Moon.

In 1802 the German astronomer Heinrich Olbers discovered another faint, starlike object that moved against the background stars. He called it Pallas, after the Greek goddess of wisdom. Like Ceres, Pallas orbits the Sun in a slightly elliptical orbit between the orbits of Mars and Jupiter. Pallas is even dimmer and smaller than Ceres, with a diameter of only 600 km.

Only two more of these small bodies—Juno and Vesta—were found until the mid-1800s, when telescopes improved. Astronomers then began to stumble across many more asteroids circling the Sun between the orbits of Mars and Jupiter, at distances of between 2 and $3\frac{1}{2}$ AU from the Sun. This region of the solar system is now called the **asteroid belt** (Figure 8-2). Asteroids whose orbits lie entirely within this region are called **belt asteroids**.

The next real breakthrough came in 1891, when the German astronomer Max Wolf applied photographic techniques to the search for asteroids. A total of 300 asteroids had been found up to that time, each painstakingly discovered by scrutinizing the skies for faint, uncharted "stars" whose positions shifted slowly from one night to the next. With the advent of astrophotography, however, the floodgates were opened. Astronomers could

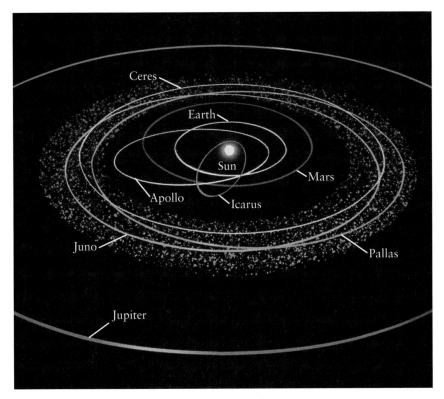

Figure 8-2 **Asteroid Orbits** Most asteroids orbit the Sun in a $1\frac{1}{2}$-AU-wide belt between the orbits of Mars and Jupiter. Of these asteroids, the orbits of Ceres, Pallas, and Juno are indicated. Some asteroids, such as Apollo and Icarus, have highly eccentric paths that cross Earth's orbit. Others, called the Trojan asteroids, are in the same orbit as Jupiter.

simply aim a camera-equipped telescope at the stars and take long exposures. If an asteroid happened to be in the field of view, it left a distinctive trail on the photographic plate (see Figure 8-3). Using this technique, Wolf alone discovered 228 asteroids.

By February 15, 1995, 6259 asteroids had been observed. That number has since increased to over 6600. Perhaps 100,000 others exist that are bright enough to appear one day on Earth-based photographs. Including the ones too small to be seen from Earth, there are probably a million or more asteroids out there.

Ceres accounts for about 30% of the mass of all the asteroids combined. Only three asteroids—Ceres, Pallas, and Vesta—have diameters greater than 300 km. Thirty other asteroids have diameters between 200 and 300 km, and 200 more are bigger than 100 km across. The vast majority of asteroids are less than 1 km across.

There is a common myth that the asteroids were once a planet that was somehow destroyed (Toolbox 8-1). This is most unlikely. If all the asteroids had once been part of a single planet, it would have had a diameter of only 1500 km, or 12% of the Earth's diameter. This is less than half the diameter of the Moon. It is much more likely that

Figure 8-3 **Discovering Asteroids** Two asteroids observed in a single photograph. Today asteroids are often discovered in long-exposure photographs.

....

AN ASTRONOMER'S TOOLBOX 8-1

Bode's Law and the Discovery of Belt Asteroids

In the late 1700s a young German astronomer, Johann Elert Bode, popularized a simple rule that describes the distances of the planets from the Sun. This rule is usually known today as **Bode's law**—an unfortunate name because it is not a physical law and was not developed by Bode. It had first been published in 1766 by Johann Titius, a German physicist and mathematician.

Bode's law is still helpful for remembering the distances of the planets from the Sun:

1   Write down the sequence of numbers 0, 3, 6, 12, 24, 48, 96, ... (Note that each number after the 3 is simply twice the preceding number.)

2   Add 4 to each number in the sequence.

3   Divide each of the resulting numbers by 10.

As shown in the table, the resulting numbers correspond remarkably well to the distances of most of the planets from the Sun measured in astronomical units. There remains, however, a gap between Mars and Jupiter. Could there be a missing planet?

Ceres orbits the Sun at an average distance of 2.77 AU. Its discovery in 1801 was in remarkable agree-

Bode's Law

Planet	Bode's prediction	Actual distance (AU)
Mercury	$(0 + 4)/10 = 0.4$	0.39
Venus	$(3 + 4)/10 = 0.7$	0.72
Earth	$(6 + 4)/10 = 1.0$	1.00
Mars	$(12 + 4)/10 = 1.6$	1.52
?	$(24 + 4)/10 = 2.8$	——
Jupiter	$(48 + 4)/10 = 5.2$	5.20
Saturn	$(96 + 4)/10 = 10.0$	9.54
Uranus	$(192 + 4)/10 = 19.6$	19.18
Neptune	$(384 + 4)/10 = 38.8$	30.06
Pluto	$(768 + 4)/10 = 77.2$	39.44

ment with Bode's law for the "missing planet." Astronomers were spurred to search for, and discover, other asteroids in the same region. They guessed mistakenly that Bode's missing planet might have somehow broken apart or exploded.

Bode's law gives increasingly poor predictions for the locations of Saturn, Uranus, Neptune, and Pluto. Astronomers today believe that Bode's law is merely a numerical coincidence. Indeed, given any set of distances between objects, it is possible to generate a "law." Not every mathematical relationship is a law of Nature.

....

the asteroids are planetesimals that have survived from the early solar system. They were prevented from ever coalescing by the powerful gravitational attraction of Jupiter.

8-2 Jupiter's gravity creates gaps in the asteroid belt

In 1867 the American astronomer Daniel Kirkwood called attention to gaps in the asteroid belt. These features, called **Kirkwood gaps,** show the influence of Jupiter's gravitational attraction, as best seen in a graph of asteroid orbital periods, like the one in Figure 8-4. Note the gaps at simple fractions ($\frac{1}{3}$, $\frac{2}{5}$, $\frac{3}{7}$, and $\frac{1}{2}$) of Jupiter's orbital period.

To understand the Kirkwood gaps, imagine an asteroid circling the Sun once every 5.93 years, exactly half of Jupiter's orbital period. On every second trip around the Sun, the asteroid is lined up between Jupiter and the Sun. In other words, the asteroid always comes nearest to Jupiter at the very same point in its orbit. Here Jupiter's pull on the asteroid is particularly strong, and over time it pulls the asteroid into an orbit further from the Sun. This periodic gravitational influence, called a *resonance*, may even eject the asteroid from the belt.

According to Kepler's third law, a period of 5.93 years corresponds to a semimajor axis of 3.28 AU. Because of Jupiter, there are no asteroids that orbit the Sun at this average distance. Similarly, there is a gap corresponding to an orbital period of one-third Jupiter's period, or 3.95 years. Additional gaps exist for other

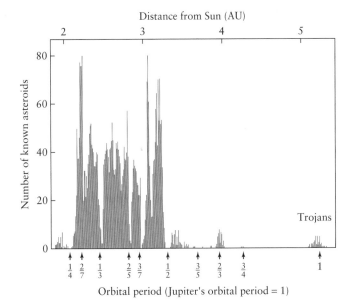

Distance from Sun (AU)

Orbital period (Jupiter's orbital period = 1)

Figure 8-4 **The Kirkwood Gaps** This graph displays the number of asteroids at various distances from the Sun. Notice that very few asteroids have orbits whose orbital periods correspond to such simple fractions as $\frac{1}{3}$, $\frac{2}{5}$, $\frac{3}{7}$, and $\frac{1}{2}$ of Jupiter's orbital period. Repeated alignments with Jupiter have deflected asteroids away from these orbits. The Trojan asteroids accompany Jupiter as it orbits the Sun.

simple relationships between the periods of asteroids and Jupiter.

Although it is likely that there are a million or more asteroids in the asteroid belt, their average separation is a staggering ten million kilometers. This is quite unlike the image that has been created by innumerable science fiction movies of asteroids so close together that you must dodge them as you fly past. (Why don't asteroids exist close together for very long? Think of what effect they have on each other.)

Nevertheless, over the 4.6 billion years that the solar system has existed, the gravitational influences of Mars and Jupiter have surely sent some asteroids careening into each other. In 1918 the Japanese astronomer Kiyotsugu Hirayama drew attention to groups of asteroids that share nearly identical orbits. These are fragments of parent asteroids.

A collision between kilometer-sized asteroids must be an awesome event. Typical collision velocities are estimated to be 3600 to 18,000 km/h (2000 to 11,000 mph), which is more than sufficient to shatter rock. In some collisions, the resulting fragments may not have enough speed to escape from each other's gravitational attraction, and thus they reassemble. Alternatively, several large fragments may end up orbiting each other. This is probably what happened to Pallas, which actually consists of a main asteroid and a large satellite. The asteroid Toutatis consists of two comparably sized pieces orbiting each other.

In the early 1990s the Jupiter-bound Galileo spacecraft passed near two asteroids—Gaspra (see Figure II-10) and Ida (Figure 8-5a)—and sent back close-up views. Both asteroids are probably fragments of larger parent bodies that were broken apart by catastrophic collisions. Because Ida's surface is more heavily cratered than Gaspra's, Ida is much older. Also, Ida has its own moon, Dactyl, a pockmarked asteroid some 1.5 km in diameter that orbits Ida at a distance of 100 km (Figure 8-5b).

8-3 There are asteroids outside the asteroid belt

While Jupiter's gravitational pull depletes certain orbits in the asteroid belt, it captures asteroids at two locations much farther from the Sun. The gravitational forces of

a

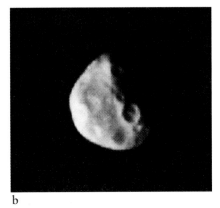

b

Figure 8-5 **Ida and Its Moon** (a) The asteroid Ida is about twice the size of Gaspra (see Figure II-10) and is considerably older. (b) Ida has its own moon, Dactyl, which is also heavily cratered.

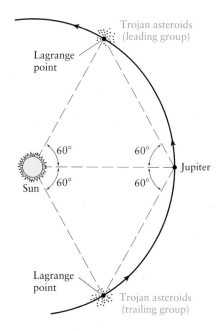

Figure 8-6 **The Trojan Asteroids** Asteroids are trapped at the two Lagrange points along Jupiter's orbit by the combined gravitational forces of Jupiter and the Sun. Asteroids at these locations are named after Homeric heroes of the Trojan War.

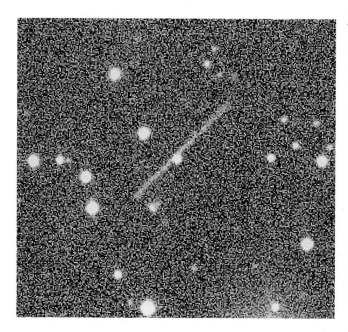

Figure 8-7 **Asteroid 1994 XM1** This image was obtained on December 9, 1994, shortly before the asteroid arrived in the Earth's vicinity. When it passed by the Earth just 12 hours later, asteroid 1994 XM1 was less than half the distance from the Earth to the Moon.

the Sun and Jupiter at these locations work together to hold asteroids in orbit. They are called **stable Lagrange points,** in honor of the French mathematician whose calculations explained them. One Lagrange point is located 60° ahead of Jupiter, and the other is 60° behind, as shown in Figure 8-6.

The asteroids trapped at Jupiter's Lagrange points are called **Trojan asteroids,** each named after a hero of the Trojan War. Approximately four dozen Trojan asteroids have been catalogued so far, and some astronomers believe that there may be several hundred rock fragments orbiting near each Lagrange point.

Still other asteroids have highly elliptical orbits that bring them into the inner regions of the solar system (see Figure 8-2). Some cross Mars' orbit, while others, called **Apollo asteroids,** even cross Earth's orbit.

The Apollo asteroid Eros passed within 23 million kilometers of our planet in 1931. On October 30, 1937, the asteroid Hermes passed within 900,000 km of Earth—only a little more than twice the distance to the Moon. On June 14, 1968, Icarus passed Earth at a distance of only 6 million kilometers. There were two other close calls quite recently. On March 23, 1989, an asteroid called 1989 FC passed within 800,000 km of the Earth, and on December 9, 1994, asteroid 1994 XM1 passed within

105,000 km. This latter asteroid is about 10 m across—the size of a small bus (Figure 8-7).

During these close encounters, astronomers can examine the details of asteroids. For example, an asteroid's brightness often varies as it rotates because different surface features reflect different amounts of light. Such data show that typical rotation periods for asteroids are between 5 and 20 hours.

Most of the Apollo asteroids will eventually strike a planet or moon—even the Earth. Some asteroids will end up heading straight into the Sun. However, the chance of the Earth being hit by one in the next few thousand years is quite remote.

Comets

Not all the early solar system consisted of rock and metal—water was present, along with other liquids and gases. We have seen in earlier chapters that water ice, along with carbon dioxide, methane, and ammonia ices, was locked up in planets and moons. Ices in the young solar system also condensed with roughly equal amounts of small rocky debris into bodies that still remain in orbit around the Sun. These dirty icebergs in space are **comets.**

8-4 Comets lack tails until they enter the inner solar system

Comets first formed in the outer reaches of the solar system at the distances of Uranus and Neptune. In that region water was plentiful and the temperature low enough for the ices to condense into chunks several kilometers across. Then gravitational tugs from Uranus and Neptune flung the comets in every direction.

There are now two reservoirs of comets. Most of the comets that eventually return to the inner solar system and develop the long tails that we usually associate with them are believed to come from a disk of comets out beyond the orbit of Pluto. This **Kuiper belt** of comets lies in the plane of the ecliptic and extends out some 500 AU from the Sun (Figure 8-8). Nearly 50 bodies in the Kuiper belt have been observed (Figure 8-9).

The vast majority of the several billion comets estimated to exist lie even farther from the Sun. Unlike the Kuiper belt comets and the rest of the solar system, these comets have a spherical distribution around the Sun called the **Oort cloud,** named after the Dutch astronomer Jan Oort, who first proposed its existence in the 1950s. Astronomers calculate that the Oort cloud extends out at least 50,000 AU, one-third the distance to the nearest stars. Most of these comets have orbits so circular that they never even get as close as Pluto is to the Sun. However, occasionally a passing star's gravitational force nudges a distant comet toward the inner solar system. As a result of its inward plunge, comets from the Oort comet cloud have highly elliptical orbits.

Because the comets in the Kuiper belt and Oort cloud are far from the Sun, they are completely frozen. These solid comet bodies, called **nuclei** (*singular* **nucleus**), are typically a few kilometers across. The first pictures of a comet's nucleus were obtained when a fleet of spacecraft flew past Comet Halley in 1986 (Figure 8-10). Halley's potato-shaped nucleus is darker than coal, probably because of carbon-rich compounds left behind after its ice evaporated.

As a comet nucleus comes within 20 AU of the Sun, solar heat begins to vaporize the ices on its surface. The

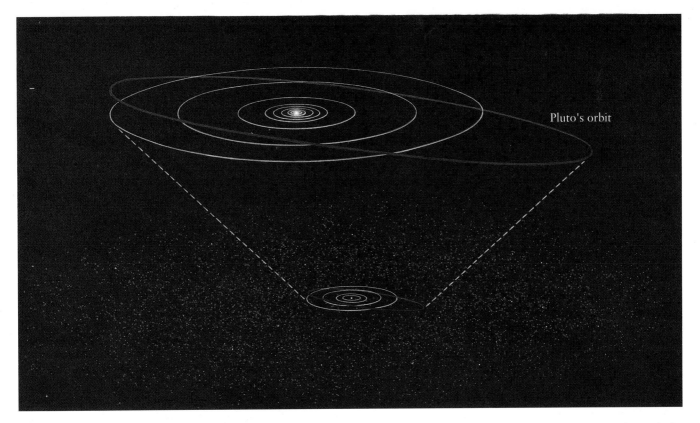

Figure 8-8 **The Kuiper Belt** The Kuiper belt of comets spreads from Pluto out 500 AU from the Sun. Most of the estimated 200 million belt comets are believed to orbit in the plane of the ecliptic. Nearly 50 of these objects have been located, the largest 320 km across. Some astronomers believe that Pluto and Charon are members of the belt.

a

b

Figure 8-9 **Kuiper Belt Objects** These 1993 images show the discovery (arrows) of one of over three dozen Kuiper belt objects. These two images of Kuiper belt object 1993 SC were taken 4.6 hours apart, during which time the object moved from (**a**) to (**b**) against the background stars.

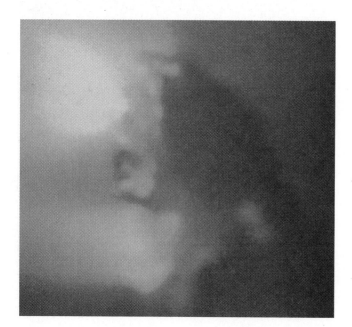

Figure 8-10 **The Nucleus of Comet Halley** This close-up picture, taken by a camera on board the Giotto spacecraft, shows the potato-shaped nucleus of the comet. Its dark nucleus measures 15 km in its longest dimension and about 8 km in its shortest. The Sun illuminates the comet from the left.

liberated gases form an atmosphere, called a **coma,** around the nucleus. Because the coma scatters sunlight, it appears as a fuzzy, luminous ball. The largest coma ever measured was over a million kilometers across—nearly as large as the Sun. Not visible to the human eye is the **hydrogen envelope,** a sphere of tenuous gas surrounding the comet's nucleus and measuring as much as 20 million kilometers in diameter (Figure 8-11).

The solar wind (see Chapters 5 and 9) and sunlight blow these luminous gases outward into one (or more) long, flowing, diaphanous **tails,** an awesome sight (Figure 8-12). The comet tails develop from coma gases and dust pushed outward from the Sun. This means that comet tails do not trail behind the nucleus, as the exhaust from a jet plane does in the Earth's atmosphere. Rather, *at the comet's nucleus, the tails always point away from the Sun* (Figure 8-13), regardless of the direction of the comet's motion. The implication that something from the Sun was "blowing" the comet's gases radially outward led Ludwig Biermann to predict the existence of the solar wind. This stream of particles from the Sun was actually discovered in 1962, a full decade later, by instruments on the spacecraft Mariner 2.

Figure 8-11 **Comet Kohoutek and Its Hydrogen Envelope** Comet Kohoutek as seen in visible light (left) and at ultraviolet wavelengths (right). The latter picture reveals a huge hydrogen cloud surrounding the comet's nucleus.

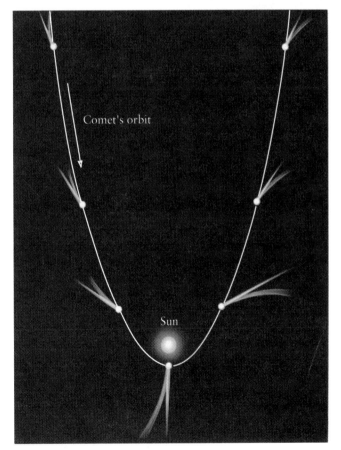

Figure 8-12 **Comet West** Astronomer Richard M. West first noticed this comet on a photograph taken with a telescope in 1975. After passing near the Sun, Comet West became one of the brightest comets in recent years. This photograph shows the comet in the predawn sky in March 1976.

Figure 8-13 **The Orbit and Tails of a Comet** The solar wind and radiation pressure from sunlight blow a comet's dust particles and ionized atoms away from the Sun. Consequently, the comet's tails always point away from the Sun.

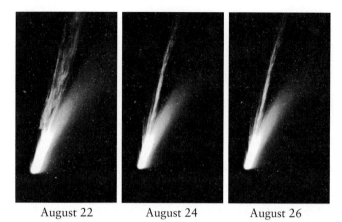

August 22   August 24   August 26

Figure 8-14 **The Two Tails of Comet Mrkos** Comet Mrkos dominated the evening sky in August 1957. These three views, taken at two-day intervals, show dramatic changes in the comet's ion tail. In contrast, the slightly curved dust tail remained fuzzy and featureless.

The Sun usually produces two comet tails: a **gas (or ion) tail** and a **dust tail**. Positively charged ions (that is, atoms missing one or more electrons) are swept directly away from the Sun by the solar wind to form the ion tail. The ions typically leave the coma at 1.4 million kilometers per hour. The relatively straight ion tail can change dramatically from night to night (Figure 8-14).

The dust tail is formed when photons strike dust particles that have been freed from the evaporating nucleus. Light exerts pressure on any object that absorbs or reflects it. This pressure, called **radiation (or photon) pressure,** is quite weak. Nevertheless, fine-grained dust particles in a comet's coma are sufficiently light to be blown away from the comet, thus producing a dust tail. However, the dust particles have enough mass so that they do not flow straight away from the Sun. Rather, the dust tail is arched in a path that lies between the ion tail and the direction from which the comet came (Figure 8-15). Comet tails can stretch to over 150 million kilometers in length.

Figure 8-15 outlines the structure of a comet. Some, like the comet shown in Figure 8-16, have a large, bright coma but a short, stubby tail. Others have an inconspicuous coma but a tail of astonishing length. The tail seen in Figure 8-17 is long enough to stretch all the way from the Earth to the Sun.

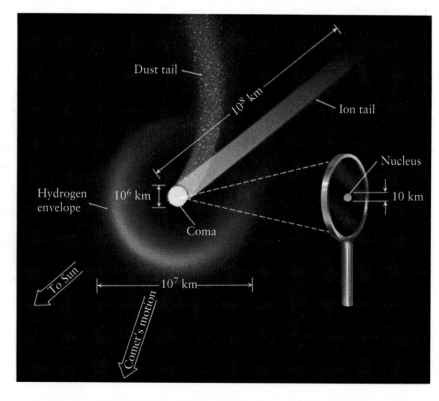

Figure 8-15 **The Structure of a Comet** The solid part of a comet (the nucleus) is roughly 10 km in diameter. The coma can be as large as 100,000 to 1,000,000 km across, and the hydrogen envelope is typically 10,000,000 km in diameter. A comet's tail can be enormous—as long as 1 AU. (This drawing is not to scale.)

EYES ON . . .   Discovering Comets

Most research astronomers do not have time to search for new comets. Like Comet Shoemaker-Levy 9, therefore, the vast majority of these objects are discovered by amateur astronomers.

To become the official discoverer of a comet takes several steps. First, you must make at least two sightings of the object to establish that the object is in motion. On each night you should determine the coordinates of the object within 1 arcsec in declination and 0.1 minute in right ascension. A good sky atlas is essential.

Next, check a listing of known comets and asteroids to verify that you are not seeing a known object. Sources of comet positions include the *International Astronomi-* *cal Union Circulars,* the *Comet Handbook of the International Comet Quarterly,* and the *Handbook of the British Astronomical Association.* Asteroids are listed in the *Minor Planet Circulars.*

Finally, report your sightings to the Central Bureau for Astronomical Telegrams. Include the celestial coordinates of the object at each observing time, your own location, the telescope you used, and the film and exposure times if you took photographs. The bureau can be reached by e-mail or the Internet (for more information, see "Astronomy and the Internet" at the back of this book).

If your comet is new, an official number will be assigned to it. It will also be named after you.

Figure 8-16 The Head of Comet Brooks This comet, named after its discoverer, had an exceptionally large, bright coma. It dominated the night skies in October 1911.

Figure 8-17 The Tail of Comet Ikeya-Seki Named after its Japanese codiscoverers, this comet dominated the predawn skies in late October 1965. Although its coma was tiny, its tail was 1 AU long.

8-5 Comets don't last forever

Astronomers, many of them amateurs, discover at least a dozen new comets in a typical year. Falling sunward from the Kuiper belt or Oort cloud, most are **long-period comets,** which move so fast that they leave the inner solar system after one pass by the Sun and take roughly 1 to 30 million years to return.

However, sometimes a comet passes so close to a planet that the planet's gravitational force changes the comet's orbit, slowing the comet down and keeping it in the inner solar system (Figure 8-18). The comet then becomes a **short-period comet,** orbiting the Sun in less than 200 years. Like Comet Halley, a short-period comet appears again and again at predictable intervals.

Comets cannot survive very many passages near the Sun. A typical comet is estimated to lose between $\frac{1}{60}$ and $\frac{1}{100}$ of its mass at each perihelion. Therefore, a comet survives around 100 close passes to the Sun before its ices evaporate completely. Figure 8-19 shows the nucleus of a comet breaking up at the distance of Mercury's orbit, shortly after it passed the Sun. Soon thereafter, its remaining dust and rock fragments spread out in a loose collection of debris that continues to circle the Sun along the comet's orbital path.

In October 1995, the comet Hale-Bopp was observed to eject some mass (Figure 8-20). This occurred when the comet was still beyond Jupiter, leading astronomers to believe that the ejection resulted from evaporation of surface ice with an assist from the comet's rapid rotation. The comet rotates within a period of about a week. The rotation is evident in pinwheel-shaped distributions of gas seen spiraling out from the comet nucleus. The ejected piece of the comet moved away from the nucleus at a speed of 109 km/hr (68 mph), disintegrating

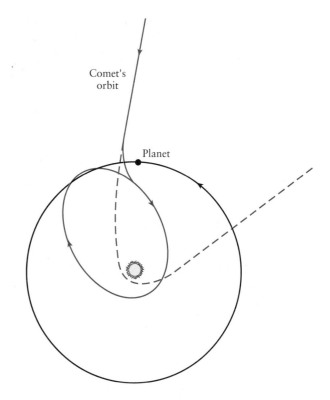

Figure 8-18 Transforming a Long-Period Comet into a Short-Period Comet The gravitational force of a planet can change a comet's orbit. Initially on highly elliptical orbits, comets are sometimes deflected into more circular paths that keep them in the inner solar system.

into a bright cloud of gas and dust and joining the general spiral structure.

A comet can also be destroyed when it comes too close to a planet or actually strikes a planet, a moon, or the Sun. A spectacular example was Comet Shoemaker-

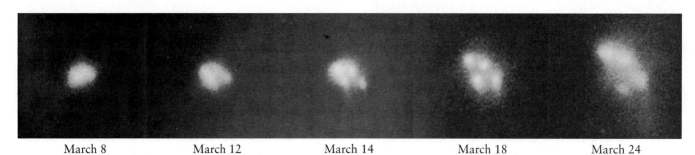

| March 8 | March 12 | March 14 | March 18 | March 24 |

Figure 8-19 The Fragmentation of Comet West Shortly after passing near the Sun in 1976, the nucleus of Comet West broke into at least four pieces. This series of five photographs clearly shows the disintegration of the comet's nucleus. Figure 8-12 shows a wide-angle view of this comet.

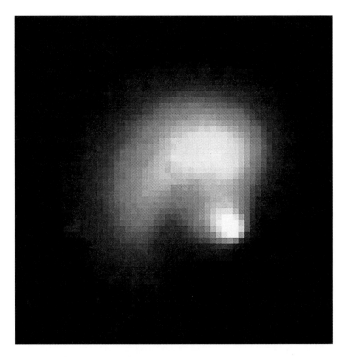

Figure 8-20 **Comet Hale-Bopp Ejects Debris** Discovered on July 23, 1995, this comet was observed to have jets of gas and debris shoot out from it at least twice between September and October of that year. This image shows the comet nucleus (lower bright region), an ejected piece of the comet's surface (upper bright region), and a spiral tail. The ejected piece eventually disintegrated, following the same spiral pattern as the tail.

Figure 8-21 **A Meteor** A meteor is produced when a piece of interplanetary rock or dust strikes the Earth's atmosphere at high speed. Exceptionally bright meteors, such as the one shown in this exposure that tracked the meteor (notice the star trails), are usually called fireballs.

Levy 9. As we discussed in Chapter 7, that comet fragmented under the tidal force from Jupiter in 1992. Two years later, with the world's astronomers watching carefully, the pieces returned and struck the planet (see Figures 7-8, 7-9, and 7-10). The fact that the comet came apart in the first place tells astronomers that its nucleus was very weakly held together. If this condition is typical of comets, then it provides an important clue to their composition and structure. For example, the ease with which Shoemaker-Levy 9 came apart suggests it may actually have been several individual pieces that had been stuck together until Jupiter's tidal force pulled them apart.

Meteoroids and Meteorites

8-6 Small rocky debris peppers the solar system

The solar system is strewn with rocky debris called **meteoroids,** which are smaller than asteroids. Although there is no official size standard, meteoroids are no more than a few hundred meters across, and the vast majority are smaller than a millimeter. The ones larger than pebbles probably broke off when asteroids collided.

Meteoroids are often pulled by gravity into the Earth's atmosphere. Air friction generates so much heat that a meteoroid's outer layer begins to vaporize. The same process creates the heat on the hull of the Space Shuttle as it reenters our atmosphere. As the meteoroid penetrates further into the air, leaving behind a trail of dusty gas, it becomes a **meteor.** Common names for these dramatic streaks of light flashing across the sky include *shooting stars, bolides,* and *fireballs.* Fireballs are meteors at least as bright as Venus (Figure 8-21); bolides are meteors that also explode, often soundlessly, in the air.

8-7 Impact craters and meteor showers mark remnants of space debris on Earth

Most meteors vaporize completely before they strike the Earth. Their dust settles to the ground, often carried by raindrops. (This is not the source of acid rain, however, which comes from natural and human-made gases ejected into the atmosphere.) Any part of a meteor that survives its fiery descent to Earth may leave an **impact crater.** Weather and water erosion are wearing away all of the 200 or more impact craters now on the Earth, and

Figure 8-22 **The Barringer Crater** An iron meteoroid measuring 50 m across struck the ground in Arizona 50,000 years ago. The result was this beautifully symmetric impact crater.

thousands more have long since been drawn into the Earth by the motion of its tectonic plates.

One of the best-preserved impact craters is the famous Barringer Crater near Winslow, Arizona (Figure 8-22). Measuring 1.2 km across and 200 m deep, it formed

Figure 8-23 **Meteor Streaks Seen During a Meteor Shower** This time exposure shows meteors streaking away from one place in the sky, in the constellation Perseus, during the Perseid meteor shower.

	TABLE 8-1		
Prominent Meteor Showers			
Shower	Date of maximum intensity	Hourly rate	Constellation
Quadrantids	January 3	40	Boötes
Lyrids	April 22	15	Lyra
Eta Aquarids	May 4	20	Aquarius
Delta Aquarids	July 30	20	Aquarius
Perseids	August 12	80	Perseus
Orionids	October 21	20	Orion
Taurids	November 4	15	Taurus
Leonids	November 16	15	Leo Major
Geminids	December 13	50	Gemini
Ursids	December 22	15	Ursa Minor

50,000 years ago when an iron-rich meteoroid some 50 m across struck the ground at 40,000 km/h (25,000 mph). The blast was like the detonation of a 20-megaton hydrogen bomb.

Although you can expect to see a meteor about every 10 minutes, there are predictable times throughout each year when the Earth is inundated with them. These **meteor showers** occur when the Earth moves through the orbit of remnant debris left from a comet. Some 30 meteor showers can be seen each year. Because the meteors in each shower appear to come from a fixed region of the sky, meteor showers are named after the constellation from which the meteors appear to radiate (Figure 8-23). For example, meteors in the Perseid shower appear to originate in the constellation Perseus. At their peaks more than one meteor can be seen each minute during such prodigious meteor showers as the Perseids. The Perseids, which take place in the summertime, are among the best free shows in astronomy. Table 8-1 lists some of the most active meteor showers. Except for the Lyrids, the meteor showers are best seen after midnight.

8-8 Meteorites are space debris that land intact

Although most meteors completely vaporize in the atmosphere, some reach the ground before totally disintegrating. Debris left on the ground is called a **meteorite,** and people have been picking them up for thousands of years. The descriptions of meteorites in historical Chi-

Figure 8-24 **A Stony Meteorite** Most meteorites that fall on the Earth are stones. Many freshly discovered specimens, like the one shown here, are coated with dark crusts. This particular stone fell in Texas.

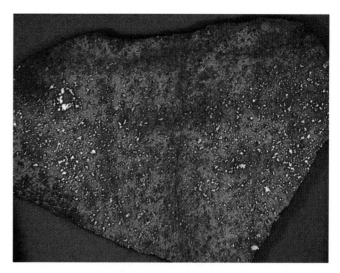

Figure 8-25 **A Cut and Polished Stone** Some stony meteorites contain tiny specks of iron, which can be seen when the stones are cut and polished. This specimen was discovered in California.

nese, Islamic, Greek, and Roman literature show that our ancestors placed special significance on these "rocks from heaven." In-falling space debris is increasing the Earth's mass by over 300 tons per day.

Meteorites are classified as stones, stony-irons, and irons. Most **stony meteorites** look much like ordinary rocks, although some are covered with a dark crust. This crust is produced when the meteorite's outer layer melts during its fiery descent through the atmosphere (Figure 8-24). When a stony meteorite is cut in two and polished, tiny flecks of iron are sometimes found in the rock (Figure 8-25).

Meteorites with a high iron content can be located with a metal detector. Consequently, the easily found iron and stony-iron meteorites dominate most museum collections, even though stony meteorites account for about 95% of all meteoritic material that falls on the Earth.

Iron meteorites (Figure 8-26) may also contain from 10% to 20% nickel by weight. Iron is moderately abundant in the universe as well as being one of the most common rock-forming elements, so it is not surprising that iron is an important constituent of asteroids and meteoroids. Another element, iridium, is common in the iron-rich minerals of meteorites but rare in ordinary rocks because most iridium settled deep into the Earth eons ago. Measurements of iridium in the Earth's crust can thus tell us the rate at which meteoritic material has been deposited on the Earth over the ages.

In 1808 Count Alois von Widmanstätten, director of the Imperial Porcelain works in Vienna, discovered a conclusive test for the most common type of iron meteorite. Most iron meteorites have a unique structure of long nickel–iron crystals called **Widmanstätten patterns,** which become visible when the meteorites are cut, polished, and briefly dipped into a dilute solution of acid

Figure 8-26 **An Iron Meteorite** Irons are composed almost entirely of iron–nickel minerals. The surface of a typical iron is covered with thumbprintlike depressions created as the meteorite's outer layers melted away during its high-speed descent through the atmosphere. This specimen was found in Australia.

Figure 8-27 **Widmanstätten Patterns** When cut, polished, and etched with a weak acid solution, most iron meteorites exhibit interlocking crystals in designs called Widmanstätten patterns. This meteorite was found in Australia.

(Figure 8-27). Because nickel–iron crystals can grow to lengths of several centimeters only if the molten metal cools slowly over many millions of years, Widmanstätten patterns are never found in counterfeit meteorites, or "meteorwrongs."

The final category of meteorites, the **stony-irons,** consist of roughly equal amounts of rock and iron. Figure 8-28, for example, shows the mineral olivine suspended in a matrix of iron. To understand why there are different types of meteorites, we consider their formation. Most meteorites were once pieces of asteroids. Heat from the rapid decay of radioactive isotopes melted newly formed asteroid interiors. Over the next few million years, differentiation occurred, just as in the young Earth. Iron sank toward the asteroid's center, while lighter rock floated up to the asteroid's surface. Iron meteorites are fragments of an asteroid's core, and stones are samples of its crust. Stony-irons presumably come from the mantle, between an asteroid's core and crust.

A class of rare stony meteorites, called **carbonaceous chondrites,** shows no evidence of ever having melted as parts of asteroids, unlike most other meteorites. Carbonaceous chondrites may therefore be primordial material from which our solar system was created. They contain complex carbon compounds and as much as 20% water. These compounds would have been broken down and the water driven out if these meteorites had been significantly heated.

Amino acids, the building blocks of proteins upon which terrestrial life is based, are among the organic compounds occasionally found inside carbonaceous chondrites, although these may be contaminants acquired after the meteoroids entered the Earth's atmosphere. Nevertheless, some scientists suspect that carbonaceous chondrites may have played a role in the origin of life on Earth.

Figure 8-28 **A Stony-Iron Meteorite** Stony-irons account for about 1% of all meteorites that fall on the Earth. This specimen, a variety of stony-iron called a pallasite, was found in Chile.

8-9 The Tunguska mystery and the Allende meteorite provide evidence of catastrophic collisions

On June 30, 1908, a spectacular explosion occurred over the Tunguska region of Siberia. The blast, comparable to a nuclear detonation of several megatons, knocked a man off his porch some 60 km away and was audible more than 1000 km away. Millions of tons of dust were injected into the atmosphere, darkening the air as far away as California.

Preoccupied with political and economic upheaval, Russia did not send a scientific expedition to the site until 1927. At that time, Soviet researchers found that trees had been seared and felled radially outward in an area about 30 km in diameter (Figure 8-29). There was no clear evidence of a crater. In fact, the trees at "ground zero" were left standing upright, although they were completely stripped of branches and leaves. Because no

Figure 8-29 **Aftermath of the Tunguska Event** In 1908 a stony asteroid traveling at supersonic speed struck the Earth's atmosphere over the Tunguska region of Siberia. Trees were blown down for many kilometers in all directions from the impact site.

significant meteorite samples were found, for many years it was assumed that a small comet had struck the Earth.

Recently, however, several teams of astronomers have argued that the Tunguska explosion was actually caused by a large meteoroid traveling at supersonic speed. The Tunguska event is consistent with an impact by a meteoroid about 80 m (260 ft) in diameter entering the Earth's atmosphere at 79,000 km/h (50,000 mph). (A small comet breaks up too high in the atmosphere to cause significant damage on the ground.)

Another chance to study catastrophic impact came shortly after midnight on February 8, 1969, when a brilliant blue-white light moved across the night sky around Chihuahua, Mexico. Hundreds of people witnessed the dazzling display. The light disappeared in a spectacular, noisy explosion that dropped thousands of rocks and pebbles over the terrified onlookers. Within hours, teams

Figure 8-30 **A Piece of the Allende Meteorite** This carbonaceous chondrite fell near Chihuahua, Mexico, in February 1969. Note the meteorite's dark color, caused by a high abundance of carbon. Geologists believe that this meteorite is a specimen of primitive planetary material. The ruler is 15 cm long.

of scientists were on their way to collect specimens of a carbonaceous chondrite, collectively named the *Allende meteorite*, after the locality (Figure 8-30).

One of the most significant discoveries to come from the Allende meteorite was evidence of the detonation of a nearby supernova 4.6 billion years ago. One of nature's most violent and spectacular phenomena, a *supernova explosion* occurs when a massive star dies. A star blows apart in a cataclysm that hurls matter outward at tremendous speeds, as we shall see in Chapter 12. During this detonation, violent collisions between nuclei produce a host of radioactive elements, including a short-lived radioactive isotope of aluminum. Scientists found unmistakable evidence that this isotope once lay within the Allende meteorite. Some astronomers interpret this as evidence for a supernova in our vicinity at about the time the Sun was born. Indeed, by compressing interstellar gas and dust, the supernova's shock wave may have triggered the birth of our solar system.

8-10 Another strike on Earth may well have killed off the dinosaurs

In the late 1970s, geologist Walter Alvarez and his physicist father, Luis Alvarez, from the University of California at Berkeley, discovered a catastrophe closer to home. Working at a site of exposed marine limestone in the Apennine Mountains in Italy that had been on the Earth's surface 65 million years ago, the Alvarez team discovered an exceptionally high abundance of iridium in a dark-colored layer of clay between limestone strata (Figure 8-31).

Since this discovery was announced in 1979, a comparable layer of iridium-rich material has been uncovered at numerous sites around the world. In every case,

Figure 8-31 **The Iridium-Rich Layer of Clay** This photograph of strata in the Apennine Mountains of Italy shows a dark-colored layer of iridium-rich clay sandwiched between white limestone (below) from the late Mesozoic era and grayish limestone (above) from the early Cenozoic era. The coin is the size of a U.S. quarter.

geologic dating reveals that this apparently worldwide layer of iridium-rich clay was deposited about 65 million years ago. Paleontologists were quick to realize the significance of this date, for it was 65 million years ago that all the dinosaurs rather suddenly became extinct. In fact, two-thirds of all the species on Earth disappeared within a brief span of time back then.

The Alvarez discovery suggests a startling explanation for the dramatic extinction of so much of the life that once inhabited our planet. Perhaps an asteroid hit the Earth at that time. An asteroid 10 km in diameter slamming into the Earth could have thrown enough dust into the atmosphere to block out sunlight for several years. As the temperature dropped drastically and plants died for lack of sunshine, the dinosaurs would have perished, along with many other creatures in the food chain that were highly dependent on vegetation. The dust eventually settled, depositing an iridium-rich layer around the world. Tiny, rodentlike creatures capable of ferreting out seeds and nuts were among the animals that managed to survive this holocaust, setting the stage for the rise of mammals and, consequently, for the evolution of humans.

In 1992 a team of geologists suggested that the hypothesized asteroid crashed into a site in Mexico. They based this conclusion on glassy debris and violently shocked grains of rock ejected from the 180-km-diameter Chicxulub Crater buried under the Yucatán Peninsula. From the known rate at which radioactive potassium decays, the scientists have pinpointed the date when the asteroid struck: 64.98 million years ago. Other geologists and paleontologists are not yet convinced that an asteroid impact led to the extinction of the dinosaurs, but most agree that this hypothesis fits the available evidence better than any other explanation that has been offered so far.

Large objects, like the one that created the Chicxulub Crater, occasionally strike the Earth with the destructive force of a nuclear weapon. A comet or small asteroid approaching our planet in the daytime sky from the general direction of the Sun would probably not be discovered by astronomers, so it could strike without warning. In spite of safeguards and diplomacy, an entirely natural phenomenon could trigger Armageddon. However, based on the frequency with which very large meteorites have struck the Earth over the past few thousand years, it seems unlikely that such an event is going to occur in our lifetimes.

WHAT DID YOU THINK?

1 *Were the asteroids a planet that was somehow destroyed?* No, the gravitational pull from Jupiter prevented a planet from ever forming there.

2 *How far apart are the asteroids on average?* The distance between asteroids averages ten million kilometers.

3 *Why do comets have tails?* Gas and dust that evaporate from the comet nucleus are pushed away from the Sun by sunlight and the solar wind.

4 *In what direction does a comet tail point?* Comet gas tails point directly away from the Sun; comet dust tails make arcs pointing away from the Sun.

5 *What is a shooting star?* A shooting star is a piece of space debris plunging through the Earth's atmosphere—a meteor.

Key Words

amino acid	comet	meteor	stable Lagrange points
Apollo asteroid	dust tail (of a comet)	meteor shower	stony meteorite
asteroid (minor planet)	gas (ion) tail	meteorite	stony-iron meteorite
asteroid belt	hydrogen envelope	meteoroid	tail (of a comet)
belt asteroid	impact crater	nucleus (of a comet)	Trojan asteroid
Bode's law	iron meteorite	Oort cloud	Widmanstätten
carbonaceous	Kirkwood gaps	radiation (photon)	patterns
chondrite	Kuiper belt	pressure	
coma (of a comet)	long-period comet	short-period comet	

Key Ideas

• Thousands of belt asteroids with diameters larger than a few kilometers circle the Sun between the orbits of Mars and Jupiter.

Gravitational attraction by Jupiter depletes certain orbits within the asteroid belt. The resulting gaps, called Kirkwood gaps, occur at simple fractions of Jupiter's orbital period.

Jupiter's and the Sun's gravity also capture asteroids in two locations, called Lagrange points, along Jupiter's orbit.

• Some asteroids move in elliptical orbits that cross the orbits of Mars and Earth. Many of these asteroids will eventually strike one of the inner planets.

An asteroid may have struck the Earth 65 million years ago, contributing to the extinction of the dinosaurs and many other species.

• Comets are fragments of ice and rock that generally move in highly elliptical orbits about the Sun at a great inclination to the plane of the ecliptic.

As a comet approaches the Sun, its icy nucleus develops a luminous coma surrounded by a vast hydrogen envelope. An ion tail and a dust tail extend from the comet, pushed away from the Sun by the solar wind and radiation pressure.

Fragments of rock from "burned-out" comets produce meteor showers.

The Kuiper belt of comets orbits the Sun beyond Pluto.

Billions of cometary nuclei probably exist in the spherical Oort cloud located far beyond Pluto.

• Boulders and small rocks in space are called meteoroids. If a meteoroid enters the Earth's atmosphere, it produces a fiery trail called a meteor. If part of the object survives the fall, the fragment that reaches the Earth's surface is called a meteorite.

Meteorites are grouped in three major classes according to their composition: iron, stony-iron, and stony meteorites.

Rare stony meteorites called carbonaceous chondrites may be relatively unmodified material from the primitive solar nebula. These meteorites often contain carbon material and may have played a role in the origin of life on Earth.

• An analysis of the Allende meteorite suggests that a nearby supernova explosion triggered the formation of the solar system 4.6 billion years ago.

Review Questions

1 Why are asteroids, meteoroids, and comets of special interest to astronomers who want to understand the early history of the solar system?

2 Describe the asteroid belt.

3 Why do you suppose that there are many small asteroids but only a few very large ones?

4 What are Kirkwood gaps and what causes them?

5 What are the Trojan asteroids, and where are they located?

6 Describe the three main classifications of meteorites. How might these different types of meteorites have originated?

7 Suppose you found a rock that you suspect to be a meteorite. Describe some of the things you could do to see if it were a meteorite or a "meteorwrong."

8 Why do astronomers believe that meteoroids come from asteroids, whereas meteor showers are related to comets?

9 With the aid of a drawing, describe the structure of a comet.

10 Why is the phrase "dirty snowball" an appropriate characterization of a comet's nucleus?

11 What is the Kuiper belt, and how might it be related to debris left over from the formation of the solar system?

Advanced Questions

12 Why do you suppose comets are generally brighter after passing perihelion (closest approach to the Sun)?

13 Can you think of another place in the solar system where there is a phenomenon similar to the Kirkwood gaps in the asteroid belt? Explain.

14 Where on Earth might you find large numbers of stony meteorites that have not been significantly changed by weather?

15 Some astronomers have recently argued that passage of the solar system through an interstellar cloud of gas could perturb the Oort cloud, causing many comets to deviate slightly from their original orbits. What might be the consequences for Earth?

* 16 Assuming a constant rate of meteor in-fall, how much mass has the Earth gained in the past 4.6 billion years?

Discussion Questions

17 Suppose it were discovered that the asteroid Hermes had been perturbed in such a way as to put it on a collision course with Earth. Describe what you would do to counter such a catastrophe using present technology.

18 From the abundance of craters on the Moon and Mercury, we know that numerous asteroids and meteoroids struck the inner planets early in the history of the solar system. Is it reasonable to suppose that numerous comets also pelted the planets 3.5 to 4.5 billion years ago? Speculate about the effects of such a cometary bombardment, especially with regard to the evolution of the primordial atmospheres of the terrestrial planets.

Observing Projects

19 Make arrangements to view an asteroid. At opposition, some of the largest asteroids are bright enough to be seen through a modest telescope. Check the "Minor Planets" section of the current issue of the *Astronomical Almanac* to see if any bright asteroids are near opposition. If so, check the current issue as well as the most recent January issue of *Sky & Telescope* for a star chart showing the asteroid's path among the constellations. You will need such a chart to distinguish the asteroid from background stars. Observe the asteroid on at least two occasions separated by a few days. On each night, draw a star chart of the objects in your telescope's field of view. Has the position of one starlike object shifted between observing sessions? Does the position of the moving object agree with the path plotted on published star charts? Do you feel confident that you have in fact seen the asteroid?

20 Make arrangements to view a comet through a telescope. Since astronomers discover roughly a dozen comets each year, there is usually a comet visible somewhere in the sky. Unfortu-nately, because most comets are quite dim, you will need to have access to a moderately large telescope. Consult recent issues of the IAU *Circular*, published by the International Astronomical Union's Central Bureau for Astronomical Telegrams, which contains predicted positions and the anticipated brightness of comets in the sky. Also, if there is an especially bright comet in the sky, the latest issue of *Sky & Telescope* might contain useful information. Is a comet visible and do you have a telescope at your disposal? If so, can you distinguish the comet from background stars? Can you see its coma? Can you see a tail?

21 Make arrangements to view a meteor shower. The date of maximum activity in Table 8-1 is the best time to observe a particular shower, although good displays can often be seen a day or two before and after the maximum. In order to see a fine meteor display, you need a clear, moonless sky. The Moon's presence above the horizon can significantly detract from the number of faint meteors you will be able to see.

9 ▶ Our Star, the Sun

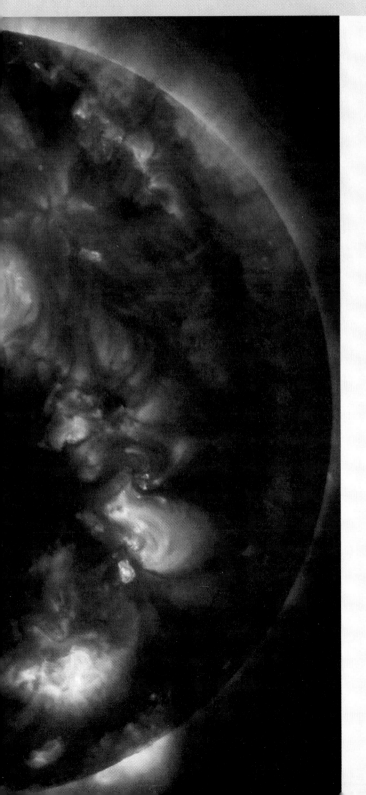

··

IN THIS CHAPTER

You will complete your exploration of the solar system by studying the Sun, a typical star. Today's technology has led to new understanding of a bewildering array of solar phenomena, including sunspots, plages, prominences, and flares. You will discover that some features of the Sun, generated by its internal heat, occur continuously. Other features, generated by the Sun's varying magnetic field, occur in 11-year cycles. You will also learn how the Sun generates the energy that makes it shine.

The X-Ray Sun Our momentary glimpses of the Sun from day to day give the impression of a blindingly bright, uniform body. This x-ray view of the Sun taken by a camera on board a rocket launched from New Mexico shows another, darker aspect of it. The x rays are emitted near the solar surface and appear in shades of yellow, orange, and red. The Sun's x-ray emissions are far less uniform than the visible light it emits. Sometimes its x rays are so intense as to be extremely dangerous to astronauts. By studying the locations and intensities of x rays from the Sun, astronomers gain an understanding of the energy transfer from the Sun's interior through its atmosphere.

1   What is the surface of the Sun like?

2   Does the Sun rotate?

3   What makes the Sun shine?

WITHOUT THE SUN'S ENERGY, we on Earth could not exist. Its ideal balance of heat, visible light, and ultraviolet radiation enabled life to form and flourish. No wonder our nearest star has been revered for millennia. No wonder, too, that astronomers have studied it intensively since Galileo first turned his telescope on the drama of sunspots. Today the Sun is our key to understanding stellar evolution, because for all its grandeur, the Sun is in fact a typical star. Table 9-1 lists the Sun's essential properties.

The Sun's Atmosphere

Although astronomers often speak of the solar surface, the Sun is so hot that it has no liquid or solid matter anywhere inside it. Moving toward the center of the Sun from space, one continually encounters ever denser gases.

9-1 The photosphere is the visible layer of the Sun

1   The Sun appears to have a surface only because most of the visible light comes from one specific layer (Figure 9-1). This region, which is about 400 km thick, is appropriately called the **photosphere** ("sphere of light"). The density of the photosphere's gas is low by Earth standards, about 0.01% as thick as the air we breathe. The photosphere has a nearly perfect blackbody spectrum corresponding to an average temperature of 5800 K (recall Figure 4-3).

The photosphere is the lowest of the three layers comprising the Sun's atmosphere. Because the upper two layers are transparent to most wavelengths of visible light, we see through them down to the photosphere. We cannot, however, see through the shimmering gases of the photosphere, and so everything below the photosphere is called the Sun's interior.

TABLE 9-1

The Sun's Vital Statistics

Statistic	Measurement
Mass	1.99×10^{30} kg
	$(3.33 \times 10^5 \ M_\oplus)$
Visual radius	6.96×10^5 km (109 $R_\oplus$)
Mean density	1410 kg/m^3
Luminosity (rate of energy emission)	3.90×10^{26} J/s
Surface temperature	5800 K
Central temperature	15.5×10^6 K
Equatorial rotation period	25 days

Under good observing conditions with a telescope and using special dark filters for protection, you can often see a blotchy pattern called *granulation* on the photosphere (Figure 9-2). Each lightly colored **granule** measures about 1000 km across and is surrounded by a darkish boundary. Time-lapse photography shows that granules form, disappear, and then re-form in cycles last-

Figure 9-1 **The Photosphere** The Sun emits most of its visible light from a thin layer of gas called the photosphere. Although it has no solid or even liquid region, we see the bottom of the photosphere as its "surface." We think of the photosphere as the lowest level of the Sun's atmosphere. Astronomers always take great care when viewing the Sun by using extremely dark filters or by projecting the Sun's image onto a screen. **Never look at the Sun directly. Doing so will destroy the retinas of your eyes.**

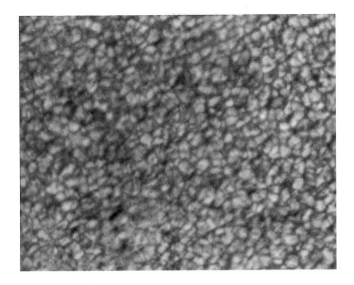

Figure 9-2 **Solar Granulation** High-resolution photographs of the Sun's surface reveal a blotchy pattern called granulation. Granules, each measuring about 1000 km across, are convection cells in the Sun's photosphere.

ing several minutes. At any single moment, several million granules cover the solar surface.

At the end of Chapter 4 we learned that radial motion of a light source affects its spectral lines through the Doppler effect (recall Figures 4-17 and 4-18 and Toolbox 4-2). By carefully measuring the wavelengths of spectral lines in various parts of individual granules, astronomers have determined that hot gas rises upward in the center of a granule. As it cools, the gas radiates its energy out into space. This is the visible light and other electromagnetic radiation we see from the Sun. The

cooled gas then spills over the edges of the granule and plunges back down into the Sun along the boundaries between granules (Figure 9-3). We saw in Chapter 7 that similar processes of convection drive the belts and zones on Jupiter and Saturn.

A granule's center is typically 100 K hotter than its edge. This difference in temperature explains why the centers of granules appear brighter than their edges. According to the Stefan–Boltzmann law (see Chapter 4), hotter regions emit more photons per square meter than do cooler regions. All these observations confirm that the Sun's surface is a vast layer of simmering gas.

The photosphere appears darker around the edge, or **limb,** than it does toward the center of the solar disk (examine Figure 9-1). This phenomenon, called **limb darkening,** arises because we see through only a certain amount of gas in the Sun's atmosphere before we reach the photosphere. However, looking at the center of the Sun's disk, we see the photosphere further down into the Sun's atmosphere, where it is hotter and brighter than it is at the limb.

9-2 The chromosphere is characterized by spikes of gas called spicules

Immediately above the photosphere is a dim layer of less dense stellar gas called the **chromosphere** ("sphere of color"). This unfortunate name suggests that it is the layer we normally see, but it is visible only when the photosphere is blocked, such as during a total solar eclipse. At that time, the chromosphere is visible as a pinkish strip some 500 km thick around the edge of the

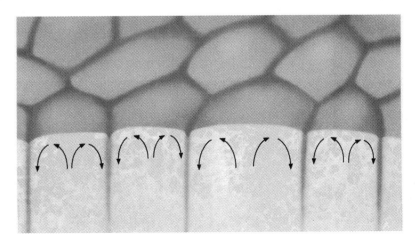

Figure 9-3 **Convection in the Photosphere** Hot gas rising upward produces bright granules. Cooler gas sinks downward along the darker, cooler boundaries between granules. This convective motion transports energy from the Sun's interior outward to the solar atmosphere.

Figure 9-4 **The Chromosphere** This photograph of the chromosphere was taken during an eclipse. It appears pinkish because the gas in the chromosphere emits only certain colors (wavelengths), among which the red ones dominate.

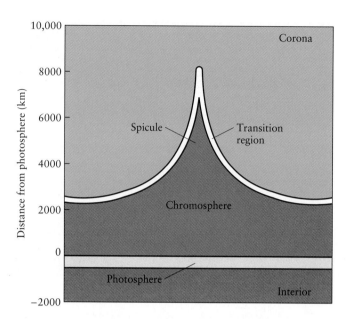

Figure 9-6 **Spicules** Small regions of the chromosphere surge upward, creating spike-shaped regions called spicules. These jets of gas last only a few minutes and cover only a few percent of the Sun's surface at any time.

dark Moon (Figure 9-4). Using special filters, however, astronomers can view the chromosphere without the benefit of an eclipse.

High-resolution images of the chromosphere reveal numerous spikes, which are jets of gas called **spicules** (Figure 9-5). A typical spicule rises for several minutes at the rate of 72,000 km/h (45,000 mph) to a height of nearly 10,000 km (Figure 9-6). Then it collapses and fades away. At any one time, roughly a third of a million spicules cover a few percent of the Sun's chromosphere.

Spicules are generally located on the boundaries of enormous regions of rising and falling chromospheric gas called **supergranules** (Figure 9-7). A typical supergranule has a diameter slightly larger than the Earth's and contains about 900 granules.

9-3 The corona ejects some of its mass into space as the solar wind

The outermost region of the Sun's atmosphere, the **corona**, extends several million kilometers from the top of the chromosphere (see Figure 9-6). The total amount of visible light we receive from the solar corona is com-

Figure 9-5 **Spicules and the Chromosphere** Numerous spicules are seen in this photograph of the Sun's chromosphere, which was taken through an H$_\alpha$ filter. The spicules are jets of

cool gas that surge upward into warmer regions of the Sun's outer atmosphere.

Figure 9-7 **Supergranules** Surrounded by spicules, supergranules are regions in the chromosphere that rise and fall. Each supergranule stretches over hundreds of granules in the photosphere below.

parable to the brightness of the full Moon—only about one-millionth as bright as the photosphere. The corona can thus be seen only when the photosphere is blocked out. This condition arises naturally during a total eclipse or artificially in a specially designed telescope called a *coronagraph*. Figure 9-8 is an exceptionally detailed photograph of the corona taken during a total eclipse.

It seems plausible that the temperature should fall as one rises through the Sun's atmosphere. After all, by moving upward, one moves farther from the heat source. Indeed, starting in the photosphere at 5800 K, the temperature drops to around 4000 K in the lower chromosphere. Surprisingly, however, the temperature then begins to rise. Around 1940 astronomers realized that the spectrum of the Sun's corona contains the emission lines of a number of highly ionized elements. For example, there is a prominent green line caused by the presence of Fe XIV (an iron atom stripped of 13 electrons). Because extremely high temperatures are required to strip that many electrons from atoms, it is clear that the corona

must be very hot. It is now known that coronal temperatures are in the range of one to two million kelvins!

Although the temperature is extremely high, the density of gas in the corona is very low, about ten trillion times less dense than the air here at sea level. The low density partly accounts for the dimness of the corona. Astronomers do not fully understand why the corona is so hot, although mounting evidence suggests that it is due to energy transported upward through the chromosphere to the corona by the Sun's complex magnetic fields.

Just as the Earth's gravity prevents most of the atmosphere from escaping into space, so, too, does the Sun's gravity keep most of its outer layers from leaving. However, some of the gas in the corona is moving fast enough—around one million kilometers per hour—to escape forever into space. As we saw in Chapter 5, such outflowing gas is called the **solar wind.**

The Sun ejects around a million tons of matter each second, mostly in the form of protons and electrons, in the solar wind. Even at this rate, this process will use up

Figure 9-8 **The Solar Corona** This extraordinary photograph was taken during the total solar eclipse of July 11, 1991. Numerous streamers are visible, extending millions of kilometers above the solar surface.

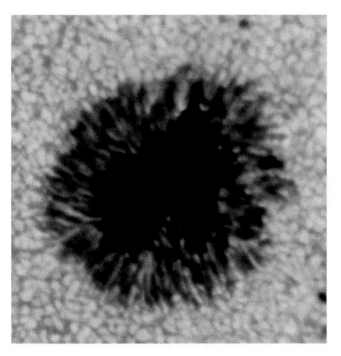

Figure 9-9 **A Sunspot** This dark region on the Sun is a typical isolated sunspot. Granulation is visible in the surrounding, undisturbed photosphere.

only a few tenths of a percent of the Sun's total mass throughout its lifetime.

The Active Sun

For all their intensity, granules, spicules, supergranules, and the solar wind will continue for as long as the Sun shines. We say that they are features of the quiet Sun. Yet the Sun's atmosphere is often disrupted by magnetic

Figure 9-10 **A Sunspot Group** This high-resolution photograph shows a sunspot group in which several sunspots overlap and others are nearby.

fields that stir things up, creating the *active* Sun. The Sun's most obvious transient features are **sunspots,** regions of the photosphere that appear dark because they are cooler than the rest of the Sun's "surface." Sometimes sunspots occur in isolation (Figure 9-9), but often they arise in clusters called sunspot groups (Figure 9-10).

9-4 Sunspots reveal the solar cycle and the Sun's rotation

Like other features of the active Sun, the number and location of sunspots vary in cycles. As shown in Figure 9-11, the average cycle lasts roughly 11 years. A time of

Figure 9-11 **The Sunspot Cycle** The number of sunspots on the Sun varies with a period of about 11 years. The most recent sunspot maximum occurred in 1990, and the most recent sunspot minimum occurred in 1986.

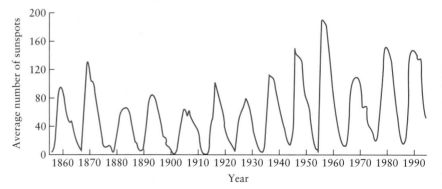

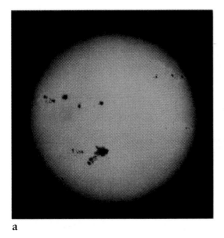

a   b

Figure 9-12 **Active and Quiet Sun** (a) The active Sun has many sunspots and other features (photo taken in 1989). (b) The quiet Sun is devoid of such features (photo taken in 1986).

many sunspots is called a **sunspot maximum.** Sunspot maxima occurred most recently in 1968, 1979, and 1990. During a **sunspot minimum,** the Sun is almost devoid of sunspots, as it was recently in 1965, 1976, and 1986 (Figure 9-12).

A typical sunspot is 10,000 km across and lasts between a few hours and a few months. Each sunspot has two parts: a dark, central region, called the *umbra,* and a brighter ring surrounding the umbra, called the *penumbra.* Seen without the surrounding, brilliant granules to outshine it, a sunspot's umbra appears red and its penumbra orange. From Wien's law (see Chapter 4), these colors indicate that the umbra is typically 4300 K and the penumbra 5000 K, both cooler than the normal photosphere.

On rare occasions a sunspot group is so large that it can be seen with the naked eye. Chinese astronomers recorded such sightings 2000 years ago, and a huge sunspot group visible to the naked eye was seen in 1979. (ALWAYS USE SPECIAL DARK FILTERS OR OTHER MEANS TO PROTECT YOUR EYES WHEN VIEWING THE SUN! *LOOKING DIRECTLY AT THE SUN CAUSES BLINDNESS!*) Of course, a telescope gives a much better view, and so it was not until Galileo that anyone had examined sunspots in detail.

By following sunspots as they moved across the solar disk (Figure 9-13), Galileo discovered that *the Sun rotates once in about four weeks.* Since a typical sunspot group lasts about two months, it can be followed for two solar rotations. Sunspot activity also lets us see the Sun's *differential rotation:* The equatorial regions rotate more rapidly than the polar regions. A sunspot near the solar

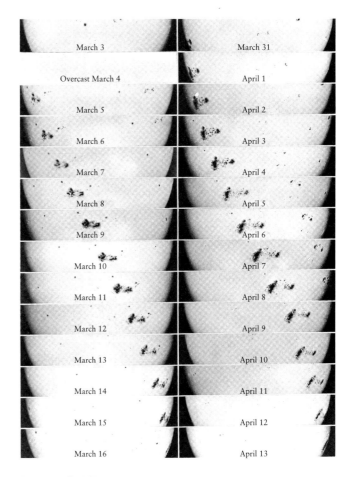

Figure 9-13 **The Sun's Rotation** By observing the same group of sunspots from one day to the next, Galileo found that the Sun rotates once in about four weeks. The equatorial regions of the Sun actually rotate somewhat faster than the polar regions. This series of photographs shows the same sunspot group over $1\frac{1}{2}$ solar rotations.

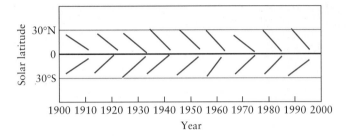

Figure 9-14 **Average Latitude of Sunspots Throughout the Sunspot Cycle** This graph shows that at the beginning of each sunspot cycle, most sunspots are at moderate latitudes, around 28° north or south. Sunspots arising later in each cycle typically form closer and closer to the Sun's equator.

equator takes 25 days to go once around the Sun, but a sunspot at 30° north or south of the equator takes about 27 days. The rotation period at 75° north or south of the equator is about 33 days, and near the poles it may be as long as 35 days.

The average latitude of new sunspots changes throughout the sunspot cycle. At the beginning of each

cycle, the sunspots are mostly at about 25° north and south latitude. Ones that form later in the cycle occur closer to the equator. Figure 9-14 shows that the average latitude at which sunspots are located varies at the same 11-year rate as does the number of sunspots.

9-5 The Sun's magnetic fields create sunspots

In 1908 the American astronomer George Ellery Hale discovered that sunspots are directly linked to intense magnetic fields on the Sun. When Hale focused a spectroscope on sunlight coming from a sunspot, he found that each spectral line in the normal solar spectrum is flanked by additional, closely spaced spectral lines not usually observed (Figure 9-15). This "splitting" of a single spectral line into two or more lines is called the **Zeeman effect;** it is named after the Dutch physicist Pieter Zeeman, who first observed it in 1896. Zeeman showed that an intense magnetic field splits the spectral lines of a light source inside the field. The more intense the magnetic field, the more the split lines are separated. For ex-

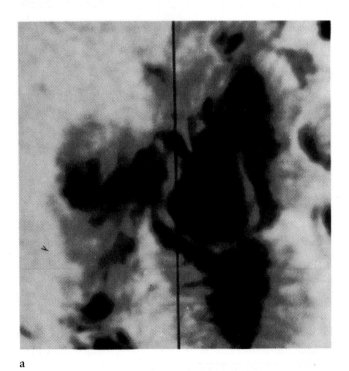

a

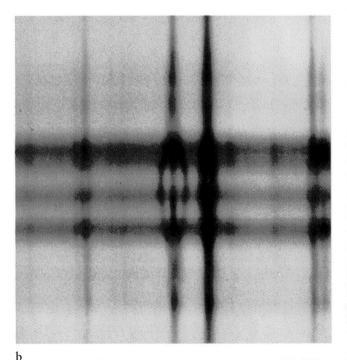

b

Figure 9-15 **Zeeman Splitting by a Sunspot's Magnetic Field** (**a**) The black line drawn across the sunspot indicates the location toward which the slit of the spectroscope was aimed. (**b**) In the resulting spectrogram, one line in the middle of the

normal solar spectrum is split into three components. The separation between the three lines corresponds to a magnetic field roughly 5000 times stronger than the Earth's natural magnetic field.

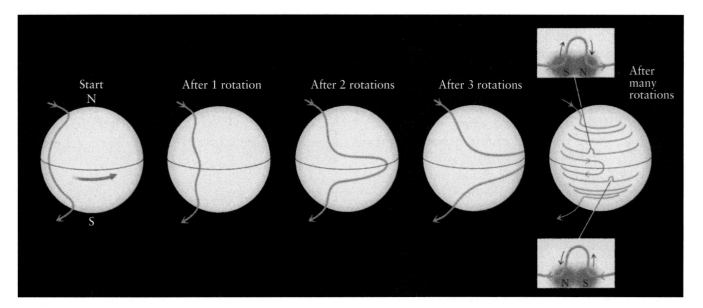

Figure 9-16 **Babcock's Magnetic Dynamo** In a possible partial explanation for the sunspot cycle, differential rotation wraps a magnetic field around the Sun. Sunspots appear where the concentrated magnetic field breaks through the solar surface.

ample, the splitting of the spectral line in Figure 9-15 into three lines corresponds to a magnetic field roughly 5000 times more intense than the Earth's natural magnetic field.

Hale's discovery demonstrates that sunspots are areas where concentrated magnetic fields project through the hot gases of the photosphere. Because of the temperature, many atoms in the photosphere are *ionized:* They lose electrons. As a result, the photosphere is a mixture of electrically charged ions and electrons called a **plasma.** A plasma is an extremely good conductor of electricity, and it interacts vigorously with magnetic fields. Specifically, the protruding magnetic field prevents hot gases inside the Sun from rising to the surface as they normally do. Such regions of the photosphere are left relatively devoid of gas and, therefore, are cooler and darker than the surrounding solar surface. These darker, cooler regions are sunspots.

On one hemisphere of the Sun, Hale found that the sunspots with a magnetic north pole always come into view before the corresponding sunspots with a magnetic south pole. On the other hemisphere, the order is reversed. The sunspots and sunspot groups in each hemisphere are connected in pairs, one where the magnetic field leaves the Sun, the other where it returns. Hale also found that this pattern reverses itself every 11 years. The hemisphere that has preceding north magnetic poles during one 11-year cycle has preceding south magnetic poles

during the next. Astronomers therefore speak of the **solar cycle,** which has a period of 22 years.

In 1960 the American astronomer Horace Babcock proposed the **magnetic dynamo** model to explain the 22-year solar cycle. As shown in Figure 9-16, the Sun's rotation generates a magnetic field by moving its electrically charged gases. This field normally lies just below the surface. However, convection under the photosphere tangles the concentrated magnetic field, and kinks or loops erupt through the solar surface. This first occurs at high latitudes; over the next 11 years it occurs closer and closer to the Sun's equator. Sunspots appear where the magnetic field projects through the photosphere.

The external loops of the magnetic field interact with the Sun's field and eventually cancel it out, causing the magnetic field to suddenly reverse itself, with the Sun's north magnetic pole becoming the south magnetic pole and vice versa. At that point all the fields creating the sunspots once again return below the surface, and the sunspot cycle is over. Because the overall magnetic field of the Sun has been reversed by the fields from the spots, the sunspots in the next cycle have reversed preceding fields, which explains why there is a 22-year cycle rather than two identical 11-year cycles.

Astronomers still do not know what holds a sunspot together week after week, even though, according to present calculations, a sunspot should break up and disperse as soon as it forms. Our understanding of sunspots is further confounded by irregularities in the solar cycle. For example, the overall reversal of the Sun's magnetic field is often piecemeal and haphazard. One pole may reverse polarity long before the other, so that for weeks or months on end the Sun may have two north poles but no south pole at all. To make matters worse, there is strong historical evidence that all traces of sunspots and the sunspot cycle have vanished for many years at a time. For example, virtually no sunspots were seen from 1645 through 1715. Such sunspot-free periods are associated with years of extreme cold on the Earth.

9-6 Magnetic fields lead to plages and filaments on the Sun

Figure 9-17 shows the active Sun's chromosphere and corona. The bright areas in this photograph are called **plages** (from the French word for "beaches"). They are hotter, and therefore brighter, than the surrounding chromosphere. Plages, which often appear just before new sunspots form, are believed to be created by magnetic fields under the photosphere crowding upward just before they emerge through the photosphere. In pushing upward, the fields compress the gases of the upper Sun.

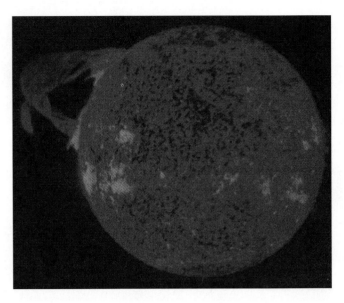

Figure 9-18 **A Prominence** A huge prominence arches above the solar surface in this Skylab photograph taken in 1973. The radiation that exposed this picture is from singly ionized helium at a wavelength of 30.4 nm, corresponding to a temperature of about 50,000 K.

This, in turn, causes this gas to become hotter and therefore glow more brightly.

The dark streaks in Figure 9-17 are features in the corona called **filaments**. They are huge volumes of gas lofted upward from the photosphere by the Sun's magnetic field. When viewed from the side, rather than from above, filaments are seen as gigantic loops or arches called **prominences** (Figure 9-18). The temperature of gas in prominences can reach 50,000 K. These features are almost always associated with sunspots. Some prominences last only for a few hours, while others persist for months.

The most violent, eruptive events on the Sun, called **solar flares**, occur in complex sunspot groups (Figure 9-19). During a solar flare, temperatures in a compact region soar to five million kelvins, blasting vast quantities of particles and radiation into space. The flare is usually over within 20 minutes. Ultraviolet and x-ray radiation from the flare take about 8 minutes to reach the Earth, while high-energy particles from it arrive a day or two later. The particles interfere with radio communications and often produce intense aurorae in the Earth's atmosphere as the Van Allen belts become overloaded with particles from the Sun. The numbers of prominences, flares, and plages vary with the same 11-year cycle as the number of sunspots.

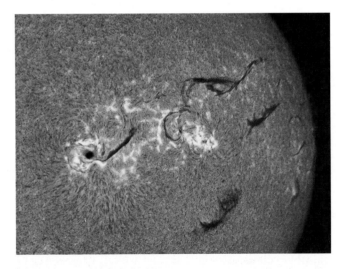

Figure 9-17 **Active Sun in H$_\alpha$** This photograph shows the chromosphere and corona during a solar maximum. The image was taken through a filter allowing only light from H$_\alpha$ emission to pass through. The hot upper layers of the Sun's atmosphere are strong emitters of H$_\alpha$ photons. Besides the sunspots, there are several other features that don't appear at the solar minimum. The snakelike features are filaments and the bright areas are plages.

Figure 9-19 **A Flare** Solar flares, which are associated with sunspot groups, produce the most energetic emission of particles from the Sun. When such flares are aimed toward the Earth, the particles often penetrate the Earth's Van Allen belts, causing aurorae, disrupting radio communications, and even causing power blackouts by disrupting power transmission.

9-7 X-ray images of the corona show a dramatically different Sun than we normally see

Recall from Chapter 4 that, according to Wien's law, the hotter a blackbody is, the shorter the dominant wavelength of the radiation it emits. The corona, at millions of kelvins, shines very brightly at x-ray wavelengths. X-ray images of the corona reveal a very blotchy, irregular inner corona (Figure 9-20). Note the large dark area, called a **coronal hole** because it is nearly devoid of the usual hot, glowing coronal gases. The coronal holes are the main corridors through which particles of the solar wind escape from the Sun. X-ray photographs also reveal numerous bright spots that are hotter than the surrounding corona. Temperatures in these bright points occasionally reach four million kelvins. Many of the bright coronal hot spots seen in x-ray pictures such as Figure 9-20 are located over sunspots.

The Sun's Interior

During the nineteenth century, geologists and biologists found convincing evidence that the Earth must have ex-

isted in more or less its present form for at least hundreds of millions of years. This fact posed severe problems for astrophysicists, because at that time it seemed impossible to explain how the Sun could have been shining for so long, radiating immense amounts of energy into space. If the Sun were shining by burning coal or gas, for example, it would be ablaze for only 5000 years before consuming all of its fuel. The answer lay deep within the Sun's interior.

9-8 Thermonuclear reactions in the core of the Sun produce its energy

In 1905 Albert Einstein provided an important key to the source of the Sun's energy with his special theory of relativity. One of the implications of this theory is that matter and energy are related by the simple equation:

$$E = mc^2$$

In other words, a mass (m) can be converted into an amount of energy (E) equivalent to mc^2, where c is the speed of light. Because c is a large number and c^2 ($c \times c$) is huge, a small amount of matter can be converted into an awesome amount of energy.

Inspired by Einstein's work, astrophysicists began to wonder if the Sun's energy output might come from the

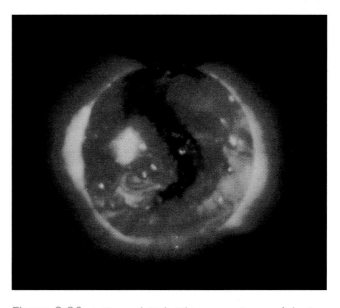

Figure 9-20 **A Coronal Hole** This x-ray picture of the Sun was taken by Skylab space station astronauts in 1973. A huge, dark, boot-shaped coronal hole dominates this view of the inner corona. Numerous bright points are also visible.

conversion of matter into energy. But exactly what would be the mechanism to make this conversion? In the 1920s the British astronomer Arthur Eddington speculated that temperatures at the center of the Sun must be much greater than had previously been thought. Under sufficiently high-temperature conditions, hydrogen nuclei fuse together to produce helium nuclei. This reaction also transforms a tiny amount of mass into a very large amount of energy—the energy of the Sun. Because of the extreme temperatures required to fuse nuclei, this process is called **thermonuclear fusion.** The central region of the Sun in which this fusion occurs is called its **core.** We will encounter other types of thermonuclear fusion (converting helium to carbon and carbon to nitrogen) in later chapters.

You may have heard the comments that mass is always conserved or that energy is always conserved. Neither of these statements is true. What is true, however, is that the total amount of mass *plus* energy is conserved. So, the destruction of mass by the Sun does not violate any laws of nature.

The thermonuclear fusion entailed in converting hydrogen into helium is called **hydrogen fusion.** Similarly descriptive expressions are used to specify the fusion of helium or other elements. The energy generated by hydrogen fusion in the Sun's core eventually escapes through the photosphere into space. That is the energy that makes the Sun shine. Hydrogen fusion is also called **hydrogen burning,** even though nothing is actually burned in the conventional sense. The ordinary burning of wood, coal, or any flammable substance is a chemical process involving only the electrons orbiting the nuclei of the atoms. Thermonuclear fusion is a far more energetic process that involves violent collisions between the nuclei themselves. Toolbox 9-1 will show you just *how* energetic.

AN ASTRONOMER'S TOOLBOX 9-1

Thermonuclear Fusion

Thermonuclear fusion occurs in the Sun's core, where the entire mass of the Sun compresses inward, creating sufficiently high pressure and temperature to fuse atomic nuclei together: The core's temperature is 15.5 million kelvins, its pressure is about 3.4×10^{11} atmospheres, and its density is 160 times greater than that of water. Under less severe conditions, the positive electric charge on each proton prevents nearby protons from penetrating it, because like charges repel each other. But in the extreme heat and pressure of the Sun's center, the protons move so fast that they can infiltrate each other's electric fields and stick or fuse together.

The nucleus of the simplest atom, hydrogen (H), consists of a single proton. The nucleus of the next simplest atom, helium (He), consists of two protons and two neutrons. In the process of four hydrogen atoms fusing into helium,

$$4H \longrightarrow He$$

two of the four protons from the hydrogen atoms change into neutrons to produce a single helium nucleus. This conversion releases two positively charged electrons from the protons, called **positrons.** When these positrons encounter regular electrons in the Sun's core, both particles are annihilated and their mass is converted into energy in the form of gamma-ray photons. In addition, massless particles called **neutrinos** are also emitted.

The mass lost during this reaction can be calculated as follows:

Mass of 4 hydrogen atoms = 6.693×10^{-27} kg

Mass of 1 helium atom = 6.645×10^{-27} kg

Mass lost = 0.048×10^{-27} kg

From Einstein's famous equation,

$$E = mc^2$$
$$= (0.048 \times 10^{-27} \text{ kg}) \times (3 \times 10^8 \text{ m/s})^2$$
$$= 4.3 \times 10^{-12} \text{ J}$$

This energy would light a 10-W lightbulb for almost $\frac{1}{2}$ trillionth of a second.

The Sun's mass, usually designated 1 $M_\odot$, is equal to 333,000 $M_\oplus$. Its total energy output per second, called its **luminosity** and denoted $L_\odot$, is 3.9×10^{26} W. To produce this luminosity, the Sun converts 600 million metric tons of hydrogen into helium within its core each second. This prodigious rate is possible because the Sun contains a vast supply of hydrogen—enough to continue the present rate of energy output for another five billion years.

9-9 Solar models describe how energy escapes from the Sun's core

A scientific description of the Sun's interior, called a **solar model,** explains how the energy from nuclear fusion gets to the photosphere. The model carefully charts the Sun's internal characteristics, such as its pressure, temperature, and density at various depths. Our model of the Sun's interior is expressed as a set of mathematical equations called the **equations of stellar structure.** These describe the conditions required inside *any* star to keep it stable, meaning that it neither expands nor collapses. Because the equations are so complex, astrophysicists today use high-speed computers to solve them. Figure 9-21 presents their results graphically. The four graphs show how the Sun's luminosity, mass, temperature, and density vary from the Sun's center to its surface. For instance, the upper graph gives the percentage of the Sun's luminosity created within that radius. The luminosity rises to 100% at about one-quarter of the way from the Sun's center to its surface. This tells us that all the Sun's energy is produced within a volume extending out to $\frac{1}{4}$ R$_\odot$.

The mass curve rises to nearly 100% at about 0.6 R$_\odot$ from the Sun's center. Almost all of the Sun's mass is therefore confined to a volume extending only 60% of the distance from the Sun's center to its surface. Indeed, the density of the gas in the photosphere is 10^4 times lower than the density of the air we breathe.

The model begins with the inward force of the Sun's gravity. This force raises the pressure and temperature in the Sun's core, as we have seen. However, because the Sun is not shrinking today, there must also be an outward force to counter the inward force of gravity. That outward force is produced by the gamma-ray photons created during fusion slamming into electrons in the core. This energy from fusion thereby raises the temperature and pressure in the core. Because the ions and electrons in the core are moving at very high speeds and are densely packed together, they don't travel far before striking other particles. In each collision the particles exert forces on each other. They then rebound and collide with other particles. The result of these frequent interactions is an outward force sufficient to counterbalance gravity. The balance between the inward force of gravity and the outward force from the gas pressure is called **hydrostatic equilibrium.**

Meanwhile, photons are slowly, randomly moving outward from the Sun's core (Figure 9-22). The outward transportation of energy by photons hitting particles, which then bounce off other particles and thereby re-

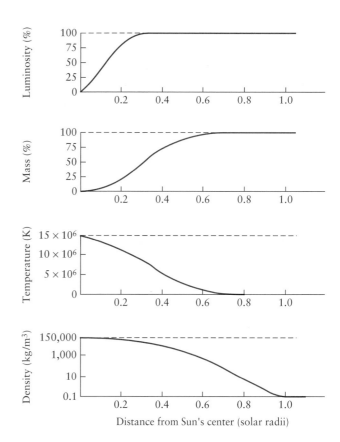

Figure 9-21 **A Theoretical Model of the Sun** The Sun's internal structure is displayed here with graphs that show how the luminosity, mass, temperature, and density vary with the distance from the Sun's center. A solar radius (the distance from the Sun's center to the photosphere) equals 696,000 km.

emit photons, is called *radiative transport* because individual photons are responsible for carrying energy outward. The emitted photons usually have less energy after the collision than they did before it. Calculations show that radiative transport is the dominant means of outward energy flow in the **radiative zone,** extending from the core to 80% of the way out to the photosphere.

Near the photosphere, bulk motion of the hot gas, rather than the flying photons, carries most of the energy the remaining distance to the surface. As we discussed earlier, this upward motion of gas blobs is called convection. We thus say that the Sun has a **convective zone.** Once the gas reaches the photosphere, it cools and settles back into the Sun. This convective flow is the origin of the granules, and the departing energy is the light that the Sun emits into space. It takes around a million years for photons to work their way from the Sun's core to its

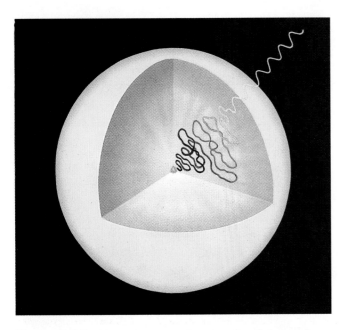

Figure 9-22 **Photons Leaving the Sun** This drawing shows the million-year odyssey of photons generated in the core of the Sun as they randomly move to the photosphere and then out into space. On their way out, the photons lose energy to ions, whose collisions help the Sun maintain hydrostatic equilibrium.

photosphere. The internal structure of the Sun is sketched in Figure 9-23.

The photons emerging from the Sun have lost energy on their way up from the core. This energy went into keeping the Sun in hydrostatic equilibrium. As a result, photon wavelengths changed. Although the photons created in the Sun are gamma rays, its most intense emission is in the visible part of the electromagnetic spectrum.

To learn more about the Sun's interior, astronomers record its vibrations, just as geologists use earthquakes to study the Earth's interior structure. Although there are no true sunquakes, the Sun does vibrate at a variety of frequencies, somewhat like a ringing bell. These vibrations, first noticed in 1960, can be detected with sensitive Doppler shift measurements. They have shown that portions of the Sun's surface move up and down by about 10 km every 5 minutes (Figure 9-24). Slower vibrations with periods ranging from 20 minutes to nearly an hour were discovered in the 1970s. More recently, longer-period oscillations have been detected. One important discovery from the study of the Sun's vibrations, called **helioseismology,** is that the convective zone is twice as thick as prior solar models predicted. Another is that be-

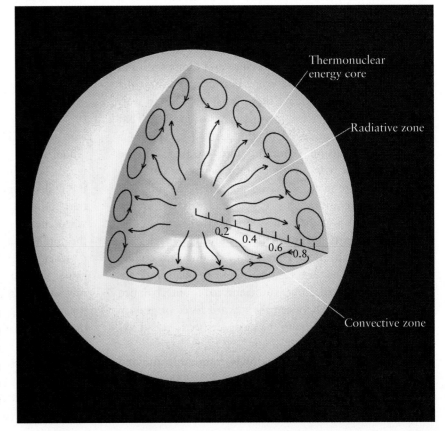

Figure 9-23 **The Sun's Internal Structure** Thermonuclear reactions occur in the Sun's core, which extends to a distance of 0.25 solar radius from the center. Energy from the core radiates outward to a distance of 0.8 solar radius. Convection is responsible for energy transport in the Sun's outer layers.

Figure 9-24 **Sound Waves Resonating in the Sun** This computer-generated image shows one of the millions of ways that the Sun vibrates because of sound waves resonating in its interior. The regions that are moving outward are colored blue; those moving inward are red. The cutaway shows how deep these oscillations are believed to extend.

low the convective zone the Sun apparently rotates like a rigid body (Figure 9-25).

9-10 The mystery of the missing neutrinos inspires speculation about the Sun's interior

For every proton that changes into a neutron during thermonuclear fusion, a neutrino is released. Neutrinos have no electric charge and were long believed to have no mass either. If so, they resemble photons and travel through space at the speed of light. However, some physicists theorize that neutrinos might have a tiny mass, probably less than 0.0001 times the mass of an electron. If neutrinos do have mass, they travel at speeds less than that of light.

Neutrinos are extraordinarily difficult to detect because they do not interact much with ordinary matter. In fact, neutrinos interact so weakly and so infrequently with matter that they easily pass through the entire Earth as if it were not there. The Sun is also largely transparent to neutrinos, allowing these particles to stream outward, unimpeded, from its core. Nearly 10^{38} neutrinos are produced at the Sun's center each second. This output is so huge that, here on Earth, roughly one hundred billion

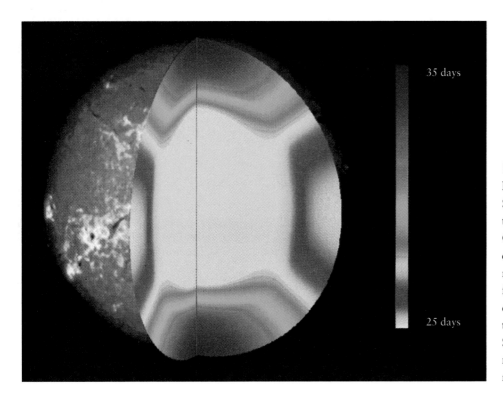

Figure 9-25 **Rotation of the Solar Interior** This cutaway picture of the Sun shows how the rate of solar rotation varies with depth and latitude. Colors represent rotation periods according to the scale given at the right. Note that the pattern of surface rotation, which varies from 25 days at the equator to 35 days near the poles, persists throughout the Sun's convective envelope. The Sun's radiative zone seems to rotate like a rigid body.

Figure 9-26 **The Solar Neutrino Experiment** This tank contains 100,000 gallons of perchloroethylene buried 1.5 km below ground at the Homestake Gold Mine in South Dakota. The Earth shields the tank from stray particles and radiation so that only solar neutrinos can convert chlorine atoms in the fluid into radioactive argon atoms. The rate at which argon atoms are created in the tank is a direct measure of the number of neutrinos coming from the Sun.

neutrinos pass through every square centimeter of your body every second!

On very rare occasions a neutrino strikes a neutron and converts it into a proton. If astronomers could detect even a few of these converted protons, it might be possible to build a "neutrino telescope" that could be used to "see" directly into the center of the Sun and the thermonuclear inferno there that is now hidden from ordinary view.

Inspired by such possibilities, Raymond Davis of the Brookhaven National Laboratory designed and built a large neutrino detector. This device consists of a huge tank containing 100,000 gallons of perchloroethylene cleaning fluid (C_2Cl_4) buried deep in a mine in South Dakota (Figure 9-26). Because matter is virtually transparent to neutrinos, most of the neutrinos from the Sun pass right through Davis' tank with no effect whatsoever. On rare occasions, though, a neutrino strikes the nucleus of one of the chlorine atoms in the cleaning fluid and

converts one of its neutrons into a proton, creating a radioactive atom of argon. The rate at which argon is produced is correlated with the number of neutrinos from the Sun arriving at the Earth.

On the average, solar neutrinos create one radioactive argon atom every three days in Davis' tank. To the continuing consternation of astronomers, this rate corresponds to only one-third of the neutrinos predicted from the "standard model" of the Sun presented in Figure 9-21. This experiment, which began in the mid-1960s, has been repeated with extreme care by other researchers around the world during the past 30 years. Even though different systems have been employed, all get comparable results. Most astronomers now agree that the solar neutrino experiments are probably sound, implying that there really are far fewer detectable neutrinos from the Sun reaching the Earth than current theory predicts. This result compels astrophysicists to reexamine their ideas about the nature of neutrinos and about the Sun.

To solve the mystery of the missing solar neutrinos, some astronomers suggest that there are other types of neutrinos that we cannot yet detect. Some theorists predict that if neutrinos have mass, they may continually change from one type to another. According to this hypothesis, many initially detectable neutrinos change into undetectable ones by the time they reach the Earth.

Another proposed explanation is that nuclear reactions in the Sun's core are exceptionally sensitive to temperature. If the Sun's center were only 10% cooler, neutrino production might drop enough to agree with current neutrino experiments. Unfortunately, obvious features of the Sun, such as its size and surface temperature, would then differ from what we observe. The missing solar neutrinos remain a mystery.

WHAT DID YOU THINK?

1 *What is the surface of the Sun like?* The photosphere is composed of hot, churning gases. There is no solid or liquid region in the Sun.

2 *Does the Sun rotate?* The Sun's surface rotates differentially. The rate varies between once every 25 and once every 35 days.

3 *What makes the Sun shine?* Thermonuclear fusion at the Sun's core is the source of the Sun's energy.

Key Words

chromosphere	helioseismology	photosphere	solar wind
convective zone	hydrogen fusion	plage	spicule
core (of the Sun)	(hydrogen burning)	plasma	sunspot
corona	hydrostatic equilibrium	positron	sunspot maximum
coronal hole	limb (of the Sun)	prominence	sunspot minimum
equations of stellar	limb darkening	radiative zone	supergranule
structure	luminosity	solar cycle	thermonuclear fusion
filament	magnetic dynamo	solar flare	Zeeman effect
granules	neutrino	solar model	

Key Ideas

• The visible surface of the Sun is a layer at the bottom of the solar atmosphere called the photosphere. The gases in this layer shine with almost perfect blackbody radiation, and convection produces features called granules there.

• Above the photosphere is a layer of hotter but less dense gas called the chromosphere. Jets of gas called spicules rise up into the chromosphere along the boundaries of supergranules.

• The outermost layer of thin gases in the solar atmosphere is called the corona, which blends into the solar wind at great distances from the Sun. The gases of the corona are very hot but at low density.

• Surface features on the Sun vary periodically in an 11-year cycle, with the magnetic fields that cause these changes actually changing over a 22-year cycle.

Sunspots are relatively cool regions produced by local concentrations of the Sun's magnetic field. The average number of sunspots increases and decreases in an 11-year cycle.

A solar flare is a brief eruption of hot, ionized gases from a sunspot group.

The magnetic dynamo model suggests that many features of the solar cycle are caused by the effects of differential rotation and convection on the Sun's magnetic field.

• The Sun's energy is produced by the thermonuclear process called hydrogen fusion, in which four hydrogen nuclei release energy when they fuse to produce a single helium nucleus. The energy released in a thermonuclear reaction comes from the conversion of matter into energy according to Einstein's equation $E = mc^2$.

A thermonuclear reaction is a nuclear reaction that occurs only at very high temperatures; hydrogen fusion, for instance, occurs only at temperatures of more than about eight million kelvins.

• A stellar model is a theoretical description of a star's interior derived from calculations based on the laws of physics.

The solar model suggests that hydrogen fusion occurs in a core that extends from the Sun's center to about 0.25 solar radius.

Throughout most of the Sun's interior, energy moves outward from the core by radiative diffusion. In the Sun's outer layers, energy is transported to the Sun's surface by convection.

• Neutrinos emitted by the Sun are detected at a lower rate than is predicted by our model of the Sun's interior.

Review Questions

1 Describe the Sun's atmosphere. Be sure to mention some of the phenomena that occur at various altitudes above the solar surface.

2 Describe the three main layers in the solar atmosphere and how you would best observe them.

* 3 When will the next sunspot maximum and minimum occur? Explain your reasoning.

4 Why is the solar cycle said to have a period of 22 years, even though the sunspot cycle is only 11 years long?

5 How do astronomers detect the presence of a magnetic field in hot gases, such as those in the solar photosphere?

6 Describe the dangers in attempting to observe the Sun. How have astronomers learned to circumvent these hazards?

7 Give an everyday example of hydrostatic equilibrium.

8 Give some everyday examples of heat transfer by convection and radiative diffusion.

9 What do astronomers mean by "a model of the Sun"?

10 Why do thermonuclear reactions in the Sun take place only in its core?

11 What is hydrogen burning? Why is hydrogen burning fundamentally unlike the burning of a log in a fireplace?

12 Describe the Sun's interior, including the main physical processes that occur at various depths within the Sun.

13 What is a neutrino, and why are astronomers so interested in detecting neutrinos from the Sun?

Advanced Questions

* 14 Using the mass and size of the Sun, calculate the Sun's average density. Compare your answer with the average densities of the outer planets. (*Hint:* The volume of a sphere of radius r is $\frac{4}{3}\pi r^3$.)

* 15 Assuming that the current rate of hydrogen fusion in the Sun remains constant, what fraction of the Sun's mass will be converted into helium over the next five billion years? How will this affect the chemical composition of the Sun?

* 16 Calculate the wavelengths at which the photosphere, chromosphere, and corona emit the most radiation. Explain how the results of your calculations suggest the best way to observe these regions of the solar atmosphere. (*Hint:* Use Wien's law and assume that the average temperatures of the photosphere, chromosphere, and corona are 5800 K, 50,000 K, and 1.5×10^6 K, respectively.)

Discussion Questions

17 Discuss the extent to which cultures around the world have worshiped the Sun as a deity throughout history. Why do you suppose there has been such widespread veneration?

18 Discuss some of the difficulties of correlating solar activity with changes in the terrestrial climate.

19 From 1645 to 1715 virtually no sunspots were seen, and northern Europe experienced a "Little Ice Age" during which record low temperatures were recorded. Could these two phenomena be related? Explain.

20 Describe some advantages and disadvantages of observing the Sun (**a**) from space and (**b**) from Earth's south pole. What kinds of phenomena and issues do solar astronomers want to explore from both Earth-orbiting and Antarctic observatories?

Observing Projects

21 Use a telescope to view the Sun by projecting the Sun's image onto a screen or sheet of white paper. **DO NOT LOOK AT THE SUN! Looking at the Sun causes blindness.** Do you see any sunspots? If so, sketch their appearance. Can you distinguish between the umbrae and penumbrae of the sunspots? Can you see limb darkening? Can you see granulation?

22 If you have access to an H_α filter attached to a telescope especially designed for viewing the Sun safely, use this instrument to examine the solar surface. How does the appearance of the Sun differ from that in white light (see the previous observing project)? What do sunspots look like in H_α? Can you see any prominences? Can you see any filaments? Are the filaments in the H_α image near any sunspots seen in white light?

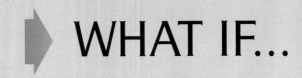
THE MOON—FULL and bright, or crescent and mysterious—is a fixture in our sky and in our culture. Throughout history, people have woven myths about the Moon and its effects on everything from childbirth to stock market activity. Countless romances have begun under a full Moon, and entire nations have dedicated themselves to reaching it.

What would the Earth be like if the Moon never existed? Obviously, there would be no eclipses or moonlight—all clear nights would be equally dark and starfilled (except where there are city lights). This would be a true boon to astronomers, who cannot see dim stars on moonlit nights. Otherwise, a moonless night would have minimal effects on humans, since all the alleged effects of moonlight on humans, such as those mentioned above, have been disproved.

In the animal world, however, nocturnal animals would be less successful at hunting, foraging, and traveling if the sky were perpetually moonless. Birds and other species that travel only on moonlit nights would have to have evolved different mechanisms for navigation.

The ocean tides, primarily caused by the gravitational attractions of the Moon and Sun, would only be one-third of what they are today. The Sun-generated tides on a moonless Earth would have constant high and low levels throughout the year—gone would be the spring and neap tides. Furthermore, the cycle of tides would take place over 12 hours (half of a solar day) instead of the approximately 12 hours and 25 minutes they are today.

We saw in Chapter 5 that the Moon is believed to have formed after a Mars-sized asteroid struck the Earth.

At that time, the day was only about six hours long. The Moon, after coalescing from impact debris, was at least ten times closer to the Earth than it is today. The monstrous tides it created back then were thousands of feet high, and they roared inland and receded seaward every three hours. This titanic flow of water pulverized the land, stripped it, and created a thick soup of elements in the oceans that, in combination with ultraviolet radiation from the Sun, enabled life to first form on Earth. Without the Moon, the tiny Sun-induced tides would not have helped enrich the oceans with land-based elements as rapidly. As a result, life would have evolved much more slowly.

The day on a moonless Earth would only be about eight hours long, rotation having slowed slightly from the original six-hour-long days because of the Sun-generated tides. The high-speed rotation of that world would lead to powerful winds, like those that whip around the giant planets today. Mountains on a moonless Earth would be lower, having been worn down by these winds.

Our physiology evolved according to a 24-hour day. This is most evident in our biological clocks, or circadian rhythms, which are the internal mechanisms that regulate sleeping and waking, eating, seeking mates, and other cyclic activities. On a world with an 8-hour day, our circadian rhythms would be hopelessly out of synch with the natural world. In order to function on such a world, all its creatures and plants would have had to evolve biological clocks based on eight hours. Considering all we have to do now, imagine what life would be like with only eight hours each day!

FOUNDATIONS

The Stars

IN THE PAGES AHEAD

Science by Starlight By analyzing starlight, an astronomer can determine stellar chemistries and masses, how far away the stars are, and how fast they are moving toward or away from us. This image shows stars in Orion and nearby constellations. These stars span virtually the entire range of star types.

The brightest star in the night sky, Sirius (part of Canis Major), is toward the lower left. It is only 8.6 ly away, making it one of the stars closest to us.

Most stars in this photograph are hundreds, or even thousands, of light-years away. The blue haze is starlight scattering off interstellar gas and dust.

You will turn your eyes to the stars. Even the distances to nearby stars are staggering compared to anything with which you are familiar. To determine these distances, you will learn how to use parallax. Knowing a star's brightness, called apparent magnitude, and its distance, you can then easily deduce its luminosity, the total energy it emits per second. You will see that stellar luminosities and spectra are clues to understanding stellar evolution, which is discussed in depth in the following four chapters. Finally, you will see that stars with different masses evolve differently. The evolution of stars shows just how special—and how ordinary—the Sun really is.

WHAT DO YOU THINK?

1 How far is the closest star other than the Sun?

2 How bright is the Sun compared to other stars?

·······························

THE *LEGEND OF THE EPHEMERA* tells of a race of remarkable insects, the Ephemera, who inhabited a great forest. These noble creatures were blessed with great intelligence, yet cursed with tragically short life spans. To the Ephemera, the forest seemed eternal and unchanging. Members of each generation lived out their brief lives without ever noticing any changes in their leafy world.

Nevertheless, careful observations and reasoning led some Ephemera to postulate that the forest was not static. They began to suspect that small green shoots grew to become huge trees and that mature trees eventually died, toppled over, and littered the forest with rotting logs, enriching the soil for future trees. Although unable to witness the transformation personally, the Ephemera became aware of life cycles stretching over the awesome periods of many years.

At first the heavens, too, seem eternal and unchanging; the views that greet us every night are virtually indistinguishable from those seen by our ancestors. This permanence, however, is an illusion. We see the stars because they emit vast amounts of radiation. As they do so, they must change, a life cycle we call **stellar evolution.**

We do have one advantage over the Ephemera, who were so small that they saw only a few trees, leaves, shoots, and rocks. Astronomers are witness to literally billions of stars in various stages of evolution. Therefore, although we can't see any one star go through more than a tiny fraction of its life, we see every stage of stellar evolution. By understanding the lives of stars, we gain insight into our place in the cosmos. We begin our journey to the stars by finding out how far we have to go across the icy void of interstellar space to reach them.

III-1 Distances to nearby stars are determined by stellar parallax

There is nothing in the human experience to prepare us for the distances between the stars. On Earth the largest distance we encounter in our daily lives is at most a few thousand kilometers. From looking at the Moon and the planets, we have some comprehension of hundreds of thousands or even millions of kilometers.

How far do you think the nearest stars are? Ask ten people and you'll probably hear answers ranging from thousands to billions of kilometers. In fact, the closest star other than the Sun, Proxima Centauri in the constellation Centaurus, is 40 trillion kilometers (24 trillion miles) away. It takes light about four years to get here from there. (Note that the constellation Centaurus is not visible from most of the northern hemisphere.) Most of the stars you see in the night sky are hundreds of times farther away than Proxima Centauri. How do we know the distance to these stars?

Astronomers know the distance to Proxima Centauri from its changing position in the sky. Throughout the year, it appears to move against the background of more distant stars, an effect called **stellar parallax.** Actually, however, it is our point of view that has changed. We experience parallax when nearby objects appear to shift their positions against a distant background as we move (see Figure 2-4).

You use parallax every day to judge distances without thinking about it. You need not even move your head; just use both eyes. As you look at a tree 30 or 40 m away, your eyeballs cross slightly. The closer the tree, the more crossed your eyes must be in order that both of them look right at it. The angle formed between your eyes and the tree tells your brain just how close the tree is. Working out distances of nearby stars is done similarly; it is just harder: It requires painstaking angle measurements and a little geometry.

As the Earth moves from one side of its orbit around the Sun to the other, a nearby star's apparent position against more distant stars shifts. The parallax angle is half the angle by which it shifts (Figure III-1) measured in arc seconds. If the parallax angle is p, the distance d to the star in parsecs is given by the equation

$$d = 1/p$$

For example, the nearest star, Proxima Centauri, has a parallax angle of 0.77 arcsec, and so its distance is 1/0.77, or approximately 1.3 pc. The first parallax measurement was made by Friedrich Wilhelm Bessel, a German astronomer and mathematician. He found the parallax angle of the star 61 Cygni to be $\frac{1}{3}$ arcsec, and so its distance is about 3 pc.

Note that 1 pc is the distance to a star whose parallax angle is exactly 1 arcsec. In fact, the parsec (from *par*allax

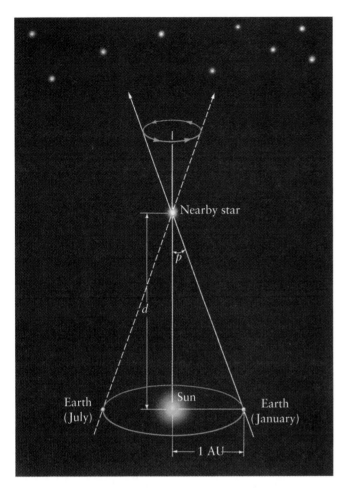

Figure III-1 Using Parallax to Determine Distance As the Earth orbits the Sun, a nearby star appears to shift its position against the background of distant stars. The angle *p* (the parallax of the star) is equal to the angular size of the radius of the Earth's orbit as seen from the star. The smaller the parallax (*p*), the larger the distance (*d*) to the star.

*sec*ond) originates in the use of parallax to measure distance. Recall from Toolbox I-2 that a parsec is equal to 3.09×10^{13} km, or 206,265 AU.

Stellar parallax angles are extremely tiny. For instance, the parallax of Proxima Centauri is comparable to the angular diameter of a dime seen from a distance of 3 km. Because parallax angles smaller than about 0.01 arcsec are difficult to measure from Earth-based observatories, the stellar parallax method gives reliable distances only for distances up to about 100 pc. Stellar parallax measurements are further limited by the seeing

disk (see Chapter 3) created for each star by the Earth's atmosphere.

Because Earth-orbiting satellites are unhampered by our atmosphere, they enable astronomers to determine the distances to stars well beyond the reach of ground-based observations. In 1989 the European Space Agency (ESA) launched a satellite called Hipparcos (an acronym for *h*igh *p*recision *par*allax *c*ollecting *s*atellite, named for Hipparchus of ancient Greece who created an early classification system for stars). Although the satellite failed to achieve its proper orbit, astronomers have used it to measure the distances to stars up to 500 pc away, including over 20,000 of the nearest stars.

Despite the information gained from stellar parallax, astronomers still want to know the distances to more remote stars for which parallax cannot be measured. We will see below how knowing the brightness of stars enables astronomers to achieve this goal.

III-2 Distances affect the brightnesses of stars as seen from Earth

Greek astronomers from Hipparchus in the second century BC through Ptolemy in the second century AD undertook the classification of stars by how bright they are. The brightest stars were originally called first-magnitude stars, and their brightness designated as +1. Those about half as bright were called second-magnitude stars (designated +2), and so forth, to sixth-magnitude stars, the dimmest ones they could see.

This scale of stellar brightness led astronomers to define **apparent magnitude.** A first-magnitude star is now defined to be 100 times brighter than a sixth-magnitude star. In other words, it would take 100 stars of apparent magnitude +6 to provide as much light as we receive from a single star of apparent magnitude +1. A magnitude difference of 1 corresponds to a factor of 2.512 in light energy, because

$$2.512^5 = 2.512 \times 2.512 \times 2.512 \times 2.512 \times 2.512 = 100$$

As another example, it takes about 2.5 third-magnitude stars to provide as much light as we receive from a single second-magnitude star.

Rigorous measurements of stellar brightness reveal that the brightest stars are actually brighter than +1.0 (first magnitude). Therefore, astronomers have had to use negative numbers for the magnitude of the very brightest

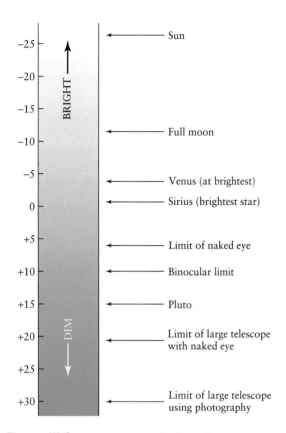

Figure III-2 **Apparent Magnitude Scale** Astronomers denote the brightness of objects in the sky by their apparent magnitude. Most stars visible to the naked eye have magnitudes between +1 and +6. Photography through a large telescope can reveal stars nearly as faint as magnitude +30.

objects. For example, Sirius—the brightest star in the night sky—has an apparent magnitude of −1.5. At its brightest, Venus shines with an apparent magnitude of −4.4, the full Moon has an apparent magnitude of −12.6, and the Sun has a magnitude of −26.7. Remember: *The smaller or more negative the apparent magnitude number for an astronomical body, the brighter the body is.*

Astronomers also extended the magnitude scale to larger positive numbers in order to describe dimmer stars visible only through their telescopes. For example, the dimmest stars visible through a pair of binoculars have a magnitude of about +10. Time-exposure photographs reveal even dimmer stars. Through some of the largest telescopes, such as the Keck telescopes, and through the Hubble Space Telescope, stars as dim as magnitude +30 can be seen. Figure III-2 illustrates the modern apparent magnitude scale.

III-3 Absolute magnitudes consider the effects of distance on brightness

Apparent magnitudes do not tell us about the actual energy emitted by the stars. A star that looks dim in the sky might be a brilliant star that just happens to be extremely far away. Imagine a source of light such as a light bulb or a star. As light moves outward from the

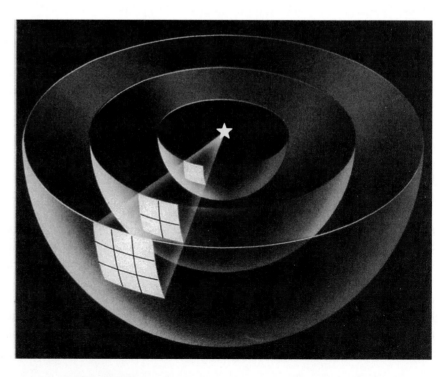

Figure III-3 **The Inverse-Square Law** This drawing shows how the same amount of radiation from a light source must illuminate an ever-increasing area as distance from the light source increases. Because the light becomes spread out as it moves away from the source, the apparent brightness of the source decreases.

The Distance–Magnitude Relationship

The closer a star, the brighter it appears. The inverse-square law leads to a simple equation for the absolute magnitude M. Suppose a star's apparent magnitude is m, and its distance from the Earth is d (measured in parsecs). Then

$$M = m - 5 \log\left(\frac{d}{10}\right)$$

where log stands for the common logarithm.

For example, consider Proxima Centauri, the nearest star in the night sky. By measuring its parallax angle we know this star is at a distance from the Earth of $d = 1.3$ pc. Its apparent magnitude is $m = +11.1$. Therefore its absolute magnitude is

$$M = 11.1 - 5 \log\left(\frac{1.3}{10}\right)$$
$$= 11.1 - (-4.4)$$
$$= 15.5$$

Compared to the Sun, with $M = 4.8$, we can see that Proxima Centauri is an intrinsically dim star.

The distance–magnitude relation is often written

$$m - M = 5 \log d - 5$$

If you know any two of d, m, and M, you can calculate the third. We shall see in a moment, for example, that we may know a star's absolute and apparent magnitude, and yet the star may be too far away for its parallax to be measured. The equation can then be used to determine its distance.

---

source, it spreads out over increasingly larger regions of space, as shown in Figure III-3. As the light spreads out, its brightness decreases, so the farther away a source of light is, the dimmer it appears. The **inverse-square law** tells us that apparent brightness decreases inversely with the square of the distance between the source and the observer. For example, double the distance to a light source and its apparent brightness decreases to $(\frac{1}{2})^2$, or $\frac{1}{4}$ its original brightness; triple the distance and the apparent brightness decreases to $(\frac{1}{3})^2$, or $\frac{1}{9}$.

To determine the total energy emitted by stars, astronomers need to know how bright they really are compared to each other; that is, how bright would stars be if they were all at the same distance from the Earth? This information is used to evaluate models of stellar evolution, which we discuss in the next few chapters. The comparison of brightnesses corrected for distance is done by calculating the apparent magnitude every star would have if it were 10 pc from the Earth. The resulting number is a star's **absolute magnitude.** For example, if the Sun were moved to a distance of 10 pc from the Earth, it would have an apparent magnitude of +4.8. Therefore the absolute magnitude of the Sun is +4.8. Absolute magnitudes range from roughly −10 for the brightest stars to +17 for the dimmest.

We cannot just move a star to exactly 10 pc from here and then remeasure its apparent magnitude. Fortunately, we can easily calculate the absolute magnitude of a nearby star. We first measure its apparent magnitude, and then we find its distance by measuring its parallax angle. Toolbox III-1 will show you how these two numbers lead to the star's absolute magnitude.

Recall from Chapter 4 that a star is very nearly a blackbody, and so the energy it emits is directly related to its brightness. The greater a star's absolute magnitude, the greater is its **luminosity,** the total amount of energy escaping from its surface each second. For convenience, stellar luminosities are expressed in multiples of the Sun's luminosity, denoted $L_\odot$, which is equal to 3.90×10^{26} W. The intrinsically brightest stars (absolute magnitude of −10) have luminosities of $10^6 \, L_\odot$. In other words, each of these stars has the energy output of a million Suns. The dimmest stars (absolute magnitude of +17) have luminosities of $10^{-5} \, L_\odot$.

The Sun is a wonderfully typical star. Both its luminosity and its absolute magnitude are about in the middle of the range for all stars, suggesting that the Sun's energy output is average. We will see in Chapter 12 that this is extremely fortunate for life on Earth: If the Sun were too much brighter (and hotter), it wouldn't have lasted long enough for life to have evolved here. And if the Sun were too much dimmer (and cooler), the Earth would not get nearly the energy needed for life to flourish.

As noted above, many stars are so remote that their stellar parallaxes cannot be measured. Knowing a star's luminosity provides astronomers with another tool for measuring the distance to stars called **spectroscopic parallax**. If we know a star's luminosity and its surface temperature, we can find its distance using the inverse-square law. Some stars, however, are so far away that we cannot accurately measure their luminosity and temperature. Distances can then still be determined from changes in the brightness of variable stars, a technique we present in Chapter 11.

III-4 The brightnesses of stars helps us classify them and study their evolution

To determine the distances and absolute magnitudes of stars, we needed only geometry and the inverse-square law, both pre-twentieth-century concepts. In the next four chapters we will use modern physics to probe the life and death of stars. You will discover that each star's absolute magnitude and surface temperature are correlated. This means that a star with a certain absolute magnitude must have a surface temperature within a limited range. For example, stars with an absolute magnitude of +4.8 (like our Sun) can only have a surface temperature between 5000 K and 6500 K.

You have seen how astronomers measure the distances to stars and their apparent magnitudes, and you know how absolute magnitudes can then be deduced. With our detailed knowledge of how the Sun operates, we can explain the evolution of other stars. You will see in Chapters 11 through 13 that stars with different masses evolve differently: Like the Sun, some end fusion when their cores are converted from hydrogen into helium, and then into carbon; other stars eventually convert their core hydrogen into neutrons; still others end up with their cores collapsing to become black holes. These different stellar histories also lead to different explosions of the outer layers of stars. Lower-mass stars like the Sun eject their outer envelopes as planetary nebulae, while more massive stars detonate as supernovae, the most powerful explosions in the universe.

We will explore some of the properties of stellar remnants, especially neutron stars and black holes. When a rotating neutron star possesses a magnetic field that does not pass through its rotation axis, the spinning magnetic fields behave just like a lighthouse, sending out a beacon of radiation. If the beam of that flickering radiation passes in our direction, we see the neutron star emit pulses of radiation. Such a star is called a pulsar, and we will study it in Chapter 12.

The most exotic of all astronomical objects, the black hole, is the object of our study in Chapter 13. You will find that black holes greatly distort the space surrounding them. They cause time to slow down and radiation in their vicinity to change wavelength. Some draw off the atmospheres of stars to which they are binary companions. Most astounding of all, you will find that black holes actually evaporate!

WHAT DID YOU THINK?

1 *How far is the closest star other than the Sun?* Proxima Centauri is 40 trillion kilometers (24 trillion miles) away. It takes light about four years to get here from there.

2 *How bright is the Sun compared to other stars?* The Sun is roughly in the middle of the range of stellar brightnesses.

Key Words

| absolute magnitude | inverse-square law | spectroscopic parallax | stellar parallax |
| apparent magnitude | luminosity | stellar evolution | |

Key Ideas

• Determining stellar distances is the first step to understanding the nature of the stars.

Distances to the nearer stars can be determined by stellar parallax, the apparent shift of a star's location against the background stars while the Earth moves along its orbit.

• The apparent magnitude of a star is a measure of how bright the star appears to Earth-based observers. The absolute magnitude of a star is a measure of the star's true brightness and is directly related to the star's energy output, or luminosity.

The absolute magnitude of a star is the apparent magnitude it would have if viewed from a distance of 10 pc. Absolute magnitudes are calculated from the star's apparent magnitude and distance.

The luminosity of a star is the amount of energy escaping from it each second.

Review Questions

1   What is stellar parallax?

2   How do astronomers use stellar parallax to measure the distances to stars?

3   Why do stellar parallax measurements work only with nearby stars?

4   What is the difference between apparent magnitude and absolute magnitude?

5   Briefly describe how you would determine the absolute magnitude of a nearby star.

6   What does a star's luminosity measure?

Advanced Questions

* 7   Van Maanen's star, named after the Dutch astronomer who discovered it, is a nearby white dwarf whose parallax angle is 0.232 arcsec. How far away is the star?

8   Discuss the advantages and disadvantages of measuring stellar parallax from a space telescope in a large solar orbit, say at the distance of Jupiter from the Sun.

9   Explain how sailors on a ship traveling parallel to a coastline at a known speed can use parallax angle measurements to determine the distance to the shore.

10 ▶ *The Nature of Stars*

····································

IN THIS CHAPTER

You will learn how astronomers analyze starlight to determine a star's temperature and chemical composition. You will also see how stellar luminosities and surface temperatures are related in the all-important Hertzsprung–Russell diagram, which reveals the evolutionary states of stars. Finally, you will explore binary star systems, in which two stars orbit each other. By observing the motions of the two stars in a binary system, astronomers can calculate their masses. All this information provides important insights into the essential nature of stars.

Structure by Starlight By analyzing starlight, an astronomer can determine such details about a star as its surface temperature, chemical composition, and luminosity. Reddish stars are comparatively cool, with surface temperatures around 3000 K. Blue-white stars have much higher surface temperatures (15,000–30,000 K). This image was made by moving the telescope, so that the stars in the tail of the constellation Scorpius appear as streaks, thereby enhancing their color differences.

WHAT DO YOU THINK?

1 What color are stars?

2 Are most stars isolated from other stars, like the Sun?

··

IN THE NINETEENTH CENTURY, astronomers realized that the Sun is just another star, one of several hundred billion in the Milky Way alone. It would be impossible to study them individually. Fortunately, however, many stars have identical properties, so by studying one we learn about millions. Astronomers look for meaningful categories in the seemingly endless variety of space. They have identified common properties of stars, including masses, chemical compositions, and rotation rates. They then developed laws to bring order to what they know—laws that will later help us to trace the lives and deaths of distant stars. The story begins with a fact we might easily overlook: Stars are not all the same color.

The Temperature of Stars

10-1 A star's color reveals its surface temperature

1 Even with the naked eye, you can see that stars have different colors. For example, in the constellation of Orion, Betelgeuse appears red and Rigel appears blue (see Figure 3-30c). Because a star behaves very nearly like a perfect blackbody, its color reveals its surface temperature. As we discussed in Chapter 4, this relationship is described by Wien's law (see Figure 4-2). Accordingly, the intensity of light from a cool star peaks at long wavelengths, and so the star looks red (Figure 10-1a); the intensity of light from a hot star peaks at shorter wavelengths, making it look blue (Figure 10-1c). The maximum intensity of a star of intermediate temperature, such as the Sun, is found near the middle of the visible spectrum (Figure 10-1b).

To measure the colors of stars accurately, astronomers have developed a technique called **photometry.** This method uses a light-sensitive device (such as a CCD; recall Figure 3-24), a standardized set of colored filters, and a telescope. The astronomer aims a telescope at a star and measures the intensity of starlight three times, each time through a different filter. The most commonly used filters are **UBV filters,** which are transparent to the ultraviolet (U), the blue (B), and the central (V) regions of the radiation spectrum, respectively (Figure 10-2). The transparency of the V filter (V for "visual") mimics the sensitivity of the human eye.

The three apparent magnitudes thus obtained, usually designated by the letter of the filter used, indicate the intensity of light received at each range of the spectrum. The astronomer then compares the intensity of starlight in the neighboring wavelength bands by subtracting one magnitude from another to form the combinations U – B and B – V, which are called the star's **color indices.**

a This star looks red

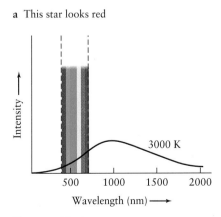

b This star looks yellow-white

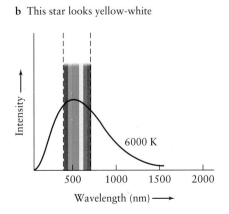

c This star looks blue

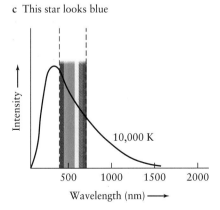

Figure 10-1 **Temperature and Color** This diagram shows the relationship between the color of a star and its surface temperature. The intensity of light emitted by three hypothetical stars is plotted against wavelength (compare Figure 4-2). The range of visible wavelengths is indicated. Where the peak of a star's intensity curve lies relative to the visible light band determines the dominant color of its visible light.

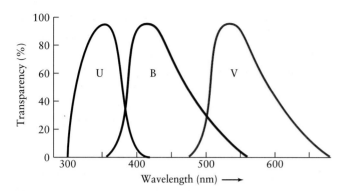

Figure 10-2 **Light Transmission Through UBV Filters** This graph shows the wavelength ranges over which standardized U, B, and V filters are transparent to light. The U filter is transparent to wavelengths from 300 to 400 nm, a range called the near-ultraviolet because it lies just beyond the violet end of the visible spectrum. The B filter is transparent from about 380 to 550 nm, and the V filter from about 500 to 650 nm.

A color index tells how much brighter or dimmer a star is in one wavelength band than in another. For example, the B − V color index indicates how much brighter or dimmer a star appears through the B filter than through the V filter. Table 10-1 gives the UBV magnitudes and color indices for several representative stars.

Color index is important because it indicates the star's surface temperature. If a star is very hot, its radiation is skewed toward the short-wavelength ultraviolet, which makes the star bright through the U filter, dimmer through the B filter, and dimmest through the V filter. Regulus (see Table 10-1) is such a star. Alternatively, if the star is cool, its radiation peaks at long wavelengths, making the star brightest through the V filter, dimmer

through the B filter, and dimmest through the U filter. As shown in Table 10-1, Aldebaran and Betelgeuse are examples of such cool stars.

Figure 10-3 graphs the relationship between the B − V color index and temperature. The red curve was determined from the blackbody curves for stars at different temperatures. If you know a star's B − V color index, you can use this graph to find the star's surface temperature. For example, the Sun's B − V index is +0.62, which corresponds to a surface temperature of 5800 K.

10-2 A star's spectrum also reveals its surface temperature

Another way to determine a star's surface temperature is through a technique called **stellar spectroscopy.** The technique was born in the 1860s, when the Italian astronomer Angelo Secchi attached a spectroscope to his telescope. A spectroscope uses a prism or finely lined glass grating to separate light into the colors of the rainbow. Using this device, Secchi discovered spectral lines in the spectra of many stars. As we saw in Chapter 4, these lines indicate an object's temperature and chemical composition.

At first glance, stellar spectra seem to come in a bewildering variety, some of which are shown in Figure 10-4. Some stellar spectra show prominent Balmer lines of hydrogen. Some exhibit many absorption lines of calcium and iron. Still others are dominated by broad absorption lines created by molecules such as titanium oxide. To cope with this diversity, astronomers since Secchi have grouped similar stellar spectra into classes, or **spectral types.** According to one classification scheme popular in

TABLE 10-1						
UBV Magnitudes and Color Indices of Selected Stars						
Star	**V**	**B**	**U**	**B − V**	**U − B**	**Apparent color**
Bellatrix	1.64	1.42	0.55	−0.22	−0.87	Blue
Regulus	1.35	1.24	0.88	−0.11	−0.36	Blue-white
Sirius	−1.46	−1.46	−1.52	0.00	−0.06	Blue-white
Megrez	3.31	3.39	3.46	+0.08	+0.07	White
Altair	0.77	0.99	1.07	+0.22	+0.08	Yellow-white
Sun	−26.78	−26.16	−26.06	+0.62	+0.10	Yellow-white
Aldebaran	0.85	2.39	4.29	+1.54	+1.90	Orange
Betelgeuse	0.50	2.35	4.41	+1.85	+2.06	Red

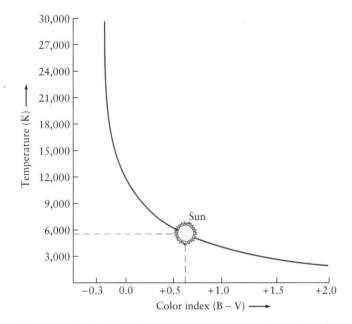

Figure 10-3 **Blackbody Temperature Versus Color Index** The B – V color index is the difference between the B and V magnitudes of a star. If the star is hotter than about 10,000 K, it is a very bluish star and its B – V index is less than zero. If a star is cooler than about 10,000 K, its B – V index is greater than zero. The Sun's B – V index is about 0.62, which corresponds to a temperature of 5800 K. After measuring a star's B and V magnitudes, an astronomer can estimate the star's surface temperature from a graph such as this one.

the late 1800s, a star was assigned a letter from A through P, according to the strength of the Balmer hydrogen lines in the star's spectrum.

After Niels Bohr explained the structure of the hydrogen atom in the early 1900s (recall Figure 4-12), astronomers realized that the temperature of the gases in a star's outer layers can be related to the strength of the lines in the star's spectrum. To see why a star's spectrum is affected by its surface temperature, consider hydrogen. Although hydrogen accounts for about three-quarters of the mass of a typical star, strong (meaning very dark) hydrogen lines do not necessarily show up in a star's visible spectrum. Recall from Chapter 4 that Balmer lines are produced when a photon excites an electron in the $n = 2$ orbit of hydrogen into a higher orbit (see Figures 4-15 and 4-16). Because a stellar surface temperature of 10,000 K maximizes such an excitation of hydrogen electrons, it produces the strongest Balmer lines. However, if the star is much hotter than 10,000 K, high-energy photons streaming through its photosphere completely strip away electrons from most of the hydrogen atoms there. Because an ionized hydrogen atom cannot produce spectral lines, a very hot star has very dim hydrogen Balmer lines even though it contains great quantities of hydrogen.

Conversely, if a star is much cooler than 10,000 K, most of the photons escaping from it possess too little

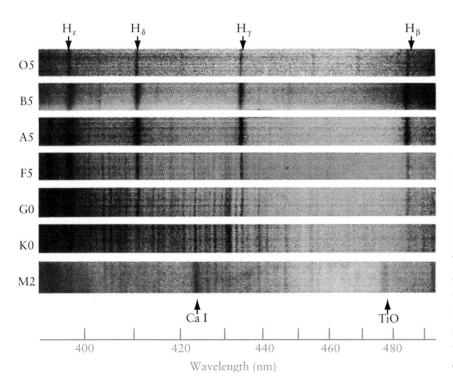

Figure 10-4 **Principal Types of Stellar Spectra** Each of these seven strips shows the spectrum of a star whose spectral type is given on the left. Spectral lines of hydrogen, calcium, and titanium oxide are indicated. The atomic hydrogen lines are strongest in A stars, which have surface temperatures of about 10,000 K. The spectra of G and K stars exhibit numerous atomic lines caused by metals, indicating temperatures from 4000 to 6000 K. The broad, dark bands in the spectrum of an M star are caused by titanium oxide molecules, which can exist only if the temperature is cooler than about 3500 K.

energy to even boost many electrons from the ground state to the $n = 2$ orbit of the hydrogen atoms, much less into more excited states. These stars also produce dim Balmer lines. In summary, to produce strong Balmer lines, a star must be hot enough to excite electrons out of the ground state, but not hot enough to ionize the atoms.

At these cooler and hotter temperatures, the spectral lines of other elements dominate a star's spectrum. For example, the spectral lines of neutral helium are pronounced at around 25,000 K because photons have enough energy to excite helium atoms without tearing away the electrons. Conversely, the spectral lines of neutral iron are especially strong at around 4500 K.

It is important to note that, unlike the ionized form of hydrogen, the ionized forms of all other elements do produce their own characteristic spectral lines. This is because neutral atoms other than hydrogen have two or more electrons; thus when only one is stripped away, other electrons remain on the atom. These remaining electrons produce a new and distinctive set of spectral lines. For instance, in stars hotter than about 30,000 K, one of the two electrons in a helium atom is torn away, creating singly ionized helium, He II. The resulting spectral lines for He II are different from the lines for neutral helium, He I. Thus, when the spectral lines of singly ionized helium appear in a star's spectrum, we know that the star has a surface temperature greater than 30,000 K.

10-3 Stars are classified by their spectra

We can thus determine a star's surface temperature from either its color index or the strength of its various spectral lines. In the early 1900s Annie Cannon and her colleagues at Harvard Observatory set up the spectral clas-

sification scheme we use today. Many of the early A through P categories were dropped because, as we just saw, the Balmer lines in a star's spectrum can be weak whether the star is very cool or very hot. By themselves, the Balmer lines don't tell us the star's surface temperature. In the late 1920s Harvard astronomer Cecilia Payne and physicist Meghnad Saha succeeded in explaining precisely how a star's Balmer spectrum is affected by its surface temperature. The remaining Balmer-based spectral types were thus reordered by stellar surface temperature into the sequence **OBAFGKM**. This sequence has traditionally been memorized with a mnemonic such as: "Oh, Be A Fine Guy, Kiss Me!" or "Oh, Be A Fine Girl, Kiss Me!"

The hottest stars, O stars, have surface temperatures over 35,000 K; their spectra are dominated by He II and Si IV (triply ionized silicon). M stars, the coolest stars, have surface temperatures around 3000 K. These latter stars are so cool that many atoms stick together in molecules such as titanium oxide, whose spectral lines are prominent. Figure 10-4 shows representative examples of each spectral type.

Astronomers have found it useful to subdivide the OBAFGKM temperature sequence further. These finer steps are indicated by adding an integer from 0 (hottest) through 9 (coolest). Thus, an A8 star is hotter than an A9 star, which is hotter than an F0 star, which is hotter than an F1 star, and so on. (What class of star is just slightly cooler than an F9?) The Sun, whose spectrum is dominated by singly ionized metals (especially Fe II and Ca II) is a G2 star.

The modern spectral classification scheme is summarized in Figure 10-5, which plots the strength of spectral lines against spectral type. Astronomers use the information in this figure to deduce the surface temperature and

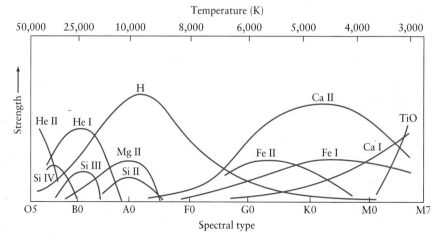

Figure 10-5 Spectral Type and Temperature The more a given wavelength is absorbed, the stronger the corresponding absorption line will be. The strengths of the absorption lines of various elements are directly related to the temperature of the star's outer layers. For example, the Sun's spectrum has strong lines of singly ionized iron and calcium (Fe II and Ca II), corresponding to a spectral type of G2 and a surface temperature of about 5800 K. Note that hydrogen lines are strongest in A stars, whereas stars cooler than about 3500 K show strong absorption caused by titanium oxide.

corresponding spectral type of a star from the intensity of the lines in its spectrum. For example, a star exhibiting especially strong Ca II and Fe I lines in its spectrum is a K3 star with a surface temperature around 4500 K.

Types of Stars

10-4 The Hertzsprung–Russell diagram identifies distinct groups of stars

At about the same time that astronomers began observing stellar spectra in the mid-1800s, the first accurate measurements of stellar parallaxes were also being made. As a result, during the next half century, observing techniques improved, and the spectral types and absolute magnitudes of many stars became known.

Around 1905 the Danish astronomer Ejnar Hertzsprung pointed out that a pattern emerges when the absolute magnitudes of stars are plotted against their color indices. Almost a decade later the American astronomer Henry Norris Russell independently discovered this regularity using spectral types instead of color indices. Plots of this latter kind are now known as **Hertzsprung–Russell diagrams**, or **H–R diagrams**, while the original absolute magnitude–color index plots are now called **color–magnitude diagrams.** Figure 10-6 is a typical Hertzsprung–Russell diagram. Each dot in Figure 10-6 represents a star whose absolute magnitude and spectral type have been determined. *Bright stars are near the top of the diagram; dim stars are near the bottom. Hot (O and B) stars are toward the left side of the graph; cool (M) stars are toward the right.*

In yet another variation, luminosity can be plotted against surface temperature. Figure 10-7 shows this last type of H–R diagram. Note that the temperature scale on the horizontal axis of the graph increases toward the left, because Hertzsprung and Russell drew their original diagrams with O stars on the left and M stars on the right. (They made this choice because of the standard sequence OBAFGKM.) Having hot stars toward the left and cool ones toward the right is a convention that no one has seriously tried to change.

The stars are not scattered randomly over an H–R diagram but are clustered in distinct regions. *Thus stars do not have random surface temperatures and absolute magnitudes; the two factors are related.* The groups into which stars fall are an essential clue to understanding their evolution, as we will see in upcoming chapters.

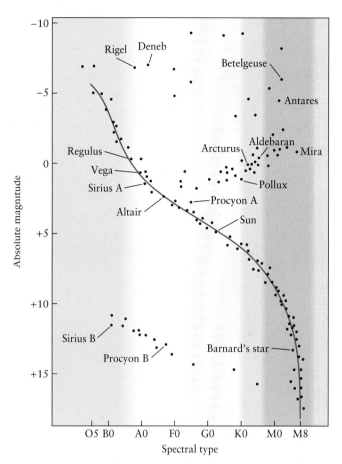

Figure 10-6 **A Hertzsprung–Russell Diagram** An H–R diagram is a graph on which the absolute magnitudes of stars are plotted against their surface temperatures or their spectral types. Each dot represents a star whose absolute magnitude and spectral type have been determined. Some well-known stars are identified. The data points reveal the existence of different types of stars grouped in specific regions in the sky: main sequence stars, giants, supergiants, and white dwarfs. The red curve indicates the location of the main sequence.

Also shown in Figure 10-7 are dashed lines indicating the radii of stars. Notice that most giants are 10 to 100 times as large as the Sun, whereas stars labeled white dwarfs are only about $\frac{1}{100}$ the size of the Sun.

The band stretching diagonally across an H–R diagram represents most of the stars we see in the nighttime sky. This band, called the **main sequence,** extends from the hot, bright, bluish stars in the upper left corner of the diagram down to the cool, dim, reddish stars in the lower right corner. A star in this region of the H–R diagram is called a **main sequence star.** About 90% of the stars surrounding the solar system are on the main sequence. Among these, the cool M stars are the most

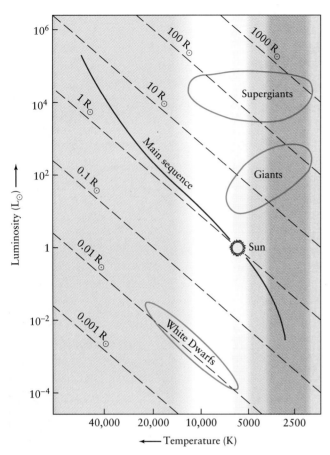

Figure 10-7 **Determining the Sizes of Stars from an H–R Diagram** On this H–R diagram, stellar luminosities are graphed against the surface temperatures of stars. Note that the various types of stars (main sequence, giants, supergiants, and white dwarfs) fall in the same regions of the graph as in Figure 10-6, where absolute magnitude is plotted against spectral type. The dashed diagonal lines indicate stellar radii. The Sun's size is midway between the largest and smallest stars we see in the sky.

common. The Sun (spectral type G2, absolute magnitude +4.8) is such a star. Notice in Figure 10-7 that most main sequence stars are roughly the same size as the Sun.

The number of main sequence stars decreases with increasing surface temperature. Therefore, the hot O stars are the rarest main sequence stars. We will explore some of the properties of main sequence stars below; we will delay a discussion of why most stars are on the main sequence until Chapter 11.

Above and to the right of the main sequence is a second major grouping of stars on the H–R diagram. Stars at these locations are bright but cool. From the Stefan–Boltzmann law, we know that a cool object radi-

EYES ON . . . Star Names

Most of the hundreds of thousands of stars that appear on typical H–R diagrams never received fanciful names such as Betelgeuse or Aldebaran. Indeed, most are so dim that they have been observed only through telescopes in the last two centuries. To study the stars yourself, you must be able to keep track of them. Astronomers have created a system of labels for all of them.

The 24 brightest stars in each constellation are assigned Greek lowercase letters:

α alpha	η eta	ν nu	τ tau
β beta	θ theta	ξ xi	υ upsilon
γ gamma	ι iota	o omicron	φ phi
δ delta	κ kappa	π pi	χ chi
ε epsilon	λ lambda	ρ rho	ψ psi
ζ zeta	μ mu	σ sigma	ω omega

A *bright* star's name is a Greek letter together with its constellation. We use the Latin possessive form of the constellations, for example:

Constellation	Possessive
Aries	Arietis
Taurus	Tauri
Gemini	Geminorum
Cancer	Cancri
Leo	Leonis
Virgo	Virginis
Libra	Librae
Scorpius	Scorpii
Ophiuchus	Ophiuchi
Sagittarius	Sagittarii
Capricornus	Capricorni
Aquarius	Aquarii
Pisces	Piscium

In most cases the brightest star in the constellation is α, the second brightest is β, the third is γ, and so on. For instance, the brightest star in the constellation Gemini is called α Geminorum. This name is more informative than its common name, Pollux.

For the millions of stars extending beyond the twenty-fourth brightest, a variety of catalogues list the stars numerically. For example, HDE 226868 is a bright, blue star, the 226,868th star in the *Henry Draper Extended Catalogue.*

ates much less light per unit of surface area than does a hot object. In order to be so bright, these stars must therefore be huge, and so they are called **giants.** By contrast, main sequence stars are often called **dwarfs.** Giants are typically 10 to 100 times as large as the Sun and have surface temperatures between 3000 and 6000 K. The cooler members of this class of stars (those with surface temperatures between 3000 and 4000 K) are often called **red giants** because they appear reddish in the nighttime sky. Aldebaran in the constellation Taurus and Arcturus in Boötes are examples of red giants that you can easily see with the naked eye.

A few rare stars are considerably bigger and brighter than typical giants. Located along the top of the H–R diagram, these superluminous stars are appropriately called **supergiants.** Betelgeuse in Orion and Antares in Scorpius are two examples that are visible in the nighttime sky. Together, giants and supergiants comprise less than 1% of the stars in our vicinity.

The remaining 9% or so of stars in our neighborhood of space fall in a final grouping toward the lower left and bottom of the Hertzsprung–Russell diagram. As their placement on the diagram shows, these stars are hot, dim, and tiny compared to the Sun. Appropriately called **white dwarfs,** they are actually remnants of stars. White dwarfs are roughly the same size as the Earth, and because of their great distances, they can only be seen with the aid of a telescope.

10-5 Stars are also classified by their luminosity

We have seen that stars can be classified into spectral types based on the most prominent lines in their spectra. However, there are subtle differences even among the spectral lines of stars having the same spectral type. Based upon these minor differences, a system of **luminosity classes** was developed in the 1930s. Luminosity class I includes all the supergiants, and luminosity class V includes the main sequence stars. The intermediate classes distinguish giants of various luminosities, as indicated in Table 10-2. We will see in Chapter 11 that the different luminosity classes correspond to different stages of stellar evolution. White dwarfs are not assigned a luminosity class because, as we noted above, they are just the remnants of stars; no fusion is taking place inside them.

Plotting luminosity classes on the H–R diagram (Figure 10-8) provides a useful subdivision of star types. Consequently, astronomers commonly describe a star by

TABLE 10-2

Stellar Luminosity Classes

Luminosity class	Type of star
I	Supergiant
II	Bright giant
III	Giant
IV	Subgiant
V	Main sequence

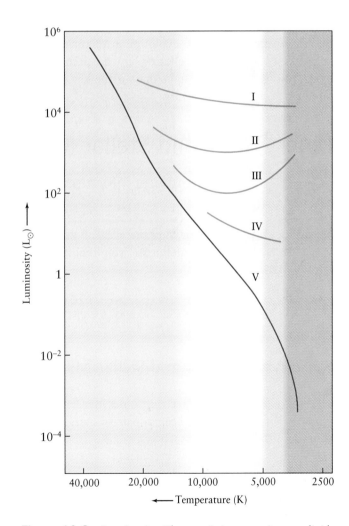

Figure 10-8 **Luminosity Classes** It is convenient to divide the H–R diagram into regions called luminosity classes. This subdivision permits finer distinctions between giants and supergiants. Luminosity class V encompasses the main sequence stars, as well as the dim red stars called red dwarfs toward the lower right side of the H–R diagram.

both its spectral type and its luminosity class; for example, the Sun is called a G2 V star. This notation supplies a great deal of information about the star, since its spectral type is correlated with its surface temperature. Thus an astronomer knows immediately that a G2 V star is a main sequence star with a luminosity of 1 L$_\odot$ and a surface temperature of 5800 K. Similarly, knowing that Aldebaran is a K5 III star tells an astronomer that it is a red giant with a luminosity of around 500 L$_\odot$ and a surface temperature of about 4000 K (see Figure 10-6).

The first important lesson to learn from the H–R diagram is that there are fundamentally different types of stars. These different kinds of stars represent different stages of stellar evolution. To truly appreciate the H–R diagram we must understand the life cycles of stars: how they are born, what happens as they mature, and how they die. We explore stellar evolution in Chapter 11, but before doing so, we need to know how much mass stars have. You will see that a star's mass is the key to determining its evolutionary history.

Binary Stars and Stellar Mass

10-6 Binary stars provide information about stellar masses

No device on Earth can measure the mass of an isolated star. However, imagine a satellite in orbit about the star.

As we saw in Chapter 2, its orbit must obey Newtonian mechanics. Newton provided the formula we need when he proved Kepler's laws. We can rewrite Kepler's third law as a relation between the masses of a star and its satellite, the satellite's orbital period, and the length of its orbit's semimajor axis:

> The sum of the masses, multiplied by the square of the orbital period, gives the cube of the semimajor axis.

By observing the separation between a star and its satellite and how long the satellite takes to complete its orbit, we can calculate the mass of the star.

We cannot send a satellite to orbit a distant star. Fortunately for astronomers, however, many stars have natural companions, and we can use their observed motions just like the orbit of our imagined satellite. About two-thirds of the stars near our solar system are members of star systems in which two or more stars orbit each other; half the objects we see as stars are actually pairs of stars so distant that they appear to us as one. Using telescopes to observe the periods of the orbits and the distances between stars, astronomers can determine stellar masses. To see how this works, look at Toolbox 10-1.

A pair of stars located at nearly the same position in the night sky is called a *double star,* and between 1782 and 1838 William Herschel and his son John catalogued thousands of them. Some double stars are not actually held by each other's gravity; in fact, they are not even

---

AN ASTRONOMER'S TOOLBOX 10-1

Kepler's Third Law and Stellar Masses

The same gravitational force that holds the Earth or a satellite in orbit can also keep a pair of stars in orbit about each other. For *any* such system, the orbits are ellipses, and Kepler's third law becomes

$$M_1 + M_2 = a^3/P^2$$

Here M_1 and M_2 are the two masses (expressed in solar masses), a is the length of the semimajor axis of the ellipse (in astronomical units), and P is the orbital period (in years). Thus, the sum of the masses can be found once the orbital period and semimajor axis are known.

For example, suppose two stars make up a double system, and one star follows an ellipse with a semimajor axis of 4 AU. We find that it takes 2.5 years to complete one orbit. Then the sum of the stars' masses is

$$M_1 + M_2 = (4)^3/(2.5)^2 = 10.2 \ M_\odot$$

The *total* mass of the system is 10.2 $M_\odot$.

As a simple test of what you have just learned, recall from Chapter 2 that we could write Kepler's third law more simply as

$$P^2 = a^3$$

Do you see why? For the Sun and a planet, the combined mass is pretty close to just one solar mass.

1908	1915	1920

Figure 10-9 **The Binary Star System Kruger 60** About one-half of the visible "stars" are actually double stars. This series of photographs shows the binary star system Kruger 60 in the constellation Cepheus. The unrelated star in the lower right corner is shown for orientation. The orbital motion of the two binary stars about each other is evident. This binary system has a period of 44.52 years, a maximum angular separation of about 3.3 arcsec, and apparent magnitudes of +9.8 and +11.4.

near each other in space. These **optical doubles** just happen to lie in the same direction as seen from Earth. For example, δ Herculis, visible with the naked eye, is an optical double with a dim star. Through a telescope these stars appear close together, but that is an optical illusion.

Other double stars are true **binary stars**—pairs in which two stars orbit each other (Figure 10-9). For example, Mizar, the middle "star" in the handle of the Big Dipper, is really two stars. Mizar A is visible to the naked eye, and Mizar B can be seen through a telescope, so they are said to form a **visual binary**. By years of pa-tient observation, astronomers can plot the orbit of one star in a visual binary about the other (Figure 10-10).

Careful observations of binaries have yielded the masses of many stars. The range of stellar masses extends from $\frac{1}{10}$ of a solar mass to about 100 solar masses. As the data accumulated, an important trend began to emerge. *On the main sequence, the more massive the star, the more luminous it is.* This **mass–luminosity relation** can be conveniently displayed as a graph (Figure 10-11). The Sun's mass lies in the middle of this range, and so we again see how ordinary our star is.

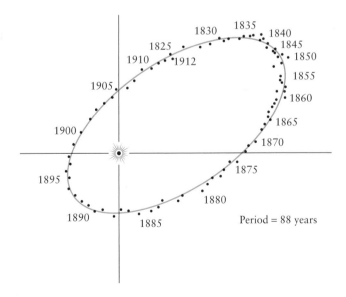

Figure 10-10 **The Orbit of 70 Ophiuchi** Once the orbit of binary stars is known, Kepler's third law can be used to deduce information about the masses of the stars. This illustration shows the orbit of a faint double star in the constellation Ophiuchus. In plotting the orbit, either star may be regarded as the stationary one—the shape and size of the orbit will be the same in either case.

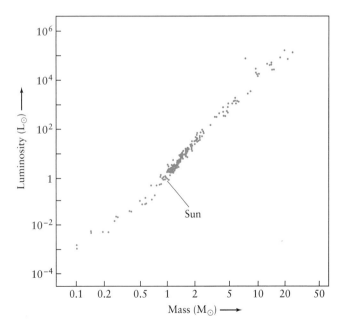

Figure 10-11 **The Mass–Luminosity Relation** For main sequence stars, mass and luminosity are directly correlated. The more massive a star, the more luminous it is. To fit them on the page, the luminosities and masses are plotted using logarithmic scales.

Figure 10-12 **A Double-Line Spectroscopic Binary** A spectroscopic binary exhibits spectral lines that shift back and forth as the two stars revolve about each other. **(a)** The stars are moving parallel to the line of sight (one star approaching Earth, the other star receding), producing two sets of shifted spectral lines. **(b)** Both stars are moving perpendicular to our line of sight.

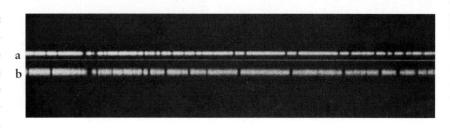

a

b

The mass–luminosity relation demonstrates that the main sequence on the H–R diagram is a progression in mass as well as in luminosity and surface temperature. The hot, bright, bluish stars in the upper left corner of the H–R diagram (see Figure 10-6) are the most massive main sequence stars in the sky. As we move down the sequence, stellar masses decrease until we reach the dim, cool, reddish stars in the lower right corner of the H–R diagram.

10-7 The orbital motion of binary stars affects the wavelengths of their spectral lines

Many binary stars are scattered throughout our Galaxy, but only those that are nearby or that are widely separated can be distinguished as visual binaries. A remote binary often presents the appearance of a single star because telescopes can't resolve the images of the individual stars. These binary systems can still be detected, however, through spectroscopy.

Spectral analysis yields incongruous spectral lines for some stars. For example, the spectrum of what appears at first to be a single star may include strong absorption lines for both hydrogen (indicating a type A star) and also titanium oxide (indicating a type M star). Because a single star cannot have both types of absorption lines prominent, such an object would have to be a binary system.

Spectroscopy can also detect the movements of stars orbiting each other because of the Doppler shift in spectral lines. As we saw in Chapter 4, a source of light coming toward you has shorter wavelengths than if the same source were stationary, and a receding source has longer wavelengths. The size of the shift is proportional to the

speed with which the source of light is moving: The greater the speed, the greater the shift.

If the two stars in a binary are orbiting at more than a few kilometers per second, they will produce two complete sets of spectral lines that shift back and forth in a regular fashion. Such stars are called **spectroscopic binary stars.** The motions of the stars revolving about their center of mass produce the periodic shifting of the spectral lines.

In many spectroscopic binaries, one of the stars is so dim that its spectral lines cannot be detected. Instead, a single set of spectral lines from the star we can see shift regularly back and forth. Such a *single-line spectroscopic binary* yields less information about its two stars than does a *double-line spectroscopic binary,* in which the lines from both stars are visible. Figure 10-12 shows two spectra of the spectroscopic binary system κ Arietis taken a few days apart. In Figure 10-12*a*, two sets of spectral lines are visible, each slightly offset in opposite directions from the normal positions of these lines. The spectral lines of the star moving toward the Earth are blueshifted; those of the other star (moving away from the Earth) are redshifted. A few days later the stars will have progressed along their orbits so that one star is moving toward the left and the other toward the right as seen from Earth. Because neither star is moving toward or away from us, their spectral lines return to their usual positions (Figure 10-12*b*).

The shifts in spectral lines can yield significant information about the orbital velocities of the stars in a spectroscopic binary. This information is best displayed as a **radial-velocity curve,** in which radial velocity is graphed over time (Figure 10-13). Recall that radial velocity is the portion of a star's motion that is directed along our line of sight to the star.

In Figure 10-13, the wavy pattern repeats with a period of about 15 days, which is the orbital period of the

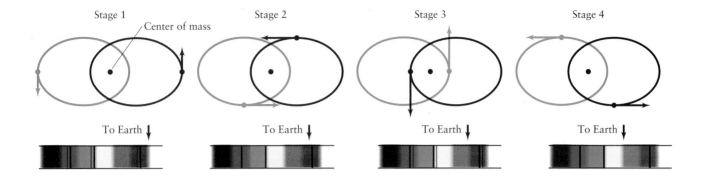

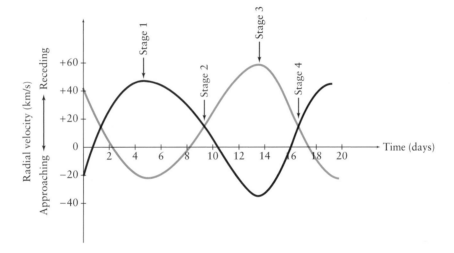

Figure 10-13 **A Radial-Velocity Curve** The graph displays the radial-velocity curve of the binary HD 171978. The draw- ings indicate the positions of the stars and their spectra at four selected moments during an orbital period.

binary. This pattern is displaced upward from the zero-velocity line by about 12 km/s, which is the overall motion of the binary system away from the Earth. Superimposed on this overall recessional motion are the periodic approaches and recessions of the two stars as they orbit around each other.

10-8 Some binary stars eclipse each other

Some binary systems are oriented so that the two stars periodically eclipse each other as seen from Earth. Such **eclipsing binaries** can be detected even when the two stars cannot be resolved as two distinct images in a telescope. The apparent magnitude of the image of the binary dims each time one star blocks out part of the other. Using a light-sensitive detector at the focus of a telescope, an astronomer can measure light intensity from binaries very accurately. From these **light curves,** such as those shown in Figure 10-14, we can see at a glance whether the eclipse is partial (Figure 10-14*a*) or total (Figure 10-14*b*).

The light curve of an eclipsing binary can yield other information as well. For example, the faintest parts of the eclipse are related to the surface temperatures of the two stars. If an eclipsing binary is also a double-line spectroscopic binary, astronomers can calculate the mass and diameter of each star from the light curves and the radial-velocity curves. These stars are rare, however, because the orbits of most spectroscopic binaries are tilted so that eclipses do not occur as seen from Earth.

Light curves can also reveal information about stellar atmospheres. Suppose that one star of a binary is a white dwarf and the other is a bloated giant. By observing exactly how the light from the bright white dwarf is gradually cut off as it begins to move behind the edge of the giant during an eclipse, astronomers can infer the

Figure 10-14 **Representative Light Curves of Eclipsing Binaries** The shape of its light curve usually reveals many details about an eclipsing binary. Illustrated here are examples of (**a**) a partial eclipse and (**b**) a total eclipse.

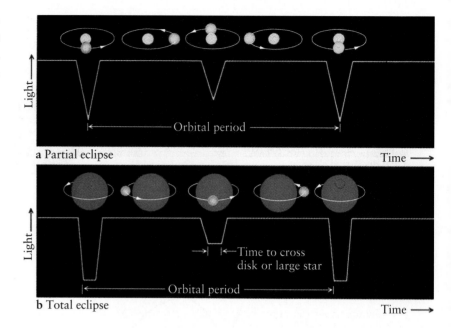

a Partial eclipse

b Total eclipse

pressure and density in the upper atmosphere of the giant. Such information is invaluable in testing models of stellar structure.

10-9 Mass transfer in close binary systems can produce unusual double stars

In a **close binary** system, only a few stellar diameters separate the stars. Such stars are so close together that the gravity of one can dramatically affect the appearance and evolution of the other. If one member of a close binary is a giant, its outer layers are pulled onto its more compact companion. Mass is transferred dramatically from one star to the other.

In the mid-1800s the French mathematician Edward Roche pointed out that the atmospheres of two stars in a binary system must remain within a pair of tear-drop-shaped regions surrounding the stars. Otherwise the gas escapes from the binary. Cut through, these **Roche lobes** take on a figure-eight shape (Figure 10-15). The more massive star is always located inside the larger Roche lobe.

In many binaries, the stars are so far apart that even during their giant stages the stars' surfaces remain well inside their Roche lobes. Each star thus lives out its life as if it were single and isolated, and the system is referred to as a **detached binary.** If two stars are relatively close together, however, one star may fill or overflow its

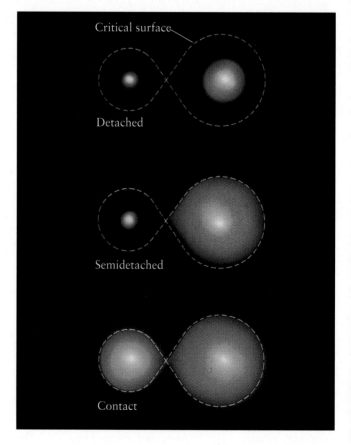

Figure 10-15 **Detached, Semidetached, and Contact Binaries** A double star is said to be detached, semidetached, or contact depending on whether either or both stars fill their Roche lobes. Mass transfer is often observed in semidetached binaries. The two stars in a contact binary share the same outer atmosphere.

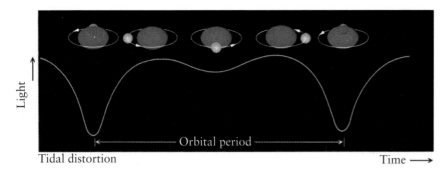

Figure 10-16 **Light Curve of a Tidally Distorted Star** Because the atmosphere of one star is pulled outward by its companion star, the light curve of this binary system is different from the light curve of a binary system in which both stars are spherical.

Roche lobe and the system is called a **semidetached binary.** Gases then flow across the point where the two Roche lobes touch and fall onto the companion star. When both stars completely fill their Roche lobes, the system is called a **contact binary** because the two stars actually touch and share a common envelope of gas.

Semidetached and contact binaries are easiest to detect if they are also eclipsing binaries. In this case, their light curves have a distinctly rounded appearance caused by these tidally distorted egg-shaped stars (Figure 10-16). The eclipsing binary called β Persei, or Algol (from an Arabic term for "demon"), is a semidetached binary that can easily be seen with the naked eye in the constellation of Perseus. From β Persei's light curve (Figure 10-17*a*) astronomers have determined that the binary contains a

star that fills its Roche lobe. Sometime in the past, as it expanded and became a giant, this star dumped a significant amount of gas onto its companion. Astronomers theorize that the *giant* was originally the more massive star, but as a result of mass transfer, the detached companion is now more massive.

Mass transfer is still occurring in a semidetached eclipsing binary called β Lyrae in the constellation of Lyra, the Harp. Like β Persei, β Lyrae contains a giant that fills its Roche lobe (Figure 10-17*b*). For many years astronomers were puzzled by the fact that the detached companion star in β Lyrae is severely underluminous, contributing virtually no light at all to the visible radiation coming from the system. Furthermore, the spectra of β Lyrae contain unusual features, some of which are

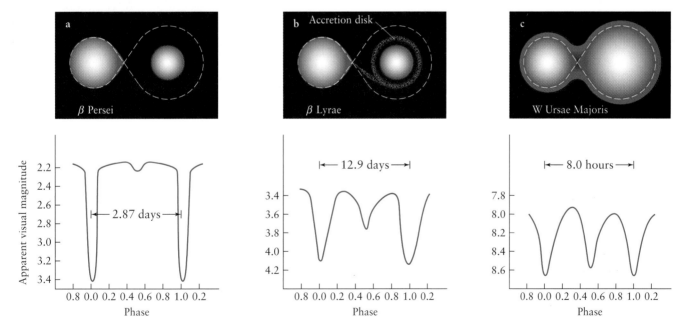

Figure 10-17 **Three Close Binaries** Light curves for and sketches of three eclipsing binaries are shown. The phase denotes the fraction of the orbital period from one primary minimum to the next. (a) β Persei, also known as Algol, is a semidetached binary. (b) β Lyrae is a semidetached binary in which mass transfer has produced an accretion disk surrounding the detached star. (c) W Ursae Majoris is a contact binary.

caused by gas flowing between the stars and around the system as a whole.

The β Lyrae system was explained in 1963 when Su-Shu Huang proposed that the underluminous star in β Lyrae is enveloped in a huge **accretion disk** of gas captured from its bloated companion. The disk is so large and thick that it completely shrouds the secondary star, making it impossible to observe at visible wavelengths. The primary star is overflowing its Roche lobe, with gases streaming onto the disk at the rate of 1 M$_\odot$ per hundred thousand years.

The fate of a semidetached system like β Persei or β Lyrae depends primarily on how fast its stars evolve. If the detached star expands to fill its Roche lobe while the companion star fills its own Roche lobe, then the result is a contact binary. An example is W Ursae Majoris, in which two stars share the same photosphere (Figure 10-17*c*). In Chapters 12 and 13 we shall see that mass transfer onto dead stars produces some of the most extraordinary objects in the sky.

WHAT DID YOU THINK?

1 *What color are stars?* Stars are found in a wide range of colors, from red through violet as well as white.

2 *Are most stars isolated from other stars, like the Sun?* In the vicinity of the Sun, two-thirds of the stars are found in pairs or larger groups.

Key Words

accretion disk	dwarf star	mass–luminosity	spectral type
binary star	eclipsing binary	relation	spectroscopic binary
close binary	giant star	OBAFGKM sequence	stellar spectroscopy
color index	Hertzsprung–Russell	optical double	supergiant
color–magnitude	(H–R) diagram	photometry	UBV filters
diagram	light curve	radial-velocity curve	visual binary
contact binary	luminosity class	red giant	white dwarf
critical surface	main sequence	Roche lobe	
detached binary	main sequence star	semidetached binary	

Key Ideas

• Stars are classified into spectral types (O, B, A, F, G, K, and M) based on their spectra. The spectral type of a star is directly related to its surface temperature.

• The Hertzsprung–Russell (H–R) diagram is a graph on which absolute magnitudes of stars are plotted against their spectral types (or, equivalently, their luminosities are plotted against surface temperatures). The H–R diagram reveals the existence of four major groupings of stars: main sequence stars, giants, supergiants, and white dwarfs.

• The mass–luminosity relation expresses a direct correlation between a main sequence star's mass and the total energy it emits.

• Binary stars are surprisingly common. Those that can be resolved into two distinct star images by an Earth-based telescope are called visual binaries.

The masses of the two stars in a binary system can be computed from measurements of the orbital period and orbital dimensions of the system.

Some binaries can be detected and analyzed, even though the system may be so distant (or the two stars so close together) that the two star images cannot be resolved with Earth-based telescopes.

A spectroscopic binary is a system detected from the periodic shift of its spectral lines. This shift is caused by the Doppler effect, as the orbits of the stars carry them alternately toward and away from the Earth.

An eclipsing binary is a system whose orbits are viewed nearly edge-on from the Earth, so that one star periodically eclipses the other. Detailed information about the stars in an eclipsing binary can be obtained by studying its light curve.

• Mass transfer, which occurs when one star in a close binary overflows its Roche lobe, can affect the appearance and evolution of both stars.

Review Questions

1 How and why is the spectrum of a star related to its surface temperature?

2 Describe UBV filters and how an astronomer uses them to measure a star's surface temperature.

3 Explain why the color index of a star is related to its surface temperature.

4 What is the primary chemical component of most stars?

5 Which is the hottest star listed in Table 10-1? Which is the coolest?

6 Draw an H–R diagram and sketch the regions occupied by main sequence stars, red giants, and white dwarfs. Briefly discuss the different ways in which you could have labeled the axes of your graph.

7 How can observations of a visual binary lead to information about the masses of its stars?

8 What is a radial-velocity curve? What kinds of stellar systems exhibit such curves?

9 What is the difference between a single-line and a double-line spectroscopic binary?

10 What is meant by the light curve of an eclipsing binary? What sorts of information can be determined from such a light curve?

11 What is the mass–luminosity relation? To what kind of stars does it apply?

12 What is a Roche lobe and what is its significance in close binary systems?

13 What is the difference between detached, semidetached, and contact binaries?

Advanced Questions

14 Sketch the radial-velocity curve of a binary whose stars are moving in nearly circular orbits that are (a) perpendicular and (b) parallel to our line of sight.

15 Sketch the light curve of an eclipsing binary whose stars are moving along highly elongated orbits with the major axes of the orbits (a) pointed toward the Earth and (b) perpendicular to our line of sight.

16 Estimate the mass of a main sequence star that is 10,000 times as luminous as the Sun. What is the luminosity of a main sequence star whose mass is $\frac{1}{10}$ that of the Sun?

17 What temperature and classification would you assign to a star with equal H and Ca II line strengths?

Discussion Question

18 How might a star's rotation affect the appearance of its spectral lines?

Observing Projects

19 Locate the star Betelgeuse in Orion. Observe it both by eye and through a small telescope. What color is it in each viewing? It helps to compare its color to those of its neighbors.

20 Locate one or more of the following double star systems: Regulus in Leo, Algieba in Leo, Mirak in Boötes, Ras Algethi in Hercules, Albireo in Cygnus, Vega in Lyra, Polaris in Ursa Minor, Rigel in Orion, Antares in Scorpius, Sirius in Canis Major, and Schedar in Cassiopeia. View them with and without a telescope. What are their colors?

21 Observe the eclipsing binary Algol (β Persei) using nearby stars to judge its brightness during the course of an eclipse. Algol has an orbital period of 2 days, 20 hours, and 53 min-

utes. When the eclipse begins, its apparent magnitude drops from 2.1 to 3.4. It remains this faint for about 2 hours. The entire eclipse, from start to finish, takes about 10 hours. Consult the "Celestial Calendar" section of the current issue of *Sky & Telescope* for the predicted dates and times of minimum brightness of Algol. Note that the schedule is given in Universal Time (the same as Greenwich Mean Time), so you will have to convert to your own time zone. Algol is normally the second brightest star in the constellation Perseus. Because of its northerly position (right ascension = 3^h 08.2^m, declination = $+40°$ 57'), Algol is readily visible from northern latitudes during the fall and winter months.

11 ▶ The Lives of Stars

IN THIS CHAPTER

You will witness star formation from cold clouds of interstellar dust and gas. Using stellar properties from Chapters 9 and 10, you will learn about a star's life on the main sequence, and then follow its remarkable transformation after it exhausts its store of hydrogen. Old stars expand into giants. Some then eject material for future generations of stars. Others pulsate, acting as beacons that enable astronomers to pinpoint distant galaxies. The H–R diagram will be your guide to all these stars—and to the other dying stars you will meet in Chapter 12.

Reflection and Emission Nebulae The two main bluish objects (toward the upper left side of this photograph) are reflection nebulae surrounding two young, hot, main sequence stars. Interstellar dust around these two stars efficiently reflects their bluish light. Several smaller reflection nebulae are scattered around the large reddish patch of ionized hydrogen gas. Dust mixed with the gas dilutes the intense red emission of the hydrogen atoms with a soft bluish haze.

WHAT DO YOU THINK?

1 Where do stars come from?

2 Do stars with greater or lesser mass last longer?

STARS EMIT HUGE AMOUNTS of radiation. As we saw in Chapter 9 on the Sun, the price for emitting energy is change: Stars lose mass and their chemical makeup evolves. The stars seem unchanging to us only because of the colossal time scales over which these transformations occur. Major stages in the life of a star can last for millions or even billions of years. By observing stars with different temperatures, brightnesses, and chemical compositions, we have come to understand **stellar evolution.** Our theory explains how stars form, how they mature, and how they grow old.

Protostars and Pre-Main-Sequence Stars

11-1 Stars form out of enormous volumes of gas and dust

1 Stars condense from clouds of gas and dust lying between existing stars. Radio telescopes have revealed that this **interstellar medium** contains about 10% of all the known mass in our Galaxy. From painstaking observations of the spectra of the interstellar medium and the stars, astronomers have determined the relative abundances of the 10 most common elements in the universe (Table 11-1). We call these numbers the standard "cosmic abundances" of matter.

Besides individual atoms, the cold depths of space allow many types of molecules to exist, including molecular hydrogen (H_2), carbon monoxide (CO), water (H_2O), ammonia (NH_3), and formaldehyde (H_2CO). These molecules can be identified by their unique spectral emissions, just as we can identify elements from atomic spectra.

Astronomers map interstellar gas in order to determine where new stars are forming. Because many interstellar molecules emit radio photons, searches for interstellar gas are primarily done in this part of the electromagnetic spectrum. We know from studying the Sun that stars are composed mostly of hydrogen. However, star-forming hydrogen gas is hard to detect in

TABLE 11-1

Relative Abundances of the Most Common Elements

Atomic number	Element	Symbol	Relative abundance
1	Hydrogen	H	1.0
2	Helium	He	0.07
6	Carbon	C	0.0004
7	Nitrogen	N	0.00009
8	Oxygen	O	0.0007
10	Neon	Ne	0.0001
12	Magnesium	Mg	0.00004
14	Silicon	Si	0.00004
16	Sulfur	S	0.00002
26	Iron	Fe	0.00003

space. Rather, radio astronomers often search for carbon monoxide (CO), which emits copious photons at a wavelength of 2.6 mm. Calculations based on the known abundances of elements reveal that there are about 10,000 H_2 molecules in a typical gas cloud for every CO molecule there. Consequently, wherever astronomers detect strong emission of CO, they know that an enormous amount of hydrogen gas must also be present.

In mapping the locations of CO emission, astronomers came to realize that vast amounts of interstellar gas are concentrated in **giant molecular clouds.** In some cases these regions appear as dark areas silhouetted against a glowing background light, such as the famous Horsehead Nebula in Figure 11-1. In other cases they appear as dark blobs that obscure the background stars (Figure 11-2). Some 6000 of these clouds are known, with masses ranging from 10^5 to 2×10^6 $M_\odot$ and diameters ranging from 50 to 300 light-years (ly). The chemical composition of a giant molecular cloud is about 74% (by mass) hydrogen, 25% helium, and 1% heavier elements. The density inside one of these clouds is about 200 hydrogen molecules per cubic centimeter, which is several thousand times greater than the average density of the gas dispersed throughout interstellar space.

The constellations of Orion and Monoceros encompass one of the most accessible regions of the sky for studying star formation and the interaction of young stars with the interstellar medium. Figure 11-3a shows a map of this region made with a radio telescope tuned to a wavelength of 2.6 mm. Note the extensive areas covered by giant molecular clouds. Such comprehensive maps of CO emission help astronomers understand how

Figure 11-1 **The Horsehead Nebula** Dust grains block the light from the background nebulosity whose glowing gases are excited by ultraviolet radiation from young, massive stars. The giant molecular cloud with the distinctive horsehead outline is located in Orion at a distance of roughly 1600 ly from Earth. The bright star to the left of center is Alnitak (ζ Orionis), the easternmost star in the belt of Orion.

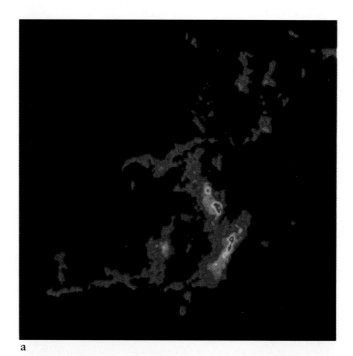

a

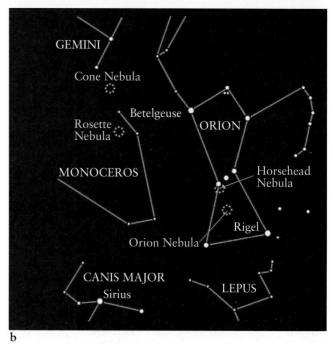

b

Figure 11-2 **A Dark Nebula** This dark nebula, Barnard 86, is located in Sagittarius. It is visible in this photograph simply because it blocks out light from the stars beyond it. The cluster of bluish stars to the left of the dark nebula is NGC 6520.

Figure 11-3 **A Map of Carbon Monoxide Features in Orion** (a) This color-coded map of a large section of the sky shows the extent of giant molecular clouds in Orion and Monoceros. The intensity of carbon monoxide (CO) emission is displayed by colors in the order of the rainbow, from violet for the weakest to red for the strongest. Black indicates no detectable emission. (b) This star chart covers the same area as the CO map in (a). The locations of four prominent star-forming nebulae are indicated. Note that the Orion and Horsehead nebulae are sites of intense CO emission.

the large-scale structure of the interstellar medium is related to the formation of stars.

11-2 Supernova explosions in cold, dark nebulae trigger the birth of stars

As we shall see in detail in Chapter 12, a supernova is a violent detonation that ends the life of a massive star. In a matter of seconds the core of the doomed star collapses, releasing vast quantities of particles and energy that blow the star apart. The star's outer layers are blasted into space at speeds of several thousand kilometers per second.

Astronomers find many remains of such dead stars across the sky. Such nebulae, like the Cygnus Loop shown in Figure 11-4, are called **supernova remnants.** Many supernova remnants have a distinctly arched appearance, as would be expected for a shell of gas expanding at supersonic speeds. As it passes through the surrounding interstellar medium, the supernova remnant excites the atoms, causing the gases to glow. If the expanding shell of a supernova remnant encounters a giant molecular cloud, it can squeeze the cloud, stimulating

star birth. As we learned in Chapter 8, there is evidence that the Sun was created in this fashion.

A simple collision between two interstellar clouds can also create new stars, since compression must occur at the boundary between the clouds. Radiation from an especially bright star or group of stars may also compress the surrounding interstellar medium enough to give birth to stars (Figure 11-5).

Once a giant molecular cloud compresses and cools enough, gravitational attraction causes small regions of gas and dust in it to collapse until they become stars. This gas must be cool because the higher the temperature, the faster its atoms and molecules move. As they move, they collide with other particles, driving them apart. In other words, the collisions between atoms and molecules create the pressure in the cloud. If the temperature is too high, the pressure overcomes any gravitational attraction between the particles, preventing them from drawing close enough together to form stars.

Each small region of an interstellar cloud behaves almost like a toy balloon. Just as gravity draws together particles of cosmic gas and dust, the gas in a balloon is compressed by the elastic walls. When heated, the balloon expands, because the gas particles move faster and create

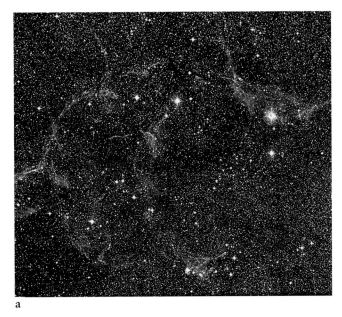

a

b

Figure 11-4 **A Supernova Remnant** (a) The Veil nebula is part of the Cygnus Loop, remnant of a supernova explosion that occurred about 20,000 years ago. The expanding spherical shell of gas now has a diameter of about 120 ly. The entire Cygnus loop fills an area of our sky six times larger than the

Moon. (b) This close-up image (of part of the Cygnus loop), taken by the Hubble Space Telescope, shows emission from different atoms: blue from oxygen, red from sulfur, and green from hydrogen.

Figure 11-5 **The Core of the Rosette Nebula** The Rosette Nebula is a large, circular nebula near one end of a sprawling giant molecular cloud in the constellation of Monoceros. Radiation from young, hot stars has blown gas away from the center of this nebula. Some of this gas has become clumped in dark globules that appear silhouetted against the glowing background gases.

greater pressure. When cooled, the balloon shrinks, because the pressure decreases and the walls are more effective in compressing the gas. Similarly, when part of an interstellar cloud cools, its gravitational attraction allows the cloud to shrink. When a region of a cloud is sufficiently cold and dense, the gravitational attraction completely overwhelms the pressure and pulls the gas together to form a new star. This collapse is called the **Jeans instability** after the British physicist James Jeans, who in 1902 calculated the conditions for it to occur.

Infrared observations show compact regions called **dense cores** inside many interstellar clouds. Their temperatures, around 10 K, are so low that the dense cores are destined to form stars. Often a giant molecular cloud will

EYES ON . . .   Nebulae

Binoculars and the naked eye are enough to let you "get your hands dirty" exploring star dust. Distant nebulae—clusters of stars and glowing gases—are among the most impressive objects in the night sky.

During the winter the Great Nebula of Orion can be observed even with the naked eye. In all likelihood, you've seen it dozens of times without knowing it. To locate the Great Nebula, find the constellation Orion in the night sky (using, for example, the star charts at the end of the book). Locate Orion's belt. Due south of the belt are three stars in a row making up Orion's sword. Examine the sword very carefully with your naked eye. Do any of the stars in it look at all odd? Now look at them through a pair of binoculars. Which one is different from the others? That one is the Great Nebula of Orion, not a

star at all! How does what you see compare with Figure 11-10?

The North American Nebula and the Pelican Nebula in Cygnus are best spotted in the fall. Pick a dark, moonless night, and use binoculars rather than a telescope. Higher magnification reveals too small a region of the sky for you to see the entirety of these vast, dim nebulae. To find them, first locate the bright star Deneb on the tail of Cygnus (using, for example, the star charts at the end of the book). The North American Nebula is located 3° east of Deneb, while the Pelican Nebula is located 2° southeast of it. These are both small angles, so sweep around the sky east of Deneb. If your binoculars are powerful enough, you should be able to see the outlines that give these nebulae their names.

have several hundred or even a few thousand dense cores. In this case, hundreds or thousands of stars form together. Such stellar nurseries are called **open clusters** of stars.

At first, a collapsing dense core is just a cool, dusty region thousands of times larger than our solar system. The dense core actually collapses from the inside out: The inner region of the dense core falls in rapidly, leaving the outer layers to drift in at a more leisurely rate. This process of increasing mass in the central region is called accretion, and the newly forming object at the center is called a **protostar.** Although fusion has not begun, a protostar glows from the heat generated by the compression of the gas it contains.

Figure 11-6 shows views of an interstellar cloud in which stars are forming. The view at visible wavelengths (Figure 11-6*a*) is a familiar sight to many telescopic observers, but it tells only part of the story. Visible light from the protostar never reaches us, because it is absorbed by the surrounding shell of infalling dust. Only the infrared picture (see Figure 11-6*b*) shows the hundreds of protostars.

If a dense core is not spinning, it collapses into a sphere, which ultimately becomes an isolated star. If it is spinning, it collapses into a disk, which may then condense into a few stars. Or, if the disk has a low enough mass, it may become a single star but with orbiting protoplanets. One such disk of gas and dust has been observed around β Pictoris (see Figure II-4). It was the first sign of planetary development outside our own solar system.

11-3 When a protostar ceases to accumulate mass, it becomes a pre-main-sequence star

After about 10^5 years of mass accretion, the protostar builds up to the mass of the Sun. A protostar of 1 M$_\odot$ is about five times larger in diameter than the Sun. Its large size makes it brighter than the Sun, and it can be seen as an intense point of infrared light. Much matter is still slowly falling inward from the dense core's outer shell. However, the radiation and particles flowing off the protostar exert outward forces on this remaining gas and dust, preventing it from ever reaching the protostar. This outward force is analogous to the force exerted on comets by the solar wind. Now mass accretion stops, and the protostar is called a **pre-main-sequence star.**

A pre-main-sequence star contracts slowly, unlike the rapid collapse of a protostar. When the temperature

a

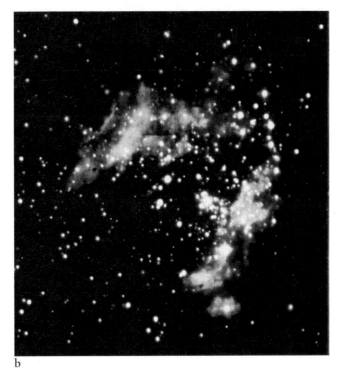

b

Figure 11-6 **Newborn Stars in the Swan Nebula** (a) This image at visible wavelengths shows the Swan (or Omega) Nebula, so named because of its characteristic shape. (b) This infrared view, constructed from images taken at wavelengths of 1.2, 1.6, and 2.2 μm, reveals hundreds of stars that do not appear in (a). A comparison of the visible and infrared views demonstrates that interstellar dust obscures many of the stars in this star-forming region.

at its core reaches ten million kelvins, hydrogen fusion begins there. As we saw in Chapter 9, this thermonuclear process releases enormous amounts of energy. The outpouring of energy from hydrogen fusion creates pressure inside the pre-main-sequence star sufficient to finally halt its gravitational contraction.

The more massive a pre-main-sequence star is, the more rapidly it ignites hydrogen fusion in its core. For example, a 5-$M_\odot$ pre-main-sequence star ignites only 100,000 years after it first forms from a protostar, whereas a 1-$M_\odot$ pre-main-sequence star takes a few tens of millions of years to do the same. By astronomical standards, these intervals are so brief that pre-main-sequence stars are considered quite transitory.

In the final stages of pre-main-sequence evolution, the outer shell of gas and dust finally dissipates. For the first time the star is directly revealed to the outside universe.

11-4 The evolutionary track of a pre-main-sequence star depends on its mass

Astrophysicists use high-speed computers and the equations of stellar structure (described in Chapter 9) to model the evolution of a pre-main-sequence star. By calculating changes in the energy that the contracting star emits, computer models can follow its changing position on a Hertzsprung–Russell diagram (Figure 11-7). Such an **evolutionary track** represents changes in a star's temperature and luminosity, not its motion in space.

Pre-main-sequence stars are relatively cool when they begin to shine at visible wavelengths. Hence the evolutionary tracks of pre-main-sequence stars begin near the right side of the H–R diagram along a curve called the **birth line** (see Figure 11-7). A star's exact location on this curve depends only on its mass. As a pre-main-sequence star of less than 4 $M_\odot$ contracts, its diminishing surface area causes its luminosity to drop. On the H–R diagram, therefore, the track of the star drops below the birth line. Eventually its surface temperature increases, and the star's track moves to the left in the diagram.

Pre-main-sequence stars more massive than 4 $M_\odot$ become hotter without much change in overall luminosity. The evolutionary tracks of these pre-main-sequence stars thus traverse the H–R diagram horizontally, from right to left. A star more massive than about 7 $M_\odot$ has no pre-main-sequence phase at all. Its gravitational compression is so great that it begins to fuse hydrogen in its protostellar phase.

Pre-main-sequence stars less massive than about 0.08 $M_\odot$ do not have enough gravitational force compressing their cores to initiate fusion. Instead, these small bodies contract to become planetlike objects, sometimes called **brown dwarfs.** As we saw in Chapter 7, Jupiter has 75 times too little mass to be a star (i.e., to have fusion in its core), so a brown dwarf is like a super-Jupiter. Confirmation of the existence of brown dwarfs was first made in 1995, when the Hubble Space Telescope imaged Gliese 229B, a brown dwarf located in the constellation Lepus 18 ly from Earth (Figure 11-8).

Protostars with masses greater than about 100 $M_\odot$ rapidly develop such extremely high temperatures that their outer layers are expelled into interstellar space, thereby reducing their masses. Stars therefore have masses between about 0.08 and 100 $M_\odot$, with lower-mass stars being most common.

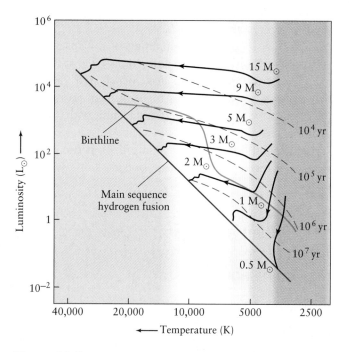

Figure 11-7 Pre-Main-Sequence Evolutionary Tracks The evolutionary tracks of seven stars having different masses are shown in this H–R diagram. The dashed lines indicate the stage reached after the indicated number of years of evolution. The birth line, shown in blue, is the location where each protostar stops accreting matter and becomes a pre-main-sequence star. Stars more massive than about 7 $M_\odot$ start fusing so rapidly that they go directly from protostars to the main sequence. Note that all tracks terminate on the main sequence at points agreeing with the mass–luminosity relation.

Figure 11-8 **A Brown Dwarf** Gliese 229B (center) is the first confirmed brown dwarf ever observed. Its spectrum is like that of Jupiter. Gliese 229B is in a binary star system. Part of its companion, Gliese 229, is seen on the left. The two stars are separated by about 43 AU. Gliese 229B has from 20 to 50 times the mass of Jupiter, but the brown dwarf is compressed to the same size as the giant planet. The spike of light was produced when Gliese 229 overloaded part of the telescope's electronics.

11-5 Young star clusters are found in H II regions

The most massive pre-main-sequence stars, those of spectral types O and B, are exceptionally hot. Because their surface temperatures are typically 15,000 to 35,000 K, they emit vast quantities of ultraviolet radiation. This energetic radiation easily ionizes any surrounding gas. Photons from an O5 star can ionize hydrogen atoms up to 500 ly away.

Because of this ionization, we can detect the formation of a cluster of stars as a magnificent glow in the nebula. While some hydrogen atoms are being knocked apart by ultraviolet photons, some of the free protons and electrons manage to get back together. As these new hydrogen atoms assemble, their electrons return to their ground state. This downward cascade through each atom's energy levels is what makes the nebula glow. Particularly prominent is the transition from $n = 3$ to $n = 2$, which produces H_{α} photons at 656 nm in the red portion of the visible spectrum (review the emission line spectrum in Figure 4-11). Thus, the nebulosity around a newborn star cluster often shines with a distinctive reddish hue. Figure 11-9 shows one of these **emission nebulae**. Because these nebulae are predominantly ionized hydrogen, they are also called **H II regions.**

Figure 11-9 **An H II Region** Because of its shape, this emission nebula is called the Eagle Nebula. It surrounds the star cluster called M16 in the constellation of Serpens Cauda (serpent's tail) at a distance of 6500 ly from Earth. Several bright, hot O and B stars are responsible for the ionizing radiation that causes the gases to glow.

Figure 11-10 **The Orion Nebula** This famous H II region can be seen with the naked eye. It is 1600 ly from Earth and has a diameter of roughly 16 ly. This nebula's mass is about 300 $M_\odot$. Four bright, massive stars at the center of the nebula produce the ultraviolet radiation that causes the gases to glow. These four stars, called the Trapezium, are separated from each other by only 0.13 ly.

An H II region is a small, bright "hot spot" in a giant molecular cloud. The collection of a few hot, bright O and B stars near the core of the nebula that produces the ionizing ultraviolet radiation is called an **OB association**. The famous Orion Nebula (Figure 11-10) is an example. Four hot, massive O and B stars at the heart of the Orion Nebula are responsible for the ionizing radiation that causes the surrounding gases to glow. The Orion Nebula is embedded in a giant molecular cloud whose mass is estimated at 500,000 $M_\odot$.

The OB association at the core of an H II region affects the rest of the giant molecular cloud (Figure 11-11). Vigorous stellar winds, along with ionizing ultraviolet radiation from the O and B stars, carve out a cavity in the cloud. Much of this outflow is supersonic. Supersonic flow always leads to the formation of a shock wave, like the sonic boom created by fast-flying aircraft. In this case a shock wave forms where the outer edge of the expanding H II region impinges on the rest of the giant molecular cloud. This shock wave compresses the hydrogen gas through which it passes, stimulating a new

Figure 11-11 **Exposing Young Stars** At the center of the Eagle Nebula, star formation is occurring in these dark pillars of gas and dust in the star cluster M16. However, intense ultraviolet radiation from existing massive stars in M16 is evaporating the dense cores in the columns, thereby prematurely terminating star formation there. Newly revealed stars are visible at the tops of the columns. These three gaseous pillars can be seen, tilted clockwise about 45°, at the center of Figure 11-9.

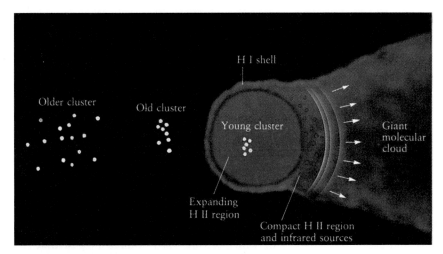

Figure 11-12 **The Evolution of an OB Association** Ultraviolet radiation from young O and B stars produces a shock wave that compresses gas further into the molecular cloud, stimulating new star formation deeper into the cloud. Meanwhile, older stars are left behind.

round of star birth. The new O and B stars continue to power the expansion of the H II region still farther into the giant molecular cloud. Meanwhile, the older O and B stars left behind begin to disperse (Figure 11-12). In this way, an OB association "eats into" a giant molecular cloud, "spitting out" stars in its wake.

Infrared observations may have revealed protostars in the swept-up layer immediately behind the shock wave from an OB association. For instance, Figure 11-13 shows both visible and infrared views of the core of the Orion Nebula. Besides the four O and B stars and glowing gas and dust that dominate the center of the view at

a

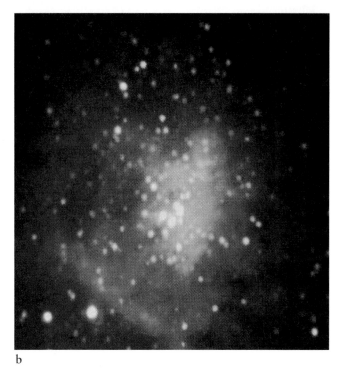

b

Figure 11-13 **The Core of the Orion Nebula** (a) This view at visible wavelengths shows the inner regions of the Orion Nebula. At the center are four massive stars, called the Trapezium, which cause the nebula to glow. (b) This infrared view, also centered on the Trapezium, covers the same area as (a). Infrared radiation can penetrate interstellar dust that absorbs visible photons. Numerous infrared objects, many of which are probably stars in the early stages of formation, can be seen.

visible wavelengths, infrared observations reveal dozens of infrared objects that may be cocoons of warm dust still enveloping pre-main-sequence stars.

11-6 Plotting a star cluster on an H–R diagram reveals its age

A young open cluster is a rich source of information about stars in their infancy. Figure 11-14 shows a beautiful emission nebula surrounding the cluster called NGC 2264 (NGC stands for *New General Catalogue*). By measuring each star's magnitude, color index, and distance, an astronomer can deduce its luminosity and surface temperature. The data for all the stars in the cluster can then be plotted on an H–R diagram, as shown in Figure 11-14. Note that the hottest stars, with surface temperatures around 20,000 K, are on the main sequence. These hot stars are the rapidly evolving, massive ones whose radiation causes the surrounding gases to

glow. Most of the stars cooler than about 10,000 K have not yet arrived at the main sequence. These less massive stars, which are in the final stages of pre-main-sequence contraction, are just now beginning to ignite thermonuclear reactions at their centers. The locations of data points on Figure 11-14 suggest that the cluster is roughly two million years old.

Spectroscopic observations of the cooler stars in NGC 2264 show that many are vigorously ejecting gas, a very common phenomenon in most stars just before they reach the main sequence. Gas-ejecting stars in spectral classes G and cooler (i.e., G, K, and M) are called **T Tauri stars,** after the first example discovered in the constellation of Taurus. Some astronomers suggest that the onset of hydrogen fusion is preceded by vigorous chromospheric activity marked by enormous spicules and flares that propel the star's outermost layers back into space. In fact, an infant star going through its T Tauri stage can lose as much as 0.4 $M_\odot$ of matter and also shed its shell while still a pre-main-sequence star.

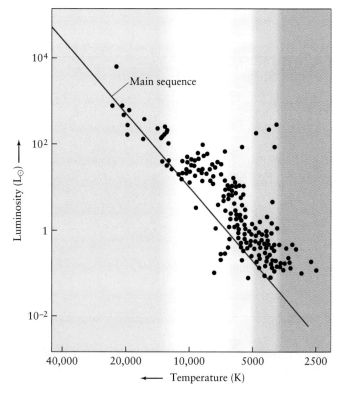

Figure 11-14 **A Young Star Cluster and Its H–R Diagram** The photograph shows an H II region and the young star cluster NGC 2264 in the constellation of Monoceros. The nebulosity is located about 2600 ly from Earth and contains numerous stars that are about to begin hydrogen fusion in their cores.

Each dot plotted on the H–R diagram represents a star in this cluster whose luminosity and surface temperature have been measured. Note that most of the cool, low-mass stars have not yet arrived at the main sequence. This star cluster probably started forming only two million years ago.

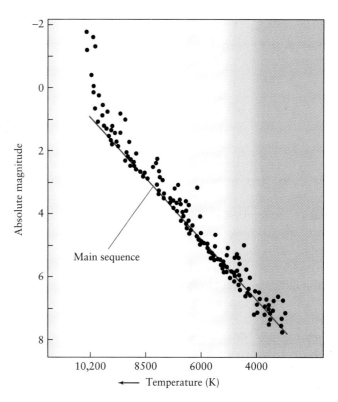

Figure 11-15 **The Pleiades and Its H–R Diagram** This open cluster, called the Pleiades, can easily be seen with the naked eye in the constellation of Taurus. It lies about 400 ly from Earth. Each dot plotted on the H–R diagram represents a star in the Pleiades whose absolute magnitude and surface temperature have been measured. Note that most of the cool, low-mass stars have arrived at the main sequence, indicating that hydrogen fusion has begun in their cores. The cluster has a diameter of about 5 ly and is about 100 million years old.

Figure 11-15 shows a young star cluster, easily visible to the unaided eye, called the Pleiades in the constellation of Taurus. In contrast to the H–R diagram for NGC 2264, nearly all the stars in the Pleiades have completed their pre-main-sequence stages. The cluster's age is about 100 million years, which is how long it takes for the least massive stars to finally begin hydrogen fusion in their cores.

Note the distinctly bluish color of the nebulosity around the Pleiades. This haze, called a **reflection nebula,** is caused by fine grains of interstellar dust that efficiently scatter and reflect blue light. Indeed, reflection nebulosity is blue for the same reason Earth's sky is blue: Particles scatter short-wavelength light much more efficiently than longer-wavelength radiation. Blue light is therefore bounced around and scattered back toward us much more intensely than is light of any other color.

Open clusters, such as the Pleiades and NGC 2264, possess barely enough mass to hold themselves together.

A star moving faster than the average speed will occasionally escape from such a cluster. In a few billion years, after its stars have separated from each other and mixed with the rest of the stars in the galaxy, the cluster will no longer exist.

Main Sequence and Giant Stars

Once the outward force caused by thermal pressure is balanced by the inward force of gravity (a condition known as *hydrostatic equilibrium*), a pre-main-sequence star ceases to collapse and a star is born. At this stage, the pre-main-sequence star's evolutionary track reaches the main sequence, as shown in Figure 11-7. *Main sequence stars are those stars fusing hydrogen into helium in their cores.*

11-7 Stars spend most of their lives on the main sequence

The **zero-age main sequence (ZAMS)** is the location on the H–R diagram where a pre-main-sequence star fusing hydrogen in its core first becomes a stable object, neither shrinking nor expanding. As was the case with the birth line, the location of a ZAMS star on the main sequence depends solely on its mass. Note that the evolutionary tracks in Figure 11-7 end at locations along the main sequence that agree with the mass–luminosity relation (recall Figure 10-11): The most massive main sequence stars are the most luminous, while the least massive stars are the least luminous.

The more massive a star is, the faster it goes through its main sequence phase. Because of the greater mass, gravity presses down with greater force on the star's core. This tremendous pressure creates such an accelerated rate of fusion that some O and B stars consume all their core hydrogen in a few million years, as shown in Table 11-2. Conversely, stars of very low mass take hundreds of billions of years to convert their cores from hydrogen into helium.

It is because fusing hydrogen into helium in their cores takes so long compared to any other stage of stellar evolution that the vast majority of stars on an H–R diagram are located on the main sequence. The Sun's total lifetime on the main sequence will be about ten billion years.

11-8 When core hydrogen fusion ceases, a main sequence star becomes a giant

When the hydrogen in a star's core is completely converted into helium, fusion ceases and the star can no longer support the weight of its outer layers. Recall from Chapter 9 that these layers are supported by thermal pressure, with energy resupplied by fusion in the core. The enormous weight pressing inward from all sides compresses the star's helium core, which causes it to heat further. As the core contracts, the hydrogen gas in a shell immediately surrounding it is drawn downward, compressed, and heated enough to begin fusing into helium. This is called **shell hydrogen fusion.** The hydrogen-fusing shell slowly works its way outward from the original core, thereby dumping more helium onto the core, which continues to contract and heat up.

TABLE 11-2

Main Sequence Lifetimes

Mass ($M_\odot$)	Surface temperature (K)	Luminosity ($L_\odot$)	Time on main sequence (10^6 years)
25	35,000	80,000	3
15	30,000	10,000	15
3	11,000	60	500
1.5	7,000	5	3,000
1.0	6,000	1	10,000
0.75	5,000	0.5	15,000
0.50	4,000	0.03	200,000

Because the core is contracting, you might think that the outer envelope of the star also shrinks, but this does not happen. In fact, the star expands to become a **giant.** With the hydrogen-fusing region now outside the core, photons created by fusion do not have as far to travel to reach the photosphere and so they lose less energy on the way up. Consequently, they heat the outer layers of the star more, thereby causing the star's outer atmosphere to swell farther and farther into space. Far from the fusing shell, the surface gases cool and soon the temperature of the star's bloated surface falls to between 3500 and 6000 K depending on the star's total mass. The cooler, less massive ones, glowing red, are often called *red giants*. Our own Sun will swell to a giant with a diameter of about 1 AU, vaporizing Mercury and perhaps causing Venus to spiral into the Sun. The Earth will be scorched to a cinder.

Giant stars are so enormous that their bloated outer layers constantly leak gases into space. At times, this *mass loss* is quite significant (Figure 11-16). Mass loss can be detected spectroscopically. Escaping gases coming toward us exhibit narrow absorption lines that are slightly blueshifted due to the Doppler effect. This small shift corresponds to a speed of 10 km/s, typical of the expansion velocities with which gases leave the tenuous outer layers of giants. A typical rate of mass loss for a giant is roughly 10^{-7} $M_\odot$ per year. For comparison, the Sun's mass loss rate is only 10^{-14} $M_\odot$ per year.

Although a giant is cooler than the main sequence star from which it evolved, the giant is brighter simply because it has more surface area from which to emit photons. As a full-fledged giant (Figure 11-17), our Sun will shine 2000 times brighter than it does today.

Figure 11-16 **A Mass Loss Star** Old stars become giants whose bloated outer atmospheres shed matter into space. This star is losing matter at a high rate and is surrounded by a reflection nebula caused by starlight reflected from dust grains. These dust grains may have condensed from material shed by the star. A typical giant can lose 10^{-7} $M_\odot$ per year, and many are surrounded by circumstellar shells of matter they have shed.

Figure 11-17 **The Sun Today and as a Giant** In about five billion years, when the Sun expands to become a giant, its diameter will increase while its core becomes more compact. Today the Sun's energy is produced in a hydrogen-fusing core whose diameter is about 300,000 km. When the Sun becomes a giant, it will draw its energy from a hydrogen-fusing shell surrounding a compact helium-rich core. The helium core will have a diameter of only 30,000 km.

11-9 Helium fusion begins at the center of a giant

Helium is the ash of hydrogen fusion. When a star first becomes a giant, its hydrogen-fusing shell surrounds a small, compact core of almost pure helium. In a moderately low-mass giant, the dense helium core is about twice the size of the Earth.

At first no thermonuclear reactions occur in the helium-rich core of a giant because the temperature there is too low to fuse helium nuclei. As the hydrogen-fusing shell moves outward in the star, however, it adds mass to the helium core. The core slowly contracts, forcing the star's central temperature to climb.

When the central temperature reaches 100 million kelvins, **core helium fusion** is ignited at the star's center.

First, two helium nuclei combine. Then a third helium nucleus is added to this combination. The entire transformation results in a carbon nucleus and the release of energy that can be written as follows:

$$^4\text{He} + {}^4\text{He} + {}^4\text{He} \longrightarrow {}^{12}\text{C} + \gamma$$

where γ denotes energy emitted as photons. Some of the carbon created can then fuse with another helium nucleus to produce oxygen:

$$^{12}\text{C} + {}^4\text{He} \longrightarrow {}^{16}\text{O} + \gamma$$

Again energy is released as gamma rays. The aging star thus has a central energy source for the first time since it left the main sequence.

The energy released in core helium fusion re-establishes thermal equilibrium—but only briefly. A mature giant fuses helium in its core for about 20% of the time that it spent fusing hydrogen as a main sequence star. While fusion is again occurring in the core, further gravitational contraction of the star ceases. Starting in the distant future the Sun will consume the helium it is now creating in its core for about two billion years.

How helium fusion begins at a giant's center depends on the mass of the star. In high-mass stars (those with masses greater than about 2 $M_\odot$), helium fusion begins gradually as temperatures in the star's core approach 100 million kelvins. In low-mass stars (those with less than 2 $M_\odot$), helium fusion begins explosively and suddenly, in an event called the **helium flash**. The helium flash occurs because of unusual conditions that develop in the core of a low-mass star on its way to becoming a giant. To appreciate these conditions we must first understand how an ordinary gas behaves, then explore how the densely packed electrons at the star's center alter this behavior.

When an ordinary gas is compressed, it heats up, and when it expands, it cools down. A high-mass star behaves the same way. If energy production overheats its core, the core expands, cooling the gases and slowing the rate of thermonuclear reactions. Conversely, if too little energy is being created to support the star's overlying layers, they move inward, compressing the core. The resulting increase in temperature speeds up the thermonuclear reactions and hence increases the energy output, which stops the contraction. Either way, the star has a "safety valve" to keep it from exploding.

In a low-mass giant, the core must undergo considerable gravitational compression to drive temperatures high enough to ignite helium fusion. At the extreme pressures deep inside the star, the atoms are completely torn apart into nuclei and electrons. In the star's highly compressed core, the free electrons are so closely crowded together that a law of physics called the **Pauli exclusion principle** becomes important. This principle, formulated in 1925 by the Austrian physicist Wolfgang Pauli, explains that two indentical particles cannot exist in the same place at the same time.

Just before the onset of helium fusion, the electrons in the core of a low-mass star are so closely crowded together that any further compression would violate the Pauli exclusion principle. Astronomers say that the helium-rich core of a low-mass giant is supported by **electron degeneracy pressure.** *Degeneracy pressure, unlike the pressure of an ordinary gas, does not depend on temperature.* Without the "safety valve" of increasing pressure, the star's core cannot expand and cool. As helium fusion begins, temperatures rise to over 100 million kelvins, and they can only continue to rise. Within a few seconds, the explosively rising temperature causes the helium to fuse at an ever-increasing rate: *the helium flash.*

In the helium flash, pressures become so high that the electrons begin to move again; the helium becomes an ordinary gas. Suddenly the usual safety valve operates once again. In a matter of hours, the star's core expands and cools. Fusion continues after the helium flash is over.

After the helium flash, the star's energy output declines, and so its outer layers again contract. The low-mass giant is left smaller, dimmer, and hotter. It has left the main sequence forever.

11-10 As stars evolve, they move on the H–R diagram

It is very enlightening to plot the evolutionary tracks of post-main-sequence stars on the Hertzsprung–Russell diagram (Figure 11-18). After each star reaches the main sequence, its evolutionary track slowly inches away from the ZAMS location as the hydrogen-fusing core grows in search of fresh fuel. The dashed line in Figure 11-18 shows the locations of the stars when all the core hydrogen has been consumed. As noted earlier, it takes a few million years for the most massive main sequence stars to reach this point, while a 1-$M_\odot$ star takes ten billion years. Lower-mass stars take tens or even hundreds of billions of years to get there.

After core hydrogen fusion ceases, the points representing high-mass stars move rapidly from left to right

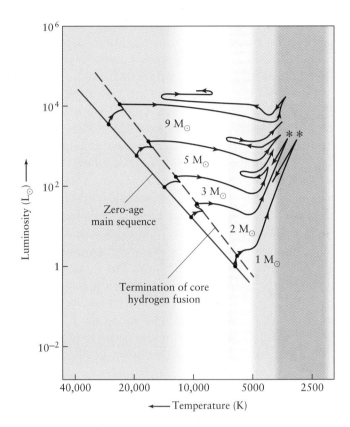

Figure 11-18 **Post-Main-Sequence Evolution** The evolutionary tracks of five stars are shown on this H–R diagram. In the high-mass stars, core helium fusion ignites smoothly where the evolutionary tracks make a sharp turn upward into the giant region of the diagram. The red stars on the 1-$M_\odot$ and 2-$M_\odot$ curves indicate where the helium flash in these stars occurs.

11-11 Globular clusters are bound groups of old stars

Post-helium-flash stars are often found in old star clusters, called **globular clusters,** so named because of their spherical shapes. A typical globular cluster, like the one shown in Figure 11-19, contains up to a million stars in a volume 300 ly across. Unlike open clusters, globular clusters are gravitationally bound groups of stars that do not disperse. Astronomers know that such clusters are old because they contain no high-mass main sequence stars: If you measure the luminosity and surface temperature of many stars in a globular cluster and plot the data on a color–magnitude diagram, as shown in Figure 11-20, you will find that the upper half of the main sequence is missing. A color–magnitude diagram is the observational equivalent of an H–R diagram. All the high-mass main sequence stars have evolved long ago into giants, leaving behind only lower-mass, slowly evolving stars still undergoing core hydrogen fusion.

The color–magnitude diagram of a globular cluster typically shows a horizontal grouping of stars to the left of the center portion of the diagram (see Figure 11-20). These stars, called **horizontal branch stars,** are post-helium-flash stars that have luminosities of about 50 $L_\odot$.

across the H–R diagram. This is when each star's core contracts and its outer layers expand. Although the stars' surface temperatures are decreasing, their surface areas are increasing, so that their overall luminosities remain roughly constant.

Just before core helium fusion begins, the evolutionary tracks of high-mass stars turn upward into the giant region of the H–R diagram. After the core helium fusion begins, however, the evolutionary tracks back away from these peak luminosities. The tracks then wander back and forth in the giant region while the stars readjust to their new energy sources.

We saw that, after the helium flash, low-mass stars become dimmer but hotter. On the H–R diagram in Figure 11-18, two low-mass stars can be found to move down and to the left.

Figure 11-19 **A Globular Cluster** A globular cluster is a spherical cluster that typically contains a few hundred thousand stars. This cluster, called M13, is located in the constellation of Hercules, roughly 25,000 ly from Earth.

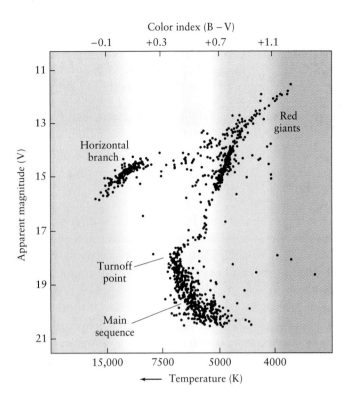

Color index (B – V)

Figure 11-20 **A Color–Magnitude Diagram of a Globular Cluster** Each dot on this graph represents the apparent magnitude and surface temperature of a star in the globular cluster called M55. Since all the stars are at virtually the same distance from Earth, their apparent magnitudes are measured rather than their absolute magnitudes. Note that the upper half of the main sequence is missing. The horizontal branch stars are low-mass stars that recently experienced the helium flash in their cores.

In years to come, these stars will move back toward the giant region as core helium fusion and shell hydrogen fusion devour their fuel.

An H–R diagram of a cluster can also be used to determine the age of the cluster. In the diagram for a very young open cluster (review Figure 11-14), all the stars are on the main sequence. As a cluster gets older, like the Pleiades, stars begin to leave the main sequence (see Figure 11-15). The high-mass, high-luminosity stars are the first to become giants as the main sequence starts to burn down like a candle. Over the years, the main sequence in a cluster gets shorter and shorter. The top of the surviving portion of the main sequence is called the *turnoff point* (see Figure 11-20). Stars at the turnoff point are just beginning to exhaust the hydrogen in their cores, and *their main sequence lifetime is equal to the age of the cluster* (recall Table 11-2). Consider the cluster M55

(see Figure 11-20), so named because it was 55th in the *Messier Catalogue* of astronomical objects. In M55, 0.8-$M_\odot$ stars have just left the main sequence, and so the cluster's age is roughly 15 billion years.

Data for several star clusters are plotted on Figure 11-21, along with turnoff point times from which the ages of the clusters can be estimated. The youngest clusters (those with their main sequences still intact) are open clusters in the disk of our Galaxy, where star formation is an ongoing process. Stars in these young clusters are said to be *metal-rich* because their spectra contain many prominent spectral lines of heavy elements. This material originally came from dead stars that exploded long ago, enriching the interstellar gases with the heavy elements formed in their cores. The young clusters are therefore formed from the debris of older generations of stars. The Sun is an example of a young, metal-rich star. Such stars are also called **population I** stars.

The oldest clusters are globular clusters. Such clusters are generally located above or below the disk of our

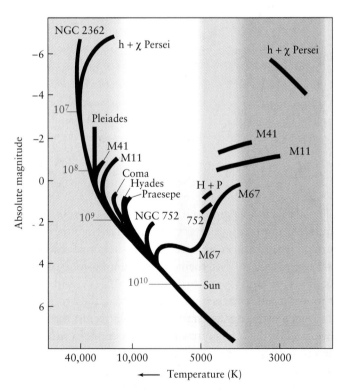

Figure 11-21 **A Composite H–R Diagram** The black bands indicate where data from various star clusters fall on the H–R diagram. The ages of turnoff points (in years) are listed in red alongside the main sequence. The age of a cluster can be estimated from the location of the turnoff point, where the cluster's most massive stars are just now leaving the main sequence.

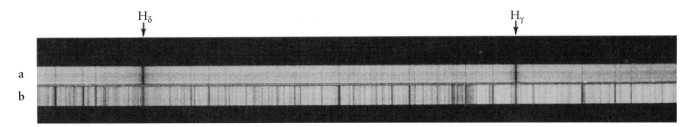

H_δ $\qquad$ H_γ

a

b

Figure 11-22 **Spectra of a Metal-Poor and a Metal-Rich Star** These spectra compare (**a**) a metal-poor star and (**b**) a metal-rich star (the Sun). Numerous spectral lines prominent in the solar spectrum are caused by elements heavier than hydrogen and helium. Note that corresponding lines in the metal-poor star's spectrum are weak or absent. Both spectra cover a wavelength range that includes two strong hydrogen absorption lines labeled H_γ and H_δ.

Galaxy, that band of light sweeping majestically across the night sky. Because their spectra show only weak lines of heavy elements, these ancient stars are said to be *metal-poor*. They were created long ago from interstellar gases that had not yet been substantially enriched with heavy elements. They are also called **population II** stars. Spectra of a population II star and of the Sun are compared in Figure 11-22.

Variable Stars

After core helium fusion begins, the evolutionary tracks of mature stars move across the middle of the H–R diagram. In Figure 11-18, we saw the evolutionary tracks of post-main-sequence stars. During these excursions across the H–R diagram, a star can become unstable and pulsate. In fact, there is a region on the H–R diagram between the main sequence and the giant branch that is called the **instability strip** (Figure 11-23). When a star's evolutionary track carries it through this region, the star slowly pulsates. As it does so, its brightness varies periodically. These so-called **variable stars** can be easily identified by their changes in brightness amid a field of stars of constant luminosity.

Low-mass, post-helium-flash stars pass through the lower end of the instability strip as they move in the horizontal branch along their evolutionary tracks. These stars become **RR Lyrae variables,** named after the prototype in the constellation of Lyra. RR Lyrae variables all have periods shorter than one day, and all have roughly the same average brightness as stars on the horizontal branch. High-mass stars pass back and forth through the upper end of the instability strip on the H–R diagram. These stars become **Cepheid variables,** often simply called Cepheids.

11-12 A Cepheid pulsates because it is alternately expanding and contracting

A Cepheid variable is characterized by the way in which its light output varies: rapid brightening followed by gradual dimming. A Cepheid variable brightens and

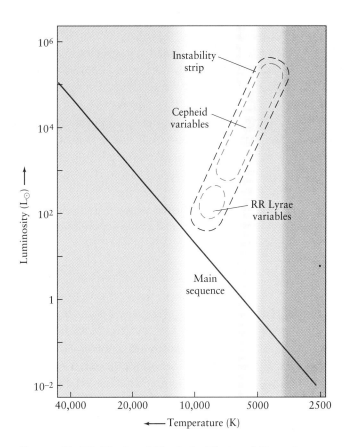

Figure 11-23 **The Instability Strip** The instability strip occupies a region between the main sequence and the giant branch on the H–R diagram. A star passing through this region along its evolutionary track becomes unstable and pulsates.

fades because the star's outer layers cyclically expand and contract. This behavior is deduced from spectroscopic observations. Spectral lines in the spectrum of δ Cephei shift back and forth with the same 5.4-day period that characterizes the variations in magnitude. According to the Doppler effect, these shifts mean that the star's surface is alternately approaching and receding from us.

When a Cepheid variable pulsates, the star's surface oscillates up and down like a spring. Consequently, the star's gases alternately heat up and cool down. Thus the characteristic light curve of a Cepheid variable results from changes in both size and surface temperature.

Just as a bouncing ball eventually comes to rest, a pulsating star would soon stop pulsating without some sort of mechanism to keep its oscillations going. In 1941 the British astronomer Arthur Eddington explained that a Cepheid variable feeds energy into its pulsations by a valvelike action involving the periodic ionization and recombination of gas in the star's outer layers. According to this theory, a star is more opaque, or "light tight," when compressed than when expanded. When the star is compressed, trapped heat pushes the star's surface outward. When the star is expanded, the heat escapes and so the star's surface, which is no longer supported, falls inward.

The details of a Cepheid's pulsation depend on the abundance of heavy elements in its atmosphere. The average luminosity of metal-rich Cepheids is roughly four times greater than the average luminosity of metal-poor Cepheids having the same period. Thus there are two classes: **Type I Cepheids**, which are the brighter, metal-rich stars, and **Type II Cepheids**, which are the dimmer, metal-poor stars. The period–luminosity relation for both types of variables is shown in Figure 11-24.

11-13 Cepheids enable astronomers to estimate vast distances

Cepheids are very important to astronomers because there is a direct relationship between a Cepheid's period of pulsation and its average luminosity. This relationship is called, appropriately enough, the **period–luminosity relation.** Dim Cepheid variables pulsate rapidly with periods of one to two days and have average brightnesses of a few hundred Suns. The most luminous Cepheids have the longest periods of all Cepheids, with variations occurring over 100 days and average brightnesses equal-

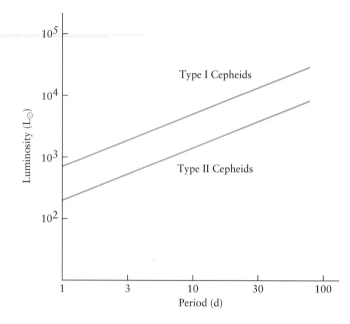

Figure 11-24 **The Period–Luminosity Relation** The period of a Cepheid variable is directly related to its average luminosity. Metal-rich (Type I) Cepheids are brighter than the metal-poor (Type II) Cepheids.

ing 10,000 $L_\odot$. Because the changes in brightness of Cepheids can be seen even in distant galaxies where other techniques for measuring distance fail, the period–luminosity relation plays an important role in determining the overall size and structure of the universe, as we shall see in Chapter 15.

In rare cases, stellar pulsations can be quite substantial. In some instances the expansion velocity is so high that the star's outer layers are ejected completely. As we shall see in the next chapter, significant mass ejection accompanies the death of stars in violent processes that renew and enrich the interstellar medium for future generations of stars.

WHAT DID YOU THINK?

1 *Where do stars come from?* Stars form from gas and dust inside giant molecular clouds

2 *Do stars with higher or lower mass last longer?* Lower-mass stars last longer because the lower gravitational force inside them causes fusion to take place at slower rates compared to the fusion inside higher-mass stars.

Key Words

birth line
brown dwarf
Cepheid variable star
core helium fusion
dense core
electron degeneracy
   pressure
emission nebula
evolutionary track
giant molecular cloud

giant star
globular cluster
H II region
helium flash
horizontal branch star
instability strip
interstellar medium
Jeans instability
OB association
open cluster

Pauli exclusion
   principle
period–luminosity
   relation
population I star
population II star
pre-main-sequence star
protostar
reflection nebula
RR Lyrae variable star

shell hydrogen fusion
stellar evolution
supernova remnant
T Tauri stars
Type I Cepheid
Type II Cepheid
variable stars
zero-age main sequence
   (ZAMS)

Key Ideas

• Enormous cold clouds of gas, called giant molecular clouds, are scattered about the Galaxy.

• Star formation begins when gravitational attraction causes protostars to coalesce within a giant molecular cloud. As a protostar contracts, its gases begin to glow. When the contraction slows down, the protostar becomes a pre-main-sequence star. When the pre-main-sequence star's core temperature becomes high enough to begin hydrogen fusion, it becomes a main sequence star.

   The most massive pre-main-sequence stars take the shortest time to become main sequence stars (O and B stars). They emit strong ultraviolet radiation that ionizes hydrogen in the surrounding cloud, creating reddish emission nebulae called H II regions.

   In the final stages of pre-main-sequence contraction, when hydrogen fusion is about to begin in the core, the star may undergo vigorous chromospheric activity that ejects large amounts of matter into space. Such gas-ejecting stars are called T Tauri stars.

• Ultraviolet radiation and stellar winds from the OB association at the core of an H II region create shock waves that compress the gas cloud, triggering the formation of more protostars.

   Supernova explosions also compress gas clouds and trigger star formation.

• A collection of a few hundred or a few thousand newborn stars is called an open cluster. Occasionally a rapidly moving star will escape, or "evaporate," from such a cluster.

• The more massive a star, the shorter its main sequence lifetime. The Sun has been a main sequence star for about five billion years and should remain so for about another five billion years. Less massive stars evolve more slowly and have longer lifetimes.

• Core hydrogen fusion ceases when hydrogen is exhausted in the core of a main sequence star, leaving a core of nearly pure helium surrounded by a shell where hydrogen fusion continues.

   Shell hydrogen fusion adds more helium to the star's core, which contracts and becomes hotter. The outer atmosphere expands considerably and the star becomes a giant.

   When the central temperature of a giant reaches about 100 million kelvins, the thermonuclear process of helium fusion begins. This process converts helium to carbon and oxygen.

   In a massive giant, helium fusion begins gradually. In a less massive giant, it begins suddenly in a process called a helium flash.

• The age of a stellar cluster can be estimated by plotting its stars on an H–R diagram. The upper portion of the main sequence will be missing because more massive main sequence stars will have become giants.

   Relatively young stars are metal-rich; ancient stars are metal-poor.

• Giants undergo extensive mass loss, sometimes producing circumstellar shells of ejected material around the stars.

• When a star's evolutionary track carries it through a region called the instability strip in the H–R diagram, the star becomes unstable and begins to pulsate.

   Cepheid variables are high-mass pulsating variables exhibiting a regular relationship between the period of pulsation and luminosity. RR Lyrae variables are low-mass pulsating variables with short periods.

Review Questions

1 What is a giant molecular cloud, and what role do these clouds play in the birth of stars?

2 Why are low temperatures necessary for protostars to form inside dense cores?

3 What is an H II region?

4 Explain why thermonuclear reactions don't occur on the surface of a main sequence star.

5 What is an evolutionary track, and how can such tracks help us interpret the H–R diagram?

6 Why do you suppose most stars we see in the sky are main sequence stars?

7 Draw the pre-main-sequence evolutionary track of the Sun on an H–R diagram. Briefly describe what was probably occurring throughout the solar system at various stages along this track.

8 On what grounds are astronomers able to say that the Sun has about five billion years remaining in its main sequence stage?

9 What will happen inside the Sun five billion years from now when it begins to turn into a giant?

10 Draw the post-main-sequence evolutionary track of the Sun on an H–R diagram up to the point when the Sun becomes a helium-burning giant. Briefly describe what might occur throughout the solar system as the Sun undergoes this transition.

11 What does it mean when an astronomer says that a star "moves" from one place to another on an H–R diagram?

12 What are Cepheid variables and how are they related to the instability strip?

13 What is the helium flash?

14 Explain how and why the turnoff point on the H–R diagram of a cluster is related to the cluster's age.

15 Why do astronomers believe that globular clusters are made of old stars?

Advanced Questions

16 How is a degenerate gas different from an ordinary gas?

17 If you took a spectrum of a reflection nebula, would you see absorption lines, emission lines, or no lines? Explain your answer.

18 Why is it useful to plot the apparent magnitudes of stars in a single cluster on an H–R diagram, rather than their absolute magnitudes?

19 What observations would you make of a star to determine whether its primary source of energy was fusing hydrogen or helium?

20 What might happen to the massive outer planets when the Sun becomes a giant?

∗ 21 How many 1.5-$M_\odot$ main sequence stars would it take to equal the luminosity of one 15-$M_\odot$ star?

∗ 22 How many times longer does a 1.5-$M_\odot$ star fuse hydrogen at its core than does a 15-$M_\odot$ star?

23 Speculate on why a shock wave from a supernova seems to produce relatively few high-mass O and B stars compared to the lower-mass A, F, G, and K stars.

24 How would you distinguish a newly formed protostar from a giant, given that they occupy the same location on the H–R diagram?

25 What observational consequences would we find in H–R diagrams for star clusters if the universe had a finite age? Could we use these consequences to establish constraints on the possible age of the universe? Explain.

Discussion Questions

26 What do you think would happen if the solar system passed through a giant molecular cloud? Do you think that the Earth has ever passed through such a cloud?

27 Speculate about the possibility of life forms and biological processes occurring in giant molecular clouds. In what ways might conditions favor or hinder biological evolution?

Observing Projects

28 Use a telescope to observe at least two of the following interstellar gas clouds. You can easily locate them with the aid of star charts published during the summer months. (*Hint:* M8 and M20 are discussed in some detail in the August 1995 issue of *Sky & Telescope*.) If you are familiar with navigating around the sky, the epoch 2000 coordinates of the objects given below will be useful:

Nebula	Right ascension	Declination
M42 (Orion)	5^h 35.4^m	$-5°$ $27'$
M43	5^h 35.6^m	$-5°$ $16'$
M20 (Trifid)	18^h 02.6^m	$-23°$ $02'$
M8 (Lagoon)	18^h 03.8^m	$-24°$ $23'$
M17 (Omega)	18^h 20.8^m	$-16°$ $11'$

In each case, can you guess which stars are responsible for the ionizing radiation that causes the nebula to glow? Draw a picture of what you see through the telescope and compare it with a photograph of the object. Which portions of the nebula are not visible through your telescope? Why are they obscured?

29 Several of the clusters included in Figure 11-21 can be seen quite well with a good pair of binoculars. Observe as many of these clusters as you can. Look at them through a telescope, if available. Note the overall distribution of stars in each cluster. Can you see any of these clusters with the naked eye? What difference do you note between binocular and telescopic images of individual clusters? You can easily locate these clusters with the aid of star charts published in such magazines as *Sky & Telescope* and *Astronomy*. If you are familiar with navigating around the sky, the epoch 2000 coordinates of the objects given below will be useful:

Nebula	Right ascension	Declination
Pleiades	3^h 47.0^m	$+24°$ $07'$
Hyades	4^h 27.0^m	$+16°$ $00'$
Praesepe	8^h 40.1^m	$+19°$ $59'$
Coma	12^h 25.0^m	$+26°$ $00'$
M11	18^h 51.1^m	$-06°$ $16'$

12 ▶ *The Deaths of Stars*

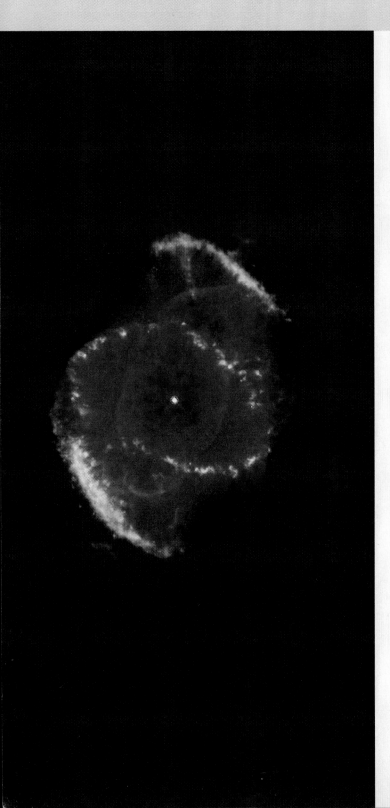

IN THIS CHAPTER

You will explore what happens to stars after they finish fusing helium in their cores. Many of them die shortly thereafter, while a few evolve further. As stars die, they eject matter far into space. You will discover that lower-mass stars expel their outer layers relatively gently, but higher-mass stars explode violently. The exposed cores of dead low-mass stars contract to become white dwarfs, whereas the remnant cores of intermediate-mass stars collapse to become neutron stars. You will then see how neutron stars have been observed as pulsars, pulsating x-ray sources, and bursters. The corpses of the highest-mass stars are black holes, to which we turn in Chapter 13.

The Cat's Eye—A Planetary Nebula Dying stars often eject their outer layers. A low-mass star can lose half its mass in a comparatively gentle process that produces a planetary nebula. The exposed stellar core typically has a surface temperature of about 100,000 K and is roughly one-tenth the size of the Sun. Ultraviolet radiation from the hot stellar core causes the surrounding gases to glow. In this image of the Cat's Eye Nebula, NGC 6543, the red glow is from hydrogen, the blue is from oxygen, and the green is from nitrogen. The central star in this nebula has exhausted all its nuclear fuel and is contracting to become a white dwarf.

WHAT DO YOU THINK?

1 Will the Sun explode? If so, what is the explosion called?

2 Where did carbon, silicon, oxygen, iron, uranium, and other heavy elements on Earth come from?

3 What is a pulsar?

4 What is a nova?

....

CARBON AND OXYGEN ARE the ashes of helium fusion. They build up in a giant star's core until the helium there is completely gone. What happens next depends on the star's mass.

Low-Mass Stars

12-1 Low-mass stars expand into the giant phase twice before becoming planetary nebulae

Recall from Chapter 11 that when shell hydrogen fusion first begins, the outpouring of energy causes the star to expand and become a giant. At this stage, a low-mass star (under 2 $M_\odot$) ascends the giant branch on the H–R diagram for the first time (Figure 12-1a). Then comes the helium flash, when the star shrinks and then moves onto the horizontal branch (Figure 12-1b).

Eventually the core is converted entirely into carbon and oxygen. Fusion there ceases once again, photon production drops off, and the inner regions of the star again contract, compressing and heating the shell of helium-rich gas just outside the core. As a result, **shell helium fusion** takes place outside the core; this shell is itself surrounded by a new hydrogen-fusing shell. All of this takes place within a volume roughly the size of the Earth. After a while, thermonuclear reactions in the hydrogen-fusing shell cease temporarily, leaving the aging star with the structure shown in Figure 12-2.

Once shell helium fusion commences, the new outpouring of energy pushes the outer envelope of the star out again. A low-mass star therefore ascends the giant branch for a second time (see Figure 12-1c), becoming brighter than it ever was before. Such **asymptotic giant branch stars**, or **AGB stars** for short, typically have diameters as big as the orbit of Mars and shine with the brightness of 10,000 $L_\odot$. At this furious rate of energy loss, the star cannot live much longer.

The death of a low-mass star begins with a thermal

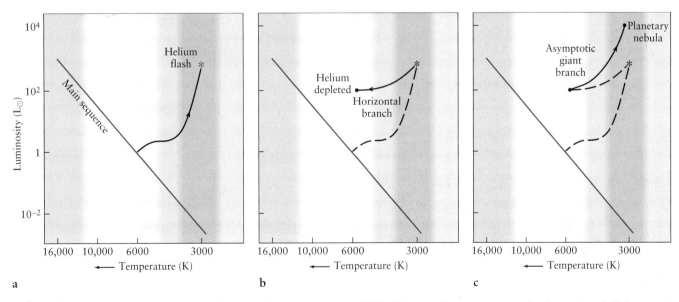

Figure 12-1 Post-Main-Sequence Evolution of Low-Mass Stars (a) The H–R track as a star makes the transition from the main sequence to the giant phase. The asterisk shows where a low-mass star is when the helium flash in it occurs. (b) After the helium flash, the star converts its helium core into carbon. While doing so, its core re-expands, decreasing shell fusion. As a result, the star's outer layers recontract. (c) After the helium core is completely transformed into carbon and oxygen, the core recollapses, and the outer layers re-expand.

Figure 12-2 **The Structure of an Old Low-Mass Star** Near the end of its life, a low-mass star becomes a giant, with a diameter almost as large as the diameter of the Earth's orbit. The star's core, the dormant hydrogen-fusing shell, and an active helium-fusing shell are contained within a volume roughly the size of the Earth.

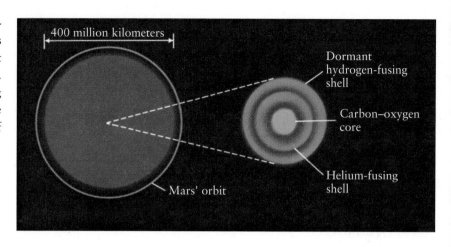

400 million kilometers

Dormant hydrogen-fusing shell

Carbon–oxygen core

Helium-fusing shell

Mars' orbit

runaway in the helium shell like the helium flash in its core earlier in its life. Now, however, the gas is not compressed enough to be degenerate because the shell is too thin. The slight increase in energy output from the thin helium-fusing shell, called a **helium shell flash,** does little to relieve the pressure from the star's overlying layers; it only drives up the temperature even more. A star has several helium shell flashes as its helium shell grows. This thermal runaway ends only after the helium-fusing shell becomes thick enough to relieve the pressure of the star's outer layers.

During each flash, the helium shell's energy output jumps a thousandfold. These brief outbursts are separated by relatively quiet intervals lasting about 300,000 years during which the helium shell becomes thicker. Finally, in one of these outbursts, the dying star's outer layers receive enough outward pressure from the helium shell flash to separate from the core completely. The layers are ejected into space.

As the ejected material expands and cools, dust grains condense and are propelled further outward by radiation pressure from the star's hot, burned-out core. The star loses as much as 60% of its mass by shedding its outer layers.

Now the hot core is exposed directly to space, emitting ultraviolet radiation intense enough to ionize the ejected gases. We call the glowing shell of ionized gases a **planetary nebula.** Planetary nebulae have nothing to do with planets. This unfortunate term was coined by astronomer William Herschel in the eighteenth century when these glowing objects, often green in appearance, were thought to look like distant planets when viewed through small telescopes.

Many planetary nebulae, such as the one shown in Figure 12-3, are distinctly circular because the gases

were ejected at roughly the same rate in all directions. In other cases, however, as in the opening figure for this chapter, the shape may resemble an hourglass, a dumbbell, or something more complex. These shapes are due to magnetic fields and to collisions with pre-existing interstellar gas.

Because planetary nebulae are actually complete shells of expanding gas and dust, why do they often look like rings (see Figure 12-3) rather than disks? A nebula is composed of a shell of gas and dust that is so thin that we see directly through most of it. Only where it is thickest,

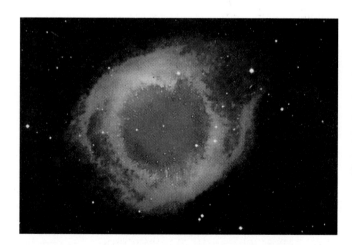

Figure 12-3 **The Planetary Nebula NGC 7293** This planetary nebula, often called the Helix Nebula, is located in the constellation of Aquarius (the Water Bearer). The star that ejected these gases is seen at the center of the glowing shell. The bluish color comes from oxygen ions, while the pink and red comes from nitrogen ions and hydrogen atoms. This nebula, located about 700 ly from Earth, has an angular diameter equal to about half that of the full Moon.

around the edge, do we see it as a ring of brightly shining gas.

Planetary nebulae are quite common in our Galaxy. Astronomers estimate that there are 20,000 to 50,000 of them. Indeed, such will be the fate of our Sun. Spectroscopic observations of these nebulae show bright emission lines of hydrogen, carbon, neon, magnesium, oxygen, and nitrogen. From the Doppler shifts of these lines, we conclude that the expanding shell of gas is moving outward from the dying star at speeds of 10 to 30 km/s. A typical planetary nebula has a diameter of roughly 1 ly, which means that it began expanding about 10,000 years ago.

Stars with masses between 2 $M_\odot$ and 4 $M_\odot$ also create planetary nebulae. Their evolution differs in one way from that of the lower mass stars just described: The pressure exerted by the more massive stars initiates core helium fusion so rapidly that a core helium flash does not occur in them.

By astronomical standards, a planetary nebula is very short-lived. After about 50,000 years, the nebula spreads over distances so far from the cooling central star that its nebulosity simply fades from view. The gases then mingle and mix with the surrounding interstellar medium, enriching it with heavy elements, such as carbon. Astronomers estimate that all the planetary nebulae in the Galaxy return a total of 5 $M_\odot$ to the interstellar medium each year. This amounts to about 15% of all matter expelled by all types of stars each year. Planetary nebulae therefore play an important role in the chemical evolution of the Galaxy as a whole.

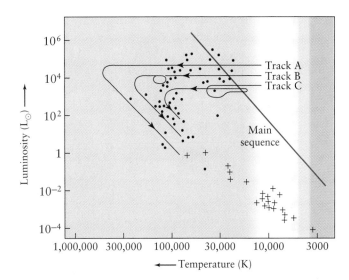

Evolutionary track	Mass ($M_\odot$)		
	Giant	Ejected nebula	White dwarf
A	3.0	1.8	1.2
B	1.5	0.7	0.8
C	0.8	0.2	0.6

Figure 12-4 **Evolution from Giants to White Dwarfs** The evolutionary tracks of three low-mass giants are shown as they eject planetary nebulae. The table gives the original mass, the amount of mass lost, and the remaining (white dwarf) mass. The dots on this graph represent the central stars of planetary nebulae whose surface temperatures and luminosities have been determined. The crosses are white dwarfs for which similar data exist.

12-2 The burned-out core of a low-mass star becomes a white dwarf

Stars less massive than about 4 $M_\odot$ never develop the central pressures or temperatures necessary to ignite thermonuclear reactions of the carbon or oxygen in their cores. Instead, as we have seen, the process of mass ejection strips away the star's outer layers and exposes the carbon–oxygen core, which simply cools off. Such burned-out hulks are called **white dwarfs.**

The evolutionary tracks of three burned-out stellar cores are shown in Figure 12-4. These particular white dwarfs evolved from main sequence stars of between 0.8 and 3.0 $M_\odot$. During the ejection phase, the appearance of these stars changes rapidly. Consequently, these objects race along their evolutionary tracks on the H–R diagram, sometimes executing loops corresponding to thermal pulses (tracks B and C in Figure 12-4). Finally,

as the ejected planetary nebulae fade and the stellar cores cool, the evolutionary tracks of the dying stars take a sharp turn downward toward the white dwarf region of the diagram.

Because there is not enough pressure inside a white dwarf to ignite additional nuclear fuels, its remaining mass severely compresses the stellar corpse. Electron degeneracy pressure supports the crushing weight of the carbon and oxygen, creating a stable stellar remnant roughly the size of the Earth. The density of matter in one of these stellar corpses is typically 10^9 kg/m^3. In other words, a teaspoonful of white dwarf matter brought to Earth would weigh five tons.

Many white dwarfs are found in our solar neighborhood, but all are too faint to be seen with the naked eye. One of the first white dwarfs to be discovered is a companion to the bright star Sirius. The binary nature of Sirius was first deduced in 1844 by the German astronomer

Figure 12-5 **Sirius and Its White Dwarf Companion** Sirius, the brightest-appearing star in the night sky, is actually a double star. The secondary star is a white dwarf, seen here at the five o'clock position in the glare of Sirius. The spikes and rays around Sirius are created by optical effects within the telescope.

Friedrich Bessel, who noticed that the star was moving back and forth, as if orbited by an unseen object. This companion, called Sirius B, was first glimpsed in 1862 (Figure 12-5). Recent satellite observations at ultraviolet wavelengths—where white dwarfs emit most of their light—demonstrate that the surface temperature of Sirius B is about 30,000 K.

As a white dwarf cools, both its luminosity and its surface temperature decline. As billions of years pass, white dwarfs get dimmer and dimmer as their surface temperatures drop toward absolute zero. This condition will be the fate of our Sun: a cold, dark, dense sphere of degenerate gases rich in oxygen and carbon, about the size of the Earth.

A white dwarf can have a mass of no more than 1.4 $M_\odot$. This mass is called the **Chandrasekhar limit** after Subrahmanyan Chandrasekhar, who received a Nobel prize for his pioneering theoretical studies of white dwarfs. Main sequence stars with more than 4 $M_\odot$ create carbon–oxygen cores above this limit. Their evolutionary cycles are very different; no planetary nebulae occur and gravity compresses the carbon–oxygen core sufficiently so that fusion continues.

High-Mass Stars

A high-mass star (greater than 4 $M_\odot$) has enough gravitational force pressing on its interior to ignite a host of additional thermonuclear reactions deep within. Even though larger nuclei repel one another more forcefully because of their greater electric charge, the nuclei in the cores of massive stars still move quickly enough to penetrate one another.

12-3 A series of different types of fusion leads to the creation of supergiants

When helium fusion ends in the core of a massive star, gravitational compression collapses the core once again, driving the star's central temperature beyond 600 million kelvins. Now carbon fusion begins, producing such elements as neon and magnesium. When a star compresses itself enough to drive its central temperature to 1.2 billion kelvins, neon fusion occurs. When the central temperature of the star then reaches 1.5 billion kelvins, oxygen fusion begins. As the star consumes increasingly heavier nuclei, it produces sulfur and isotopes of silicon and phosphorus, among other elements.

Each successive thermonuclear reaction occurs more rapidly than the last. For example, detailed calculations for a 25-$M_\odot$ star demonstrate that carbon fusion occurs for 600 years, neon fusion for 1 year, and oxygen fusion for only 6 months. After half a year of core oxygen fusion, gravitational compression forces the central temperature up to 2.7 billion kelvins and silicon fusion begins. This thermonuclear process proceeds so furiously that the entire core supply of silicon in a 25-$M_\odot$ star is used up in 1 day.

Each stage of fusion adds a new shell of matter outside the core (Figure 12-6), creating a structure resembling the layers of an enormous onion. Together, the tremendous numbers of fusion-generated photons created in all these shells as well as in the core heat the outer layers of the star, so that it expands to become even larger than a giant. Such **supergiants** are almost as big across as Jupiter's orbit around the Sun.

Supergiant stars, which are brighter than 10^5 $L_\odot$, emit winds throughout most of their existence, with mass loss rates exceeding those of giants (see Chapter 11). Figure 12-7 shows a supergiant star losing mass. Betelgeuse, which is 470 ly away in the constellation of Orion, is another good example of a supergiant experiencing mass loss. Recent spectroscopic observations show that this

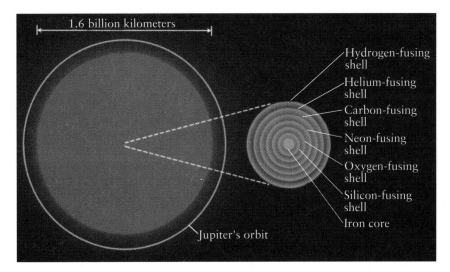

1.6 billion kilometers

Hydrogen-fusing shell
Helium-fusing shell
Carbon-fusing shell
Neon-fusing shell
Oxygen-fusing shell
Silicon-fusing shell
Iron core

Jupiter's orbit

Figure 12-6 **The Structure of an Old High-Mass Star** Near the end of its life, a high-mass star becomes a supergiant with a diameter almost as big as the orbit of Jupiter. The star's energy comes from six concentric fusing shells, all contained within a volume roughly the same size as the Earth.

star is losing mass at the rate of 1.7×10^{-7} $M_\odot$ per year and is surrounded by a huge shell of ejected gas called a *circumstellar shell* that is expanding at 10 km/s. These escaping gases have been detected at distances of 10,000 AU from the star. Consequently, the expanding circumstellar shell has an overall diameter of $\frac{1}{3}$ ly.

Silicon fusion involves many hundreds of nuclear reactions, but its major final product is an inert iron core. Surrounding this iron core, successive layers of shell fusion consume the star's remaining reserves of fuel (see Figure 12-6). Whereas the star's enormously bloated atmosphere is nearly as big as the orbit of Jupiter, the entire energy-producing region of the star is again contained in a volume the size of the Earth. The buildup of an inert, iron-rich core signals the impending violent death of a massive star.

12-4 High-mass stars die violently by blowing themselves apart in supernovae

Unlike lighter elements, iron cannot fuel further thermonuclear reactions. The protons and neutrons inside iron nuclei are already so tightly bound together that no further energy can be extracted by fusing still more nuclei together. The sequence of fusion stages therefore ends.

Because iron atoms do not fuse, the electrons in the core must now support the star's outer layers by the brute strength of electron degeneracy pressure alone. Soon, however, the continued deposition of fresh iron from the silicon-fusing shell causes the core's mass to exceed the Chandrasekhar limit. Electron degeneracy suddenly becomes unable to support the star's enormous weight, and the core collapses.

Figure 12-7 **A Mass Loss Supergiant Star** Beautiful nebulosity surrounds a supergiant star that is experiencing significant mass loss. The ejected material collides and interacts with the surrounding interstellar gas and dust, thereby producing the cosmic bubble seen here. This supergiant and its nebulosity are located in the constellation of Canis Major.

TABLE 12-1

Evolutionary Stages of a 25-$M_\odot$ Star

Stage	Central temperature (K)	Central density (kg/m³)	Duration of stage
Hydrogen burning	4×10^7	5×10^3	7×10^6 yr
Helium burning	2×10^8	7×10^5	5×10^5 yr
Carbon burning	6×10^8	2×10^8	600 yr
Neon burning	1.2×10^9	4×10^9	1 yr
Oxygen burning	1.5×10^9	1×10^{10}	6 mo
Silicon burning	2.7×10^9	3×10^{10}	1 d
Core collapse	5.4×10^9	3×10^{12}	0.2 s
Core bounce	2.3×10^{10}	4×10^{17}	Milliseconds
Explosion	About 10^9	Varies	10 s

Any isolated giant star with a mass greater than 4 $M_\odot$ develops an iron core whose mass will eventually exceed the Chandrasekhar limit. When this limit is exceeded, a rapid series of cataclysms is triggered that tears the star apart in a few seconds. Let us see how this happens in the death of a 25-$M_\odot$ star according to supercomputer simulations (Table 12-1).

In a 25-$M_\odot$ star, electron degeneracy pressure fails when the density is sufficiently high inside the iron core. The core then collapses immediately. In roughly a tenth of a second, the central temperature exceeds five billion kelvins. Gamma-ray photons associated with this intense heat have so much energy that they begin to break apart the iron nuclei in a process called **photodisintegration.** Although it took millions of years to create the iron core, it takes less than a second to convert it all back into elemental protons, neutrons, and electrons.

Within another tenth of a second, as the density continues to climb, the electrons are forced to combine with protons to produce neutrons, and the process releases a flood of neutrinos. As we saw in Chapter 9, neutrinos have no electric charge and are believed to have very little, if any, mass; thus they resemble photons and travel at or near the speed of light. Despite their ability to pass through the Earth or Sun without interaction, the newly created neutrinos in a collapsing high-mass star do not immediately escape from the star's core because the matter there is so fantastically dense.

About one-quarter second after the collapse begins, the density of the entire core reaches 4×10^{17} kg/m³, which is *nuclear density,* the density at which neutrons and protons are normally packed together inside nuclei. (Compare this to the density of water, 10^3 kg/m³.) At nu-

clear density, matter is virtually incompressible. Thus, when the neutron-rich material of the core reaches this density, it suddenly becomes very stiff. The collapse of the core then comes to a halt so abruptly that it rebounds and begins rushing back out. This is called *core bounce.*

During this critical stage, the star's unsupported layers of shell-fusing matter are plunging inward at up to 15% of the speed of light. As this material crashes onto the rebounding core, shock waves generate enormous temperatures and pressures, causing the falling material to bounce back out. In just a fraction of a second, a tremendous volume of matter begins to move back up toward the star's surface. This wave accelerates rapidly as it encounters less and less resistance, and soon it forms an outgoing shock wave. After a few hours, this shock wave reaches the star's surface. Combined with the outward force created by the neutrinos slamming into the dense matter, the shock wave lifts the star's outer layers away from the core in a mighty blast. The star becomes a **supernova.**

As the outer layers of a massive dying star are blasted into space, they are compressed so much by the neutrinos and the shock wave that fusion actually occurs during the explosion, creating a broad assortment of elements. Indeed, this is where most of the elements on the Earth and other terrestrial planets comes from, including most of the atoms in our bodies.

As the supernova expands, the star's luminosity suddenly increases by a factor of 10^8. The energy emitted by a supernova is truly staggering—as much as all 200 billion stars in the Milky Way Galaxy combined! In other words, for a few days following the explosion, a super-

Figure 12-8 **The Supernova SN 1987A** In 1987 a supernova was discovered in a nearby galaxy called the Large Magellanic Cloud (LMC). This photograph shows a portion of the LMC that includes the supernova and a huge H II region called the Tarantula Nebula. At maximum brightness observers at southern latitudes saw the supernova without a telescope.

nova shines as brightly as an entire galaxy. Our model 25-$M_\odot$ star ejects 24 $M_\odot$ of its mass. Supernovae of less massive stars return proportionately less mass to the interstellar medium.

12-5 Supernova SN 1987A gave us a close-up look at the death of a massive star

On February 23, 1987, a supernova was discovered in the Large Magellanic Cloud, a galaxy near our own Milky Way. The supernova was designated SN 1987A, the first to be observed that year (Figure 12-8). It occurred in 30 Doradus, an enormous H II region often called the Tarantula Nebula because of its spiderlike appearance. The supernova was so bright that it could be seen with the naked eye.

Supernovae are often seen in remote galaxies with the aid of a telescope. In 1885 a supernova in the Andromeda Galaxy was just barely visible to the unaided eye. We have to go back to 1604 to find another supernova bright enough to be seen without a telescope. But SN 1987A gave astronomers the unique opportunity to study the death of a nearby massive star using modern equipment.

At first SN 1987A reached only a tenth of the luminosity typical of an exploding massive star (see Table

12-1). For the next 85 days, it gradually brightened before slowly dimming as is characteristic of an ordinary supernova. Fortunately, the doomed star had been observed before it became a supernova. The Large Magellanic Cloud is about 160,000 ly from Earth—near enough to us that many of its stars have been individually observed and catalogued. The doomed star had been identified as a B3 I supergiant (Figure 12-9). Could this explain SN 1987A's unusually slow brightening?

When this star was on the main sequence, its mass was about 20 $M_\odot$, although by the time it exploded it had probably shed a few solar masses. The evolutionary track for an aging 20-$M_\odot$ star wanders back and forth across the top of the H–R diagram, and so the star alternates between being a hot (blue) supergiant and cool

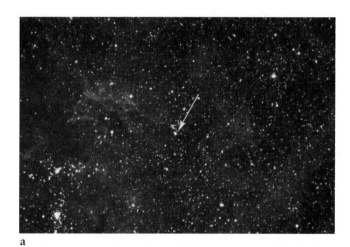

a

b

Figure 12-9 **The Doomed Star and SN 1987A** (a) A small section of the Large Magellanic Cloud before the outburst of SN 1987A. The doomed star, a blue supergiant, is identified by an arrow. (b) The supernova a few days after the explosion.

(red) supergiant. The star's size changes significantly as its surface temperature changes. A blue supergiant is only 10 times larger in diameter than the Sun, but a red supergiant of the same luminosity is 1000 times larger. Because the doomed star was relatively small when it exploded, it reached only a tenth of the brightness that it would have attained had it been a red supergiant.

Ordinary telescopic observations of a supernova explosion can show us only the expanding outer layers of the dying star. Even x-ray or radio observations fail to see through the hot gases being blasted into space. Thus, using ordinary techniques we cannot observe the extraordinary events occurring in and around the doomed star's core. However, most of the energy of a supernova explosion is carried away in a characteristic form: Neutrinos are produced in great profusion as the star's core collapses. By detecting these neutrinos and measuring their properties, astronomers can learn many details about the star's collapsing core, especially about its bounce. However, neutrinos are very difficult to detect. As we saw in Chapter 9, they seldom interact with ordinary matter. Fortunately, several neutrino detectors, including the original perchlorethylene detector described in Chapter 9, were operating when the neutrinos from SN 1987A reached Earth.

Other, water-based neutrino detectors were also active at the time. These detectors use a different principle to detect neutrinos than do those using perchlorethylene. Water contains many protons, namely, the nuclei of hydrogen atoms. When a high-energy neutrino strikes a proton, it produces a positron, which is like an electron but with a positive charge. The positron in turn emits a flash of light called **Cerenkov radiation**. As the Russian physicist Pavel A. Cerenkov first observed, the flash occurs whenever a particle moves through water faster than light can. Such motion does not violate the tenet that the speed of light *in a vacuum* (3×10^8 m/s) is the ultimate speed limit in the universe. Light is slowed considerably as it passes through water, and high-energy particles can exceed this reduced speed without violating the laws of physics. Thus scientists detect neutrinos indirectly by observing Cerenkov flashes with light-sensitive devices called *photomultipliers* mounted in the water (Figure 12-10).

Nearly a day before SN 1987A was first observed in the sky, teams of scientists at neutrino detectors in Japan and the United States reported finding Cerenkov flashes from a burst of neutrinos. The Kamiokande II detector in Japan detected 12 neutrinos at about the same time that

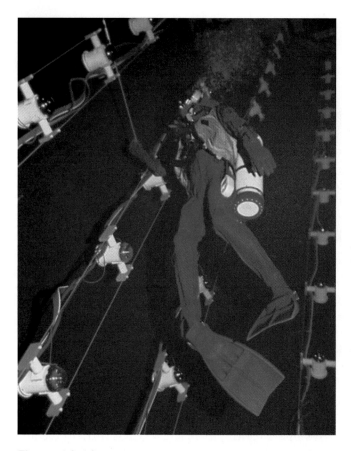

Figure 12-10 **Inside a Neutrino Detector** A diver is shown servicing one of the photomultiplier tubes in the Irvine-Michigan-Brookhaven (IMB) neutrino detector. This apparatus detected brief flashes of light caused by neutrinos from SN 1987A when they struck protons in the water.

8 were found by the IMB (Irvine-Michigan-Brookhaven) detector in a salt mine under Lake Erie. Neutrinos preceded the visible outburst because they escaped from the dying star before the shock wave from the collapsing core reached the star's surface. They were detected in the Earth's northern hemisphere, where the supernova is always below the horizon, after having passed through the Earth.

About three and a half years after SN 1987A detonated, astronomers using the Hubble Space Telescope obtained a picture showing several rings of glowing gas around the exploded star (Figure 12-11). This gas was emitted by the star 10,000 years before the supernova when a hydrogen-rich stellar envelope was ejected in an hourglass shape by gentle stellar winds from the doomed star. The star was then a red supergiant. The relic gas is

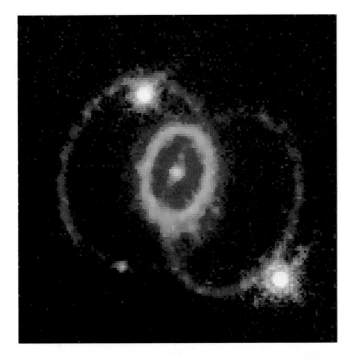

Figure 12-11 **Shells of Gas Around SN 1987A** Intense radiation from the supernova explosion caused part of an hourglass-shaped shell of gas surrounding SN 1987A to glow. The shell was ejected from the star 10,000 years before it detonated. The bright ring is glowing gas around the hourglass's waist. One dim ring is above the waist, while a second dim ring is behind it. The white spots are unrelated stars.

now being illuminated by photons from the supernova. Astronomers predict that the gas ejected from the supernova itself will strike the circumstellar shell sometime between 1997 and 2004. This collision will cause the shell of gas to brighten considerably, eventually illuminating the entire hourglass.

Astronomers have been rather lucky with SN 1987A. The doomed star had been studied and its distance from Earth was known. The supernova was located in an unobscured part of the sky, and neutrino detectors happened to be operating at the time of the outburst. Astronomers will be monitoring the progress of this supernova for years to come.

12-6 Accreting white dwarfs in close binary systems can also become supernovae

Supernovae such as SN 1987A, which are the death throes of massive stars, are known as **Type II supernovae.** Observations reveal that others, known as **Type I supernovae,** begin as white dwarfs.

A supernova's spectrum determines its type: Hydrogen lines are prominent in Type II supernovae but absent in Type I. Both types begin with a sudden rise in brightness (Figure 12-12). A Type I supernova typically reaches

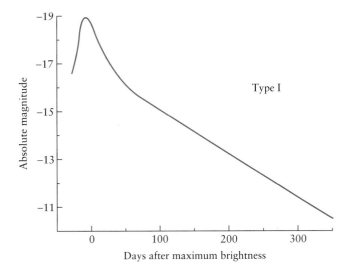

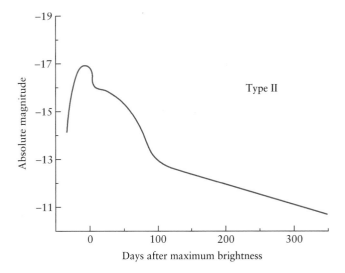

Figure 12-12 **Supernova Light Curves** A Type I supernova, which gradually declines in brightness, is caused by an exploding white dwarf in a close binary system. A Type II supernova usually has alternating intervals of steep and gradual declines in brightness. A Type II supernova is caused by the explosive death of a massive star.

an absolute magnitude of –19 at peak brightness, while a Type II supernova usually peaks at –17. Type I supernovae then decline gradually for more than a year, whereas Type II supernovae alternate between periods of steep and gradual declines in brightness. Type II light curves therefore have a steplike appearance.

A Type I supernova begins with a carbon–oxygen-rich white dwarf in a close, semidetached binary system. As we saw in Chapter 10, mass transfer can occur in a close binary if one star overflows its Roche lobe (recall Figures 10-15 and 10-17). To trigger a Type I supernova, a swollen giant companion star dumps gas onto a white dwarf. When the white dwarf's mass gets close to the Chandrasekhar limit, the increased core pressure created by the additional mass enables carbon fusion to begin. In a catastrophic runaway process reminiscent of the helium flash, the rate of carbon fusion skyrockets and the star blows up.

A Type I supernova is powered by nuclear energy. What we see is simply the fallout from a thermonuclear explosion, which produces a wide array of radioactive isotopes. Especially abundant is an unstable isotope of nickel that decays into a radioactive isotope of cobalt. Most of the electromagnetic display of a Type I supernova, including the smooth decline of its light curve, results directly from the radioactive decay of nickel and cobalt.

12-7 The remnants of a supernova explosion can be detected for centuries afterward

Astronomers find the debris of supernova explosions scattered across the sky. A beautiful example is the Veil Nebula in the Cygnus Loop, (see Figure 11-4). The doomed star's outer layers were blasted into space so violently that they are still traveling at supersonic speeds. As this expanding shell of gas plows through the interstellar medium, it collides with atoms and molecules, making the gases glow.

Many supernova remnants cover sizable fractions of the sky. The largest is the Gum Nebula, with a diameter of 60° (Figure 12-13). This nebula looks so big because it is so close: Its near side is only about 300 ly from Earth. Studies of the nebula's expansion rate suggest that the supernova exploded around 9000 BC. People then living in Egypt and India could have witnessed it. At maximum brilliance, the exploding star probably was as bright as the Moon at first quarter.

Figure 12-13 **The Gum Nebula** The Gum Nebula is the largest known supernova remnant, spanning 60° of the sky, roughly centered on the southern constellation of Vela. The nearest portions of this expanding nebula are only 300 ly from the Earth. The supernova explosion occurred about 11,000 years ago, and its remnant now has a diameter of about 2300 ly. Only the central regions of the nebula are shown here.

Many supernova remnants can only be detected at nonvisible wavelengths, ranging from x rays through radio waves. For example, Figure 12-14 shows both x-ray and radio images of the supernova remnant Cassiopeia A. Visible-light photographs of this part of the sky reveal only a few small, faint wisps. Thus radio searches for supernova remnants are more fruitful than visual searches. Only two dozen supernova remnants have been found on photographs, but more than 100 remnants have been discovered by radio astronomers.

From the expansion rate of the nebulosity in Cassiopeia A, astronomers conclude that the supernova explosion occurred about 300 years ago. Although telescopes were in wide use by the late 1600s, no one saw the outburst. In fact, the last supernova of a star in our Galaxy near its maximum brightness was seen by Johannes Kepler in 1604. In 1572 Tycho Brahe also recorded the sudden appearance of an exceptionally bright star in the sky. To find any other accounts of supernova explosions, we must delve into astronomical records that are almost a thousand years old.

Astronomers have seen more than 600 supernovae in distant galaxies. These observations suggest that in a typical galaxy like the Milky Way, Type I supernovae

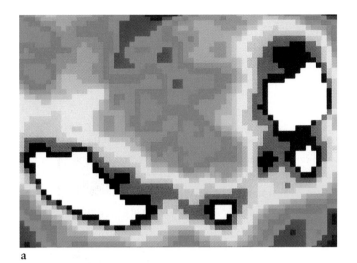

a

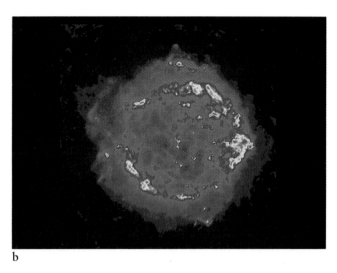

b

Figure 12-14 **Cassiopeia A** Supernova remnants such as Cassiopeia A are typically strong sources of x rays and radio waves. **(a)** An x-ray picture of Cassiopeia A taken from the Exosat satellite, a European Space Agency mission. **(b)** A corre- sponding radio image produced by the Very Large Array (VLA). The supernova explosion that produced this nebula occurred 300 years ago, about 10,000 ly from Earth.

occur roughly once every 36 years, while Type II supernovae occur about once every 44 years. Thus there should be about five supernovae exploding in our Milky Way Galaxy each century. Where are they?

Vigorous stellar evolution in our Galaxy occurs primarily in the Galaxy's disk, where the giant molecular clouds are located. Our Galaxy's disk is therefore the place where massive stars are born and supernovae explode. However, this region of our Galaxy is so filled with interstellar gas and dust that we simply cannot see very far into it. Supernovae probably do erupt every few decades in remote parts of our Galaxy, but their detonations are hidden by interstellar debris.

Neutron Stars and Pulsars

12-8 The cores of many Type II supernovae become neutron stars

When stars of between 4 and 9 $M_\odot$ explode as supernovae, their remnant cores are highly compressed clumps of neutrons called **neutron stars.** In other words, there are at least two types of stellar corpses: white dwarfs and neutron stars.

The neutron was discovered during laboratory experiments in 1932. Within a year, two astronomers had predicted the existence of neutron stars. Inspired by the realization that white dwarfs are supported by electron degeneracy pressure, Fritz Zwicky and Walter Baade proposed that a highly compact ball of neutrons could also produce a powerful pressure that would support a stellar corpse. They suggested that this **neutron degeneracy pressure** might be even greater than electron degeneracy pressure (Chapter 11) and thus allow stellar remnants with masses beyond the Chandrasekhar limit. Zwicky and Baade wrote, "We advance the view that supernovae represent the transition from ordinary stars into neutron stars, which in their final stages consist of extremely closely packed neutrons."

Most scientists ignored Zwicky and Baade's prophetic proposal for years. After all, a neutron star would have to be a rather weird object. In order to transform protons and electrons into neutrons, the density in the star would have to be equal to nuclear density, about 10^{17} kg/m^3. Thus, a thimbleful of neutron star matter brought back to Earth would weigh 100 million tons. Furthermore, an object compacted to nuclear density would be very small. A 2-$M_\odot$ neutron star would have a diameter of only 8 km, making it only as large as Jupiter's smallest known moons. The surface gravity on one of these neutron stars would be so strong that the escape velocity would equal one-half the speed of light. All these conditions seemed so outrageous that few astronomers paid any serious attention to the subject of neutron stars—until 1968.

Figure 12-15 **A Recording of the First Pulsar** The intensity of radio emission from the first pulsar varies, but the spacing between the pulses is exactly 1.3373011 seconds.

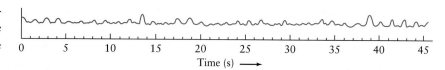

Time (s) ⟶

As a young graduate student at Cambridge University, Jocelyn Bell had spent many months helping to construct an array of radio antennas covering $4\frac{1}{2}$ acres in the English countryside. By the fall of 1967 the instrument was completed, and Bell and her colleagues began detecting radio emissions from various celestial sources. In November, while scrutinizing data from the new telescope, Bell noticed that the antennas had detected regular beeps from one particular location in the sky. Careful repetition of the observations demonstrated that the radio pulses were arriving with a regular period of 1.3373011 seconds (Figure 12-15).

The regularity of this pulsating radio source was so striking that the Cambridge team suspected that they might be detecting signals from an advanced alien civilization. This possibility was soon discarded as several more of these pulsating radio sources, which soon came to be know as **pulsars,** were discovered across the sky. In all cases, the periods were extremely regular, ranging between 0.2 second and 1.5 seconds.

When the discovery of pulsars was officially announced in early 1968, astronomers around the world began proposing all sorts of explanations. Many of these theories were bizarre, and arguments raged for months. Most astronomers could not accept that pulsars might be associated with dead stars. There seemed to be a sufficient number of white dwarfs in the sky to account for all the stars that have died since our Galaxy was formed. It was thus generally assumed that all dying stars somehow manage to eject enough matter so that their corpses do not exceed the Chandrasekhar limit. However, by late 1968, all controversy was laid to rest with the discovery of a pulsar in the middle of the Crab Nebula.

In AD 1054 Chinese astronomers recorded the appearance of a supernova (they called it a "guest star") in the constellation of Taurus. When we turn a telescope toward this location, we find the Crab Nebula, shown in Figure 12-16. It looks like the residue of an explosion and is, in fact, a supernova remnant. The pulsar at the center of the Crab Nebula is called the Crab pulsar.

The Crab pulsar is spinning too fast to be a white dwarf. Its period is 0.033 second, which means that it pulses 30 times each second. At that speed, something as

bulky as a white dwarf would immediately fly apart. Since that time, astronomers have discovered numerous other pulsars, including PSR 1257+12, a stellar corpse that pulses 888 times each second. Clearly pulsars must be incredibly compact—as small and dense as a neutron star.

12-9 A rotating magnetic field explains the pulses from a neutron star

A neutron star must be rotating rapidly, since virtually all stars rotate. The Sun takes nearly a full month to rotate once about its axis. But collapsing stars speed up as they shrink, just as an ice skater doing a pirouette speeds up when he pulls in his arms. (Recall that the total amount of angular momentum in a system always remains constant.) Even an ordinary star rotating once a month would be spinning faster than once a second if compressed to the size of a neutron star.

In addition to rapid rotation, most neutron stars have intense magnetic fields. It is probably safe to say

Figure 12-16 **The Crab Nebula** This beautiful nebula, named for the armlike appearance of its filamentary structure, is the remnant of a supernova seen in AD 1054. The distance to the nebula is about 6000 ly, and its present angular size (4 by 6 arcmin) corresponds to linear dimensions of about 7 by 10 ly.

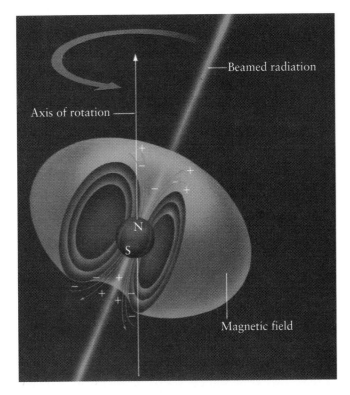

Figure 12-17 **A Rotating, Magnetized Neutron Star** It is reasonable to suppose that a neutron star rotates rapidly and possesses a powerful magnetic field. Charged particles are accelerated near the star's magnetic poles and produce two oppositely directed beams of radiation. As the star rotates, the beams sweep around the sky. If the Earth happens to lie in the path of the beams, we see a pulsar.

that almost every star has some kind of magnetic field. In an average star like our Sun, the magnetic field is spread out over millions upon millions of square kilometers just under the star's surface (see Chapter 9). However, if a star of solar dimensions collapses down to a neutron star, its magnetic field becomes very concentrated. The strength of the magnetic field increases a billionfold.

The axis of rotation of a typical neutron star is not the same as the axis connecting its north and south magnetic poles (Figure 12-17), much as the magnetic and rotation axes of the planets and Sun are different axes that are inclined to each other. As the neutron star rotates, its powerful magnetic field therefore rapidly changes direction. Like a giant electric generator, the star creates intense electric fields, which act on protons and electrons near its surface. The powerful electric fields channel these charged particles, causing them to flow out from the neutron star's polar regions, as sketched in Figure 12-17. As the particles stream along the field, they accelerate and emit energy. The result is two very thin beams of radiation pouring out of the neutron star's north and south magnetic polar regions—a pulsar.

A rotating, magnetized neutron star is somewhat like a lighthouse beacon. As the star rotates, its beams of radiation sweep around the sky. If the Earth happens to be located in the right direction, a brief flash can be observed each time a beam whips past our line of sight. This explanation for pulsars is often called the **lighthouse model**. Indeed, the center of the Crab Nebula is actually flashing on and off 30 times each second (Figure 12-18). Also visibly flashing is the Vela pulsar at the core of the Gum Nebula (see Figure 12-13). The Vela pulsar, with a period of 0.089 second, is the slowest pulsar ever detected at visual wavelengths. Only the very youngest pulsars are energetic enough to emit visible flashes along with their radio pulses.

The Crab pulsar is one of the youngest pulsars, its creation having been observed some 900 years ago. The Vela pulsar is also quite young: It was created about

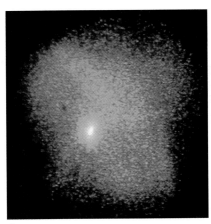
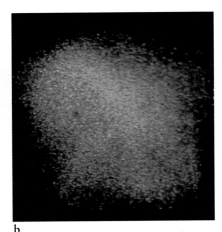

a   b

Figure 12-18 **The Crab Pulsar** A pulsar is located at the center of the Crab Nebula. These two photographs show the Crab pulsar in its (**a**) on and (**b**) off states. Like the radio pulses, these visual flashes have a period of 0.033 second.

11,000 years ago. Because they use some of their energy of rotation via the pulses they emit, *pulsars slow down as they get older.* Astronomers are observing SN 1987A with the hope of seeing through the supernova's ejected gases to search for pulses that would indicate a rapidly spinning neutron star.

While most pulsars appear to be isolated bodies, two have been discovered with companions. One, labeled PSR 1957+20, but more commonly called *the binary pulsar,* has a massive companion, which is probably another neutron star. The two neutron stars orbit each other with a period of about 8 hours. Because they are so close together and so incredibly dense, these two bodies follow orbits predicted by Einstein's general theory of relativity (Chapter 13) rather than Newton's laws of motion (Chapter 2). Indeed, their behavior is a strong confirmation that general relativity is a more accurate model of the behavior of matter. The second pulsar with a companion, labeled PSR B1257+2, is even more unusual than the binary pulsar in that its Doppler shift reveals the presence of at least two planets in orbit around it.

12-10 Pulsating x-ray sources are neutron stars in close binary systems

During the 1960s astronomers obtained tantalizing x-ray views of the sky during short rocket and balloon flights that briefly lifted x-ray detectors above the Earth's atmosphere. Several strong x-ray sources were discovered, and each was named after the constellation in which it was located. For example, Scorpius X-1 is the first x-ray source found in the constellation of Scorpius.

Astronomers were so intrigued by these preliminary discoveries that they built and launched Explorer 42, an x-ray-detecting satellite that could make observations 24 hours a day. The satellite was launched in 1970 from Kenya to place it in an orbit above the Earth's equator. In recognition of the hospitality of the Kenyan people, Explorer 42 was renamed Uhuru, which means "freedom" in Swahili.

Uhuru gave us our first comprehensive look at the x-ray sky. As the satellite slowly rotated, its x-ray detectors swept across the heavens. Each time an x-ray source came into view, signals were transmitted to receiving stations on the ground. Before its battery and transmitter failed in early 1973, Uhuru had succeeded in locating 339 x-ray sources.

The discovery of pulsars was still fresh in everyone's mind when the Uhuru team discovered x-ray pulses coming from Centaurus X-3 in early 1971. Figure 12-19 shows data from one sweep of Uhuru's detectors across Centaurus X-3. The pulses have a regular period of 4.84 seconds. A few months later similar pulses were discovered coming from a source called Hercules X-1, which has a period of 1.24 seconds. Because the periods of these two x-ray sources are so short, astronomers began to suspect that they had found rapidly rotating neutron stars.

It soon became clear, however, that systems such as Centaurus X-3 and Hercules X-1 are not ordinary pulsars like the Crab or Vela pulsars. Centaurus X-3 completely turns off periodically. Every 2.087 days, it turns off for almost 12 hours. This fact suggests that Centaurus X-3 is an eclipsing binary and that the x-ray source takes nearly 12 hours to pass behind its companion star.

The case for the binary nature of Hercules X-1 is even more compelling. It has an off state corresponding to a 6-hour eclipse every 1.7 days, and careful timing of the x-ray pulses shows a periodic Doppler shift every 1.7 days. This information provides direct evidence of

Figure 12-19 **X-Ray Pulses from Centaurus X-3** This graph shows the intensity of x rays detected by Uhuru as Centaurus X-3 moved across the satellite's field of view. The successive pulses are separated by 4.84 seconds. The variation in the height of the pulses was a result of the changing orientation of Uhuru's x-ray detectors toward the source as the satellite rotated.

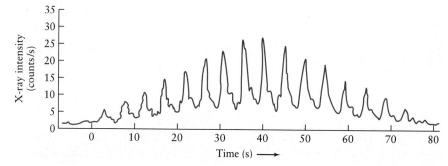

orbital motion about a companion star: When the x-ray source is approaching us, its pulses are separated by slightly less than 1.24 seconds. When the source is receding from us, slightly more than 1.24 seconds elapse between the pulses.

Careful visual searches around the location of Hercules X-1 soon revealed a dim star named HZ Herculis. The apparent magnitude of this star varies between +13 and +15, with a period of 1.7 days. Because this period is exactly the same as the orbital period of the x-ray source, astronomers conclude that HZ Herculis is the companion star around which Hercules X-1 orbits.

Astronomers now realize that systems such as Centaurus X-3 and Hercules X-1 are examples of binary stars in which one is a neutron star. All these binaries have very short orbital periods. Consequently, the distance between the ordinary star and its neutron star must be very small. This proximity enables the neutron star to capture gas escaping from the ordinary companion star.

To explain the pulsations of x-ray sources such as Centaurus X-3 or Hercules X-1, astronomers assume that the ordinary star either fills or nearly fills its Roche lobe. Either way, matter escapes from the star. If the star fills its lobe, as in the case of Hercules X-1, mass loss results from direct overflow through the Roche lobe; if the star's surface lies just inside its lobe, as with Centaurus X-3 (Figure 12-20), a stellar wind carries off the mass. A typical rate of mass loss for the ordinary star is roughly 10^{-9} $M_\odot$ per year.

The neutron star in a pulsating x-ray source, like an ordinary pulsar, rotates rapidly and has a powerful magnetic field inclined to the axis of rotation (recall Figure 12-17). Because of its strong gravity, the neutron star easily captures much of the gas escaping from its companion star. As the gas falls toward the neutron star, the magnetic field funnels the incoming matter down onto its magnetic polar regions. The neutron star's gravity is so strong that the gas is traveling at nearly half the speed of light by the time it crashes onto the star's surface. Because this violent impact creates hot spots at both poles with temperatures of about 10^8 K, these hot spots emit abundant x rays with a luminosity nearly 100,000 times brighter than the Sun. As the neutron star rotates, two beams of x rays from the polar caps sweep around the sky. As with other pulsars, if the Earth happens to be in the path of one of the beams, we can observe a pulsating x-ray source. The pulse period is thus equal to the neutron star's rotation period. For example, the neutron star in Hercules X-1 is spinning at the rate of once every 1.24 seconds.

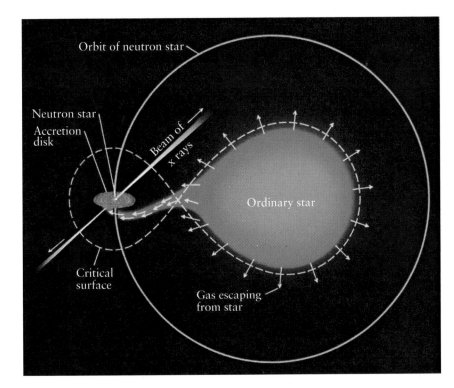

Figure 12-20 **A Model of a Pulsating X-Ray Source** Gas escaping from an ordinary star is captured by the neutron star. The infalling gas is funneled down onto the neutron star's magnetic poles, where it strikes the star with enough energy to create two x-ray-emitting hot spots. As the neutron star spins, beams of x rays from the hot spots sweep around the sky.

12-11 Other neutron stars in binary systems emit powerful jets of gas

In a binary system, gas captured by a neutron star's gravity goes into orbit around the neutron star, as shown in Figure 12-20. Recall from Chapter 10 that the result is a rotating disk of material called an *accretion disk.* Accretion disks have been detected in many close binary systems where mass transfer is occurring (recall Figure 10-17*b*).

With pulsating x-ray sources like Hercules X-1, the rate at which gas falls onto the neutron star from the inner edge of the accretion disk is low enough to allow the resulting x rays to escape. If the companion star is dumping vast amounts of material onto the neutron star, however, the resulting energy cannot escape easily. Instead, tremendous pressures build up in the gases crowding down onto the neutron star. In the plane of the accretion disk, the newly arrived gases are constantly spiraling in toward the neutron star. Their escape lies perpendicular to the accretion disk. The result should be two powerful jets of high-velocity hot gases. That is just what was seen for the weird star known as SS433.

In the autumn of 1978, Bruce Margon and his colleagues were observing SS433, which had been noted for strong emission lines in its spectrum. To everyone's sur-prise, the spectrum of SS433 could be broken into two sets of spectral lines, identical except that one set is very redshifted away from its usual wavelengths while another set is comparably blueshifted. Somehow, SS433 is coming and going at the same time. To make matters even more puzzling, the wavelengths of these redshifted and blueshifted lines change dramatically from one night to the next.

Astronomers had never seen anything like this, and soon many were observing SS433. By mid-1979, it was clear that the system's redshifted and blueshifted lines are actually moving back and forth across the spectrum of SS433 with a period of 164 days. Astrophysicists were quick to point out that the two sets of spectral lines could be generated by two oppositely directed jets of gas, one tilted toward us and the other away from us. Furthermore, the 164-day variation could be explained by a precession, or "wobble," of the axis of rotation of the accretion disk and its two jets. As the two jets circle about the sky every 164 days, we see a periodic variation in the Doppler shift.

All these features come together in the model sketched in Figure 12-21. To explain the large redshifts and blueshifts discovered by Margon, gas in the two oppositely directed jets must have a speed of 78,000 km/s, roughly one-quarter the speed of light. In addition, the accretion

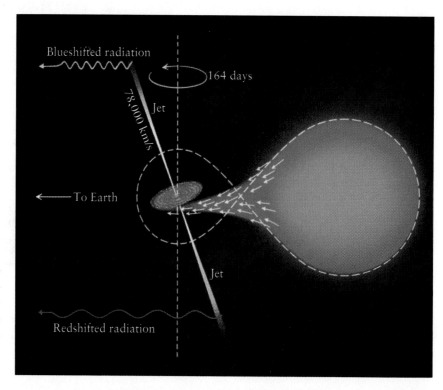

Figure 12-21 A Model of SS433 Gas from a normal star is captured into an accretion disk about a neutron star. Two high-speed, oppositely directed jets of gas are ejected from the faces of the disk. Because the disk is tilted, the gravitational pull of the normal star causes the jets to precess with a period of 164 days.

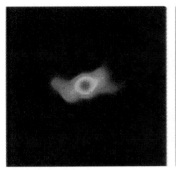 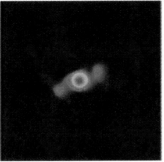 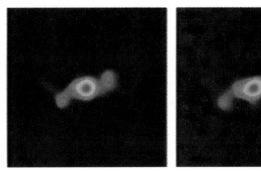

Figure 12-22 **Four Views of SS433** These four radio views, taken in early 1981, show jets of gas extending out to one-sixth of a light-year on either side of SS433. Three-quarters of the radio emission comes from SS433 itself (the red central blob), which is located at the center of a supernova remnant 13,000 ly from Earth.

disk must be tilted with respect to the orbital plane of the two stars of the binary system. Just as the Earth's axis is tilted with respect to the plane of the ecliptic, causing the Earth to precess, the tilt of the accretion disk results in the 164-day precession of the two jets.

Figure 12-22 shows four high-resolution radio views of SS433. Note the two oppositely directed appendages emerging from the central source. As we shall see in Chapter 16, many so-called quasars have a similar radio structure, though on a much larger scale. Because most quasars are very far away, they are difficult to study. The real significance of SS433 may be that it gives us a miniature quasarlike object in our own celestial backyard.

Novae and Bursters

Because low-mass stars are far more common than high-mass stars, white dwarfs are far more common than neutron stars. With all the bizarre and fascinating phenomena associated with neutron stars, you might be wondering if white dwarfs do anything more dramatic than simply cool off. The answer is definitely yes.

Occasionally a star in the sky suddenly brightens by a factor of between 10^4 and 10^6. This phenomenon is called a **nova** (not to be confused with a supernova, which involves a much greater increase in brightness). Novae are fairly common. Their abrupt rise in brightness is followed by a gradual decline that may stretch over several months or more (Figures 12-23 and 12-24).

Painstaking observations strongly suggest that novae occur in close binary systems containing a white dwarf. The ordinary companion star presumably fills its Roche lobe, so it gradually deposits fresh hydrogen onto the white dwarf. This new mass becomes a dense layer cov-

ering the hot surface of the white dwarf. As more gas is deposited and compressed, the temperature in the hydrogen layer continues to increase. Finally, at about 10^7 K, hydrogen fusion ignites throughout the layer, blowing it into interstellar space. This explosion is the nova. After a nova, fusion ceases. The companion star, however, may retain enough mass to supply a new layer of surface hydrogen, enabling some novae to reoccur.

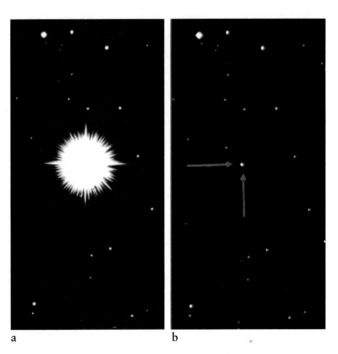

a   b

Figure 12-23 **Nova Herculis 1934** These two pictures show a nova (**a**) shortly after peak brightness as a magnitude +3 star and (**b**) two months later, when it had faded to magnitude +12. Novae are named after the constellation and year in which they appeared.

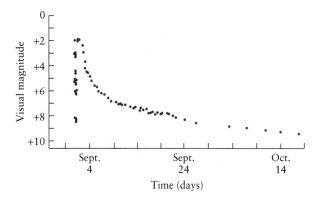

Figure 12-24 **The Light Curve of a Nova** This graph shows the history of Cygni 1975, a nova that blazed forth in the constellation of Cygnus in September 1975. The rapid rise in magnitude followed by a gradual decline is characteristic of all novae.

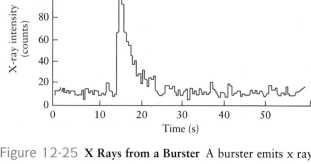

Figure 12-25 **X Rays from a Burster** A burster emits x rays with a constant low intensity interspersed with occasional powerful bursts. This burst was recorded in September 1975 by an x-ray telescope that was pointed toward the globular cluster NGC 6624. About one-third of all known bursters are located in globular clusters.

Neutron stars can also gather new mass from a companion star and flare up. Beginning in late 1975, astronomers analyzing data from x-ray satellites detected sudden, powerful bursts of radiation. The record of a typical burst is shown in Figure 12-25. The source, called an **x-ray burster,** emits x rays at a constant low level; then suddenly, without warning, there is an abrupt increase followed by a gradual decline. A typical burst lasts for only 20 seconds. Several dozen x-ray bursters have been located, most lying in the plane of the Milky Way Galaxy. Their locations on the celestial sphere indicate that they are actually in the Milky Way.

X-ray bursters, like novae, are believed to arise from mass transfer in binary star systems. With a burster, how-

ever, the stellar corpse is a neutron star rather than a white dwarf. Gases escaping from the ordinary companion star fall onto the neutron star. The energy released as this gas crashes down onto the neutron star's surface produces the low-level x rays that are continuously emitted by the burster.

Most of the gas falling onto the neutron star is hydrogen, which becomes compressed against the hot surface of the star by the star's powerful surface gravity. In fact, temperatures and pressures in this accreting layer are so high that the arriving hydrogen is promptly converted into helium by the hydrogen-fusing process. Constant hydrogen fusion soon produces a layer of helium that covers the entire neutron star.

Figure 12-26 **Distribution of Gamma-Ray Bursts** Gamma-ray bursters have been observed everywhere in the sky, indicating that, unlike x-ray bursters, they do not originate in the disk of the Milky Way Galaxy. This map of the entire sky "unfolded" onto the page shows bursts detected by the Compton Gamma Ray Observatory. The size of each dot indicates the relative strength of each burst. The colors indicate the x-ray wavelengths from longest to shortest wavelength; they are coded red, orange, green, light blue, dark blue, violet.

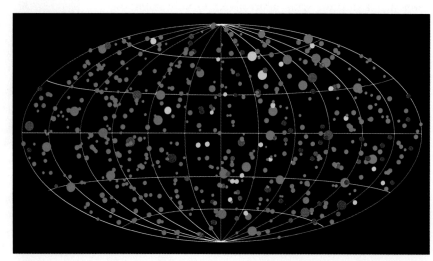

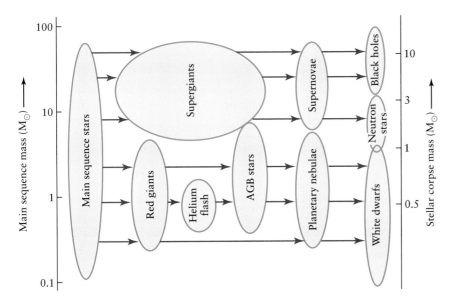

Figure 12-27 **A Summary of Stellar Evolution** The evolution of an isolated star depends on its mass. The scale at the left indicates the mass of a star when it is on the main sequence. The scale at the right gives the mass of the resulting stellar corpse. Stars less massive than about 4 $M_\odot$ can eject enough mass to become white dwarfs. High-mass stars can produce Type II supernovae and become neutron stars or black holes.

When the helium layer is about 1 m thick, helium fusion ignites explosively, and we observe a sudden burst of x rays. In other words, whereas explosive hydrogen fusion on a white dwarf produces a nova, explosive helium fusion on a neutron star produces an x-ray burster. In both cases, the fusion is explosive because the fuel is so strongly compressed against the star's surface that it is degenerate, like the star itself. As we saw with the helium flash inside giants, ignition of a degenerate thermonuclear fuel always involves a sudden thermal runaway because the usual safety valve between temperature and pressure is not operating.

There are also **gamma-ray bursters.** Unlike x-ray bursters, these objects have bursts lasting for only a few seconds; in addition, they are located all over the celestial sphere, indicating that they lie outside the Milky Way Galaxy (Figure 12-26). Over nine hundred gamma-ray bursters have been discovered. Their origins are still being debated. Some astronomers believe they are associated with neutron stars, while others believe they are related to supernovae.

In life as well as in death, the mass of a star determines its fate (Figure 12-27). Just as there is an upper limit to the mass of a white dwarf (called the Chandrasekhar limit), there is also an upper limit to the mass of a neutron star, called the *Oppenheimer-Volkov limit*. Above this limit, neutron degeneracy pressure cannot support the overbearing weight of the star's matter pressing in from all sides. The Chandrasekhar limit for a white dwarf is 1.4 $M_\odot$, and the Oppenheimer-Volkov limit for a neutron star is about 3 $M_\odot$. What might happen if a dying massive star failed to eject enough matter to get below the Oppenheimer-Volkov limit?

The gravity associated with a neutron star is so strong that the escape velocity is roughly one-half the speed of light. With a stellar corpse greater than 3 $M_\odot$, there is so much matter crushed into such a small volume that the escape velocity exceeds the speed of light. Because nothing can travel faster than light, nothing—not even light—can leave the dead star. The star therefore disappears, its powerful gravity leaving a hole in the fabric of the universe. The discovery of neutron stars thus inspired astrophysicists to examine seriously one of the most bizarre and fantastic objects ever predicted by modern science—the black hole.

WHAT DID YOU THINK?

1 *Will the Sun explode? If so, what is the explosion called?* The Sun will explode as a planetary nebula in about five billion years.

2 *Where did carbon, silicon, oxygen, iron, uranium, and other heavy elements on Earth come from?* These elements are created by supernovae.

3 *What is a pulsar?* A pulsar is a rotating neutron star in which the magnetic field does not pass through the rotation axis.

4 *What is a nova?* A nova is a relatively gentle explosion of hydrogen gas on the surface of a white dwarf in a binary star system.

Key Words

asymptotic giant branch (AGB) star	lighthouse model	planetary nebula	Type I supernova
Cerenkov radiation	neutron degeneracy pressure	pulsar	Type II supernova
Chandrasekhar limit	neutron star	shell helium fusion	white dwarf
gamma-ray burster	nova (*plural* novae)	supergiant	x-ray burster
helium shell flash	photodisintegration	supernova	

Key Ideas

• A low-mass main sequence star becomes a giant when shell hydrogen fusion begins. It becomes a horizontal branch star when core helium fusion begins.

  Thermal pulses in the helium-fusing shell can eject the star's outer layers.

• The burned-out core of a low-mass star becomes a dense carbon–oxygen sphere about the size of the Earth called a white dwarf.

  The maximum mass of a white dwarf (the Chandrasekhar limit) is 1.4 $M_\odot$.

• After exhausting its central supply of hydrogen and helium, a high-mass star undergoes a sequence of other thermonuclear reactions in its core. These are carbon fusion, neon fusion, oxygen fusion, and silicon fusion. The star eventually develops an iron-rich core.

• A high-mass star dies in a supernova explosion that ejects most of the star's matter into space at very high speeds. This so-called Type II supernova is triggered by the gravitational collapse of the doomed star's core.

  The star's core becomes a neutron star or even a black hole. A neutron star is a very dense stellar corpse consisting of closely packed neutrons in a sphere roughly 8 km in diameter.

  Neutrinos were detected from the supernova SN 1987A, which was visible to the naked eye.

• An accreting white dwarf in a close binary system can also become a supernova when carbon fusion ignites explosively throughout such a degenerate star. Such a detonation is called a Type I supernova.

• A pulsar is a rapidly rotating neutron star with a powerful magnetic field, making it a source of periodic radio and other electromagnetic pulses.

  Energy pours out of the polar regions of the neutron star in intense beams that sweep around the sky.

  Some x-ray sources exhibit regular pulses. These objects are thought to be neutron stars in close binary systems with ordinary stars.

• Material from the ordinary star in a binary pair can fall onto the surface of its companion white dwarf or neutron star to produce a surface layer in which thermonuclear reactions can occur.

  Explosive hydrogen fusion may occur in the surface layer of a companion white dwarf, producing the sudden increase in luminosity that we call a nova.

  Explosive helium fusion may occur in the surface layer of a companion neutron star, producing the sudden increase in x-ray radiation called a burster.

Review Questions

1 What is the difference between a giant and a supergiant?

2 Why is the temperature in a star's core so important in determining which nuclear reactions can occur there?

3 How is a planetary nebula formed?

4 What is a white dwarf?

5 What is the significance of the Chandrasekhar limit?

6 What is a neutron star?

7 Compare a white dwarf and a neutron star. Which of the two types of stellar corpses is more common?

8 On an H–R diagram, sketch the evolutionary track that the Sun will follow between the time it leaves the main sequence and when it becomes a white dwarf. Approximately how much mass will the Sun have when it becomes a white dwarf? Where will the rest of the mass have gone?

9 Why do you suppose that all the white dwarfs known to astronomers are relatively close to the Sun?

10 Why have radio searches for supernova remnants been more fruitful than searches at visible wavelengths?

11 Why do astronomers believe that pulsars are rapidly rotating neutron stars?

12 What is the difference between Type I and Type II supernovae?

13 Compare a nova with a Type I supernova. What do they have in common? How are they different?

14 Compare a nova and a burster. What do they have in common? How are they different?

15 What is SS433?

16 Describe what radio pulsars, x-ray pulsars, and bursters have in common. How are they different manifestations of the same type of astronomical object?

Advanced Questions

17 What prevents thermonuclear reactions from occurring at the center of a white dwarf? If no thermonuclear reactions take place in its core, why doesn't the star collapse?

18 Suppose you wanted to determine the age of a planetary nebula. What observations would you make, and how would you use the resulting data?

19 What reasons can you think of to explain why the rate of expansion of the gas shell in a planetary nebula is not uniform in all directions?

20 What kinds of stars would you monitor if you wished to observe a supernova explosion from its very beginning? Look up tabulated lists of the brightest and nearest stars. Which, if any, of these stars are possible supernova candidates? Explain.

21 To determine accurately the period of a pulsar, astronomers must take the Earth's orbital motion about the Sun into account. Explain why.

*** 22** The distance to the Crab Nebula is about 2000 pc. When did it actually explode?

Discussion Questions

23 Suppose that you discover a small glowing disk of light while searching the sky with a telescope. How would you decide just from your observations if this object was a planetary nebula? Could your object be something else? Explain.

24 Immediately after the first pulsar was discovered, one explanation offered was that the pulses were signals from an extraterrestrial civilization. Why do you suppose astronomers discarded this idea?

Observing Projects

25 With the help of star maps, locate and observe as many planetary nebulae as possible on clear, moonless nights, using the largest telescope at your disposal. Some of the more notable ones include: Little Dumbbell (M76), NGC 1535, Eskimo, Ghost of Jupiter, Owl (M97), Ring (M57), Blinking Planetary, Dumbbell, Saturn Nebula, and NGC 7662. Note and compare the various shapes of the different nebulae.

26 Two supernova remnants can be seen through modest telescopes, one in the winter sky and the other in the summer sky. Both are quite faint, however, so you should schedule your observations for a moonless night. The winter sky contains the Crab Nebula, which is discussed in detail in this chapter. It is located near the star marking the eastern horn of Taurus (the Bull). Its coordinates are R.A. = 5^h 34.5^m and Dec. = +22° 00'. Whereas the entire Crab Nebula easily fits into the field of view of an eyepiece, the Veil (or Cirrus) Nebula in the summer sky is so vast that you can see only a small fraction of it at a time. The easiest way to find the Veil Nebula is to aim the telescope at the star 52 Cygni (R.A. = 20^h 45.7^m and Dec. = +30° 43'), which lies on one of the brightest portions of the nebula. If you then move the telescope slightly north or south until 52 Cygni is just out of the field of view, you should see giant wisps of glowing gas.

13 ▶ *Black Holes*

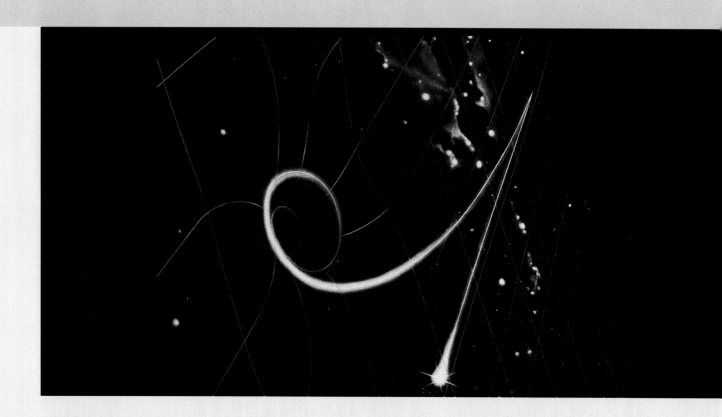

IN THIS CHAPTER

You will discover regions of space and time that are severely distorted by the extremely dense mass they contain. Black holes arise from the collapse of neutron stars more massive than 3 $M_\odot$, from enormous clumps of matter at the centers of some galaxies, and, probably, from the Big Bang explosion. You will find that Einstein's general theory of relativity predicts their surprisingly simple properties. Finally, you will follow a probe into the heart of a black hole and learn of a black hole's bizarre fate.

The Behavior of Light near a Black Hole Black holes cause nearby spacetime to become distorted. In this artist's rendition, two light beams fired in slightly different directions from the upper right travel on very different paths. They both move in straight lines, but the meaning of "straight" near a black hole defies our intuition. The beam on the right, aimed slightly away from the black hole, feels virtually no effect from it. The other beam, inclined toward the black hole, travels through space that is so distorted that it twists the light into a spiral, which is swept down into the black hole. Don't let this painting (or similar images in science fiction movies) mislead you into believing that black holes have a "top" or a "bottom;" they can be entered from any direction.

WHAT DO YOU THINK?

1. Are black holes just holes in space?

2. What is at the surface of a black hole?

3. What power or force enables black holes to draw things in?

4. Do black holes last forever?

. . . .

BLACK HOLES INSPIRE AWE, fear, and uncertainty. Many people harbor the belief that black holes are destined to "swallow up" all the matter in the universe. Happily, black holes are more benign than that, but they are no less strange.

The gravitational force acting on a neutron star smaller than 3 $M_\odot$ is counterbalanced by an equal, outward force from neutron degeneracy pressure. Prevented from collapsing further, the star is incredibly compact, barely 40 km across. But not even neutron degeneracy pressure can stand up to masses greater than 3 $M_\odot$ and their inward gravitational force.

The Evidence for Black Holes

Once a massive star overcomes even neutron degeneracy pressure, what prevents it from collapsing until its mass approaches infinite density and infinitesimal volume? The answer is "Nothing"! No wonder the dead star and the space around it come to defy even Newton's laws. Nothing—not even light—can manage to escape from its gravity: It becomes a **black hole**. We must turn to Einstein's powerful special and general theories of relativity in order to understand these exotic objects.

13-1 Special relativity changes our conception of space and time

In 1905 Albert Einstein began a revolution in physics with his **special theory of relativity.** He was guided by two innovative ideas. While its implications proved revolutionary, the first notion seems simple: *Your description of physical reality is the same regardless of the velocity at which you move.* In other words, as long as you are moving in a straight line at a constant speed, you experience the same laws of physics as anyone else moving at any other constant speed or in any other direction. To illustrate, suppose you were inside a closed boxcar moving in a straight line at 100 km/h. Any scientific measurements you take in the moving boxcar would yield the same results as the measurements taken while the boxcar was at rest (or at any other speed).

The second idea seems more bizarre: *Regardless of your speed or direction, you always measure the speed of light to be the same.* Suppose that you are in a car moving toward a distant street lamp. Even if you are moving at 95% the speed of light, you will still measure the speed of the oncoming light photons to be the same as if you were standing by the roadside.

Einstein's special theory of relativity expresses these two assumptions mathematically. Its results have been confirmed in innumerable experiments. The first result is that the length of an object decreases as its speed increases; the faster an object moves past you, the shorter its dimension is in the direction of its motion. In other words, if a boxcar moves past you, it is actually shorter, from your perspective, than when it is at rest (Figure 13-1). However, if you measure the length of the moving boxcar while you are inside it, then you are moving with it, and your measurement of its length will be the same as when it was at rest. The word *relativity* emphasizes the importance of the relative speed between the observer and the measured object.

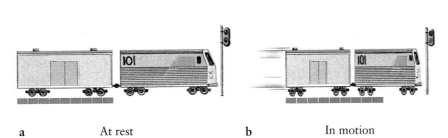

a At rest **b** In motion

Figure 13-1 **Movement and Space** According to the special theory of relativity, the faster an object moves, the shorter it becomes in its direction of motion, becoming infinitesimally small as its speed approaches the speed of light. The dimensions perpendicular to the object's motion are unchanged compared to the same dimensions when the object is at rest.

Einstein's second result is just as strange: Clocks passing by you run more slowly than do clocks at rest. In fact, the faster a clock passes by, the slower it appears to tick from your perspective. For example, pilots and airline attendants actually age more slowly than they would if they didn't fly. Note, however, that their activities slow down at the same rate as the clocks around them: They don't feel time moving more slowly. Only an observer moving more slowly sees that the clock of the high-speed travelers has slowed. These connections between motion and clocks mean that space and time cannot be considered as two separate concepts. Relativity requires us to consider them as a single entity, thus creating the concept of **spacetime.**

Finally, Einstein showed that the mass of an object increases as it moves faster. A body approaching the speed of light becomes infinitely massive. To push mass faster than the speed of light would therefore require an infinite amount of energy. No wonder the speed of light is the universal speed limit.

13-2 General relativity predicts black holes

First published in 1915, Einstein's **general theory of relativity** extends his special theory of relativity. It describes how matter affects not only the law of gravity but also the very fabric of space and time. He found that matter affects the nature of space and the rate at which time passes. Newton's laws of motion and his universal law of gravitation are strictly accurate only for small masses and low densities. Especially near the enormous density of a collapsing neutron star, only general relativity correctly predicts how objects will move. Even light changes its course!

In general relativity, unlike in Newtonian physics, the path of light is affected by nearby mass (Figure 13-2): A massive object actually curves space and even slows the flow of time. The more massive or denser the object, the more it curves nearby space and the more it slows down time in its vicinity. Just by being near matter, clocks run slower than they would in empty space. Furthermore, photons leaving the vicinity of a star lose energy in climbing out of the star's gravitational field. They show this change by shifting to longer wavelengths. This is called **gravitational redshift.**

Einstein's description of gravity is radically different from Newton's. According to Newtonian mechanics, space is perfectly flat and extends infinitely far in all di-

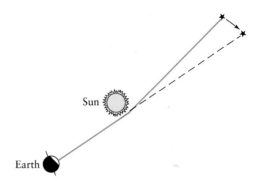

Figure 13-2 Warped Space and the Curved Path of Light That the warping of space by matter deflects light was one of the first predictions of general relativity to be confirmed, in 1919. This confirmation came when stars near the Sun were observed during an eclipse. Starlight was deflected around the Sun by up to 1.75 arcsec, a small but measurable amount.

rections. Similarly, Newtonian clocks monotonously tick at an unchanging rate, never speeding up or slowing down. In this rigid, unalterable framework of space and time, gravity is described as a "force" that acts at a distance. The planets literally pull on each other across empty space with a strength described by Newton's universal law of gravitation.

The general theory of relativity does not treat gravity as a force at all. Instead, gravity causes space to become curved and time to slow down. A planet or spacecraft passing near the Sun is deflected from a straight-line path because space itself is curved.

Several predictions of general relativity have been tested:

The perihelion position of Mercury as seen from the Sun shifts or precesses by 43 arcsec more than is predicted by Newtonian gravitational theory (Figure 13-3).

Light is measurably deflected by the Sun's gravitational curving of space (see Figure 13-2).

Extremely accurate clocks run more slowly when being flown in airplanes than identical clocks remaining on the ground.

The spectra of some stars are observed to have gravitational redshifts.

These observations are consistent with the theory of general relativity but not with the laws of Newtonian physics. General relativity has been confirmed again and again.

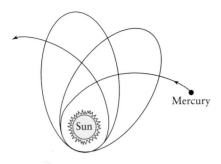

Figure 13-3 **Mercury's Orbit Explained by General Relativity** The location of Mercury's perihelion (its position closest to the Sun, greatly exaggerated in this diagram) changes with each orbit due to the gravitational influences of the other planets as well as effects predicted by Einstein's theory of general relativity. The amount of this change is inconsistent with the predictions made by Newton's theory of gravity alone.

Applying general relativity to collapsing neutron stars, we learn that sufficiently dense matter actually causes nearby space to curve so much that it closes in on itself (Figure 13-4). Photons flying outward at an angle

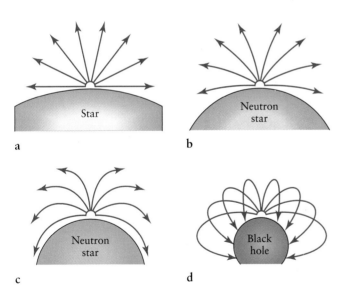

Figure 13-4 **Trapping of Light by a Black Hole** (a) Light departing from a main sequence, giant, or supergiant star is affected very little by the star's gravitational force. (b) However, as a neutron star collapses, its gravitational force increasingly curves the paths of departing light until (c) some of the photons actually return to the star's surface. (d) When the neutron star becomes a black hole, all light leaving its surface remains trapped. Most photons curve back in, except those flying straight upward, which become infinitely redshifted, thereby disappearing.

from such a collapsing star arc back inward; photons flying straight outward lose all their energy, and they cease to exist.

If light, the fastest moving of all known things, can't escape from such a region around the dense matter, then nothing can! Such regions out of which no matter or any form of electromagnetic radiation can escape are called black holes. Plummeting in on itself, a collapsing neutron star becomes so dense that it ceases to consist of neutrons. Indeed, general relativity predicts that it compresses to infinite density, a state called a **singularity**. But when matter gets sufficiently dense, even general relativity ceases to be valid; a more comprehensive theory must still be developed. Astrophysicists hope that combining the laws of quantum mechanics with general relativity will explain the final form of matter in a black hole. All we can say with confidence is that the matter in black holes becomes incredibly dense and compact.

13-3 Several binary star systems contain black holes

Black holes are more than fine points of relativity theory: They are real, and more of them are being located all the time. To find evidence for black holes, we look first to binary star systems. When a massive star in a binary system becomes a black hole, it may vanish from sight, but it has dramatic effects on its companion star.

As we saw in Chapter 10, most stars are in binary star systems. When a star in a close binary becomes a black hole, it draws off some of the companion star's atmosphere. The gas swirls into the black hole like water going down a bathtub drain. The inward swirling gas forms an accretion disk (see Chapter 12), which is compressed so much that it gives off x rays (Figure 13-5). If a visible star has a sufficiently tiny, sufficiently massive x-ray-emitting companion, we have located a black hole.

Shortly after the Uhuru x-ray satellite was launched in the early 1970s, astronomers found a promising candidate—an x-ray source called Cygnus X-1. This source is highly variable and irregular. Its strong x-ray emission flickers on time scales as short as a hundredth of a second. If different parts of an x-ray source grew bright at different times, its emission would be a continuous stream. In order for Cygnus X-1 to flicker, the entire star must brighten and dim as a unit. Therefore light must have time to travel across Cygnus X-1 between each pulse. Since light travels 3000 km in a hundredth of a second, Cygnus X-1 must be smaller than the Earth.

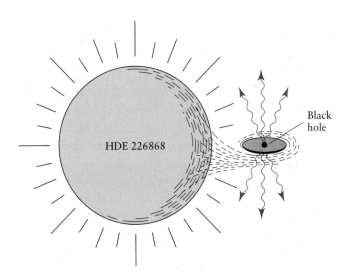

Figure 13-5 X Rays Generated by Accretion of Matter near a Black Hole Stellar-remnant black holes are detected in close binary star systems, such as Cygnus X-1, LMC X-3, A0620-00, and V404 Cygni. This drawing of Cygnus X-1 shows how gas from the companion star transfers to the black hole, thereby creating an accretion disk. As the gas spirals inward, friction and compression heat it so much that the gas emits x rays, which can be detected.

Cygnus X-1 occasionally emits radio radiation, and in 1971 radio astronomers succeeded in associating it with the visible star HDE 226868 (Figure 13-6). Spectroscopic observations revealed that HDE 226868 is a B0 supergiant with a surface temperature of about 31,000 K. Because such stars do not emit significant amounts of x rays, HDE 226868 alone cannot be Cygnus X-1.

Further spectroscopic observations soon showed that the lines in the spectrum of HDE 226868 shift back and forth with a period of 5.6 days. This behavior is characteristic of a single-line spectroscopic binary; HDE 226868's companion is too dim to produce its own set of spectral lines. The clear implication is that HDE 226868 and Cygnus X-1 are the two components of a binary star system.

The B0 supergiant HDE 226868 is estimated to have a mass of about 30 $M_\odot$, like other B0 supergiants. As a result, Cygnus X-1 must have a mass of about 7 $M_\odot$; otherwise, it would not exert enough gravitational pull to make the B0 star wobble by the amount deduced from the periodic Doppler shift of its spectral lines. Cygnus X-1 can't be a white dwarf or a neutron star, because its mass is too large for either of these objects. There is only one remaining possibility: It must be a fully collapsed neutron star—a black hole.

In the early 1980s a binary system similar to Cygnus X-1 was identified in the nearby galaxy called the Large Magellanic Cloud (see Chapter 12). The x-ray source, called LMC X-3, exhibits rapid fluctuations, just like those of Cygnus X-1. LMC X-3 orbits a B3 main sequence star every 1.7 days. From its orbital data, astronomers conclude that the mass of LMC X-3 is probably about 6 $M_\odot$, which would make it a black hole. This conclusion is still tentative. Some astronomers argue that we may have overestimated the object's mass.

Another black hole candidate is a spectroscopic binary in the constellation of Monoceros that contains the flickering x-ray source A0620-00. The visible companion of A0620-00 is an orange dwarf star of spectral type K, which orbits the x-ray source every 7.75 hours. From orbital data, astronomers conclude that the mass of A0620-00 must be greater than 3.2 $M_\odot$ and more probably about 9 $M_\odot$.

Perhaps the most convincing black hole candidate of all is the spectroscopic binary called V404 Cygni, which consists of an x-ray source orbited by a low-mass G or K star. Doppler shift measurements reveal that the line-of-sight velocity of the visible star varies by more than 400 km/s as it orbits its unseen companion every 6.47 days. These data give a firm lower limit for the mass of the companion of at least 6.26 $M_\odot$.

Figure 13-6 X-Ray Source Cygnus X-1 and Its Companion Black Hole Cygnus X-1 contains a 7-$M_\odot$ black hole in orbit with a B0 blue giant star, HDE 226868. This star system is located about 8000 ly from Earth. This photograph was taken with the 200-in. telescope at Mount Palomar. The slightly dimmer star nearby is an optical double that is not part of a binary system with HDE 226868.

a

b

Figure 13-7   **Galaxies with Supermassive Black Holes**   Notice the bright regions in the centers of each of these galaxies, where stars and gas are held in tight orbits by black holes. (a) M87's bright nucleus is so small (about the size of the solar system) and pulls on the nearby stars with so much force that astronomers believe it is a 3×10^9 M$_\odot$ black hole. (b) The nucleus of the nearby spiral galaxy Andromeda (M31) may harbor a black hole of a few million solar masses.

13-4 Black holes exist at the centers of some galaxies and may also have formed in the early universe

The fate of massive neutron stars led to the idea of black holes back in 1939. Could black holes have been created by other mechanisms? Yes, they could have formed from the clumping of mass at the centers of some newborn galaxies. In addition, the Big Bang itself may have compressed matter sufficiently to form black holes. However, so far, the only firm evidence we have is of galactic black holes.

As we will explore in more detail in Chapter 17, galaxies were created from condensing gas in the early universe. When a galaxy was forming, some of the gas plunged straight inward, colliding with similar gas coming in from the other direction. As it piled up at the center, much like the mass in a protostar, this matter compressed itself into a **supermassive black hole.**

In May 1994 the newly repaired Hubble Space Telescope obtained compelling evidence for a black hole at the very center of the galaxy M87. At the heart of M87 is a tiny, bright source of light. Spectra showed that nearby gas and stars are orbiting extremely rapidly. They can be held in place only if the bright object contains some three *billion* solar masses (Figure 13-7*a*). Given that its size is only slightly larger than the solar system, it can only be a black hole.

A second supermassive black hole has tentatively been identified at the center of the nearby Andromeda galaxy (Figure 13-7*b*). Estimates put this second galactic black hole at a few million solar masses. Another compelling candidate is galaxy M106. As we shall see in the next chapter, a similar black hole may even be lurking at the center of our Milky Way, only 25,000 ly from the Earth.

Even more exotic black holes may have formed along with the universe itself. The Big Bang explosion from which the universe emerged may have been chaotic and powerful enough to have compressed tiny knots of matter into **primordial black holes.** Their masses may have ranged from a few grams to greater than the mass of the Earth. Astronomers have not yet observed evidence of primordial black holes.

Inside a Black Hole

A black hole has a complicated birth, but its nature is surprisingly simple. It has a boundary shaped like a sphere; it either rotates or it doesn't; and it slowly disappears. That is almost all there is to know.

AN ASTRONOMER'S TOOLBOX 13-1

The Size of Black Holes

According to Einstein's general theory of relativity, the Schwarzschild radius R_{Sch} of any black hole is related to its mass M. Its radius can be determined by the equation

$$R_{Sch} = \frac{2GM}{c^2}$$

where c is the speed of light, 3×10^8 m/s^2, and G is the gravitational constant, 6.67×10^{-11} m^3/kg$\cdot$s^2.

A 5-$M_\odot$ main sequence star, for example, has a radius of 3×10^6 km, while the above equation reveals that a black hole with the same mass has a 15-km Schwarzschild radius. Even a galactic black hole containing three billion solar masses has a Schwarzschild radius of only 60 AU (which is just twice the size of Neptune's orbit around the Sun). A primordial black hole with the mass of Mount Everest would have a Schwarzschild radius of just 1.5×10^{-15} m!

13-5 Matter in a black hole becomes much simpler than out in the universe

A black hole is separated from the rest of the universe by a boundary called the **event horizon.** Within this imagined shell, even light cannot escape. The event horizon is not like the surface of a solid body. There is no matter at this location except for the instant it takes all infalling mass to cross the event horizon and enter the black hole.

We cannot look inside a black hole since no electromagnetic radiation escapes from it. Our understanding of its structure comes from the equations of general relativity. According to Einstein's theory, the event horizon is a sphere. The distance from the center of the black hole to its boundary is called the **Schwarzschild radius** (abbreviated R_{Sch}). Toolbox 13-1 shows how to calculate this distance, which depends only on the mass. The more massive the black hole, the larger is its event horizon.

When a stellar remnant collapses to a black hole, it loses its magnetic field. The field's energy radiates away in the form of **gravitational waves,** which are ripples in the very fabric of spacetime. Wherever the gravitational wave passes, spacetime itself becomes momentarily distorted. Scientists have designed antennas to detect these bursts of radiation emitted by massive, dying stars.

Besides its magnetic field, matter within a black hole loses almost all trace of its composition and origin. It retains only three properties that it had before entering the black hole: its *mass,* its *angular momentum,* and its *electrical charge.* All other attributes of matter vanish inside a black hole, and familiar concepts such as proton, neutron, electron, atom, and molecule no longer apply. In addition, because few large bodies appear to have a net charge, it is doubtful that many black holes do either. We therefore find that there are only two different types of black holes: those that rotate and those that do not.

If the mass creating a black hole is not rotating, it has no angular momentum, and the black hole it forms does not rotate either. We call these **Schwarzschild black holes.** General relativity predicts that all the mass in such a black hole collapses to a singularity at its center (Figure 13-8).

When the matter creating a black hole does possess angular momentum, the matter collapses instead to a *ring* located inside the black hole between its center and the event horizon (Figure 13-9). Such rotating black holes are called **Kerr black holes** in honor of the New Zealand mathematician Roy Kerr, who first calculated their structure in 1963. Most Kerr black holes should be spinning thousands of times every second, even faster than the pulsars we studied in Chapter 12. The mass of any isolated black hole is concentrated in a singularity, while all the rest of its interior is empty space.

Unlike Schwarzschild black holes, Kerr black holes also possess a doughnut-shaped region directly *outside* their event horizons in which objects cannot remain at rest without falling into the black hole. Called **ergoregions,** they are regions of spacetime that the rotating black hole drags around, like so much batter in a blender. If it is moving fast enough, an object entering this region can fly out of it again; if it stops in the ergoregion, however, it will fall into the black hole.

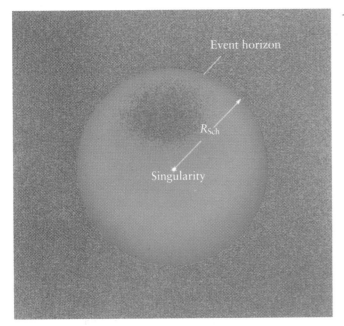

Structure of a Schwarzschild (Nonrotating) Black Hole A nonrotating black hole has a very simple structure. In fact, it has only two notable parts: its singularity and its boundary. Its mass, called a singularity because it is so dense, collects at its center. The spherical boundary between the black hole and the outside universe, R_{Sch}, is called the event horizon. There is no concrete surface at the event horizon. In fact, except for its location at the boundary of the black hole, an event horizon lacks any features at all.

13-6 Falling into a black hole is an infinite voyage

Imagine being in a spacecraft orbiting only 100 Schwarzschild radii (1500 km) from an isolated 5-$M_\odot$ black hole. You are held in orbit by the black hole's gravitational force. Even at that short distance, the only effect the black hole has on you is from its gravitational attraction. It is only when you get very, very close to the event horizon that bizarre things begin to happen. To investigate these changes, you send a cube-shaped probe toward the black hole, with one side of the cube always facing "downward," toward the black hole. The probe emits a blue glow so that you can follow its progress. What happens to the cube as it approaches the black hole?

From the time you eject it until it reaches about 10 Schwarzschild radii (150 km), you see the probe descend as if it were falling toward a planet or the Moon (Figure 13-10a). At 10 Schwarzschild radii, however, the probe begins to feel a severe tidal effect from the black hole. The closest face of the probe feels more gravitational pull from the black hole than do its parts farther away, and it begins to be stretched apart.

By the time the probe comes within a few Schwarzschild radii of the event horizon, the tidal forces on it are so great that it violently elongates. The part of the probe closest to the black hole accelerates downward and away from the rest of the probe. Furthermore, the sides of the probe are drawn together: They are falling in straight lines toward a common center (Figure 13-10b). The net gravitational effect of moving close to the event horizon is for the probe to be pulled long and thin. From a practical perspective, this means that the probe would be violently torn apart, since it is not composed of perfectly elastic material.

As the probe nears the black hole, blue photons leaving it must give up more and more energy to escape the

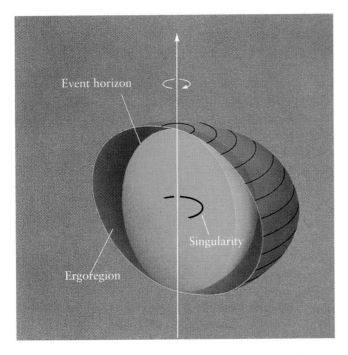

Figure 13-9 **Structure of a Kerr (Rotating) Black Hole** Rotating black holes are only slightly more complex than nonrotating ones. The singularity of a Kerr black hole is located in an infinitely thin ring around the center of the hole. It appears as an arc in this cutaway drawing. The event horizon is again a spherical surface. There is also a donut-shaped region, called the ergoregion, just outside the event horizon, in which nothing can remain at rest. Space in the ergoregion is being curved or pulled around by the rotating black hole.

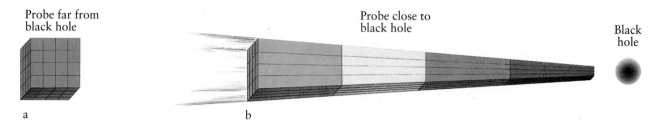

Probe far from
black hole

Probe close to
black hole

Black
hole

a   b

Figure 13-10 **Effect of a Black Hole's Tidal Force on In-falling Matter** (a) A cube-shaped probe is 1500 km from a black hole. (b) Near the Schwarzschild radius, the probe is pulled long and thin by the difference in the gravitational forces felt by its different sides. This is a tidal effect, a greatly magnified version of the Moon's gravitational force on the Earth. The probe changes color as photons from it experience extreme gravitational redshift.

increasing gravitational force. However, unlike a projectile fired upward, photons cannot slow down. Rather, they lose energy by increasing their wavelengths (see Toolbox 3-1). This is another example of the gravitational redshift predicted by general relativity. The closer the probe gets to the event horizon, the more its light is redshifted—first to green, then yellow, then orange, then red, then infrared, and finally radio waves.

Stranger still is the black hole's effect on time. General relativity predicts that when the probe approaches within a few Schwarzschild radii of the black hole, its infall rate will slow down as seen from far away. Also, signals from the probe show that its clocks are running more slowly than they did when it left your spacecraft. Time dilation is so great near the event horizon that the probe will appear to stop and hover above it.

Nevertheless, the probe itself actually crosses the event horizon and continues falling toward the black hole's singularity. Unlike the images we see in the movies, your probe can enter the black hole from any direction. Pulled apart by tidal effects, it disintegrates as it falls inward. It could not survive passage through a black hole, even if there were a way to come out somewhere else.

Nonetheless, could a black hole be connected to another part of spacetime or even some other universe? General relativity predicts such connections, called **wormholes**, for Kerr black holes, but astrophysicists are skeptical. Their conviction is called **cosmic censorship:** Nothing can enter or leave a singularity except by crossing the event horizon.

13-7 Black holes evaporate

4 In exploring the fate of black holes, astrophysicists find that, again, these objects confound common sense. With

its mass effectively cloaked behind the event horizon and presumably collapsed into a singularity, it seems plausible that there is no way of getting mass from the black hole back out into the universe again. Not possible, that is, until one recalls that mass and energy are two sides of the same coin. What if there were a way of converting the mass into a form of energy that *could* get out of the black hole, such as gravitational energy? The conversion path is Einstein's equation $E = mc^2$, and it does occur according to an idea first proposed by Stephen Hawking, the British theoretical physicist.

The reason that black holes convert their mass into energy, which then returns to the universe, is a process called *virtual particle production.* Quantum mechanics (see Chapter 4) allows pairs of particles to form spontaneously. Each pair is always two antiparticles that form, come back together, and annihilate each other, all within the incredibly short time of about 10^{-21} s. For example, an electron and its antiparticle, a positron, might form and destroy each other in this process. This is occurring everywhere in the universe, not just near black holes. Because the particles annihilate each other so rapidly, we normally never know of their existence.

However, particle production has been observed in high-energy accelerators. These machines accelerate electrons, protons, or other bodies to nearly the speed of light. Occasionally one of these high-speed particles encounters a pair of virtual particles *before* the two virtual particles annihilate each other. The collision forces the virtual particles apart, preventing them from annihilating each other, and making the two particles real and observable.

Among other places, virtual particles form *just outside* the event horizon of a black hole. If one particle is created slightly farther from the hole than its companion, the two virtual particles feel a *tidal* force from the

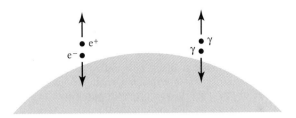

Figure 13-11 **Evaporation of a Black Hole** Throughout the universe, pairs of virtual particles spontaneously appear and disappear so quickly that they do not violate any laws of physics. The tidal force just outside the event horizon of a black hole is strong enough to tear two virtual particles apart before they destroy each other, and they become real. Some fall into the black hole, but some new particles escape, causing the black hole to shrink and eventually to evaporate completely. Here we see just a few of the exotic particles in the making: an electron (e$^-$) and positron (e$^+$), and a pair of photons (γ).

black hole's tremendous gravitational pull (recall the blue probe discussed earlier). When the gravitational tidal force is strong enough, it can pull the two virtual particles apart, making them real (Figure 13-11).

To conserve momentum, at least one of the newly formed real particles always falls into the black hole. But the other particle will sometimes have enough energy to escape from the vicinity of the black hole. These particles fly free into space. In this case, the black hole has effectively emitted energy equal to $E = mc^2$, where m is the mass of the freed particle. Where the virtual particles became real is a temporary void in space outside the event horizon. The black hole fills the void with energy that comes from its mass; *some of the black hole's mass is converted into gravitational energy and then transmitted outside the event horizon to replace the energy taken away by the escaped particle!*

This is called the **Hawking process,** named after its proposer. The time it takes a black hole to completely evaporate by the Hawking process increases with the mass of the hole. More massive black holes have lower tidal forces at their event horizons than do less massive, and therefore smaller-diameter, black holes. Therefore, the rate at which virtual particles are converted into real particles around bigger black holes is actually slower than it is around smaller ones. The faster the real particles are produced, the faster they wear down the mass of the black hole; thus lower-mass black holes evaporate more rapidly.

A stellar black hole of 5 M$_\odot$ would take more than 10^{62} years to evaporate (much longer than the present age of the universe), while a 10^{10} kg primordial black hole (equivalent to the mass of Mount Everest) would take only about 15 billion years. Depending on the precise age of the universe, a black hole of this size and age should be in the final throes of evaporating. Astronomers are trying to identify events in space corresponding to the violent, particle-generating deaths of primordial black holes.

WHAT DID YOU THINK?

1 *Are black holes just holes in space?* No, black holes contain highly compressed matter—they are not empty.

2 *What is at the surface of a black hole?* The surface of a black hole, called the event horizon, is empty space—there is no stationary matter there.

3 *What power or force enables black holes to draw things in?* The only force that pulls things in is the gravitational attraction of the matter in the black hole.

4 *Do black holes last forever?* No, black holes evaporate.

Key Words

black hole	gravitational redshift	Schwarzschild black	special theory of
cosmic censorship	gravitational waves	hole	relativity
ergoregion	Hawking process	Schwarzschild radius	supermassive black
event horizon	Kerr black hole	singularity	hole
general theory of	primordial black hole	spacetime	wormhole
relativity			

Key Ideas

• According to general relativity, mass causes space to become curved and time to slow down. These effects are significant only near very large masses or very compact objects.

• A black hole is an object so dense that the escape velocity from it exceeds the speed of light.

　If a stellar corpse is more massive than 3 $M_\odot$, gravitational compression overcomes neutron degeneracy pressure and forces it to collapse further and become a black hole.

　Observations indicate that some binary star systems harbor black holes. In such systems, gases captured by the black hole from the companion star heat up and emit detectable x rays.

　Supermassive black holes originated in the cores of some galaxies. Other black holes may have formed at the beginning of the universe.

• The event horizon of a black hole is a spherical boundary where the escape velocity equals the speed of light. No matter or electromagnetic radiation can escape from inside the event horizon. The distance from the center of the black hole to the event horizon is called the Schwarzschild radius.

The matter inside a black hole collapses to a singularity. The singularity for nonrotating matter is a point at the center of the black hole. For rotating matter, the singularity is a ring inside the event horizon.

　A black hole has only three physical properties: mass, angular momentum, and electric charge.

• Nonrotating black holes are called Schwarzschild black holes. Rotating black holes are called Kerr black holes. The event horizon of a Kerr black hole is surrounded by an ergoregion in which all matter must be moving to avoid being pulled into the black hole.

• As matter approaches the singularity of a black hole, it is torn by extreme tidal forces, light from it is redshifted, and time appears to slow. Einstein's general theory of relativity may be refined to forbid wormholes.

• Black holes can evaporate by the Hawking process, in which virtual particles near the black hole become real. This transition decreases the mass of a black hole.

Review Questions

1 Under what conditions do all outward pressures on a collapsing star fail to stop its inward motion?

2 In what way is a black hole blacker than black ink or a black piece of paper?

3 If the Sun suddenly became a black hole, how would the Earth's orbit be affected?

4 What is the law of cosmic censorship?

5 What are the differences between rotating and nonrotating black holes?

6 Why are all the black hole candidates that are stellar remnants members of close binary systems?

7 If light cannot escape from a black hole, how can we detect x rays from such objects?

Advanced Questions

* **8** What is the event horizon, measured in kilometers, of a black hole containing 3 $M_\odot$? 30 $M_\odot$?

9 If more massive stars evolve and die before less massive ones, why do some black hole candidates have lower masses than their normal stellar companions?

10 Under what circumstances might a white dwarf or a neutron star in a binary star system become a black hole?

11 Which type of black hole do science fiction writers (implicitly) use in sending spaceships from one place to another through the hole? Why would the other type not be suitable?

Observing Project

12 As you well know, you cannot see a black hole with a telescope. Nevertheless, you might want to observe the visible companion of Cygnus X-1. The epoch 2000 coordinates of this ninth-magnitude star are R.A. = $19^h 58.4^m$ and Dec. = +35° 12′. This location is quite near the bright star η Cygni (the star in Cygnus, the Swan that represents the swan's long neck). This constellation appears high in the night sky in the summer and appears in the shape of a cross. Compare what you see with Figure 13-6.

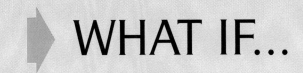

WHAT IF...

The Earth Orbited a 1.5-M_☉ Sun?

THE EARTH IS at a perfect distance from a wonderful star. The Sun provides just enough heat so that liquid water, necessary to sustain life, can exist here. Would our planet still be suitable for the evolution of life if the Sun were 1.5 times more massive than it actually is?

Kepler's third law tells us that the period of Earth's orbit around the new Sun at the same distance we are today would be only 298 days. A 20-year-old student today would be $24\frac{1}{2}$ years old if the Earth orbited the bigger Sun.

Recall from Chapter 4 that the peak of the Sun's blackbody radiation is in the blue-green part of the visible spectrum because of its surface temperature of 5800 K. Because our new Sun's surface temperature would be 8400 K, it would appear blue-white in our sky. The sky itself would appear more intensely blue than it does today because the new Sun would emit many more blue photons than does our present one. Indeed, the new Sun would give off seven times as much energy per second as our present Sun. The new Sun's radius would also be 20% larger than that of our present Sun.

The effect of the new Sun's increased energy emission on the Earth would be profound. Because of the new Sun's higher infrared (heat) output, the Earth would be hotter than it is today. Just that extra heat would raise the average global temperature by about 10 K (about 20°F). This does not seem like a lot, but the impact of that slight increase would actually boost the atmospheric temperature much higher.

The extra heat from the new Sun would cause more ocean water to evaporate into the atmosphere. Since water is a greenhouse gas (i.e., it traps infrared radiation), the air temperature would rise, causing even more water to evaporate from the oceans, which, in turn, would cause the air to heat even more. This vicious cycle, called a *runaway greenhouse effect*, would make the Earth's surface so hot and dry that it would be uninhabitable.

By moving the Earth about 2.6 times farther away from the more massive Sun, the temperature would become suitable for life. However, that would solve only the least of our problems. Just by increasing the Sun's mass by 50%, the ultraviolet radiation emitted would be several *thousand* times stronger. This is because the energy output of stars with different surface temperatures varies with wavelength. While the output of visible light would change slightly, the output of ultraviolet would be vastly higher. Therefore, even though the Earth's surface temperature would be suitable farther from the new Sun, the flood of ultraviolet radiation would be so strong that the ozone layer would be overwhelmed and the level of ultraviolet radiation at the Earth's surface would be much higher than it is today. This would be so even with the greater concentration of ozone created by the increased ultraviolet from the more massive Sun.

Life would have to evolve greater protection from ultraviolet than it has today. And if this were not a great enough challenge, suppose humans were evolving on the Earth orbiting a 1.5-M_☉ Sun some 4.6 billion years after the solar system formed. They would discover that their star had evolved so rapidly that it was just about to expand into the red giant phase and consume the Earth!

IV → FOUNDATIONS

The Universe

The Realm of the Galaxies Beyond the solar system, beyond the Milky Way, lies the realm of the galaxies. Just as stars orbit in their respective galaxies, so, too, do many galaxies orbit each other. This image shows hundreds of galaxies grouped together some 325 million light-years from Earth. This computer-enhanced photograph shows only a tiny sliver of the universe, yet it provides a valuable glimpse at the variety and staggering numbers of galaxies that exist. This region of the universe is part of the so-called Great Wall of galaxies stretching across more than one-third of the sky.

IN THE PAGES AHEAD

You will discover that the Milky Way is a galaxy—billions of stars bound together by their gravitational attraction. Until the twentieth century, few astronomers believed that myriads of galaxies lie beyond the Milky Way. You will see how the period–luminosity law for Cepheid variables settled the debate by giving the distances to other galaxies. In the rest of this book you will explore these galaxies, other distant objects, and the large-scale structure of the universe. You will ponder the ultimate fate of the universe, and you will estimate whether advanced life has formed on distant planets orbiting other stars.

WHAT DO YOU THINK?

1 Are all the stars members of the Milky Way?

2 How many galaxies are there?

· · · · · · · · · · · · · · · · · · · ·

SO FAR IN THIS BOOK, we have discussed stars singly, in binary systems, and in the relatively small groups called open and globular clusters. In binary systems and in globular clusters, the stars are *gravitationally bound:* They are held together by their mutual attraction. Now we shall see that the same forces create far vaster groupings of stars. The remainder of this book explores how matter is distributed throughout the universe, what the universe is, where it came from, and where it is going.

IV-1 The Milky Way is only one among billions of galaxies

For those of us fortunate enough to live away from bright outdoor lights, the Milky Way appears as a filmy band of white, overlain with the glow of individual stars. As children we are all told—correctly, it turns out—that

our solar system is part of the **Milky Way Galaxy,** an enormous assemblage of stars. But just how are we a part of it? Perhaps, we might imagine, all the stars in the universe belong to the Milky Way. If so, then "the galaxy" and "the universe" would mean the same thing. Indeed, that is precisely what most astronomers believed until the early decades of this century.

Today we know that the universe contains myriad **1** galaxies. Each galaxy is a grouping of millions, billions, or even trillions of stars, all gravitationally bound together. These stars are sometimes accompanied by huge quantities of interstellar gas and dust. As we saw in Chapter 10, new stars are forming in giant molecular clouds every day.

As early as 1755, the famous German philosopher Immanuel Kant suggested that vast collections of stars lie far beyond the confines of the Milky Way. Less than a century later, an Irish astronomer observed the structure of some of those "island universes." William Parsons was the third Earl of Rosse in Ireland. He was rich, he liked machines, and he was fascinated with astronomy. Accordingly, he set about building gigantic telescopes. In February 1845, his pièce de résistance was finished. This telescope's massive mirror measured 6 ft in diameter and was mounted at one end of a 60-ft tube controlled by cables, straps, pulleys, and cranes (Figure IV-1*a*). For

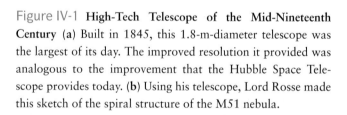

Figure IV-1 High-Tech Telescope of the Mid-Nineteenth Century (**a**) Built in 1845, this 1.8-m-diameter telescope was the largest of its day. The improved resolution it provided was analogous to the improvement that the Hubble Space Telescope provides today. (**b**) Using his telescope, Lord Rosse made this sketch of the spiral structure of the M51 nebula.

a

b

Figure IV-2 **The Spiral Galaxy M51** Also called NGC 5194, this spiral galaxy in the constellation of Canes Venatici is known as the Whirlpool galaxy because of its distinctive appearance. Its distance from Earth is about 20 million light-years. The blob at the end of one of the spiral arms is a companion galaxy.

many years, this triumph of nineteenth-century engineering enjoyed the reputation of being the largest telescope in the world.

With this new telescope, Lord Rosse examined many of the nebulae discovered and catalogued by William Herschel. Recall from Chapter 10 that William Herschel and his son John, among others, discovered and recorded details of many astronomical objects including double star systems and, as is relevant here, fuzzy objects they called **nebulae** (singular **nebula**). Lord Rosse observed that some of these nebulae have a distinct spiral structure. Perhaps the best example is M51 (also called NGC 5194).

Lacking photographic equipment, Lord Rosse made drawings of what he saw. Figure IV-1*b* shows his drawing of M51. Views like this inspired Lord Rosse to echo Kant's proposal of "island universes." Figure IV-2 shows a modern photograph of M51.

Most astronomers of Rosse's day did not agree. A considerable number of the "nebulae" listed in the *New General Catalogue* were, in fact, interstellar clouds and star clusters scattered throughout the Milky Way. It seemed likely that these intriguing spiral nebulae were also members of our Galaxy.

The astronomical community became increasingly divided over the nature of spiral nebulae. Finally, in April 1920, a debate was held at the National Academy

of Sciences in Washington, D.C. On one side was Harlow Shapley, a young, brilliant astronomer renowned for his recent determination of the size of the Milky Way Galaxy. Shapley believed the spiral nebulae to be relatively small, nearby objects scattered around our Galaxy like the globular clusters he had studied. Opposing Shapley was Heber D. Curtis of the Lick Observatory near San Jose, California. Curtis championed the island universe theory, arguing that each of these spiral nebulae is a separate rotating system of stars much like our own Galaxy.

The **Shapley–Curtis debate** generated much heat but little light. Nothing was decided, because no one had any firm evidence to demonstrate exactly how far away the spiral nebulae were. Astronomy desperately needed a definitive determination of the distances to a spiral nebula. Such a measurement was the first great achievement of a young lawyer who abandoned a Kentucky law practice and moved to Chicago to study astronomy. His name was Edwin Hubble.

IV-2 Studies of Cepheid variables by Henrietta Leavitt helped Edwin Hubble discover the distances to galaxies

Edwin Hubble joined the staff of the Mount Wilson Observatory in Pasadena, California, and in 1923 he took a historic photograph of the Andromeda nebula, M31, one of the spiral nebulae around which controversy raged. A modern photograph of it appears in Figure IV-3. Hubble carefully examined his photographic plate and discovered what he first thought to be a nova. Referring to previous plates of that region, he soon realized that the object was actually a Cepheid variable star. As we saw in Chapter 11, these pulsating stars vary in brightness periodically. Further scrutiny over the next several months revealed many other Cepheids, two of which are identified in Figure IV-4.

Only a decade before, in 1912, the American astronomer Henrietta Leavitt had published her important study of Cepheid variables. Leavitt studied numerous Cepheids in the Small Magellanic Cloud, a small galaxy very near the Milky Way. (On a clear night, this galaxy is visible to the naked eye in the southern hemisphere as a thin cloud near the Large Magellanic Cloud.) As we learned in Chapter 11, there are two kinds of Cepheid variables: the metal-rich Type I Cepheids, which Leavitt studied, and the slightly dimmer, metal-poor Type II Cepheids. The latter were not discovered until the 1940s, when U.S. cities were

Figure IV-3 **The Andromeda Galaxy (M31 or NGC 224)**
This nearby galaxy covers an area of the sky roughly five times as large as the full Moon. Under good observing conditions, the galaxy's bright central bulge can be glimpsed with the naked eye in the constellation of Andromeda. The distance to the galaxy is 2.2 million light-years. The white rectangle outlines the area shown in Figure IV-4.

Figure IV-4 **Cepheid Variables in the Andromeda Galaxy**
Two Cepheid variables are identified in this view of the outskirts of the Andromeda galaxy. Because these stars appear so faint (although they are in fact quite luminous), Hubble successfully demonstrated that the Andromeda "nebula" is extremely far away.

blacked out during World War II and the sky was, therefore, especially dark.

Leavitt's study led her to the period–luminosity law for Type I Cepheids: As we saw in Figure 11-24, there is a direct relationship between a Cepheid's luminosity (or absolute magnitude) and its period of oscillation. By observing the period, Hubble was therefore able to determine the absolute magnitude of the Cepheid. We have also seen a relationship between apparent magnitude, absolute magnitude, and distance (see Toolbox III-1). Hubble could calculate the distance to the Cepheid star and therefore the distance to the nebula containing it. Suppose you find a Cepheid variable, measure its period, and use a graph such as Figure IV-5 to determine the star's average luminosity. The star's true brightness can be expressed as an absolute magnitude. Meanwhile, you

observe the star's apparent magnitude. Because you now know both the apparent and the absolute magnitudes, you can calculate the star's distance.

Straightforward calculations using modern data on the distance of Cepheids demonstrates that M31 is some 2.2 million light-years *beyond* the Milky Way. This proves that M31 is not an open or globular cluster (the traditional nebulae) in our Galaxy, but rather an enormous separate stellar system: a separate galaxy. Today M31 is called the Andromeda galaxy. It is the only object not part of the Milky Way that can be seen with the naked eye from the Earth's northern hemisphere.

Hubble's results, which were presented at the end of 1924, settled the Shapley–Curtis debate once and for all. The universe was recognized to be far larger and populated with far bigger objects than most astronomers

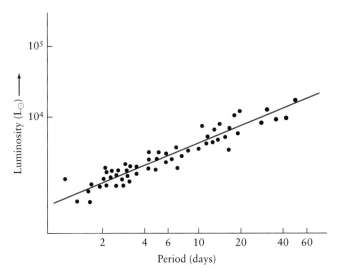

Figure IV-5   The Period–Luminosity Relation This graph shows the relationship between the periods and luminosities of classical (Type I) Cepheid variables. Each black dot represents a Cepheid whose brightness and period have been measured. The line is the best fit to the data.

challenging for astronomers to completely survey the Galaxy's contents.

Continued observations have revealed that there are **2** literally billions and billions of galaxies. In Chapter 15 we learn about their different types. By looking at neighboring galaxies not concealed by clouds, we learn about the activity in their centers. We also discover that galaxies are themselves often found in groups, and the large-scale structure of the universe does not stop there. Most groups of galaxies are themselves parts of larger groups, distributed as if on the surfaces of enormous, empty bubbles. These groupings and their movements away from Earth offer us clues to the structure and origin of the universe.

Quasars and active galaxies, exceptionally powerful sources of radiation, are examined in Chapter 16. We learn of their structure and the causes of their emissions. In Chapter 17 we consider the beginning, evolution, and possible endings of the whole universe. Because this subject involves the most distant observations, it entails the greatest uncertainty of all the topics we have covered. Finally, in Chapter 18 we take a brief look at the possible existence of other life in the universe.

had imagined. Hubble had discovered the realm of the galaxies.

In Chapter 14 we shall apply Hubble's method to find our place in the Milky Way Galaxy. Recall that the Sun's proximity, analogously, makes it our best understood star. It might seem that the nearness of the stars and clouds in the Milky Way would make it the best understood galaxy. However, we will see that the clouds of gas and dust surrounding the solar system make it very

WHAT DID YOU THINK?

1 *Are all the stars members of the Milky Way?* While all the stars we can see without the aid of telescopes are in the Milky Way, most stars in the universe are in other galaxies.

2 *How many galaxies are there?* There are billions and billions of galaxies.

Key Words

Milky Way Galaxy   nebula (*plural* nebulae)   Shapley–Curtis debate

Key Ideas

• At the beginning of this century, astronomers were divided as to whether all stars and nebulae are members of the Milky Way Galaxy.

• The Shapley–Curtis debate was the first major discussion of whether the Milky Way contains all the stars in the universe.

• As Henrietta Leavitt showed, Cepheid variable stars are important in determining the distance to other galaxies.

• Edwin Hubble first determined that there are groups of stars far outside the Milky Way Galaxy.

Review Questions

1 What was the Shapley–Curtis debate all about? Was a winner declared at the end of the debate? Whose ideas turned out to be correct?

2 How did Edwin Hubble prove that Andromeda is not a nebula in our Milky Way Galaxy?

Observing Project

3 On a clear night, locate and observe the Andromeda galaxy. It is "up" in the early evening between August and March in the northern hemisphere. If you are not conversant with the constellations, take a star map with you. Locate the **W** made by five bright stars in Cassiopeia. Also locate Polaris, the north pole star. Polaris is a medium-bright star found by following the line from the two stars in the Big Dipper farthest from the handle upward until you reach a medium-bright star. Draw a mental line from Polaris to the second star on the right side of Cassiopeia if you see it as the letter **W** or the second star on the left of Cassiopeia if you see it as the letter **M**. Keep moving south along that line one-half as far again and you should see a fuzzy "star," which is Andromeda. Describe what you see. Examine Andromeda through binoculars and again through the largest telescope you have available. Describe what you see.

14 ▶ *The Milky Way Galaxy*

···

IN THIS CHAPTER

You will explore our Milky Way Galaxy, learning about its mass, composition, and structure. You will find that it is a dynamic environment, with stars continually forming and dying. Observations show that mass exists in the Milky Way that astronomers have yet to identify. By exploring our Galaxy, you will gain important insights into the properties of galaxies in general, thereby widening your perspective on the universe and enabling you to ask fundamental questions about the cosmos.

Our Galaxy This wide-angle photograph, taken from Australia, spans 180° of the Milky Way, from the Southern Cross at the left to Cygnus at the right. The center of the Galaxy is in the constellation Sagittarius, in the middle of this photograph. Figure 14-1 is a comparable photograph showing the northern Milky Way.

WHAT DO YOU THINK?

1 Where in the Milky Way is the solar system located?

2 How fast is the Sun moving in the Milky Way?

3 How many stars are in the Milky Way Galaxy?

• • • • • • • • • • • • • • • • •

LIKE THE CURTAIN IN a theater, the Milky Way raises our expectations. What breathtaking natural displays lie beyond the milky white glow of stars and gas? How will we account for them using scientific theories such as Kepler's laws and Einstein's general relativity, which were originally based on worlds near our Sun? Starting with the Milky Way, we shall begin to comprehend the overall structure of the universe, just as the Sun gave us insight into other stars.

The Spiral Structure of Our Galaxy

The Milky Way stretches all the way around the sky in a continuous band that is almost perpendicular to the plane of the ecliptic. Galileo, the first person to look at the Milky Way through a telescope, immediately discov-

ered that it contains countless dim stars. Figure 14-1 is a wide-angle photograph showing roughly half of the Milky Way.

Because it completely encircles us, astronomers in the eighteenth century first began to suspect that the Sun and all the stars in the sky are part of the enormous disk-shaped assemblage we call the Milky Way Galaxy. In the 1780s William Herschel attempted to deduce the Sun's location in the Galaxy by counting the number of stars in 683 regions of the sky. He reasoned that the greatest density of stars should be seen toward the Galaxy's center and a lesser density seen toward the edge. However, *Herschel found roughly the same density of stars all along the Milky Way*. As a result, he concluded that we are at the center of the Galaxy. His conclusion was wrong because he misinterpreted his observations.

14-1 Interstellar dust hides the true extent of the Milky Way

The reason for Herschel's mistake was discovered in the 1930s by R. J. Trumpler: Herschel didn't know about the interstellar gas and dust that were affecting the starlight he saw. While studying star clusters, Trumpler discovered that remote clusters appear dimmer than would be expected just from their distance alone. Trumpler therefore concluded that interstellar space is not a perfect vacuum: It contains dust that absorbs light from

Figure 14-1 **The Milky Way** This wide-angle photograph shows the Milky Way extending from Sagittarius on the left to Cassiopeia on the right. Note the dark lanes and blotches. This mottling is caused by interstellar gas and dust that obscure the light from background stars.

distant stars. Like the stars themselves, this obscuring material is concentrated in the plane of the Galaxy.

Great patches of this interstellar dust are clearly visible in wide-angle photographs, such as Figure 14-1. There is so much dust that the center of our Galaxy is totally obscured from view. Any visual photons from there are absorbed or scattered before they reach us. Herschel therefore was seeing only nearby stars, and he measured apparent magnitudes that were dimmer than they would have been had there been no interstellar dust. Without adjusting for the effects of the dust, he concluded the stars were farther away than they really are. He also had no idea of either the enormous size of the Galaxy or the vast number of stars concentrated around the galactic center and invisible to us.

Because interstellar dust is concentrated in the plane of the Galaxy, the absorption of starlight is strongest in those parts of the sky covered by the Milky Way. Above or below the plane of the Galaxy our view is relatively unobscured. Knowledge of our true position in the Galaxy eventually came from observations of globular clusters in these unobscured portions of the sky.

As we saw in Foundations IV, Henrietta Leavitt published her studies of Type I Cepheids in 1912. Ever since, these variable stars have been important tools in determining distances to the stars. Shortly after Leavitt's discovery, Harlow Shapley, who later debated the existence of other galaxies with Heber Curtis, became very interested in the family of pulsating stars known as RR Lyrae variables, which are quite similar to Cepheid variables. RR Lyrae variables are commonly found in globular clusters (Figure 14-2). We saw in Chapter 11 that RR Lyrae variable stars also have a period–luminosity relationship. Therefore they, too, can be used to determine distances.

Shapley used the period–luminosity relation to determine the distances to the then-known 93 globular clusters in the sky. From their directions and distances, he mapped out the distribution of these clusters in three-dimensional space. By 1915 Shapley had noticed a peculiar property of globular clusters. Most are located in one-half of the sky, widely scattered around the constellation Sagittarius. Figure 14-3 shows two globular clusters in this part of the sky.

By 1917 Shapley had discovered that the globular clusters form a huge spherical system that is not centered on the Earth. The clusters are instead centered about a point in the Milky Way toward Sagittarius. Shapley then made the bold conjecture: *The globular clusters orbit the*

Figure 14-2 **The Globular Cluster M55** The arrows indicate three RR Lyrae variables in this globular cluster located in the constellation of Sagittarius. From the average apparent brightness (as seen in this photograph) and the average true brightness of these stars (known to be roughly 100 $L_\odot$), astronomers have deduced that the distance to this cluster is 20,000 ly.

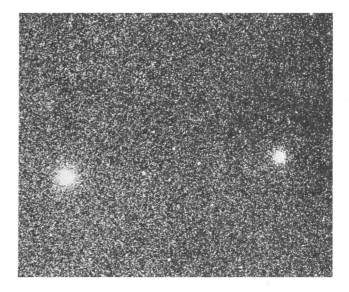

Figure 14-3 **A View Toward the Galactic Center** More than a million stars fill this view, which covers a relatively clear window just 4° south of the galactic nucleus in Sagittarius. Two prominent globular clusters are also seen. Although most regions of the sky toward Sagittarius are thick with dust, there is surprisingly little obscuring matter in this tiny section of the sky.

center of the Milky Way. They therefore outline the true size and extent of the Galaxy. The result of his pioneering research was the realization that the Earth is not at the center of the Galaxy.

14-2 Radio observations help map the galactic disk

Decades after Shapley indirectly found the Galaxy's center, astronomers were able to measure the distance from the Sun to the **galactic nucleus** at the very heart of our Galaxy by detecting radio waves through the interstellar gas and clouds. Observations of gas clouds orbiting the galactic center indicate that this distance is 28,000 ly. Because of their long wavelengths, radio waves easily penetrate the interstellar medium without being scattered or absorbed. (Recall that the Magellan spacecraft used radio waves to pierce the veil of clouds around Venus to create a detailed map of that planet's surface.)

The distance to the galactic nucleus establishes a scale from which the dimensions of other galactic features can be determined. The galactic nucleus is surrounded by a flattened sphere of stars, called the **nuclear bulge,** that is about 20,000 ly in diameter. Swirling out from the nuclear bulge in the plane of the disk are sev-

eral **spiral arms.** The visible **disk** of our Galaxy is about 100,000 ly in diameter and about 2000 ly thick (Figure 14-4). The spherical distribution of globular clusters defines the **halo** of the Galaxy. If we could view our Galaxy edge-on from a great distance, it would probably look somewhat like NGC 4565, shown in Figure 14-5.

Radio observations also reveal details about the concentrations of gas and dust in the Galaxy's spiral arms. Because hydrogen is by far the most abundant element in the universe, astronomers looked for radio emissions from concentrations of hydrogen gas in the disk of the Galaxy. This is useful for detecting gas devoid of the carbon monoxide that usually highlights interstellar regions. Unfortunately, the major electron transitions in the hydrogen atom (see Figure 4-16) produce photons at ultraviolet and visible wavelengths that do not penetrate the interstellar medium. How can radio telescopes detect all this hydrogen? The answer lies in atomic physics.

In addition to mass and charge, particles such as protons and electrons possess a tiny amount of angular momentum commonly called **spin.** An electron or a proton can be crudely visualized as a tiny spinning sphere. According to the laws of quantum mechanics, the electron and proton in a hydrogen atom can only spin in either the same or opposite directions (Figure 14-6); they can have no other spin orientations. If the electron in a

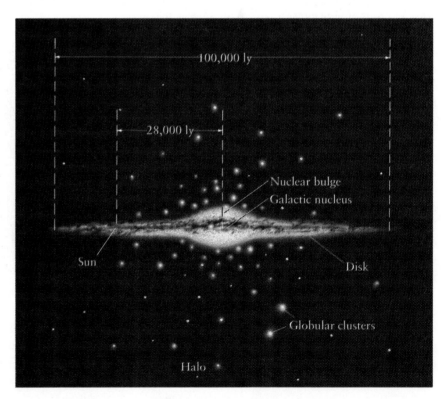

Figure 14-4 **Edge-on View of Our Galaxy** There are three major components to our Galaxy: a thin disk, a nuclear bulge, and a halo. The disk contains gas and dust along with population I (young, metal-rich) stars. The halo is composed almost exclusively of population II (old, metal-poor) stars.

Figure 14-5 **Edge-on View of a Spiral Galaxy** If we could view the Milky Way edge-on from a great distance, it would look like this spiral galaxy in the constellation of Coma Berenices. A thin layer of dust and gas is clearly visible in the plane of the galaxy. Also note the reddish color of the bulge that surrounds the galaxy's nucleus.

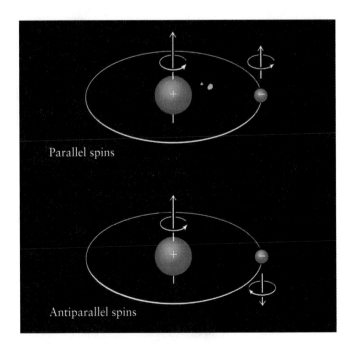

Figure 14-6 **Electron Spin and the Hydrogen Atom** In the lowest orbit of the hydrogen atom, the electron and the proton can spin in either the same or opposite directions. When the electron flips over (reverses its direction of rotation), the atom either gains or loses a tiny amount of energy. This energy is either absorbed or emitted as photons with wavelengths of 21 cm.

hydrogen atom flips from one orientation to the other, the atom must gain or lose a tiny amount of energy. In particular, when going from parallel to opposite spins, the atom emits a low-energy photon whose wavelength is 21 cm. In 1951 a team of astronomers succeeded in detecting the faint hiss of 21-cm radio static from spin flips of interstellar hydrogen.

The detection of **21-cm radio radiation** was a major breakthrough in mapping the galactic disk. To see why, suppose that you aim your radio telescope across the Galaxy as sketched in Figure 14-7. Your radio receiver picks up 21-cm emission from hydrogen clouds at points 1, 2, 3, and 4 (point S is the location of the Sun). However, the radio waves from these various clouds are Doppler shifted by slightly different amounts because they are moving at different speeds as they travel around the Galaxy. These various Doppler shifts smear the 21-cm radiation over a range of wavelengths. Because these radio waves from gas clouds in different parts of the Galaxy arrive at our radio telescopes with slightly different wavelengths, it is possible to identify which radio signals come from which gas clouds and thus produce a map of the Galaxy, such as that shown in Figure 14-8.

Our map reveals numerous arched lanes of neutral hydrogen gas but gives only a vague hint of spiral structure. We need to use even more of the photons from space in order to improve our understanding of the

Figure 14-7 **A Technique for Mapping the Galaxy** Hydrogen clouds at different locations along our line of sight are moving at different speeds. Radio waves from the various gas clouds are therefore subjected to slightly different Doppler shifts, permitting astronomers to sort out the gas clouds and map the Galaxy.

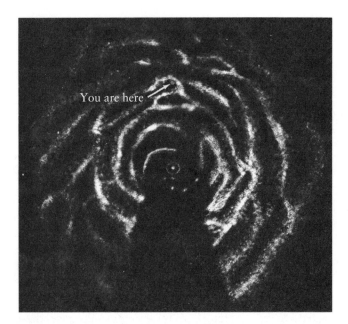

Figure 14-8 **A Map of the Galaxy** This map, based on radio telescope surveys of 21-cm radiation, shows the distribution of hydrogen gas in a face-on view of the Galaxy. Many hints of spiral structure are seen. The Sun's location is indicated by a yellow arrow near the top of the map. The galactic nucleus is marked with a dot surrounded by a circle. Details in the large, blank, wedge-shaped region toward the bottom of the map are unknown because gas in this part of the sky is moving perpendicular to our line of sight and thus does not exhibit a detectable Doppler shift.

Galaxy's spiral arms. Note that photographs of other galaxies (Figure 14-9) show spiral arms outlined by "spiral tracers"—bright, population I stars and emission nebulae. As we saw in Chapter 11, these features indicate active star formation. Thus, another useful way to chart the spiral structure of our Galaxy is to map the locations of star-forming complexes marked by H II regions, giant molecular clouds, and massive, hot, young stars in OB associations.

Dust absorption limits the range of visual observations in the plane of the Galaxy to less than 10,000 ly from the Earth. Nevertheless, there are enough OB associations and H II regions visible in the sky to plot the spiral arms near the Sun. Furthermore, as we saw in Chapter 11 (recall Figure 11-3a), carbon monoxide is a good tracer of molecular clouds. Radio observations of this molecule have recently been used to chart remote regions of the Galaxy. Taken together, all these observations indicate that our Galaxy has at least four major spiral arms and several short arm segments (Figure 14-10a).

a

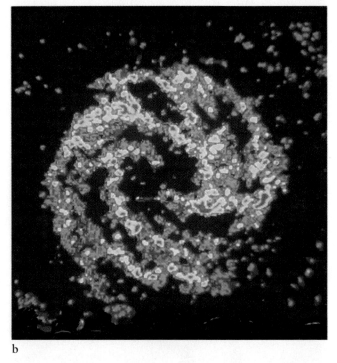

b

Figure 14-9 **A Spiral Galaxy** This galaxy, called M83, is in the southern constellation of Centaurus about 12 million light-years from Earth. (**a**) This photograph at visible wavelengths clearly shows the spiral arms illuminated by young stars and glowing H II regions. (**b**) This radio view at a wavelength of 21 cm shows the emission from neutral hydrogen gas. Note that the spiral arms are more clearly demarcated by hot stars and H II regions than by 21-cm radio emission.

a

b

Figure 14-10 **Our Galaxy Seen Face-on** (a) Our Galaxy has four major spiral arms and several shorter arm segments. The Sun is located near the Orion arm, between two major spiral arms. The Galaxy's diameter is about 100,000 ly, and the Sun is about 28,000 ly from the galactic center. (**b**) If the Milky Way has a bar of stars crossing its nuclear bulge as some recent observations suggest, then it probably looks more like the galaxy NGC 1073 in the constellation Leo.

Recent observations suggest there may also be a bar of stars and gas crossing the nuclear bulge. If so, the Milky Way looks more like the galaxy in Figure 14-10b.

The Sun is located near a relatively short arm segment called the Orion arm, which includes the Orion nebula (see Figure 11-10) and neighboring sites of vigorous star formation in that constellation. Two major spiral arms border either side of the Sun's position. On the side toward the galactic center is the Sagittarius arm, which you see in the summer when you look at the portion of the Milky Way stretching across Scorpius and Sagittarius (see the chapter opening figure). Directed away from the galactic center is the Perseus arm, which is visible in the winter. The remaining two major spiral arms are usually referred to as the Centaurus arm and the Cygnus arm, neither of which can be seen at visible wavelengths because of obscuring dust in the interstellar medium.

14-3 The Galaxy is rotating

Just as the orbital motion of the planets keeps them from falling into the Sun, the motion of the stars and clouds around the galactic center keeps these bodies apart. If the stars and clouds in our Galaxy were not in relative motion, their mutual gravitational forces would have caused them to fall into one massive lump billions of years ago. We would not be here. However, just as detecting the positions of the stars and clouds has been difficult, so, too, has been measuring the orbital motion of the stars and gas, collectively called the *galactic rotation*.

Radio observations of 21-cm radiation from hydrogen gas give important clues about our Galaxy's rotation. By measuring Doppler shifts, astronomers can determine the speed of objects toward or away from us across the Galaxy. These observations clearly indicate that our Galaxy does not rotate like a rigid body but rather exhibits *differential rotation:* Stars at different distances from the galactic center orbit the Galaxy at different speeds.

Further clues to the rotation of the Galaxy come from the motions of stars. Of course, the Sun itself is moving. Because of the stars' differential rotation, the Sun is like a car on a circular freeway with the fast lane on one side and the slow lane on the other. As sketched in Figure 14-11, stars in the fast lane are passing the Sun and thus appear to be moving in one direction, while stars in the slow lane are being overtaken by the Sun and therefore appear to be moving in the opposite direction.

2 Unfortunately, like the 21-cm observations, this study reveals only how fast stars and gas are moving relative to the Sun. To get a complete picture of the Galaxy's rotation, we must find out how fast the Sun itself is traveling around the center of the Galaxy. The Swedish astronomer Bertil Lindblad proposed a method of computing this speed. He noted that not all the stars in the sky move in the orderly pattern shown in Figure 14-11. Globular clusters and stars in the halo of our Galaxy do not participate in the general rotation of the Galaxy, but instead have more or less random motions. Using the average of these random motions as a background, astronomers have estimated that the Sun moves along its orbit about the galactic center at a speed of 230 km/s, or about 828,000 km (half a million miles) per hour.

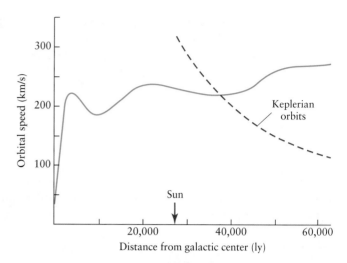

Figure 14-12 **The Galaxy's Rotation Curve** This graph plots the orbital speeds of stars and gas in the Galaxy out to 60,000 ly from the galactic center. The dashed line labeled "Keplerian orbits" indicates how the rotation curve should decline beyond the edge of the Galaxy. The rotation curve continues to rise, proving that vast quantities of dark matter surround the Galaxy.

Given the Sun's speed and its distance from the galactic center, astronomers can calculate the Sun's orbital period. Traveling at 828,000 km per hour, our Sun takes around 230 million years to complete one trip around the Galaxy. This result demonstrates how vast our Galaxy is.

Because we know the true speed of the Sun, we can use the Doppler shifts measured by radio astronomers to determine the actual speeds of the stars. This computation gives us a **rotation curve,** a graph showing the orbital speed of stars and interstellar clouds at various distances from the center of the Milky Way (Figure 14-12).

Knowing the Sun's velocity around the Galaxy from the rotation curve, we use Kepler's third law to estimate the mass of the Galaxy. Putting in the numbers, we obtain a mass of about 1.1×10^{11} $M_\odot$, and that is only the mass *inside* the Sun's orbit around the Galaxy. What could lie beyond?

Mysteries at the Galactic Fringes and Nucleus

Looking at the Milky Way all around the Earth, we know that stars and other matter extend beyond the Sun's orbit. Unfortunately, because this matter does not

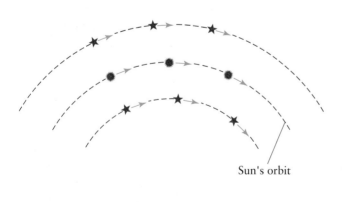

Figure 14-11 **Differential Rotation of the Galaxy** Stars at different distances from the galactic center move at different speeds. As a result, stars on one side of the Sun's orbit about the Galaxy seem to be overtaking us, while stars on the other side seem to be lagging behind.

affect the Sun's motion, its mass cannot be calculated from Kepler's law. In recent years, astronomers have been astonished to discover just how much matter apparently lies beyond the orbit of the Sun.

14-4 Most of the matter in the Galaxy has not yet been identified

According to Kepler's third law, the farther a star or cloud is from the center of the Milky Way, the slower it should be moving, just as the orbital speeds of the planets decrease with increasing distance from the Sun (see the dashed red line on Figure 14-12). In reality, however, galactic orbital speeds continue to *climb* well beyond the visible edge of the galactic disk. This means there must be more gravitational force from the Galaxy acting on the distant stars and clouds than we can see or have taken into account.

A surprising amount of matter must therefore lie beyond the Sun's orbit in the Galaxy. Nearly 90% of the mass of our Galaxy has not been located; the Milky Way's mass could exceed 1×10^{12} M$_\odot$. This total galactic mass translates into roughly 200 billion stars. To make matters even more mysterious, this outlying matter is dark and does not show up on photographs. Some astronomers suspect that this **dark matter** is spherically distributed all around the Galaxy along with the globular clusters in a massive halo. The nature of this dark matter—whether black holes, neutrinos, gas, Jupiterlike bodies, or dim stars—is, so far, a complete mystery. This matter is sometimes called the **missing mass.** This is a poor name since the matter isn't missing; we just don't yet know its location or nature.

14-5 The galactic nucleus is also still poorly understood

The nucleus of our Galaxy is an active, crowded place. The number of stars in Figure 14-3 gives a hint of the stellar congestion there. If you lived on a planet near the galactic center, you could see a million stars as bright as Sirius, the brightest star in our own night sky. The total intensity of starlight from all those nearby stars would be equivalent to 200 of our full Moons. Night would never really fall.

Because of the interstellar absorption of light at visible wavelengths, most of our knowledge about the center of the Galaxy comes from infrared and radio observations. Figure 14-13 shows three infrared views looking toward the nucleus of the Galaxy. Figure 14-13a is a wide-angle view covering a 50° segment of the Milky Way through Sagittarius and Scorpius. The prominent band across this image is a thin layer of dust in the plane of the Galaxy. The numerous knots and blobs along the dust layer are interstellar clouds heated by young O and B stars. Figure 14-13b is an IRAS view of the galactic center. Numerous streamers of dust (in blue) surround it. The strongest infrared emission (in white) comes from **Sagittarius A,** which is a grouping of several powerful sources of radio waves. One of these sources, called Sagittarius A*, is believed to be the galactic nucleus. Figure 14-13c shows stars within 1 ly of Sagittarius A*, with resolution of 0.02 ly.

Radio observations give a different picture of the center of our Galaxy. In 1960 Doppler shift measurements of 21-cm radiation revealed two enormous arms of hydrogen. One arm, which is located between Earth and the galactic nucleus, is approaching us at a speed of 53 km/s. The other arm, on the far side of the galactic nucleus, is receding from us at a rate of 135 km/s. The total amount of hydrogen in these expanding arms is at least several million solar masses. Something quite extraordinary must have happened about 10 million years ago to expel such an enormous amount of gas from the central region of the Galaxy.

In addition to 21-cm radiation from neutral hydrogen gas, astronomers have also detected radio noise coming from the galactic center. This radio emission, which is produced by high-speed electrons spiraling around a magnetic field, is called **synchrotron radiation.** In spite of its small size, Sagittarius A is one of the brightest sources of synchrotron radiation in the entire sky.

Some of the most detailed radio images of the galactic nucleus come from the Very Large Array (VLA). Figure 14-14a is a wide-angle view of Sagittarius A, covering an area about 250 ly across. Huge filaments perpendicular to the plane of the Galaxy stretch 200 ly northward of the galactic disk, then abruptly arch southward toward Sagittarius A. The orderly arrangement of these filaments suggests that a magnetic field may be controlling the distribution and flow of ionized gas, just like magnetic fields on the Sun funnel such gas to create solar prominences.

The inner core of Sagittarius A, shown in Figure 14-14b, covers an area about 30 ly across. A pinwheel-like feature surrounds Sagittarius A* at the center of this view. One arm of this pinwheel is part of a ring of gas and dust orbiting the galactic center.

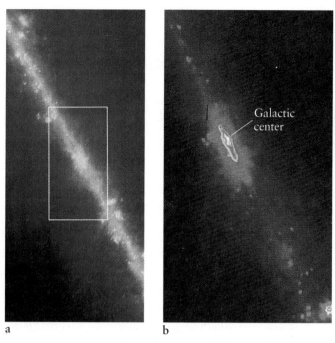

a

b

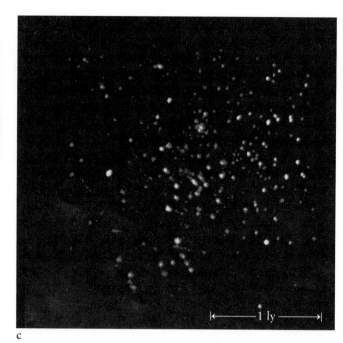

c

Figure 14-13 **The Galactic Center** (a) This wide-angle view at infrared wavelengths shows a 50° segment of the Milky Way centered on the nucleus of the Galaxy. Black represents the dimmest regions of infrared emission, with blue the next dimmest, followed by yellow and red; white represents the strongest emission. The prominent band across this photograph is a layer of dust in the plane of the Galaxy. Numerous knots and blobs along the plane of the Galaxy are interstellar clouds of gas and dust heated by nearby stars. (b) This close-up infrared view of the galactic center covers the area outlined by the white rectangle in (a). (c) This infrared image shows about 300 of the brightest stars less than 1 ly from Sagittarius A*, which is at the center of the picture. The distribution of stars around the galactic center implies a very high density (about a million solar masses per cubic light-year) of less luminous stars.

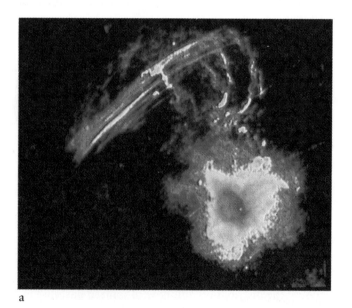

a

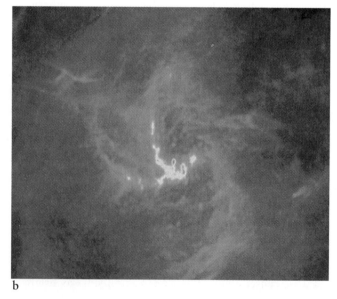

b

Figure 14-14 **Two Radio Views of the Galactic Nucleus** These two pictures show the appearance of the center of our Galaxy at radio wavelengths. The strongest radio emission is shown in red, weaker emission being colored green through blue. (a) This view covers an area of the sky about the same size as the full Moon, corresponding to a distance of 250 ly across. The parallel filaments may be associated with a magnetic field. The galactic nucleus is toward the lower right, at the center of the strongest emission. (b) This high-resolution view shows details of the galactic center covering an area 30 ly across. The pinwheel-like structure is centered on Sagittarius A*.

The motion of the gas clouds orbiting near the galactic nucleus can be deduced from the Doppler shifts of infrared spectral lines. In the late 1970s, for example, astronomers discovered that neon emission lines are severely broadened, perhaps from the orbital speed of the gas around the galactic nucleus. Because the spectra of receding and approaching gases are shifted in opposite directions, the spectral lines are smeared, with a range of wavelengths corresponding to a range of line-of-sight velocities. On one side of the galactic nucleus, gas is coming toward us at speeds up to 200 km/s, but on the other side it is rushing away from us at the same speeds.

Something must be holding this high-speed gas in such tight orbits about the galactic nucleus. Using Kepler's third law, astronomers estimate that 10^6 $M_\odot$ is needed to prevent this gas from flying off into interstellar space. The observed broadening of spectral lines suggests that an object with the mass of a million Suns is concentrated at Sagittarius A*. This object must be extremely compact—much smaller than a few light-years across. Many astronomers argue that an object so massive and compact must be a supermassive black hole. As we saw in Chapter 13, extraordinary activity is also occurring in the nuclei of many other galaxies, which indicates the presence of supermassive black holes at their centers, as well.

However, some astronomers disagree about the presence of a supermassive black hole in the Milky Way. High-resolution infrared views of the galactic center show no indications of the presence of a supermassive black hole. During the coming years, observations from Earth-orbiting satellites as well as from radio telescopes on the ground may elucidate the mysterious core of the Milky Way.

WHAT DID YOU THINK?

1 *Where in the Milky Way is the solar system located?* The solar system is about 28,000 ly from the center of the Galaxy, near the Orion spiral arm.

2 *How fast is the Sun moving in the Milky Way?* The Sun orbits the center of the Milky Way Galaxy at a speed of 828,000 km per hour.

3 *How many stars are in the Milky Way Galaxy?* The Milky Way has about 200 billion stars.

Key Words

dark matter (missing mass)
disk (of a galaxy)
galactic nucleus

halo (of a galaxy)
nuclear bulge
rotation curve (of a galaxy)

Sagittarius A
spin (of an electron or proton)

spiral arm
synchrotron radiation
21-cm radio radiation

Key Ideas

• Our Galaxy has a disk about 100,000 ly in diameter and about 2000 ly thick, with a high concentration of interstellar dust and gas. It contains around 200 billion stars.

• The nucleus of the Milky Way is surrounded by a flattened sphere of stars called the nuclear bulge; the entire Galaxy is surrounded by a spherical distribution of globular clusters called the halo of the Galaxy.

• OB associations, H II regions, and molecular clouds in the galactic disk outline huge spiral arms.

• From studies of the rotation of the Galaxy, astronomers estimate that the total mass of the Galaxy is about 1×10^{12} $M_\odot$, with much of this mass presently undetectable.

• The Sun is located about 28,000 ly from the galactic nucleus, between two major spiral arms. The Sun moves in its orbit at a speed of about 828,000 km per hour and takes about 230 million years to complete one orbit about the center of the Galaxy.

• Interstellar dust obscures our view into the plane of the galactic disk at visual wavelengths. Hydrogen clouds can be detected despite the intervening interstellar dust by the 21-cm radio waves emitted by changes in the relative spins of electrons and protons.

• The galactic nucleus has been studied at infrared and radio wavelengths, which pass readily through intervening interstellar dust and H II regions that illuminate the spiral arms. These observations have revealed many details of the galactic nucleus, but much remains unexplained.

A supermassive black hole with a mass of about 10^6 $M_\odot$ may exist at the galactic center.

Review Questions

1 Why do you suppose that the Milky Way is far more prominent in July than in December?

2 How would the Milky Way appear to us if our solar system were located at the edge of the Galaxy?

3 What observations led Harlow Shapley to conclude that we are not at the center of the Galaxy?

4 Explain why globular clusters spend most of their time in the galactic halo, even though their eccentric orbits take them across the disk of the Galaxy.

5 How do hydrogen atoms generate 21-cm radiation? What do astronomers learn about our Galaxy from observations of that radiation?

6 Why do astronomers believe that vast quantities of dark matter surround our Galaxy?

7 What is synchrotron radiation?

8 Why are there no massive O and B stars in globular clusters?

9 How would you estimate the total number of stars in the Galaxy?

10 What evidence suggests that a supermassive black hole might be located at the center of our Galaxy?

Advanced Questions

11 Why don't astronomers detect 21-cm radiation from the hydrogen in giant molecular clouds?

12 Describe the rotation curve you would get if the Galaxy rotated like a rigid body.

* 13 Approximately how many times has the solar system orbited the center of the Galaxy since the Sun and planets were formed?

14 Compare the apparent distribution of open clusters, which contain young stars, with the distribution of globular clusters relative to the Milky Way. Can you think why open clusters were originally referred to as galactic clusters?

* 15 The visible disk of the Galaxy is about 100,000 ly in diameter and 2000 ly thick. If about five supernovae explode in the Galaxy each century, how often on the average would you expect to see a supernova within 1000 ly of the Sun?

16 Speculate on the reasons for the rapid rise in the Galaxy's rotation curve (see Figure 14-12) at distances close to the galactic center.

Discussion Question

17 What observations would you propose to determine the nature of the hidden mass in our Galaxy's halo?

Observing Project

18 Observe the Milky Way on a clear, moonless night with a high-quality pair of binoculars. During the summer months, look toward the galactic center in Sagittarius and follow the Milky Way arching across the sky through the constellations of Aquila and Cygnus. In the winter, follow the Milky Way as it sweeps across such recognizable constellations as Canis Major, Auriga, and Perseus. Note the details in the structure of the Milky Way. For instance, toward the galactic center you should be able to see a few stretches of the Milky Way that clearly display dark patches and mottling where clouds of interstellar gas and dust have blocked the light from background stars. Enjoy the spectacle.

15 ▶ *Galaxies*

··

IN THIS CHAPTER

You will discover that galaxies are categorized by their shapes. You will learn why some galaxies have spiral arms and how some galaxies devour others in dramatic collisions. You will discover that galaxies are found in clusters and are surrounded by mysterious halos of dark matter. Finally, you will explore the large-scale structure of the universe and its rate of expansion.

A Spiral Galaxy This galaxy, called NGC 2997, is a beautiful example of a two-armed spiral. The yellow color of the nucleus shows that this part of the galaxy is dominated by old, relatively cool stars. The blue color of the arms is caused by light from young, hot stars and from glowing interstellar gas clouds.

WHAT DO YOU THINK?

1 Do galaxies all have spiral arms?

2 Are most of the stars in a spiral galaxy in its arms?

3 Are galaxies isolated objects?

4 Are all galaxies moving away from the Milky Way?

· · · · · · · · · · · · · · · · · · · ·

OFF IN A GALACTIC suburb, far from the center of the action, the Sun is one of several hundred billion stars in the Milky Way. And the Milky Way is but one of billions of galaxies separated by inconceivably vast stretches of intergalactic space. We now move on to develop a global perspective on the universe—the realm of the galaxies and beyond.

Types of Galaxies

There are billions of galaxies beyond our Milky Way. **1** Unlike the Milky Way, most do not have spiral arms. However, despite their incredible number, there is surprising consistency in the overall shapes of most galaxies. Edwin Hubble began cataloging galaxies by their appearance in the 1920s after his measurements of Cepheid variables proved they lie far outside the Milky Way. The **Hubble classification** of galaxies—spirals, barred spirals, ellipticals, irregulars, and their subclasses—is still used today.

Figure 15-1 **A Spiral Galaxy** This galaxy, NGC 1566, is a typical spiral galaxy. It is a highly flattened disk, with spiral arms and a bright nuclear bulge.

15-1 The tightness of a spiral galaxy's arms is correlated to the size of its nuclear bulge

Spiral galaxies (Figure 15-1) are characterized by a nuclear bulge and by arched lanes of stars and glowing interstellar clouds, which appear as spiral arms. The closest spiral galaxy to the Milky Way is the Andromeda galaxy, M31, which is visible to the naked eye as a fuzzy blob in the northern hemisphere (recall Figure IV-3).

Hubble noted that how tightly the spiral arms are wound varies among different spiral galaxies (Figure 15-2). Further, the size of the nuclear bulge varies. Hub-

Sa   **Sb**   **Sc**

Figure 15-2 **Various Spiral Galaxies (Face-on Views)** Edwin Hubble classified spiral galaxies according to the winding of the spiral arms and the size of the nuclear bulge. Three examples are shown here.

a   b   c

Figure 15-3 **Spiral Galaxies Tilted with Respect to the Milky Way** (a) Because of its large nuclear bulge, this galaxy is classified as an Sa. If we could see it face-on, the spiral arms would be tightly wound around a voluminous bulge. (b) Note the smaller nuclear bulge in this Sb galaxy. (c) Note the interstellar dust obscuring the relatively insignificant nuclear bulge of this Sc galaxy.

ble also observed that these two variations are correlated: The tighter the spiral, the larger the nuclear bulge.

Spirals with tightly wound spiral arms (and fat nuclear bulges) are called *Sa galaxies*. Those with moderately wound spiral arms (and a moderate nuclear bulge) are *Sb galaxies*. Finally, loosely wound spirals (with tiny nuclear bulges) are *Sc galaxies*. A typical spiral galaxy contains 100 billion stars and measures 100,000 ly in diameter.

Because not all spiral galaxies are oriented face-on to the Earth, their spiral arms are not always evident. However, we can still classify many of them just from the size of their central bulges. For instance, M104 (Figure 15-3a) must be an Sa galaxy with tightly wound arms because of its huge central bulge. An Sb galaxy (Figure 15-3b) has a smaller central bulge. The tiny central bulge of an Sc (Figure 15-3c) is hardly noticeable at all in this galaxy, which is steeply tilted to our view.

Besides the degree of winding, the overall appearance of individual spiral arms varies from galaxy to galaxy. In some galaxies, called **flocculent spirals** from the word meaning "fleecy," the spiral arms are broad, fuzzy, chaotic, and poorly defined (Figure 15-4a). Other galaxies, called **grand-design spirals**, exhibit beautiful

a   b

Figure 15-4 **Variety in Spiral Arms** The differences in spiral galaxies suggest that at least two mechanisms can create spiral arms. (a) This galaxy has fuzzy, poorly defined spiral arms probably created by self-propagating star formation. (b) This galaxy has thin, well-defined spiral arms probably created by a spiral density wave.

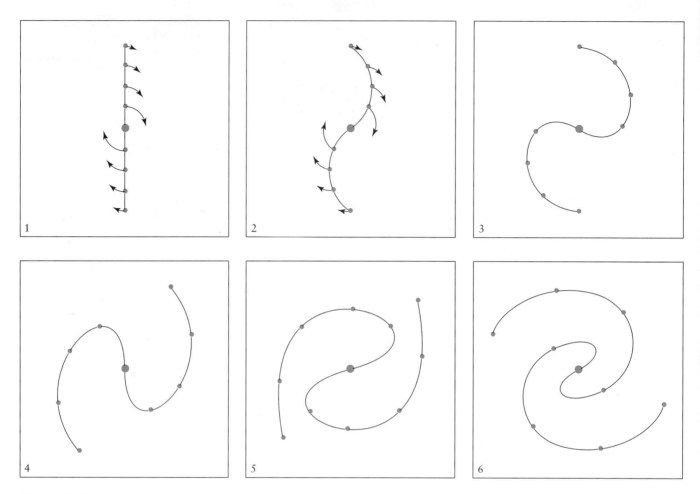

Figure 15-5 **The Winding Dilemma** The rotation curve of our Galaxy indicates that stars farther from the center take longer to go around than do stars closer in. Therefore, if the spiral arms were created by the motion of stars, eventually the arms would become tightly wound. Such winding is not observed in our Galaxy or in other galaxies.

arching arms outlined by brilliant H II regions and OB associations. In these latter galaxies the spiral arms are thin, delicate, graceful, and well-defined (Figure 15-4*b*). These two different types of spiral structures suggest that more than one mechanism gives rise to the creation of a spiral galaxy.

For many years the very existence of spiral arms confounded astronomers. As Figure 14-12 demonstrated, the velocities of stars and gases are fairly constant over a large portion of a galaxy's disk. Consequently, stars farther from the center moving at the same speed as stars closer in should take longer to orbit the galaxy. As a result, the spiral arms should eventually "wind up," wrapping themselves tightly around the nucleus (Figure 15-5). After a few galactic rotations, the spiral structure should disappear altogether. However, there is no evidence of

this happening either in our Galaxy or in other galaxies. This "winding dilemma" suggests that the spiral arms are not the result of the orbiting motions of stars.

15-2 Self-sustaining star formation and spiral density waves produce spiral arms

Imagine stars forming in a dense interstellar cloud somewhere in the disk of a galaxy. The galaxy does not yet have spiral arms. As soon as hot, massive stars form, their radiation compresses nearby interstellar gas, triggering the formation of additional stars in that gas. Massive stars quickly explode as supernovae and produce shock waves, which further compress the surrounding interstellar medium, and so the star-forming region grows. This process of *self-propagating star formation* helps account for spiral arms. The galaxy's differential rotation drags the inner edges of the young region ahead of the outer edges. The newly formed stars soon spread out in the form of a spiral arm highlighted by bright O and B stars and glowing nebulae.

Bursts of star formation come and go more or less at random across a galaxy. Bits and pieces of spiral arms therefore appear only to disappear as older, massive stars die off. This process creates chaotic spiral arms, such as those seen in flocculent galaxies (see Figure 15-4a), but it doesn't create smooth spirals, such as those in grand-design galaxies. So, self-propagating star formation cannot be the whole story. We must seek a completely different explanation for the more stately grand-design spirals. Our search for them begins in a pond.

If you throw a rock into the water, waves are created. As the waves move across the pond's surface, individual water molecules simply bob up and down. No water actually travels with the wave pattern. As you know from experience, the waves move outward in concentric rings from where the rock struck. Suppose the pond is rotating when you throw in a rock. *Now* how do the waves move? According to astronomer Bertil Lindblad, working in the 1920s, they should be spiral-shaped.

Similar waves travel through spiral galaxies. Like water in a pond, interstellar gas sustains ripples in galaxies called **spiral density waves**. They never wind up; instead, they orbit in a rigidly defined spiral pattern. Spiral density waves move more slowly than the gas and stars in the galaxy.

Russian scientists in the 1970s tested Lindblad's theory. Lacking the computer power to simulate galaxies numerically, they threw rocks into pie pans of water rotating on phonograph turntables! The resulting water wave patterns were indeed spirals.

In the mid-1960s the American astronomers C. C. Lin and Frank Shu explained how galactic spiral density waves travelling through the interstellar medium lead to spiral arms. Lin and Shu argued that density waves passing through the disk of a galaxy cause the faster moving interstellar gas and dust to pile up temporarily. A spiral arm is therefore simply a temporary compression like a galactic traffic jam. Imagine workers painting a line down a busy freeway. The cars normally cruise at 65 mph, but the crew of painters causes a bottleneck. The cars slow down temporarily to avoid hitting other cars and the slowly moving paint truck. As seen from the air, there is a noticeable congestion of cars around the painters (Figure 15-6). An individual car spends only a few moments

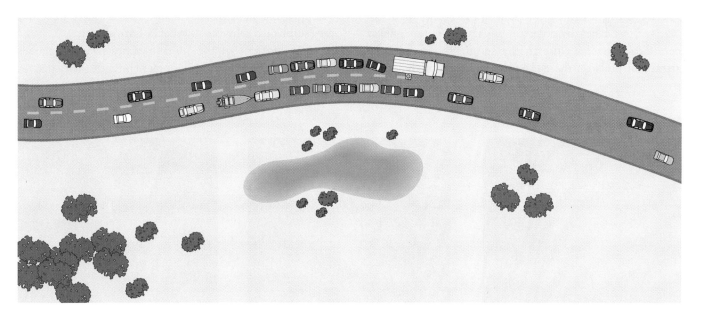

Figure 15-6 **Compression Wave in Traffic Flow** When the normal traffic flow is slowed down, cars bunch together. In a grand-design galaxy, a density wave moves through the stars and gas. The wave is merely a region of slightly denser matter, which, in turn, creates more gravitational force. This compresses the gas and enhances star formation, which outlines the spiral density wave.

in the moving traffic jam before resuming its usual speed, but the traffic jam itself lasts all day long. In much the same way, any interstellar gas and dust is disturbed relatively briefly, but the density wave maintains well-defined spiral arms for much longer.

As the interstellar medium sweeps through the spiral density waves, old gas and dust left behind from dead stars are compressed into new clouds, which then form new metal-rich stars. The sprawling dust lanes in Figure 15-4b attest to the recent passage of a compressional shock wave.

2 We see spiral arms because they contain numerous bright O and B stars and copious quantities of dust and gas that these stars illuminate (see Figure 15-4b). Stars as dim as the Sun contribute virtually nothing to the brightness of spiral arms, which explains why the winding dilemma does not occur in the presence of a spiral density wave: The massive stars highlighting it explode before they finish passing through it. The remaining, longer-lived, lower-mass stars, like our Sun, fill the space between the spiral arms without emitting much light. Indeed, as pronounced as spiral arms may appear, *only about 5% more stars are found there than are found in between the spiral arms.*

It takes an enormous amount of energy to compress the interstellar gas and dust. After a billion years or so, even spiral density waves would begin to fade away. There must be a driving mechanism that keeps them going, such as a nearby companion galaxy. As this galaxy periodically passes close to the spiral, its gravitational attraction pulls on the gas, stars, and dust of the spiral galaxy to generate new density waves. Indeed, grand-design galaxies are usually found in the presence of a companion galaxy.

15-3 Bars of stars run through the nuclear bulges of barred spiral galaxies

Most astronomers believe that the Milky Way is a spiral galaxy. As we have noted, however, some evidence suggests that it is a **barred spiral,** a spiral galaxy with a bar of stars crossing through the nuclear bulge. The arms in barred spirals extend from the ends of the bar rather than from the nuclear bulge itself. Computer models suggest that the bars may arise in galaxies with less dark matter than in normal, or unbarred, spirals.

Hubble found that the winding of the spiral arms in barred spirals again correlates with the size of the nuclear bulge (Figure 15-7). An *SBa galaxy* has a large central bulge (and tightly wound spiral arms). Likewise, a barred spiral with moderately wound spiral arms (and a moderate central bulge) is an *SBb galaxy,* and an *SBc galaxy* has loosely wound spiral arms (and a tiny central bulge). Ordinary spiral galaxies outnumber barred spirals by about 2 to 1.

Although most spiral galaxies have two spiral arms, a sizable minority have more arms. The Milky Way, for example, has at least four arms. Astronomers have not yet established why the numbers of arms vary.

The motions of the stars in many nearby galaxies have been measured from Doppler shifts of their spectra.

| SBa | SBb | SBc |

Figure 15-7 Various Barred Spiral Galaxies As with spiral galaxies, Edwin Hubble classified barred spirals according to the winding of their spiral arms and the size of the nuclear bulge. Three examples are shown here.

E0   E3   E6

Figure 15-8 **Various Elliptical Galaxies** Hubble classified elliptical galaxies according to how round or elongated they are. An E0 galaxy is round; a very elongated elliptical galaxy is an E7. Three examples are shown here.

All spiral and barred spiral galaxies discovered to date have their arms trailing around behind them as they rotate; they are appropriately called **trailing-arm spirals.**

15-4 Elliptical galaxies have a variety of sizes and masses

Elliptical galaxies, named for their distinctive shapes, have no spiral arms. Hubble subdivided elliptical galaxies according to how round or oval they look. The roundest elliptical galaxies are called *E0 galaxies,* while the most elongated ones are *E7 galaxies.* Elliptical galaxies with intermediate amounts of flattening are numbered E1 to E6 (Figure 15-8).

The Hubble scheme classifies galaxies solely by their appearance from our Earth-bound view. Yet whenever we see anything in the sky, we are seeing a two-dimensional view of a three-dimensional object. In the case of most elliptical galaxies, we have no way of knowing anything about the third dimension. What looks like an E0 galaxy (essentially circular) might actually be egg-shaped when viewed from another side. Conversely, an elongated E7 galaxy might look circular when viewed face-on. There is observational evidence that many elliptical galaxies do, indeed, have different appearances from virtually every direction.

Elliptical galaxies look far less dramatic than their spiral and barred spiral cousins, because they contain relatively little interstellar gas and dust. Because stars form in interstellar clouds, few stars should be forming in ellipticals. Observations of spectra confirm that these galaxies contain primarily population II, low-mass, long-lived stars.

Elliptical galaxies exist in an tremendous range of sizes and masses—from the biggest to the smallest galaxies in the universe. Figure 15-9 shows an example of a **giant elliptical galaxy.** This huge galaxy, about 20 times larger than the Milky Way, is located near the middle of

Figure 15-9 **A Giant Elliptical Galaxy** This huge elliptical galaxy sits near the center of a rich cluster of galaxies in the constellation of Coma Berenices. Many normal-sized galaxies surround this giant elliptical, which is about two million light-years in diameter.

Figure 15-10 **A Dwarf Elliptical Galaxy** This nearby galaxy, called Leo I, about 1 million light-years from Earth, is a satellite of the Milky Way and a member of the cluster to which we belong.

a large cluster of galaxies in the constellation of Coma Berenices (Berenice's Hair).

Giant ellipticals, containing some ten trillion solar masses, are rare compared to other types of galaxies, while **dwarf elliptical galaxies** are extremely common. Dwarf ellipticals are only a fraction the size of an average elliptical galaxy and contain so few stars—only a few million—that these galaxies are nearly transparent. Because you can actually see straight through the center of a dwarf elliptical galaxy and out the other side, as shown in Figure 15-10, distant ones are hard to detect. Consequently, many more dwarf ellipticals undoubtedly exist than have been identified.

15-5 The regular types of galaxies can be represented in a tuning fork diagram

Edwin Hubble connected the three regularly shaped types of galaxies—spirals, barred spirals, and ellipticals—in a diagram shaped like a tuning fork (Figure 15-11). According to this scheme, S0 galaxies, called **lenticulars,** are

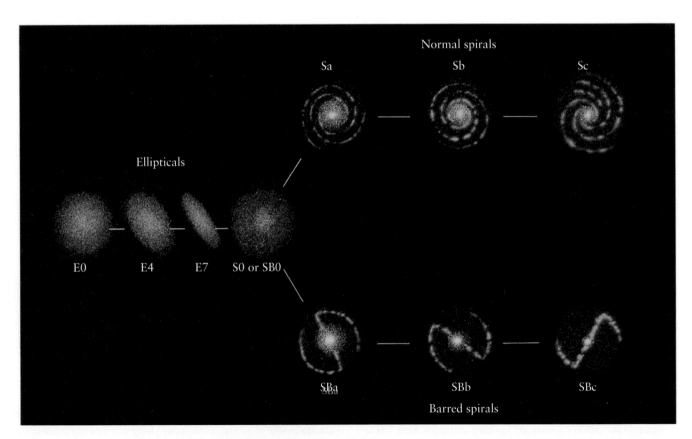

Figure 15-11 **Hubble's Tuning Fork Diagram** Hubble summarized his classification scheme for regular galaxies with this tuning fork diagram. An S0 galaxy is a transitional type between ellipticals and spirals.

Figure 15-12 **The Large Magellanic Cloud (LMC)** At a distance of only 160,000 ly, this irregular galaxy is a nearby companion of our Milky Way Galaxy. Note the huge H II region (called the Tarantula Nebula or 30 Doradus) toward the left side of this image. Its diameter of 800 ly and mass of 5 million Suns makes it the largest known H II region.

Figure 15-13 **The Small Magellanic Cloud (SMC)** The SMC is only slightly farther away from us than the LMC. Because of its sprawling, asymmetrical shape, the SMC is classified as an irregular galaxy. Note that the SMC is rich in young blue stars.

a transition type between ellipticals and the two kinds of spirals. Although they look somewhat like ellipticals, S0 galaxies have both a central bulge and a disk like spiral galaxies, but they lack spiral arms.

Hubble found some galaxies that cannot be classified as spirals, barred spirals, or ellipticals. He called these **irregular galaxies.** Examples include the Large Magellanic Cloud (LMC) and the Small Magellanic Cloud (SMC), both of which can be seen with the naked eye from southern latitudes and are among the nearest galaxies to the Milky Way. For want of any better idea, the irregular galaxies are sometimes placed between the ends of the tuning fork tines.

Telescopic views of the Magellanic clouds easily distinguish individual stars (Figures 15-12 and 15-13). Note that the SMC does not exhibit any of the geometric symmetry characteristic of spirals or ellipticals and is therefore a true irregular. However, the LMC does have a vague barlike structure.

The Hubble diagram suggests that one type of galaxy may change into another. This idea has been in and out of favor with astronomers for decades. As we shall see later in this chapter, extensive observations by the Hubble Space Telescope and by ground-based telescopes now suggest that interactions between galaxies do, indeed, lead to changes in their structures.

Clusters and Beyond

15-6 Galaxies are grouped in clusters and superclusters

Galaxies are not scattered randomly throughout the universe but are grouped in **clusters.** One typical cluster, called the Fornax cluster because it is located in the constellation of Fornax (the Furnace), is seen in Figure 15-14. Observations of galactic motion reveal that members of a cluster of galaxies are *gravitationally bound* together: They orbit each other and occasionally even collide.

Clusters of galaxies are themselves grouped together in huge associations called **superclusters.** A typical supercluster contains dozens of individual clusters spread over a volume up to 100 million light-years across. In Figure 15-15, which covers many hundreds of square degrees as seen from Earth, note the delicate, filamentary structure spread across the sky.

The arrangement of superclusters in space was clarified in the early 1980s when astronomers began discovering enormous **voids** between superclusters where exceptionally few galaxies are found. These voids are roughly spherical and measure 100 million to 400 million light-years in diameter. Recent surveys reveal that

Figure 15-14 **A Cluster of Galaxies** This cluster of galaxies, called the Fornax cluster, is about 60 million light-years from Earth. Both elliptical and spiral galaxies are easily identified. The barred spiral galaxy at the lower left is NGC 1365, the largest and most impressive member of the cluster.

Figure 15-15 **Two Million Galaxies** This map shows the distribution of roughly two million galaxies over about 10% of the sky. Each dot is shaded according to the number of galaxies it contains. White dots indicate more than 20 galaxies, blue dots between 1 and 19 galaxies; there are no galaxies in dark blue areas. The clusters form a lacy, filamentary structure. The larger elongated bright areas are superclusters and filaments, which generally surround darker voids containing few galaxies. Statistical analysis of this map shows that galaxies clump together on distance scales up to about 150 million light-years. Over distances greater than this, the distribution of galaxies in the universe appears to be roughly uniform.

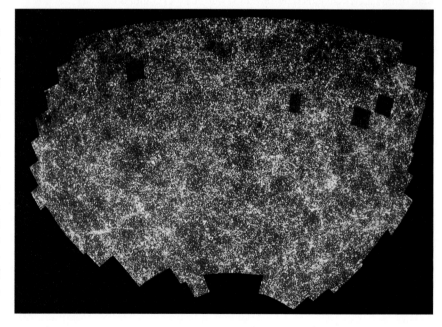

galaxies are concentrated on the surfaces between these voids (Figure 15-16). One notable sheet of galaxies, shown as a ribbon in the two-dimensional Figure 15-16, has been dubbed the **Great Wall.** The arc of this "wall of galaxies" runs for some 650 million light-years. Like a sheet of paper seen edge-on, the Great Wall also has depth, extending toward us at least 225 million light-years. The distribution of clusters of galaxies throughout the universe is therefore said to be "sudsy" because it resembles a collection of giant soap bubbles. Galaxies surround voids in the same way that bubbles are concen-

trated on the surface of soap film. Many astronomers suspect that this sudsy pattern contains important clues about conditions shortly after the Big Bang that led to the formation of superclusters of galaxies.

15-7 Clusters of galaxies can be densely or sparsely populated

A cluster of galaxies is said to be either **poor** or **rich,** depending on how many galaxies it contains. For example,

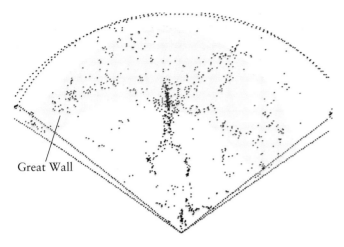

Figure 15-16 **A Slice of the Universe** This map shows the locations of 1099 galaxies in a thin slice of the universe. To produce this map, a team of astronomers painstakingly measured the redshifts of galaxies out to a distance of nearly a billion light-years from Earth in a strip of sky 6° wide and 153° long. Note that most galaxies surround voids that are roughly circular. This distribution of galaxies is like a slice cut through the suds in a kitchen sink.

the Milky Way Galaxy, the Andromeda galaxy, and the Large and Small Magellanic clouds belong to a poor cluster called the **Local Group.** The Local Group contains some 30 galaxies, over a third of which are dwarf ellipticals. Figure 15-17 shows a map of the Local Group; Figure 15-18*a* shows a close-up view of one of these galaxies, which was discovered in 1994. This

galaxy had been hidden from view until now by the obscuring gas and dust of our own Galaxy. Located in the constellation of Sagittarius just 50,000 ly from the center of the Milky Way, this newly discovered member of the Local Group is our closest known neighbor—closer even than the Large Magellanic Cloud. The new galaxy appears to be coming apart under the tidal forces from our own Galaxy, which will absorb it over the next hundred million years.

Another galaxy discovered in 1994 in the same region of the sky is 10 million light-years from the Milky Way. Named Dwingeloo 1 for the radio telescope that located it, this galaxy is in a nearby cluster (Figure 15-18*b*). It is highly likely that other nearby galaxies will be discovered in the coming years.

The nearest fairly rich cluster is the Virgo cluster. It is a sprawling collection of over 1000 galaxies covering a 10° × 12° area of the sky. Cepheids in the spiral galaxy M100 in Virgo indicate that this cluster is about 50 million light-years away from the Milky Way. The overall diameter of the Virgo cluster is about 7 million light-years.

Three giant elliptical galaxies dominate the center of the Virgo cluster; two of them appear in Figure 15-19. These enormous galaxies are two million light-years in diameter, 20 times as large as an ordinary elliptical or spiral galaxy. In other words, one of these giant ellipticals is nearly the same size as the entire Local Group!

We also categorize clusters of galaxies as regular or irregular, depending on their shapes. A **regular cluster** is distinctly spherical, with a marked concentration of

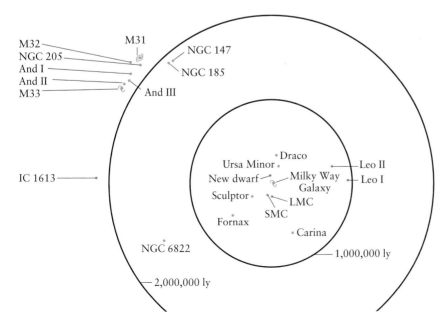

Figure 15-17 **The Local Group** Our Galaxy belongs to a poor, irregular cluster consisting of about 30 galaxies, a dozen of which are dwarf ellipticals. The Andromeda galaxy (M31) is the largest and most massive galaxy in the Local Group. The second largest is the Milky Way itself. M31 and the Milky Way are each surrounded by a dozen satellite galaxies. The newly discovered dwarf galaxy near the Milky Way has yet to be named.

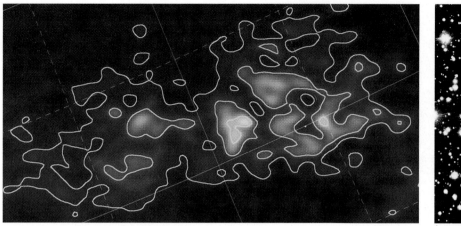

a

b

Figure 15-18 **Recently Discovered Galaxies** Hidden by the gas and dust in the Milky Way, these two galaxies were only recently discovered. (a) This optical image shows a dwarf galaxy in the Local Group that is being consumed by the Milky Way. It is presently 50,000 ly from the Milky Way, making it the closest known companion to our Galaxy. (The curved lines em-phasize the galaxy's structure and highlight areas of equal brightness. The grid lines orient the galaxy in right ascension and declination.) (b) This visible light image shows Dwingeloo 1, a spiral galaxy located 10 million light-years from the Milky Way in a nearby cluster of galaxies. Unlike the galaxy in (a), Dwingeloo 1 is too far away to be destroyed by the Milky Way.

galaxies at its center. Their numerous gravitational interactions over the ages have spread the galaxies into their distinctive spherical distribution. The Virgo cluster, in contrast, is an **irregular cluster** because its galaxies are more randomly scattered about a sprawling region of the sky.

Figure 15-19 **The Center of the Virgo Cluster** This fairly rich cluster lies about 50 million light-years from us. Only the center of this huge cluster appears in this view. Note the two giant elliptical galaxies (M84 and M86).

Figure 15-20 **The Coma Cluster** This rich, regular cluster containing thousands of galaxies is about 300 million light-years from Earth. Such regular clusters are composed mostly of elliptical and S0 galaxies and are common sources of x rays. This Hubble Space Telescope image is a mosaic of 16 photographs (part of the field is missing because of the way the cameras are arrayed on the telescope). The brightest member of the Coma cluster shown here is the elliptical galaxy NGC 4881.

Figure 15-21 **The Hercules Cluster** This irregular cluster, which is about 700 million light-years from Earth, contains a high proportion of spiral galaxies, often associated in pairs and small groups.

The nearest example of a rich, regular cluster is the Coma cluster, located 300 million light-years from us in the constellation of Coma Berenices (Figure 15-20). Despite its great distance, more than 1000 bright galaxies within it are easily visible on photographic plates. Certainly the Coma cluster must contain many thousands of dwarf ellipticals soon to be detected by the Hubble Space Telescope—maybe as many as 10,000 galaxies overall. The core of the Coma cluster is dominated by two giant ellipticals surrounded by many normal-sized galaxies.

Rich, regular clusters like the Coma cluster contain mostly elliptical and S0 galaxies. Only 15% of the Coma cluster's galaxies are spirals and irregulars. Irregular clusters, such as the Virgo cluster and the Hercules cluster (Figure 15-21), have a more even mixture of galaxy types. Two-thirds of the 200 brightest galaxies in the Hercules cluster are spirals, 19% are ellipticals, and the rest are irregulars.

15-8 Galaxies in a cluster can collide and combine

The galaxies in a cluster all orbit a common center of mass. Occasionally two galaxies pass through one another, and their stars pass by each other (Figure 15-22). However, *there is so much space between the stars that the probability of two stars crashing into each other is extremely small*. On the other hand, the galaxies' huge clouds of interstellar gas and dust are so large that they do collide, slamming into each other and producing

strong shock waves. The colliding interstellar clouds are stopped in their tracks. The collision can bring galaxies together—or hurl their stars far into space.

A really violent collision can strip both galaxies of their interstellar gas and dust, while the stars of each keep right on going. The violence of the collision heats the gas stripped from these galaxies to extremely high temperatures. This process may be a major source of the hot **intergalactic gas** often observed in rich, regular clusters.

In a less violent collision between two galaxies or in a near miss, the compressed interstellar gas may have enough time to cool sufficiently to allow many stars to form. Such collisions can thus stimulate prolific star formation, which may account for the **starburst galaxies** that blaze with the light of numerous newborn stars. These galaxies are characterized by bright centers surrounded by clouds of warm interstellar dust, indicating a recent, vigorous episode of star birth (Figure 15-23). The warm dust is so abundant that starburst galaxies are among the most luminous objects in the universe at infrared wavelengths.

The starburst galaxy M82 in Figure 15-23 is one member of a nearby cluster that includes the beautiful

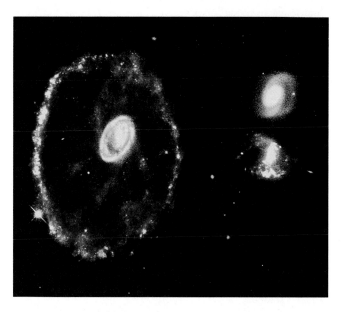

Figure 15-22 **Cartwheel Galaxy** This ring-shaped galaxy 500 million light-years from Earth is the result of another galaxy, probably the blue-white one on the right, having passed through the middle of the larger one. The passage sent shock waves through the Cartwheel, creating the circular structure. The shock waves also stimulated a burst of star formation, creating many bright blue and white stars. The spiral galaxy on the upper right of the image is probably in the background.

Figure 15-23 **A Starburst Galaxy** Prolific star formation is occurring at the center of this galaxy, called M82, located about 120 million light-years from Earth. This activity was probably triggered by tidal interactions with neighboring galaxies. Note the turbulent appearance of the interstellar gas and dust around the galaxy's center.

spiral galaxy M81 and a fainter companion called NGC 3077 (Figure 15-24a). A recent radio survey of that region of the sky revealed enormous streams of hydrogen gas connecting the three galaxies (Figure 15-24b). The loops and twists in these streamers suggest that the three galaxies have had several close encounters over the ages. Incidentally, a similar stream of hydrogen gas connects our Galaxy with its neighbor, the Large Magellanic Cloud.

Gravitational interaction between colliding galaxies can also hurl thousands of stars out into intergalactic space along huge, arching streams. These shapes have been dramatically illustrated in computer simulations like the one shown in Figure 15-25. Note the remarkable similarity between this simulation and the colliding galaxies shown in Figure 15-26.

In a rich cluster there must be many near-misses between galaxies. If galaxies are surrounded by extended halos of dim stars, these near-misses could strip the

a

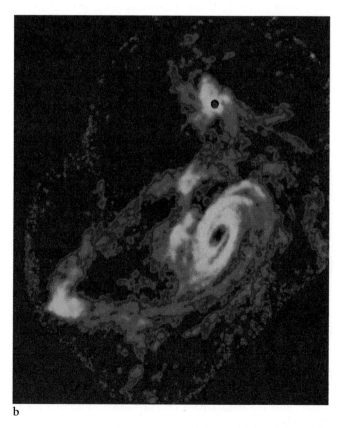

b

Figure 15-24 **The M81/M82/NGC 3077 Cluster** The starburst galaxy M82 is in a nearby cluster whose three members are connected by streamers of hydrogen gas. (a) This photograph shows the three galaxies at visual wavelengths. (b) This radio image, created from data taken by the Very Large Array, shows the streamers of hydrogen gas that connect the three galaxies.

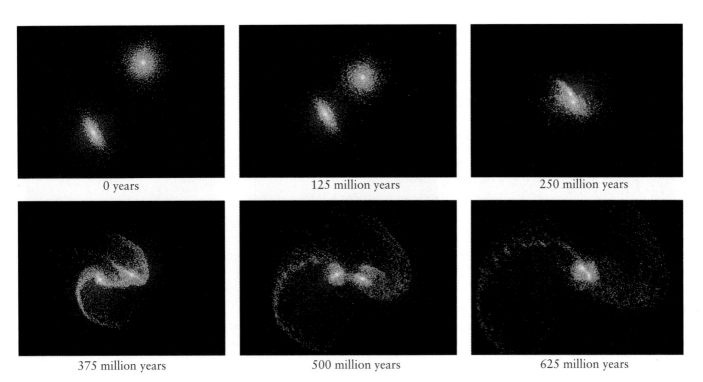

0 years

125 million years

250 million years

375 million years

500 million years

625 million years

Figure 15-25 A Simulated Collision Between Two Galaxies These frames from a computer simulation show the collision and merging of two disk-shaped galaxies accompanied by ejection of many stars into intergalactic space. Stars in the disk of each galaxy are colored blue, while stars in their central bulges are yellow. Red indicates dark matter that surrounds each galaxy. Compare the frame at 625 million years with the photograph of NGC 2623 in Figure 15-26.

Figure 15-26 A Colliding Pair of Galaxies with Antennas Many pairs of colliding galaxies exhibit long antennas of stars ejected by the collision. This particular system, called NGC 2623, is also a significant source of radio radiation. Supercomputer simulations, like the one shown in Figure 15-25, give important insights into possible histories of such systems.

galaxies of these outlying stars. In this way, a loosely dispersed sea of dim stars might come to populate the space between galaxies in a cluster. One of the projects assigned to the Hubble Space Telescope is to search for these dim stars in extended halos and in intergalactic space.

After galaxies collide, some stars are flung far and wide, scattering material into intergalactic space. However, other stars slow down, causing the remnants of the galaxies to merge. Several dramatic examples of **galactic mergers** have been recently discovered (Figure 15-27). Astronomers also speak of **galactic cannibalism,** which occurs when a large galaxy captures and "devours" a smaller one. Cannibalism differs from mergers in that the dining galaxy is significantly bigger than its dinner, whereas merging galaxies are about the same size.

Many astronomers suspect that giant ellipticals are the product of galactic cannibalism. As we have seen, giant galaxies typically occupy the centers of rich clusters. Often, smaller galaxies are located around these giants (see Figure 15-9). As they pass through the extended

Figure 15-27 **Merging Galaxies** This contorted object in the constellation of Ophiuchus consists of two spiral galaxies in the process of merging. The collision between the two galaxies has triggered an immense burst of star formation.

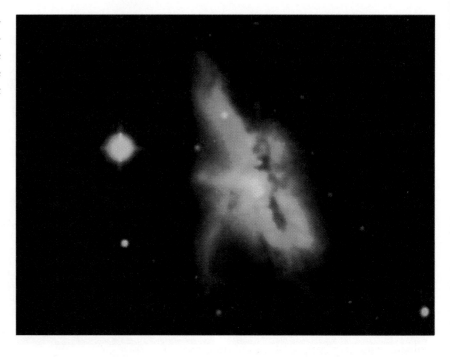

halo of a giant elliptical, these smaller galaxies slow down and are eventually consumed by the larger galaxy.

Figure 15-28 shows a computer simulation in which a large, disk-shaped galaxy devours a small satellite galaxy. The large galaxy consists of 90% stars (in blue) and 10% gas (in white) by mass. It is surrounded by a halo of dark matter having a mass about 3.3 times that of the disk. The satellite galaxy, which has a tenth of the

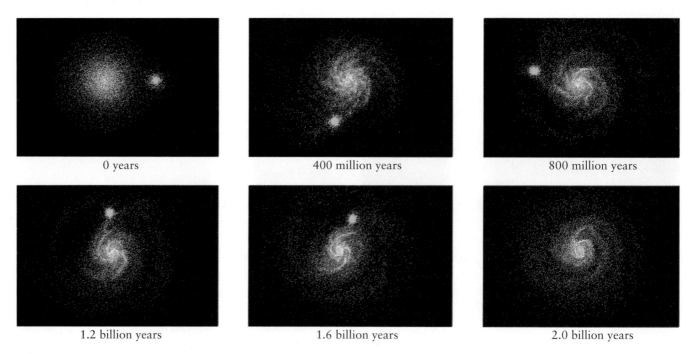

| 0 years | 400 million years | 800 million years |

| 1.2 billion years | 1.6 billion years | 2.0 billion years |

Figure 15-28 **Simulated Galactic Cannibalism** This simulation, performed at the Pittsburgh Supercomputing Center, shows a small galaxy (stars in orange) being devoured by a larger, disk-shaped galaxy (stars in blue, gas in white). The pictures display progress at 400-million-year intervals. Note how spiral arms are induced in the disk galaxy by its interaction with the satellite galaxy.

mass of the large galaxy, contains only stars (in orange). Initially the satellite is in circular orbit about the large galaxy. Note that spiral arms appear in the large galaxy as the collision proceeds. Two billion years elapse as the satellite spirals in toward the core of the large galaxy. Although much material is stripped from the satellite, most of its stars plunge into the nucleus of the large galaxy.

15-9 Galactic halos may account for some "missing mass" in the universe

What prevents the rapidly moving galaxies in clusters and superclusters from wandering away from each other? There must be sufficient matter to provide the necessary gravitational force to bind galaxies together in clusters, and clusters into superclusters. However, *no cluster or supercluster of galaxies contains enough visible matter to stay bound together.* Recall that a similar problem arose in Chapter 14 regarding individual galaxies. A lot of nonluminous matter must be scattered about each cluster of galaxies, or else the galaxies would have long ago wandered away in random directions and the clusters would no longer exist today. Recall that this unexplained material is called dark matter, and the puzzle of its existence is known as the missing-mass problem. Analyses demonstrate that the total mass needed to bind a typical rich cluster is 10 times greater than the mass of material that shows up on visible-light photographs.

Astronomers using x-ray telescopes have recently solved part of this mystery. Satellite observations have revealed x rays pouring from the space between galaxies in rich clusters. This radiation is emitted by substantial amounts of hot intergalactic gas at temperatures between 10 and 100 million kelvins. The mass of this hot intergalactic medium is typically as great as the combined mass of all the visible galaxies in a rich cluster.

Extensive evidence supports the idea of extended halos surrounding galaxies. Many galaxies have rotation curves similar to that of our Milky Way (recall Figure 14-12). These rotation curves remain remarkably flat out to surprisingly great distances from the galaxies' centers. For example, Figure 15-29 shows the rotation curves of four spiral galaxies. In all cases, the orbital speed is fairly constant, even out where stars and nebulae are too dim and widely scattered for reliable measurements. According to Kepler's third law, we should see a decline in orbital speed in the outer portions of a galaxy. The implication is that we have still not detected the true edge of

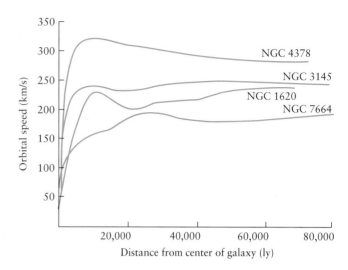

Figure 15-29 **The Rotation Curve of Four Spiral Galaxies** This graph shows the orbital speed of material in the disks of four spiral galaxies. Many galaxies have flat rotation curves, indicating the presence of extended halos of dark matter.

these and many similar galaxies. Because this decline has not been observed in many galaxies, astronomers conclude that there must be a considerable amount of dark matter extending well beyond the visible portion of a galaxy's disk.

Identifying the nature of this unseen matter is one of the most important goals in modern astronomy. As we will see in Chapter 17, the total amount of this matter will determine the fate of the universe.

15-10 The redshifts of remote superclusters indicate that the universe is expanding

Whenever an astronomer finds an object in the sky that can be photographed, one of the first tasks is to study its composition by attaching a spectrograph to a telescope and recording its spectrum. As long ago as 1914, V. M. Slipher, working at the Lowell Observatory in Arizona, took spectra of "spiral nebulae." He was surprised to discover that the spectral lines of 11 of the 15 spiral nebulae he studied were substantially redshifted. This redshift indicated that they are moving away from us at significant speeds. This marked dominance of redshifts was presented by Curtis in the Shapley–Curtis debate as evidence that the "spiral nebulae" could not be part of our Milky Way Galaxy.

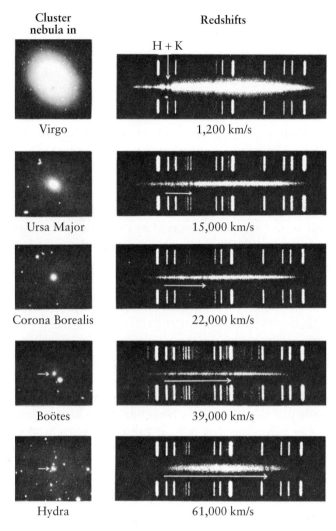

Cluster nebula in   **Redshifts**

H + K

Virgo — 1,200 km/s

Ursa Major — 15,000 km/s

Corona Borealis — 22,000 km/s

Boötes — 39,000 km/s

Hydra — 61,000 km/s

Figure 15-30 **Five Galaxies and Their Spectra** The photographs of these five elliptical galaxies all have the same magnification. They are labeled according to the constellation in which each galaxy is located. The spectrum of each galaxy is the hazy band between the comparison spectra. In all five cases, the so-called H and K lines of calcium are seen. The recessional velocity (calculated from the Doppler shifts of the H and K lines) is given below each spectrum. Note that the fainter—and thus more distant—a galaxy is, the greater is its redshift.

During the 1920s Edwin Hubble and Milton Humason photographed the spectra of many galaxies with the 100-in. telescope on Mount Wilson. Using techniques such as the brightness of Cepheid variables, Hubble estimated the distances to a number of galaxies. Using the Doppler effect (recall Figure 4-17), Hubble calculated the speed at which each galaxy is receding from us. Five representative elliptical galaxies and their spectra are

shown in Figure 15-30. As indicated by this illustration, there seems to be a direct correlation between the distance to a galaxy and the size of its redshift: Galaxies in distant superclusters are moving away from us more rapidly than galaxies in nearby superclusters. This recessional motion pervades the universe and is now called the **Hubble flow.**

When Hubble plotted the data on a graph of distance versus speed, he found that the points lie nearly along a straight line. Figure 15-31 is a modern version of Hubble's 1929 graph. It is important to note that this result does not apply to the galaxies in the Local Group or to the other clusters in our local supercluster, because these galaxies and clusters are all bound together by mutual gravitational attraction. Hubble flow implies only that the superclusters of galaxies are moving away from each other.

This relationship between the distances to galaxies and their redshifts is one of the most important astronomical discoveries of the twentieth century. It tells us that we are living in an expanding universe. The **Hubble law,** which gives the speed of the expanding universe, is most easily stated as a formula:

$$\text{Recessional velocity} = H_0 \times \text{distance}$$

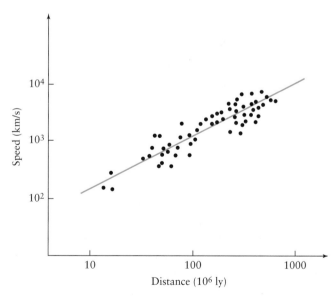

Figure 15-31 **The Hubble Law** The distances and recessional velocities of 60 Sc spiral galaxies are plotted on this graph. The straight line is the "best fit" for the data. This linear relationship between distance and speed is called the Hubble law.

The Hubble Law

The Hubble law describes our expanding universe. It is a simple formula:

$$v = H_0 r$$

where v is the recessional velocity, r is the distance, and H_0 is the Hubble constant. It is the formula for the straight line displayed in Figure 15-31, and the Hubble constant is the slope of this line.

As Figure 15-31 shows, we usually measure the velocity in kilometers per second and the distance in megaparsecs. Different techniques for estimating the distances to remote galaxies currently lead to differing estimates

for H_0, ranging from 50 km/s/Mpc (read "50 kilometers per second per megaparsec") to 90 km/s/Mpc.

Suppose that the Hubble constant is 75 km/s/Mpc, the value that we shall generally use in this book. Then a galaxy 1 Mpc from us is moving away at a rate of

$$v = (75 \text{ km/s/Mpc}) \times (1 \text{ Mpc}) = 75 \text{ km/s}$$

A galaxy 2 Mpc away is therefore receding at 150 km/s, and so on. A galaxy located 100 million parsecs from Earth should be rushing away from us with a speed of 7500 km/s! Uncertainty in the value of H_0 obviously means a very great uncertainty in the distances and motions of remote galaxies.

where H_0 is a constant, commonly called the **Hubble constant.** Toolbox 15-1 shows how the Hubble law works.

The exact value of the Hubble constant is a topic of active research and extremely heated debate among astronomers. For this book, we will use the value that is roughly the average of all present data:

$$H_0 = 75 \text{ km/s/Mpc}$$

To determine the Hubble constant, astronomers must measure the redshifts and distances to many galaxies. Although redshift measurements can be quite precise, it is very difficult to measure the distances to remote galaxies accurately. Indeed, conflicting measurements of distance are the main reason that the value of H_0 is so controversial.

Recall from the discussion in Foundations III that we can always find the distance to an object if we know both its apparent and absolute magnitudes. Astronomers use the term **standard candle** to denote any object whose absolute magnitude is known. Cepheid variables, the brightest supergiants, globular clusters, and supernovae are all useful as standard candles because astronomers know something about their absolute magnitudes. To determine the distance to a galaxy, an astronomer must measure the apparent magnitude of one or more of these standard candles in that galaxy. As soon as both the apparent and absolute magnitudes are known, the distance to the galaxy can be easily calculated.

The distances to nearby galaxies can be determined by fairly reliable methods. For example, with the Hubble Space Telescope, Cepheid variable stars can now be seen out to 200 Mly from Earth. The distances to galaxies in this nearby volume of space can thus be determined from the period–luminosity law.

Beyond 200 Mly, even the brightest Cepheid variables, which have absolute magnitudes of about −6, are not visible with current technology. Astronomers then turn to more luminous stars, such as the brightest red and blue supergiants, which can be seen out to distances of 500 million and 800 million light-years, respectively. Thus, out to these limits, we can determine the distances to galaxies from the apparent magnitudes of these luminous supergiants.

When individual stars are no longer discernible, astronomers turn to the light from entire star clusters and gas clouds. The brightest globular clusters can be seen out to 1.3 billion light-years from Earth, and the brightest H II regions out to 3 billion light-years. From the apparent brightness of these clusters and nebulae, distances to remote galaxies have been estimated that suggest a value for the Hubble constant of 55 km/s/Mpc.

One way to measure the distance to the most remote galaxies is to measure the brightness of supernova explosions within them. The brightest supernovae reach an absolute magnitude of −19 at the peak of their outbursts (Figure 15-32). These brilliant outbursts can be seen out to distances over 10 billion light-years away.

Figure 15-32 **A Supernova in M61** In 1961 a supernova was discovered in this spiral galaxy, which is a member of the Virgo cluster. Supernovae can be seen in extremely remote galaxies and are important standard candles used to determine the distances to these faraway galaxies.

An intriguing and controversial method for determining distances was developed in the 1970s by astronomers Brent Tully and Richard Fisher. It is controversial because the Hubble constant it predicts is uncomfortably large for most theories of the universe. They discovered that the width of the hydrogen 21-cm emission line (discussed in Chapter 14) of a spiral galaxy is related to the galaxy's absolute magnitude: The broader the line, the brighter the galaxy. This correlation is called the **Tully–Fisher relation.**

Since line widths can be measured quite accurately, astronomers can use the Tully–Fisher relation to determine the luminosities (absolute magnitudes) of spiral galaxies and thus their distances. Measurements of 21-cm-line widths for distant, receding galaxies yields a Hubble constant of 88 km/s/Mpc. If this value is correct, we may need to change our estimate of the age of the universe.

The major obstacle in determining the Hubble constant is that the farther we look into space, the fewer standard candles we have. Consequently, it becomes less possible to double-check estimated distances to remote objects. Unfortunately, remote galaxies are precisely the objects whose distances we must determine to find the value of the Hubble constant.

By refining the standard candles through more observations with instruments like the Hubble Space Telescope (HST) and the new, large ground-based instruments such as the Keck telescopes, we hope to determine once and for all the true value for H_0. This advance will contribute significantly to our understanding of the structure and evolution of galaxies and the universe.

WHAT DID YOU THINK?

1 *Do galaxies all have spiral arms?* No, galaxies may be either spiral, barred spiral, elliptical, or irregular. Only spirals and barred spirals have arms.

2 *Are most of the stars in a spiral galaxy in its arms?* No, the spiral arms contain only 5% more stars than the regions between the arms.

3 *Are galaxies isolated objects?* No, galaxies are grouped in clusters, and clusters are grouped in superclusters.

4 *Are all galaxies moving away from the Milky Way?* No, only the galaxies in other superclusters are necessarily receding from us.

Key Words

barred spiral galaxy	galactic cannibalism	Great Wall	intergalactic gas
cluster (of galaxies)	galactic merger	Hubble classification	irregular cluster
dwarf elliptical galaxy	giant elliptical galaxy	Hubble constant	(of galaxies)
elliptical galaxy	grand-design spiral	Hubble flow	irregular galaxy
flocculent spiral galaxy	galaxy	Hubble law	lenticular galaxy

Local Group   rich cluster   starburst galaxy   Tully–Fisher
poor cluster   (of galaxies)   supercluster   relationship
   (of galaxies)   spiral galaxy   (of galaxies)   void
regular cluster   spiral density wave   trailing-arm spiral
   (of galaxies)   standard candle

Key Ideas

• The Hubble classification system groups galaxies into four major categories: spiral, barred spiral, elliptical, and irregular.

Spiral galaxies and barred spiral galaxies are sites of active star formation.

According to the theory of self-propagating star formation, spiral arms of flocculent galaxies are caused by the births and deaths of stars over extended regions of a galaxy. Differential rotation of a galaxy stretches the star-forming regions into elongated arches of stars and nebulae we see as spiral arms.

According to the spiral density wave theory, spiral arms of grand-design galaxies are caused by density waves. The gravitational field of a spiral density wave compresses the interstellar clouds through which it passes, thereby triggering the formation of stars including OB associations, which highlight the arms.

• Elliptical galaxies contain much less interstellar gas and dust than do spiral galaxies; little star formation is occurring in elliptical galaxies.

• Galaxies are grouped in clusters rather than scattered randomly through the universe.

A rich cluster contains hundreds or even thousands of galaxies; a poor cluster may contain only a few dozen.

A regular cluster has a nearly spherical shape with a central concentration of galaxies; in an irregular cluster, the distribution of galaxies is asymmetrical.

Our Galaxy is a member of a poor, irregular cluster called the Local Group.

Rich, regular clusters contain mostly elliptical and S0 galaxies; irregular clusters contain more spiral and irregular galaxies.

Giant elliptical galaxies are often found near the centers of rich clusters.

• The observable mass of a cluster of galaxies is not large enough to account for the observed motions of the galaxies; a large amount of unobserved mass must be present between the galaxies. This unaccounted-for dark matter is known as the missing-mass problem.

• Hot intergalactic gases emit x rays in rich clusters.

• When two galaxies collide, their stars initially pass each other but their interstellar media collide violently, either stripping the gas and dust from the galaxies or triggering prolific star formation.

The gravitational effects of a galactic collision can throw stars out of their galaxies into intergalactic space.

Galactic mergers occur; a large galaxy in a rich cluster may grow steadily through galactic cannibalism, perhaps producing a giant elliptical galaxy.

• There is a simple linear relationship between the distance from the Earth to a galaxy and the redshift of that galaxy (which is a measure of the speed at which it is receding from us); this relationship is the Hubble law, recessional velocity = $H_0 \times$ distance, where H_0 is the Hubble constant.

• Standard candles—such as Cepheid variables, the brightest supergiants, globular clusters, H II regions, and supernovae in a galaxy—are used to estimate intergalactic distances. Because of difficulties in measuring the distances to remote galaxies, the value of the Hubble constant, H_0, is not known with certainty.

Review Questions

1   What is the Hubble classification scheme? Which category includes the biggest galaxies? Into which category do the smallest galaxies fall? Which type of galaxy is the most common?

2   In which types of galaxies are new stars most likely forming? Describe the observational evidence that supports your answer.

3   What is the difference between a flocculent spiral galaxy and a grand-design spiral galaxy?

4   Briefly describe how the theory of self-propagating star formation accounts for the existence of spiral arms in galaxies.

5   Briefly describe how the spiral density wave theory accounts for the existence of spiral arms in galaxies.

6   How is it possible that galaxies in our Local Group still remain to be discovered? In what part of the sky are these galaxies located? What sorts of observations might reveal these galaxies?

7   Are there any galaxies besides our own that can be seen with the naked eye? If so, which ones can you name?

8   What is the difference between a rich cluster of galaxies and a poor one? What is the difference between a regular cluster of galaxies and an irregular one?

9   How can a collision between galaxies produce a starburst galaxy?

10   Why do astronomers believe that there must be considerable quantities of dark matter in clusters of galaxies?

11   Explain why the dark matter in galaxy clusters cannot be neutral hydrogen.

12   What is the Hubble law?

13   Some galaxies in the Local Group exhibit blueshifted spectral lines. Why aren't these blueshifts violations of the Hubble law? Explain.

14   What is a standard candle? Why are standard candles important to astronomers trying to measure the Hubble constant?

15   What kinds of stars would you expect to find populating space between galaxies in a cluster?

Advanced Questions

∗ 16   Suppose a spectrum of a distant galaxy showed that its redshift corresponds to a speed of 22,000 km/s. How far away is the galaxy?

∗ 17   A cluster of galaxies in the southern constellation of Pavo (the Peacock) is located 100 Mpc from Earth. How fast, on average, is this cluster receding from us? Why do different galaxies in the cluster show different velocities?

18   How might you determine what fraction of a galaxy's redshift is caused by the galaxy's orbital motion about the center of mass of its cluster?

Discussion Questions

19   Discuss the advantages and disadvantages of using the various standard candles to determine extragalactic distances.

20   Discuss whether the various Hubble types of galaxies actually represent some sort of evolutionary sequence.

21   Discuss the sorts of phenomena that can occur when galaxies collide. Do you think that such collisions can change the Hubble type of a galaxy?

22   From what you know about stellar evolution, the interstellar medium, and the spiral density wave theory, explain the appearance and structure of the spiral arms of grand-design spiral galaxies.

Observing Projects

23   Using a telescope with an aperture of at least 30 cm (12 in.), observe as many of the spiral galaxies listed in the table below as you can. Many of these galaxies are members of the Virgo cluster, which can best be seen during the spring. Since all galaxies are quite faint, be sure to schedule your observations for a moonless night. The best view is obtained when a galaxy is near the meridian. While at the eyepiece, make a sketch of what you see. Can you distinguish any spiral structure? After completing your observations, compare your sketches with photographs found in such popular books as *Galaxies* by Timothy Ferris (Sierra Club Books, 1980) and *The Color Atlas of the Galaxies* by James Wray (Cambridge University Press, 1988).

Spiral galaxy	R.A.	Decl.	Hubble type	Spiral galaxy	R.A.	Decl.	Hubble type
M31 (NGC 224)	$0^h 42.7^m$	+41° 16′	Sb	M91 (NGC 4548)	$12^h 35.4^m$	+14° 30′	SBb
M58 (NGC 4579)	12 37.7	+11 49	Sb	M94 (NGC 4736)	12 50.9	+41 07	Sb
M61 (NGC 4303)	12 21.9	+4 28	Sc	M98 (NGC 4192)	12 13.8	+14 54	Sb
M63 (NGC 5055)	13 15.8	+42 02	Sb	M99 (NGC 4254)	12 18.8	+14 25	Sc
M64 (NGC 4826)	12 56.7	+21 41	Sb	M100 (NGC 4321)	12 22.9	+15 49	Sc
M74 (NGC 628)	1 36.7	+15 47	Sc	M101 (NGC 5457)	14 03.2	+54 21	Sc
M83 (NGC 5236)	13 37.0	−29 52	Sc	M104 (NGC 4594)	12 40.0	−11 37	Sa
M88 (NGC 4501)	12 32.0	+14 25	Sb	M108 (NGC 3556)	11 11.5	+55 40	Sc
M90 (NGC 4569)	12 36.8	+13 10	Sb				

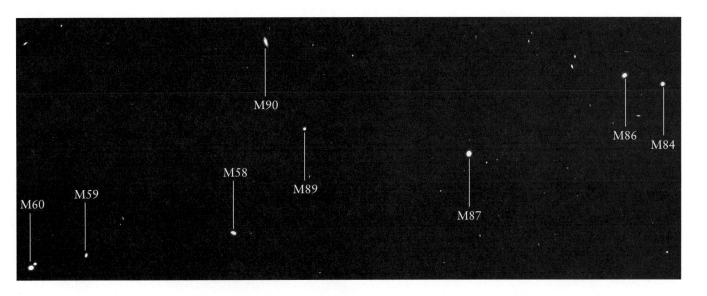

24 Using a telescope with an aperture of at least 30 cm (12 in.), observe as many of the elliptical galaxies listed in the table below as you can. Six of these galaxies are in the Virgo cluster, which is conveniently located in the evening sky from March through June. The photograph above, which covers a 2° × 5° swath of the cluster, may help you find some of these galaxies. As in the previous exercise, be sure to schedule your observations for a moonless night, when the galaxies you wish to observe will be near the meridian. Do these elliptical galaxies differ in appearance from spiral galaxies?

Elliptical galaxy	R.A.	Decl.	Hubble type
M58 (NGC 4472)	12^h 29.8^m	+8° 00′	E4
M59 (NGC 4621)	12 42.0	+11 39	E3
M60 (NGC 4649)	12 43.7	+11 33	E1
M84 (NGC 4374)	12 25.1	+12 53	E1
M86 (NGC 4406)	12 26.2	+12 57	E3
M89 (NGC 4552)	12 35.7	+12 33	E0
M110 (NGC 205)	00 40.4	+41 41	E6

25 Using a telescope with an aperture of at least 30 cm (12 in.), observe as many of the interacting galaxies listed in the table below as you can. As in the previous exercises, be sure to schedule your observations for a moonless night when the galaxies you wish to observe will be near the meridian. While at the eyepiece, make a sketch of what you see. Can you distinguish hints of interplay among the galaxies? After completing your observations, compare your sketches with photographs found in such popular books as *Galaxies* by Timothy Ferris (Sierra Club Books, 1980).

Interacting galaxies	R.A.	Decl.
M51 (NGC 5194)	13^h 29.9^m	+47° 12′
NGC 5195	13 30.0	+47 16
M65 (NGC 3623)	11 18.9	+13 05
M66 (NGC 3627)	11 20.2	+12 59
NGC 3628	11 20.3	+13 36
M81 (NGC 3031)	9 55.6	+69 04
M82 (NGC 3034)	9 55.8	+69 41
M95 (NGC 3351)	10 44.0	+11 42
M96 (NGC 3368)	10 46.8	+11 49
M105 (NGC 3379)	10 47.8	+12 35

16 ▶ Quasars and Active Galaxies

IN THIS CHAPTER

You will examine perhaps the most distant, most luminous objects in the universe: quasars. You will learn of their unusual spectra and their intense energy output within a surprisingly small volume. Like other bright objects, called active galaxies, that you will also consider, quasar centers may contain supermassive black holes, and you will see how gas spiraling into these central engines explains their power.

The Core of a Radio Galaxy This artist's rendition shows the scenario that many astronomers believe is responsible for double radio sources. A supermassive black hole at the center of a galaxy is surrounded by an accretion disk. In the inner regions of the accretion disk, matter crowding toward the hole is diverted outward along two oppositely directed beams. These beams deposit energy into two huge, radio-emitting lobes located on either side of the galaxy.

WHAT DO YOU THINK?

1 What do quasars look like?

2 What powers a quasar?

· ·

THE SUN MAY RELEASE far more energy than anything on Earth, but its brightness is reassuringly familiar. It helps us understand even more tremendous energies liberated by dying stars. But even the emissions from supernovae are as nothing, compared to the power of quasars and active galaxies. Some of these objects emit more energy each *second* than the Sun does in two hundred years. Yet our knowledge of them began in an amateur astronomer's backyard.

Quasars

The development of radio astronomy in the late 1940s opened the realm of nonvisual astronomy. The first radio telescope was built in 1936 by Grote Reber in his backyard in Illinois. By 1944 Reber had detected strong radio emissions from sources in the constellations of Sagittarius, Cassiopeia, and Cygnus. Two of these sources, nicknamed Sagittarius A (abbreviated "Sgr A") and Cassiopeia A (Cas A), happen to be in our Galaxy. The first is the galactic nucleus (see Chapter 14) and the second a supernova remnant (see Chapter 12). However, Reber's third source, called Cygnus A (Cyg A), proved hard to categorize. The mystery only deepened in 1951, when Walter Baade and Rudolph Minkowski, using the 200-in. optical telescope on Palomar Mountain, discovered a strange-looking galaxy at the same position. Figure 16-1 is a photograph of the optical counterpart of Cyg A.

The galaxy associated with Cyg A is very dim. Nevertheless, Baade and Minkowski managed to photograph its spectrum. They detected a redshift corresponding to a speed of 17,000 km/s. According to the Hubble law, this speed indicates that Cyg A lies 750 million light-years from Earth!

The enormous distance to Cyg A astounded astronomers, because it is one of the brightest radio sources in the sky. Although barely visible through the giant telescope at Palomar, Cyg A's radio waves can be picked up by amateur astronomers with backyard equipment! Its energy output must therefore be colossal. In fact, Cyg A shines with a radio luminosity 10^7 times as bright as that of an ordinary galaxy such as M31 in Andromeda yet is

Figure 16-1 **Cygnus A (3C 405)** This strange-looking galaxy was discovered at the location of the radio source Cygnus A (Cyg A). This galaxy has a redshift corresponding to a recessional speed of 6% of the speed of light, which, according to the Hubble law, corresponds to a distance of about 750 million light-years from Earth. Because Cyg A is one of the brightest radio sources in the sky, the energy output of this remote galaxy must be enormous.

far more distant. The object corresponding to Cyg A has to be something quite extraordinary.

16-1 Quasars look like stars but have huge redshifts

Cygnus A is not the only powerful radio source in the far-distant sky. Starting in the 1950s radio astronomers were busy making long lists of radio sources. One of the most famous lists, titled the *Third Cambridge Catalogue*, was published in 1959. (The first two catalogues produced by the British team were filled with inaccuracies.) Even today astronomers often refer to its 471 radio sources by their "3C numbers." Cyg A, for example, is designated 3C 405, because it is the 405th source on the Cambridge list. Because of its extraordinary luminosity, astronomers were eager to learn whether any other sources in the 3C catalogue had similarly extraordinary properties.

One interesting case was 3C 48. In 1960 Allan Sandage used the Palomar telescope to discover a "star"

Figure 16-2 **The Quasar 3C 48** For several years astronomers erroneously believed that this object (arrow) is simply a peculiar nearby star that happens to emit radio waves. Actually, the redshift of this starlike object is so great that, according to the Hubble law, it must be roughly 4 billion light-years away.

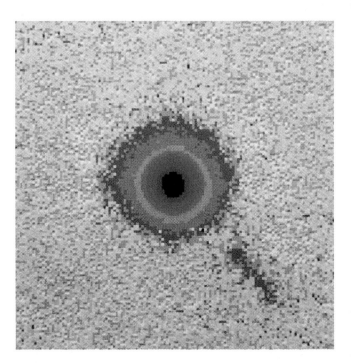

Figure 16-3 **The Quasar 3C 273** This greatly enlarged view shows the starlike object associated with the radio source 3C 273. Note the luminous jet on one side. By 1963 astronomers determined that the redshift of this quasar is so great that, according to the Hubble law, it is nearly 2 billion light-years from Earth.

at the location of this radio source (Figure 16-2). Because ordinary stars are not strong sources of radio emission, 3C 48 had to be something unusual. Indeed, its spectrum showed a series of emission lines that no one could identify. Although 3C 48 was clearly an oddball, many astronomers thought it was just another strange star in our Galaxy.

Another such "star," called 3C 273, was discovered in 1962. It was found to have a luminous "jet" of bright gas protruding from one side, as shown in Figure 16-3. Like 3C 48, this object emits a series of bright spectral lines that no one could identify.

A breakthrough was finally made in 1963, when Maarten Schmidt at the California Institute of Technology found that four of the brightest spectral lines of 3C 273 are positioned relative to one another just like four familiar spectral lines of hydrogen. However, these emission lines of 3C 273 are found at much longer wavelengths than the usual wavelengths of the hydrogen lines. In other words, the light from 3C 273 is subjected to a substantial redshift.

Spectra for stars in our Galaxy exhibit comparatively small Doppler shifts, because these stars cannot move extremely fast relative to the Sun without soon escaping from the Galaxy. Schmidt thus conjectured that 3C 273 might not be a nearby star after all. Pursuing this hunch, he promptly identified all four spectral lines as being hydrogen lines that have suffered an enormous redshift corresponding to a speed of almost 15% of the speed of light. According to the Hubble law, this huge redshift implies the incredible distance to 3C 273 of roughly two billion light-years.

Figure 16-4 shows the spectrum of 3C 273. Remember from Chapter 4 that a spectrograph produces a graph of intensity versus wavelength on which emission lines appear as peaks and absorption lines appear as valleys. The spectral lines of 3C 273 are brighter than the background radiation at other wavelengths, called the *continuum*. The emission lines are caused by excited gas atoms emitting radiation at specific wavelengths. The strong emission lines mean that something unusual is heating the gas.

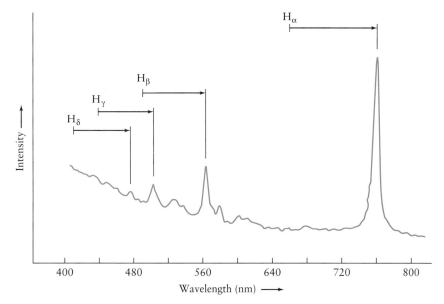

Figure 16-4 **The Spectrum of 3C 273** Four bright emission lines caused by hydrogen dominate the spectrum of 3C 273. The arrows indicate how far these spectral lines are redshifted from their usual wavelengths.

Inspired by Schmidt's success, astronomers looked again at the spectral lines of 3C 48. Their redshift corresponded to a velocity of nearly one-third the speed of light. Therefore 3C 48 must be nearly twice as far away as 3C 273, or about four billion light-years from Earth if a Hubble constant (H_0) of 75 km/s/Mpc is used. (As discussed in the previous chapter, the value of the Hubble constant is not known with certainty. If it is higher than 75, then these sources are nearer to us than the distances given in this chapter; if it is lower, then they are farther away.)

Because of their starlike appearances and strong emissions, 3C 48 and 3C 273 were dubbed **quasi-stellar radio sources.** This term was soon shortened to **quasars.** All quasars, also called **quasi-stellar objects (QSOs),** look like stars but have incredible energy output. They need not specifically be *radio* sources; most, in fact, are not. Yet the name *quasar* has stuck, and thousands have been discovered since the pioneering days of the early 1960s. Virtually all have enormous redshifts, in some cases corresponding to speeds greater than 90% of the speed of light; Figure 16-5 shows the spectrum of a quasar whose

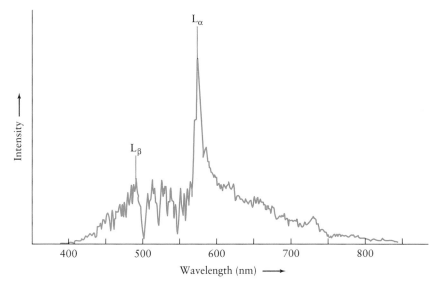

Figure 16-5 **The Spectrum of a High-Redshift Quasar** The light from this quasar, known as PKS 2000-330, is so highly redshifted that spectral lines normally in the far-ultraviolet (L_α and L_β) can be seen at visible wavelengths. Note the large number of deep absorption lines on the short-wavelength side of L_α. These lines, collectively called the "Lyman-alpha forest," are probably caused by remote clouds of gas along our line of sight to the quasar. Hydrogen in these clouds absorbs photons from the quasar at wavelengths less redshifted than the quasar's L_α line.

Figure 16-6 **The Brightness of 3C 279**
This graph shows variations in the magnitude of brightness of the quasar 3C 279. Note the large outburst in 1937. The data were obtained by carefully examining old photographic plates in the files of the Harvard College Observatory.

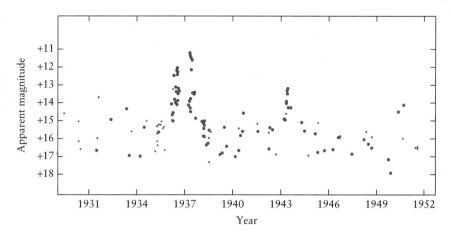

redshift corresponds to 92% of the speed of light. From the Hubble law, it follows that the distances to these high-redshift quasars are typically in the range of 10 to 13 billion light-years—their light has taken 10 to 13 billion years to reach us. When we look at these quasars, we are seeing objects as they existed when the universe was very young.

16-2 A quasar emits a huge amount of energy from a small volume

Galaxies are big and bright. A typical large galaxy, like our own Milky Way, contains several hundred billion stars and shines with the luminosity of 10 billion Suns. The most gigantic and most luminous galaxies, such as the giant ellipticals, are only 10 times brighter. Yet beyond eight billion light-years from Earth, even the brightest galaxies are too faint to be easily detected. Most ordinary galaxies are too dim to be detected at half that distance. The fact that quasars can be seen at distances where galaxies are not visible means that quasars are far more luminous than galaxies. Indeed, a typical quasar is 100 times brighter than our Milky Way.

In the mid-1960s, several astronomers discovered that some of the newly identified quasars had been photographed inadvertently in the past. For example, 3C 273 was found on numerous photographs, including one taken in 1887. By carefully examining the images of quasars on these old photographs, astronomers found that *quasars fluctuate in brightness, occasionally flaring up*. See, for example, the data from old photographs of another quasar, 3C 279, plotted in Figure 16-6. Note the prominent outbursts that occurred around 1937 and 1943. During these outbursts, the luminosity of 3C 279 increased by a factor of at least 25. Because of the enormous distance to this quasar, it must have been shining during those peak periods at least 10,000 times brighter than the entire Milky Way.

Because quasars fluctuate in brightness, astronomers have been able to place strict upper limits on the sizes of quasars. As we saw in Chapter 13, an object cannot vary in brightness faster than the time it takes light to travel across that object. For example, an object that is 1 ly in diameter cannot vary in brightness with a period of less than 1 year.

The brightnesses of many quasars vary over intervals of only a few months, weeks, days, or even hours! Recent x-ray observations reveal large variations taking place in as little as three hours. This rapid flickering means that the source of the quasar's energy must be quite small by galactic standards. In fact, the energy-emitting region of a typical quasar—the "powerhouse" that blazes with the luminosity of 100 galaxies—is less than 1 light-day in diameter. *If quasars are indeed at the huge distances indicated by their redshifts, something must be producing the luminosity of 100 galaxies in a volume roughly the same size as our solar system!*

16-3 Active galaxies bridge the energy gap between ordinary galaxies and quasars

Some astronomers in the 1960s could not accept the fact that quasars have such a huge energy output. They argued that quasars are closer and therefore need to put out much less energy. Something else, they insisted, must be causing such massive redshifts. No such mechanism has been found, and most quasars are still believed to be at great distances from us. In recent years, however, astronomers have discovered objects with luminosities between those of ordinary galaxies and remote quasars.

Some of these strange galaxies have unusually bright, starlike nuclei; others have strong emission lines in their spectra; still others are highly variable. Some have jets and beams of radiation emanating from their cores, and most of these objects are more luminous than ordinary galaxies. All are called **active galaxies**. Some, called **peculiar galaxies** (and denoted "pec"), appear to be blowing themselves apart. Any of the Hubble classes of galaxies (see Chapter 15) can be peculiar.

Carl Seyfert at the Mount Wilson Observatory discovered the first active galaxies in 1943 while surveying spiral galaxies. Now called **Seyfert galaxies,** these luminous objects have bright, starlike nuclei and strong emission lines in their spectra. For example, NGC 4151 has an extremely rich spectrum with many prominent emission lines. Some of these emission lines are produced by iron atoms with a dozen or more electrons stripped away, indicating that NGC 4151 contains some extremely hot gas. Seyfert galaxies also vary in brightness. For instance, sometimes the magnitude of NGC 4151 changes over a few days.

Another example of a Seyfert galaxy is NGC 1068, shown in Figure 16-7. At infrared wavelengths this galaxy shines with the brilliance of 10^{11} Suns. This extraordinary luminosity has been observed to vary by as much as 7×10^9 $L_\odot$ over only a few weeks. In other words, the infrared power output of the nucleus of NGC 1068 rises and falls by an amount nearly equal to the total luminosity of our entire Galaxy.

Many more Seyfert galaxies have been discovered in recent years. Approximately 10% of the most luminous galaxies in the sky are Seyfert galaxies. Some of the brightest Seyfert galaxies shine as brightly as faint quasars, which leads most astronomers to suspect that

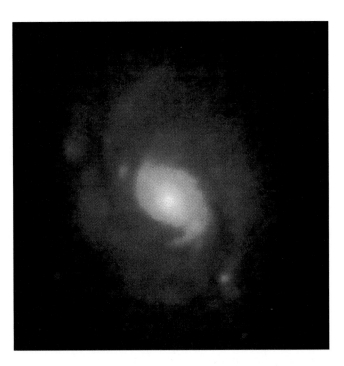

Figure 16-7 **The Seyfert Galaxy NGC 1068** This Seyfert galaxy, also called M77 or 3C 71, is renowned for its extraordinary infrared luminosity, which varies over intervals as short as a few weeks. Note how bright the inner spiral arms are compared with the outer spiral arms.

the nuclei of Seyfert galaxies are, in fact, low-luminosity quasars.

Some Seyfert galaxies exhibit vestiges of violent, explosive phenomena in their nuclei. For instance, NGC 1275 in Figure 16-8 has filaments of gas tens of thousands of light-years long protruding from its nucleus in all directions. Spectroscopic studies indicate that this gas is being blasted away from the galaxy's nucleus at 3000 km/s. In

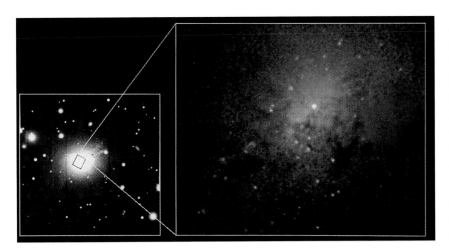

Figure 16-8 **The Active Galaxy NGC 1275 (3C 84)** On the left is a ground-based view of this peculiarly shaped Seyfert galaxy. Located in the Perseus cluster, it is a strong source of x rays and radio radiation. According to the Hubble law, the galaxy's redshift indicates a distance of roughly 300 million light-years from Earth. On the right, a close-up of the central portion of NGC 1275 reveals individual clusters of stars (bright blue dots) among the gas protruding from the nucleus. Spectroscopic observations confirm significant mass ejection from the galaxy's center.

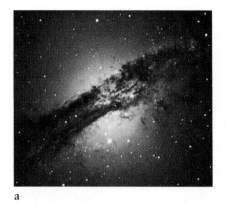

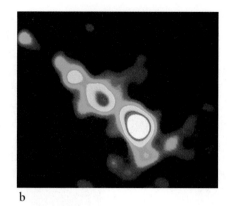

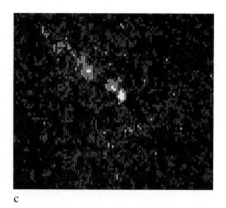

a

b

c

Figure 16-9 **The Peculiar Galaxy NGC 5128 (Centaurus A)** This extraordinary galaxy is located in the constellation of Centaurus, roughly 13 million light-years from Earth. These three views show visible, radio, and x-ray images of this galaxy, all to the same scale. (a) This photograph at visible wavelengths shows a dust lane across the face of the galaxy. (b) This false-color radio image shows that vast quantities of radio radiation pour from extended regions of the sky on either side of the dust lane. (c) This x-ray picture from the Einstein Observatory shows that NGC 5128 has a bright x-ray nucleus. An x-ray jet protrudes from the galaxy's nucleus along a direction perpendicular to the galaxy's dust lane.

1977 Vera Rubin and her colleagues reported observations demonstrating that NGC 1275 actually consists of two galaxies. As we saw in Chapter 15, a collision or close encounter between two galaxies can eject matter into intergalactic space. More and more of this intergalactic medium is being discovered between galaxies.

The high-speed ejection of matter from active galaxies is seen best at nonvisible wavelengths. For instance,

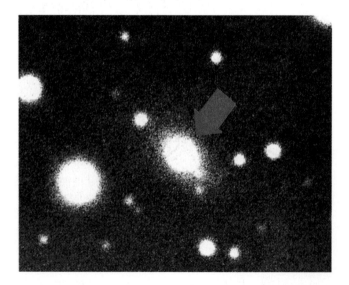

Figure 16-10 **BL Lacertae** This photograph shows fuzz around BL Lacertae. BL Lac objects appear to be giant elliptical galaxies with bright starlike nuclei, much as Seyfert galaxies are spiral galaxies with quasarlike nuclei. BL Lac objects contain much less gas and dust than do Seyfert galaxies.

Figure 16-9a is a photograph of the galaxy NGC 5128 in the southern constellation of Centaurus. An unusually broad dust lane stretches across the galaxy. NGC 5128 is classified as an "E0 (pec)." In the radio and x-ray spectra, the ejected matter from NGC 5128 is visible (Figure 16-9b and c).

Another class of active galaxies are the **BL Lacertae objects,** named after their prototype, BL Lacertae (BL Lac) in the constellation of Lacerta (the Lizard). BL Lac (Figure 16-10) was discovered in 1929, when it was mistaken for a variable star, largely because its brightness varies by a factor of 15 in only a few months. BL Lac's most intriguing characteristic is a *totally featureless spectrum,* exhibiting neither absorption nor emission lines. A BL Lac object is an elliptical galaxy with a bright starlike center, much as a Seyfert galaxy is a spiral galaxy with a quasarlike center.

16-4 Active galaxies lie at the center of double radio sources

The galaxy NGC 5128 (Centaurus A), one of the brightest sources of radio waves in the sky, was one of the first radio sources discovered when radio telescopes were erected in Australia. In part, the brightness of Centaurus A at radio wavelengths comes from its proximity to Earth, only 13 million light-years away. As shown in Figure 16-9b, radio waves pour from two regions, called **radio lobes,** on either side of the galaxy's dust lane. Far-

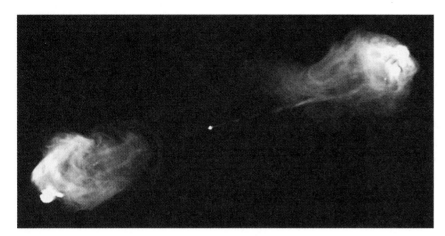

Figure 16-11 **A Radio Image of Cygnus A** This color-coded radio picture was produced at the Very Large Array. Most of the radio emission from Cygnus A comes from the radio lobes located on either side of the peculiar galaxy seen in Figure 16-1. These two radio lobes are each about 160,000 ly from the optical galaxy, and each contains a brilliant, condensed region of radio emission.

ther from the dust lane is a second set of radio lobes that spans a volume 2 Mly across. Recent x-ray pictures of NGC 5128 reveal an x-ray jet (see Figure 16-9c) sticking out of the galaxy's nucleus. This jet, which is perpendicular to the galaxy's dust lane, is aimed toward one of the radio lobes. These observations suggest that particles and energy stream out of the galaxy's nucleus toward the radio lobes.

By 1970 radio astronomers had discovered dozens of objects similar to Centaurus A; these are now called **double radio sources.** An active galaxy, usually resembling a giant elliptical, is often found between the two radio lobes. For instance, the visible galaxy associated with Cyg A (see Figure 16-1) is also located between two radio lobes, as shown in Figure 16-11.

All double radio sources seem to have some sort of central "engine" that ejects electrons and magnetic fields outward along two oppositely directed jets at very nearly the speed of light. After traveling many thousands or even millions of light-years, this ejected material slows down, allowing the electrons and the magnetic field to produce the radio radiation that we detect. Recall that a specific type of radio emission, called *synchrotron radiation*, occurs whenever energetic electrons move in a spiral within a magnetic field. The radio waves that come from the lobes of a double radio source have all the characteristics of synchrotron radiation. At radio wavelengths, the double radio sources are among the brightest objects in the universe.

The idea that a double radio source involves powerful jets of particles traveling near the speed of light is supported by the existence of **head–tail sources,** so named because each such source appears to have a head of concentrated radio emission with a weaker tail trailing behind it. A good example is the active elliptical

galaxy NGC 1265 in the Perseus cluster of galaxies. NGC 1265 is known to be moving at a high speed (2500 km/s) relative to the cluster as a whole. Figure 16-12 is a radio map of NGC 1265. Note that the radio emission has a distinctly windswept appearance. Just as smoke pouring from a steam locomotive trails a rapidly moving train, particles ejected in the two jets from this galaxy are deflected by the galaxy's passage through the sparse intergalactic medium.

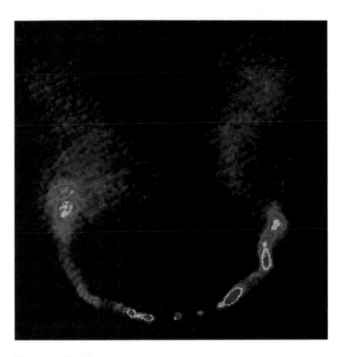

Figure 16-12 **The Head–Tail Source NGC 1265** The active elliptical galaxy NGC 1265 cannot be an ordinary double radio source, since it is moving at a high speed through the intergalactic medium. Because of this motion, the two jets trail the galaxy, giving this radio source a distinctly windswept appearance.

Supermassive Black Holes as Central Engines

How do quasars and active galaxies produce such enormous amounts of energy from such small volumes? As long ago as 1968, the British astronomer Donald Lynden-Bell suggested that the gravitational field of a supermassive black hole (see Chapter 13) could be the engine that supplies the necessary power.

16-5 Supermassive black holes lurk at the centers of some galaxies

As we saw in Chapter 13, finding black holes is a difficult business. At best, we can only see the effects of the hole's gravity and try to rule out non-black-hole explanations of the data. This general approach has been applied to several galaxies, including M31, M32, M87, M104, and the Milky Way.

The Andromeda galaxy (M31) is the largest, most massive galaxy in the Local Group (see Figures IV-3 and 15-17). At a distance of only 2.2 million light-years from Earth, M31 is close enough to us that details in its core less than 1 pc across can be resolved by the Hubble Space Telescope, revealing that this galaxy has two bright regions in its core. The dimmer one is believed to be its true nucleus, while the brighter one is thought to be just a cluster of stars.

In the mid-1980s several astronomers made careful spectroscopic observations of the core of M31. Using measurements of Doppler shifts, they determined that stars within 50 ly of the galaxy's nucleus are orbiting the nucleus at exceptionally high speeds, which suggests that a massive object is located at the galaxy's center. Without the gravity of such an object to keep the stars in their high-speed orbits, they would have escaped from the galaxy's core long ago. From such observations, astronomers estimate that the mass of the central object is about 50 million solar masses. That much matter confined to such a small volume strongly suggests the existence of a supermassive black hole.

Located near M31 is a small elliptical galaxy called M32 (Figure 16-13a). High-resolution spectroscopy indicates that stars close to the center of M32 are also orbiting this galaxy's nucleus at unusually high speeds, which could be explained by the presence of a supermassive black hole there. Furthermore, a picture taken by the Hubble Space Telescope (Figure 16-13b) shows that the concentration of stars at the core of M32 is truly remarkable. The density of stars there is more than a hun-

a

b

Figure 16-13 **The Elliptical Galaxy M32** (a) This small galaxy is a satellite of M31, a portion of which is seen at the left of this wide-angle photograph. Both galaxies are roughly 2.2 million light-years from Earth. (b) This high-resolution image from the Hubble Space Telescope shows the center of M32. Note the concentration of stars at the nucleus of the galaxy. The area covered in this view is only 175 ly on a side.

Figure 16-14 **The Sombrero Galaxy (M104)** This spiral galaxy in Virgo is nearly edge-on to our Earth-based view. Spectroscopic observations suggest that a billion-solar-mass black hole is located at the galaxy's center.

dred million times greater than the density of stars in the Sun's neighborhood. The concentration of stars and their high speeds strongly support the belief that a supermassive black hole exists at the center of M32.

Astronomers have uncovered evidence of supermassive black holes in other, more remote galaxies. John Kormendy used a 3.6-m telescope on Mauna Kea to examine the core of M104 spectroscopically (Figure 16-14). Once again high-speed gas orbiting the galaxy's nucleus was found. These observations suggest that the center of this galaxy is dominated by a supermassive black hole containing a billion solar masses. Similar spectroscopic obser-

vations of the edge-on S0 galaxy NGC 3115 also reveal a billion-solar-mass black hole at the core.

Recent observations with the Hubble Space Telescope show distinct evidence for a supermassive black hole at the center of the giant elliptical galaxy M87. Located some 50 million light-years from Earth, M87 is an active galaxy that has long been recognized as quite unusual. In 1918 Heber Curtis at the Lick Observatory reported that the center of M87 has "a curious straight ray . . . apparently connected with the nucleus by a thin line of matter." Figure 16-15*a* is a long exposure of M87 and several neighboring galaxies; M87's nucleus and jet are

a

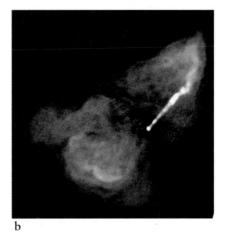

b

c

Figure 16-15 **The Giant Elliptical Galaxy M87** M87 is located near the center of the sprawling, rich Virgo cluster, which is about 50 million light-years from Earth. (a) This long exposure shows the extent of M87 and some smaller, neighboring galaxies. The numerous fuzzy spots that surround M87 are globular clusters; each contains about a million stars. (b) This radio image shows the intergalactic gas emitted by M87. These gas

clouds dwarf the giant galaxy, the bright spot near the center of the picture. (c) This recent Hubble Space Telescope image of M87 shows the gas disk in its nucleus. M87's extraordinarily bright nucleus and the gas jets result from a three-billion-solar-mass black hole, whose gravity causes huge amounts of gas and an enormous number of stars to crowd around it.

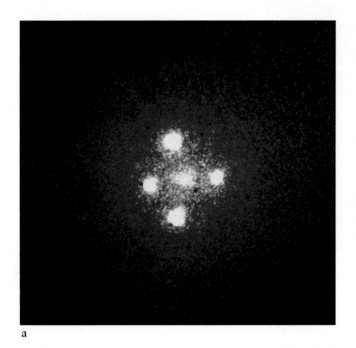

a

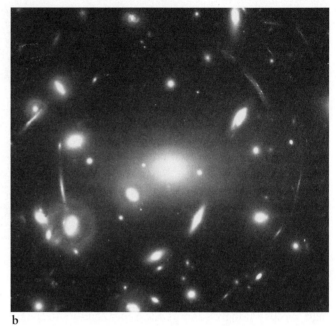

b

Figure 16-16 **An Einstein Cross and an Einstein Ring**
(a) This image from the Hubble Space Telescope shows the
gravitational lensing of a quasar in the constellation of Pegasus.
The quasar, about 8 billion light-years from Earth, is seen as
four separate images surrounding a galaxy that is only 400 million light-years away. The diffuse image at the center of this
Einstein cross is the core of the intervening galaxy. (b) In this
image from the Hubble Space Telescope, the quasar that is being lensed is behind the intervening galaxies. The quasar light
arrives here as arcs of a ring, called an Einstein ring, after having passed the galaxies.

buried in the galaxy's glare. Figure 16-15b is a radio
image showing the gas emission from M87, which is a
powerful source of radio waves and x rays.

The Hubble Space Telescope image of M87 in Figure
16-15c shows an exceptionally bright, starlike nucleus
and a surrounding disk of gas with trailing spiral arms.
To produce this fiery glow, stars must be packed so
tightly at the center of M87 that their density is at least
300 times greater than that normally found at the centers
of giant ellipticals. This dense clustering of stars indicates the presence of a black hole with a mass of nearly
three billion Suns at the center of M87.

If supermassive black holes lurk within quasars and
active galaxies, then the light they emit should be subject
to Einstein's general theory of relativity. Like the light
from stars behind the Sun, the light from distant objects
such as quasars should be deflected as it travels past
other galaxies toward us. This distortion of light, or
gravitational lensing, can lead to our receiving several
images of the distant quasar or galaxy. If the background

object is exactly behind an intervening galaxy, the light
should actually be focused in a ring, called an **Einstein
ring,** rather than as several separate images. Figure 16-16a
shows four images of a quasar in a configuration called
an **Einstein cross.** Figure 16-16b shows arcs that are part
of an Einstein ring. All of these observations help confirm the validity of Einstein's general theory of relativity.

16-6 Jets of matter from around a black hole may explain quasars and active galaxies

If supermassive black holes are commonplace at the centers of galaxies, their gravitational energy could give rise
to quasars and active galactic nuclei. Here is one convincing scenario. Because the centers of galaxies are con-
gested places, the black hole captures a massive accretion
disk of gas and dust. According to Kepler's third law,
material in the inner regions of this disk orbits the most

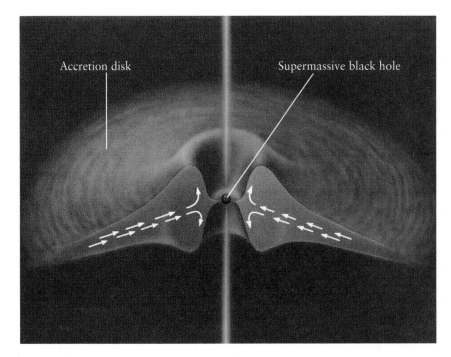

Figure 16-17 **A Supermassive Black Hole as the Central Engine** The energy output of an active galaxy or a quasar may involve an extremely massive black hole that captures matter from its surroundings. In the scenario depicted here, the inflow of material through an accretion disk is redirected to produce two powerful jets of particles traveling at nearly the speed of light. (Also see artist's rendition on the first page of this chapter.)

rapidly. This material would be constantly rubbing against and thus heating up the more slowly moving gases in the outer regions. As the gases spiral toward the black hole, they release this energy violently, giving the hole its brilliant luminosity.

The crowding of in-falling matter also leads to the formation of jets of gas, which stream out from the black hole. Although some of the incoming material is swallowed up, the pressures in the accretion disk become so great that most of the hot gases never really get close to the event horizon. Instead, the pressure there is relieved by matter being violently ejected along the perpendicular to the accretion disk, where the hot gases experience the least resistance. The result is two oppositely directed beams as sketched in Figure 16-17 and on the opening page of this chapter.

In 1992 the Hubble Space Telescope took a picture of the spiral galaxy M51 that some astronomers interpret as showing an accretion disk around a supermassive black hole. An overall view of this galaxy, sometimes called the Whirlpool galaxy, appears in Figure IV-2. Figure 16-18 is a Hubble Space Telescope image of the galaxy's nucleus. The dark, horizontal bar in the picture is our edge-on view of a ring of dust and gas about 100 ly in diameter. This bar is identified as the accretion disk, because ground-based radio and optical observations show a

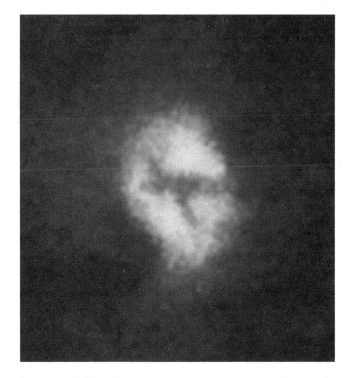

Figure 16-18 **The Nucleus of M51** This image, taken by the Hubble Space Telescope in 1992, shows a dark X silhouetted across the nucleus of the spiral galaxy M51. The dark, horizontal bar may be an edge-on view of a huge dusty ring that surrounds a supermassive black hole. The fainter, inclined bar is puzzling.

Figure 16-19 **The Orientation of the Central Engine and Its Jets** Double radio sources, quasars, and BL Lacertae objects may be the same type of object viewed from different directions. If one of the jets is aimed almost directly at the Earth, we see a BL Lac object. If the jet is somewhat tilted to our line of sight, we see a quasar. If the jets are nearly perpendicular to our line of sight, we see a double radio source.

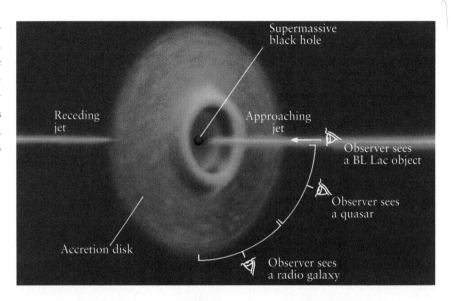

double-lobed structure that is bisected by the bar. The fainter, inclined bar in the image is still a mystery.

What confines the ejected matter to narrow jets? First, the ejecta could initially be forced into narrow jets by magnetic fields. Second, gas still falling toward the black hole could prevent escaping matter from spreading. To see why, consider water squirting out of the nozzle of an ordinary garden hose. The stream of water broadens and the spray fans out through a wide angle in the air. If the nozzle is placed in a swimming pool, however, the stream of water does not fan out as much. Similarly, as the two jets of hot gas leave the vicinity of the black hole, they must blast their way through the gas that is still crowding inward. Passage through this material causes the jets to become extremely narrow, concentrated beams. This "self-focusing" may explain not only the double radio sources but also the jets and beams we see protruding from active galaxies and some quasars.

The main difference between double radio sources, quasars, and BL Lac objects may be just the angle at which the central engine is viewed. As Figure 16-19 shows, an observer sees a double radio source when the accretion disk is viewed nearly edge-on, so that the jets are nearly in the plane of the sky. At a steeper angle, the observer sees a quasar. If one of the jets is aimed almost directly at Earth, a BL Lac object is seen.

The nature of quasars may still be eluding us. In 1994 John Bahcall and collaborators discovered eight quasars that do not appear to be embedded in galaxies (Figure 16-20). Since the gas and stars swirling around a supermassive black hole are believed to be supplied by a host galaxy, these "naked" quasars seem out of place. Furthermore, the naked quasars all have nearby companion galaxies that appear to be merging with the quasar. One possibility is that some supermassive black holes form in relative isolation and then stimulate normal stars

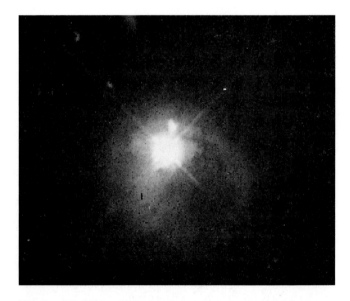

Figure 16-20 **A Naked Quasar** The quasar PKS 2349-01, located in the center of this image, is not embedded in a visible galaxy. The wisps surrounding the bright central quasar are pieces of a companion galaxy that may have been torn apart by the quasar. Some quasars may form alone and then merge with nearby galaxies, as is apparently happening here.

to form in the surrounding gas, thereby creating a galaxy. Another possibility is that the quasar pulls another galaxy around it. A third possibility is that these recent observations are being misinterpreted and the quasars are not really isolated. The next few years will be exciting as the model for quasars and active galaxies is refined and retested.

WHAT DID YOU THINK?

1 *What do quasars look like?* They look like stars but spew out immense energy.

2 *What powers a quasar?* A quasar is a galaxy powered by a supermassive black hole at its center.

Key Words

active galaxy
BL Lacertae (BL Lac) object
double radio source
Einstein cross

Einstein ring
gravitational lensing
head–tail source
peculiar galaxy

quasar (quasi-stellar radio source)
quasi-stellar object (QSO)

radio lobe
Seyfert galaxy

Key Ideas

• A quasar, or quasi-stellar radio source, is an object that looks like a star but that has a huge redshift. This redshift corresponds to an extreme distance from the Earth according to the Hubble law.

To be seen from Earth, a quasar must be very luminous, typically about 100 times brighter than an ordinary galaxy.

Relatively rapid fluctuations in the brightnesses of some quasars indicate that they cannot be much larger than the diameter of our solar system.

• An active galaxy is an extremely luminous galaxy that has one or more unusual features: an unusually bright, starlike nucleus, strong emission lines in its spectrum, rapid variations in luminosity, or jets or beams of radiation emanating from its core.

An active galaxy with a bright, starlike nucleus and strong emission lines in its spectrum is categorized as a Seyfert galaxy.

BL Lacertae objects have bright nuclei whose cores show relatively rapid variations in luminosity.

• Most double radio sources seem to have an active galaxy located between its two characteristic radio lobes.

A head–tail radio source shows evidence of jets of high-speed particles emerging from an active galaxy.

• Most quasars are probably very distant active galaxies.

• Spectroscopic observations indicate huge concentrations of matter at the centers of certain galaxies.

• The strong energy emission from quasars, active galaxies, and double radio sources may be produced as matter falls toward a supermassive black hole at the center.

Some matter spiraling in toward a supermassive black hole would be squeezed into two oppositely directed beams that eject particles and energy into intergalactic space.

Review Questions

1 Suppose you suspected a certain object in the sky to be a quasar. What sort of observations might you perform to confirm your hypothesis?

2 Explain why astronomers do not use any of the standard candles described in Chapter 15 to determine the distances to quasars.

3 Explain how the rate of variability of a source of light can be used to place an upper limit on the size of the source.

4 What is an active galaxy? How many different kinds of active galaxies can you name? How do they differ from one another?

5 Why do astronomers believe that the energy-producing region of a quasar is very small?

6 How is synchrotron radiation produced?

7 Why does it seem reasonable to suppose that quasars are extremely distant active galaxies?

8   What is a double radio source?

9   How does SS433 (discussed in Chapter 12) compare with a typical double radio source?

10   What is a supermassive black hole? What observational evidence suggests that supermassive black holes might be located at the centers of certain galaxies?

11   Why do many astronomers believe that the engine at the center of a quasar is a supermassive black hole surrounded by an accretion disk?

12   How might the orientation of the jets emanating from the center of a galaxy relative to our line of sight be related to the type of active galaxy that we observe?

Advanced Questions

13   In the 1960s it was suggested that quasars might be compact objects ejected at high speeds from the centers of nearby ordinary galaxies. Why does the absence of blueshifted quasars disprove this hypothesis?

14   When quasars were first discovered, many astronomers were optimistic that these extremely luminous objects could be used to probe distant regions of the universe. For example, it was hoped that quasars would provide high-redshift data from which the Hubble constant could be accurately determined. Why do you suppose these hopes have not been realized?

Discussion Questions

15   Speculate on the possibility that quasars, double radio sources, and giant elliptical galaxies represent an evolutionary sequence.

16   Some quasars show several sets of absorption lines whose redshifts are less than the redshift of the quasars' emission lines. For example, the quasar PKS 0237-23 has five sets of absorption lines, all with redshifts somewhat less than the redshift of the quasar's emission lines. Propose an explanation for these sets of absorption lines.

17 ▶ *Cosmology*

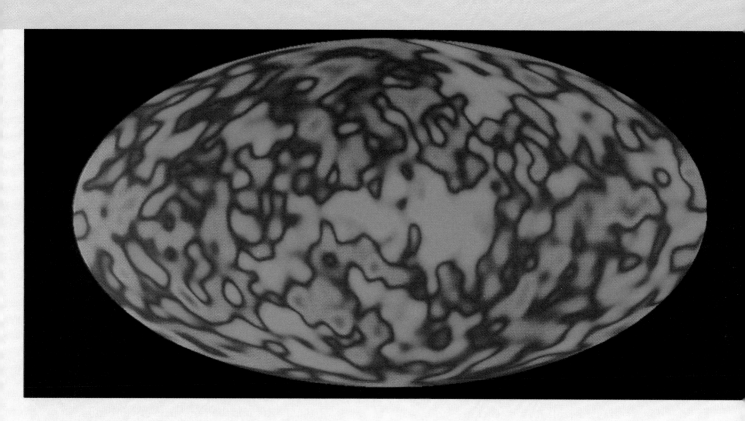

IN THIS CHAPTER

You will explore the structure and fate of the universe. You have studied its individual elements, from our home planet to distant superclusters of galaxies and quasars. Now you will step back. What is the universe? How did it begin? Is it changing? Will it ever end? Some of these questions do not have definitive answers today, but our knowledge is increasing dramatically. You will see evidence supporting our best theory for the origins of the universe—the Big Bang. You will trace the emergence of matter, the formation of galaxies, and how a primordial explosion may have given the universe its shape and determined its ultimate fate.

Structure of the Early Universe This microwave map of the sky, produced from data taken by NASA's Cosmic Background Explorer (COBE), shows temperature variations in the cosmic microwave background. Pink regions are about 0.0003 K warmer than the average temperature of 2.73 K; blue regions are about 0.0003 K cooler than the average. These tiny temperature fluctuations date back to the earliest moments of the universe and are related to the large-scale structure of the universe today. The radiation detected to make this map is from a time 300,000 years after the Big Bang.

WHAT DO YOU THINK?

1 What is the universe?

2 Did the universe have a beginning?

3 Will the universe end?

• •

"IN THE BEGINNING . . . ." People have speculated about the universe since before the dawn of recorded time, but until the development of general relativity and quantum mechanics, science had little to add to the discussion. Today astronomers have means of *testing* theories of the universe's origins, and only one theory proposed so far has survived: the Big Bang. With the aid of this theory, we can now describe scientifically the evolution of the universe—from about 10^{-43} s after the universe began until today—and we are closing in on the very first moment.

The Big Bang

Cosmology is the study of the origin, structure, and evolution of the universe. Modern cosmology got off to a shaky start in 1915, when Einstein published his general theory of relativity. His equations predicted that the universe should either be expanding or contracting. Yet if the universe includes all matter, energy, and spacetime, how can it be expanding? The answer lies in spacetime itself.

17-1 Expansion explains the cosmological redshift

The prediction of a changing universe flew in the face of the belief in an infinite, static universe, a concept promoted by Isaac Newton two centuries earlier. Newton believed that each star is fixed in place and held there under the influence of a uniform gravitational pull from every part of the sky. If the stars were not uniformly distributed, he argued, one region would have more mass than another. The denser region's gravity would then attract other stars, causing them to clump together further. Because he didn't observe this clumping, Newton concluded that the universe must be uniform. He also concluded that without any gravitational sources of change,

the universe must be infinitely old and should continue to exist forever without undergoing any major changes in structure.

Newtonian cosmology and prevailing theology made Einstein doubt the implications of his own theory, and so he missed the opportunity to propose that we live in a changing universe. Instead, he adjusted his elegant equations to yield a static cosmos. Einstein later said that by doing so he had made the biggest blunder of his career. Today we continue to learn from his equations in their original form.

Edwin Hubble is usually credited with the discovery that we live in an **expanding universe.** As we saw in Chapter 15, Hubble discovered the simple linear relationship between the distances to galaxies in other superclusters and the redshifts of those galaxies' spectral lines. This relationship, the Hubble law, states that the farther a galaxy is from Earth, the greater its redshift. Thus, remote galaxies appear to be moving away from us with speeds proportional to their distances from us (review Figure 15-31).

Years of careful observations by many astronomers have shown that no matter in which direction spectral line measurements are made, all distant galaxies are moving away from the Earth. Not only that, but *the Hubble law relating velocities and distances is the same in all directions*. That is, the Hubble constant, H_0, is the same no matter where you aim your telescope. The fact that the recession rates are the same in all directions is a condition called **isotropy.** The universe is isotropic; its global expansion is the same in every direction.

At first glance the general expansion away from us suggests that the Earth is at the center of the universe. After all, where else could we be if all the distant galaxies are moving away from us? In fact, the answer is that we could be practically anywhere in the universe! To understand why, see "Eyes on the Expanding Universe." According to general relativity, spacetime is not rigid. The universe is expanding because spacetime itself expands. As it expands, the distance between superclusters of galaxies grows. Naturally that includes the distances of other superclusters from our own. The *entire* universe is expanding, and we just happen to be a part of it.

Because spacetime is continually expanding, photons from remote galaxies are all redshifted. Imagine a photon coming toward us from a distant galaxy. As the photon travels through space, space is expanding, and so the photon's wavelength becomes stretched. When the photon reaches our eyes, we see a drawn-out wavelength, which is the redshift. The longer the photon's journey,

EYES ON . . .   The Expanding Universe

To understand the universe, scientists build models, a kind of mathematical snapshot of the world about us. To develop such a picture, it often helps to begin with a simple analogy. Since the expansion of the universe is hard to visualize, imagine for a moment a rum raisin cake baking in an oven. As the cake rises, the raisins move further apart. They do not get larger, nor do they move through the batter. Each raisin remains at rest in its own little bit of cake and is carried along as the cake spreads out.

Now think of each raisin as a supercluster of galaxies and the batter between the raisins as the rest of spacetime. As the universe expands, the distance between widely separated superclusters of galaxies grows larger and larger. The expansion of the universe *is* the expansion of spacetime.

Suppose you were on one particular raisin, say the raisin labeled 3 in the drawing below, left. You would see raisins 2 and 4 moving away from you with equal velocities. Every 10 minutes raisins 2 and 4 get one inch farther away as the cake expands as in the drawing below, right. In that same time interval, raisins 1 and 5 move twice as far from you as raisins 2 and 4. To go twice as far away in the same time, raisins 1 and 5 must be moving twice as fast as the closer raisins 2 and 4. This is exactly what Hubble's law of cosmological expansion (recessional velocity = $H_0 \times$ distance) says: *Double the distance and you double the velocity.*

To see why we need not occupy the center of the universe, put yourself now on raisin 4. You see raisins 3 and 5 moving away at the same rate—in fact, they recede at the same rate as you saw raisins 2 and 4 move away when you were on raisin 3. From raisin 4, raisins 1 and 6 are moving away twice as fast as raisins 3 and 5. From any raisin you will see all the other raisins moving away. Similarly, no matter where we are in the universe, we would still see all the other superclusters receding.

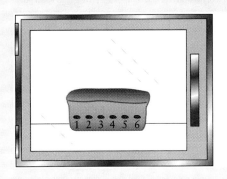
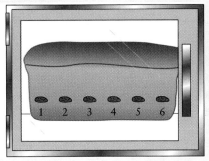

The Expanding Rum Raisin Cake Analogy
The expanding universe can be compared to a rising rum raisin cake. Just as the raisins move apart as the cake rises, all the superclusters of galaxies recede from each other as the universe expands.

the more its wavelength will have been stretched. Thus, photons from distant galaxies have larger redshifts than photons from nearby galaxies.

A redshift caused by the expansion of the universe is properly called a **cosmological redshift.** *The Doppler shifts we have studied so far are caused by an object's motion through spacetime, whereas the cosmological redshift is caused by the expansion of spacetime.* Except for the most distant galaxies and quasars, the Doppler shift and the cosmological redshift predict the same relationship between redshift and motion. That is why Hubble was justified in using the Doppler equation to calculate the recession of galaxies.

17-2 The expanding universe probably originated in an explosion called the Big Bang

The universe has been expanding for billions of years, so in the past the density of matter was much greater than it is at present. If we press far enough into the distant past, all the matter in the universe once must have been concentrated in a state of unimaginably high density. Based on Hubble's observations of an expanding universe, George Gamow in the 1940s proposed that the universe began in a colossal explosion, today called the

Big Bang. Ironically, Sir Fred Hoyle, who coined the name in 1950, is one of the very few astrophysicists who do not believe that the Big Bang occurred.

Imagine watching a movie of two superclusters receding from each other. If we then run the film backward, time runs in reverse and we observe the superclusters approaching each other. The time it would take for the galaxies to collide is the time since the Big Bang. We can use a simple equation to make our first estimate of the age of the universe:

$$\text{Time since Big Bang} = \frac{\text{separation distance}}{\text{recessional velocity}}$$

Recall that Hubble found the relationship

$$\text{Recessional velocity} = H_0 \times \text{separation distance}$$

which can be rewritten as

$$H_0 = \frac{\text{recessional velocity}}{\text{separation distance}}$$

Comparing the first and last of these equations, we see that Hubble's constant is exactly the inverse of the time since the universe began. Using a Hubble constant of 75 km/s/Mpc, we can convert megaparsecs into kilometers, so that the distances cancel out (Toolbox 17-1):

$$\frac{1}{H_0} = \frac{1}{75 \text{ km/s/Mpc}} = 13 \text{ billion years}$$

Because some stars in globular clusters appear to be 14 billion years old, either stars are older than the universe, or our value for Hubble's constant is suspect. Hubble's constant will almost surely turn out to be lower. Recall from Chapter 15 that some observations place it at 55 km/s/Mpc, which implies an age for the universe of 17 billion years. For now, however, although our observations remain incomplete and controversial, we shall continue to use the common estimate of 75 km/s/Mpc.

Because light takes time to travel to Earth, *the farther we look into space, the farther back in time we are seeing.* But if the universe has a finite age of around 13 billion years, we could not see anything farther than 13 billion light-years or so away. At that distance, the objects we see are just being formed. Therefore, all that we can see from the Earth is contained in an enormous sphere (Figure 17-1). The boundary of this sphere is called the **cosmic particle horizon.** This sphere is our en-

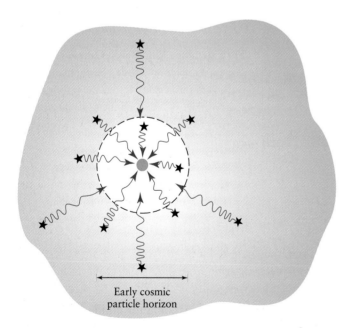

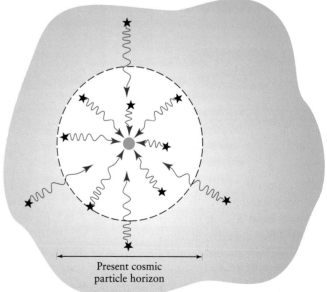

Early cosmic particle horizon

Present cosmic particle horizon

Figure 17-1 The Observable Universe These diagrams show how the universe expanded so rapidly that there are regions in it whose light has not yet reached us. We still see galaxies forming at the farthest reaches of our telescopes' resolving power, within a billion years after the Big Bang. These galaxies, formed at the same time as the Milky Way, appear young, because the light

from their beginnings is just now reaching us. The radius of the cosmic particle horizon is equal to the distance that light has traveled since the Big Bang. Because the Big Bang occurred about 13 billion years ago, the cosmic particle horizon today is about 13 billion light-years away in all directions.

H_0 and the Age of the Universe

The Hubble constant may not look like a measure of time. All we need, however, is a little practice with unit conversion. The trick is to remove all distance units from H_0 and then invert it. Recall that we are using $H_0 = 75$ km/s/Mpc, which has different units of distance in the numerator and denominator. As we saw in Chapter 15, this mix of units makes it easy to determine how fast a galaxy is receding once its distance is known. For example, a galaxy 10 Mpc away is moving away at

$$75 \text{ km/s/Mpc} \times 10 \text{ Mpc} = 750 \text{ km/s}$$

We know that

$$1 \text{ pc} = 3.0856 \times 10^{13} \text{ km}$$

and so 1 Mpc $= 3.0856 \times 10^{19}$ km. Since there are 3.156×10^7 seconds in a year, we get the conversion factors we need:

$$\frac{1}{H_0} = \left(\frac{1}{75 \text{ km/s/Mpc}} \right) \times (3.0856 \times 10^{19} \text{ km/Mpc})$$

$$\times \left(\frac{1}{3.156 \times 10^7 \text{ s/yr}} \right) \approx 13 \times 10^9 \text{ yr}$$

tire observable universe. Light from all the luminous objects in this volume has had time to reach the Earth, but light from objects beyond this sphere has not gotten here yet, and so we are unaware of objects in these regions. With our technology we can practically see light from this horizon. We are thus almost able to see back to the time before galaxies and stars first formed.

17-3 The microwave radiation that fills all space is evidence for the Big Bang

Expansion of the universe by itself is insufficient evidence for accepting the Big Bang theory. Indeed, a plausible alternative called the steady state theory was proposed in 1948 by Fred Hoyle, Herman Bondi, and Thomas Gold. In this theory the universe is constantly expanding, and as it expands, new matter is created that takes the place of the receding matter. The newly formed matter then becomes new galaxies. Where this new matter comes from is unknown, but the laws of physics do not necessarily exclude its creation. So why do virtually all astronomers now believe the Big Bang rather than the steady state? The answer lies in other observations that are consistent with the former but not with the latter.

Shortly after World War II, Ralph Alpher and George Gamow proposed that the universe immediately following the Big Bang was incredibly hot—far in excess of 10^{12} kelvins. It must therefore have been filled with high-energy, short-wavelength radiation. This radiation is called the cosmic microwave background. As the universe expanded, it cooled, and cosmological redshift stretched the wavelengths of the radiation left over from the Big Bang. Think of turning off a hot oven and opening the oven door. As heat disperses throughout the kitchen, the oven cools. Similarly, the temperature of the cosmic microwave background, the remnants of which still fill all of space, should be quite low now—only a few kelvins above absolute zero. In the early 1960s, Robert Dicke, P. J. E. Peebles, and their colleagues at Princeton began designing an antenna to detect this cosmic background radiation at microwave wavelengths.

Meanwhile, just a few miles from Princeton, two physicists had already found the cosmic radiation. Arno Penzias and Robert Wilson of Bell Telephone Laboratories were working on a new horn antenna designed to relay telephone calls to Earth-orbiting communications satellites (Figure 17-2). Penzias and Wilson were deeply puzzled: No matter where in the sky they pointed their antenna, they detected a faint background noise. Careful removal of heat-generating pigeon droppings failed to eliminate the noise completely. Thanks to a colleague, they soon learned of the then-theoretical cosmic microwave background and the work of Dicke and Peebles in trying to locate it. Communicating with the Princeton astronomers, Penzias and Wilson were able to claim the first detection.

Our most accurate cosmic microwave background measurements to date come from the Cosmic Background Explorer (COBE) satellite, which was placed in

Figure 17-2 **The Bell Labs Horn Antenna**
This horn antenna at Holmdel, New Jersey,
was used by Arno Penzias and Robert Wilson
in 1965 to detect the cosmic microwave
background.

orbit about the Earth in 1989 (Figure 17-3). Data from
COBE's spectrometer shown in Figure 17-4 demonstrate
that this ancient radiation has the spectrum of a blackbody with a temperature of 2.73 K. This background radiation, familiarly called the *3-degree background radia-*

tion, is consistent with the Big Bang theory alone. None
of the other cosmological theories, such as the steady
state theory, predict cosmic background radiation.

Like the expansion of the universe, the cosmic microwave background is almost perfectly isotropic: Its intensity is nearly the same in different directions in the
sky. Figure 17-5 shows a map of the microwave sky. It is
slightly warmer than average in the direction of the constellation of Leo and slightly cooler in the opposite direction, toward Aquarius.

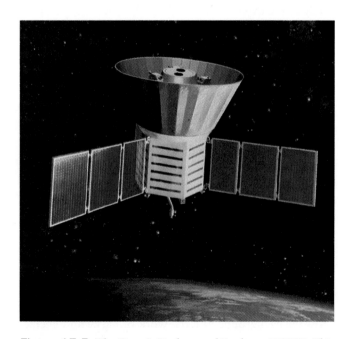

Figure 17-3 **The Cosmic Background Explorer (COBE)** This
satellite, launched in 1989, measured the spectrum and angular
distribution of the cosmic microwave background over a wavelength range of 1 μm to 1 cm. COBE (pronounced co-bee)
found local temperature variations across the sky but no overall deviation from a perfect blackbody spectrum.

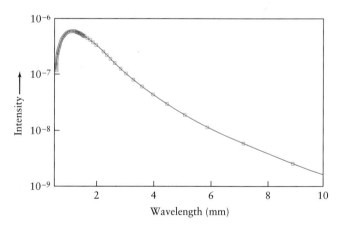

Figure 17-4 **The Spectrum of the Cosmic Microwave Background** The little squares on this graph are COBE's measurements of the brightness of the cosmic microwave background
plotted against wavelength. The data fall along a blackbody
curve for 2.73 K to a remarkably high degree of accuracy. The
peak of the curve occurs at a wavelength of 1.1 mm, in accor-
dance with Wien's law.

Figure 17-5 **The Microwave Sky** This map of the microwave sky was produced from data taken by instruments on board COBE. The galactic center is in the middle of the map, and the plane of the Milky Way runs horizontally across the map. Color indicates temperature: Magenta is warm and blue is cool. The temperature variation across the sky is caused by the Earth's motion through the microwave background.

The tiny temperature variation depicted in Figure 17-5 results from the Earth's overall motion through the cosmos. We are moving through the background radiation because of our rotation and orbit around the Sun. In addition, the Sun, the Milky Way, and even the Local Group of galaxies are themselves in motion (Figure 17-6). If the Earth were exactly at rest with respect to the microwave background, the radiation would be perfectly isotropic.

The observed temperature differences mean that the Earth is moving toward Leo at a speed of 390 km/s. Taking into account the known velocity of the Sun around the center of our Galaxy, we find that the entire Milky Way Galaxy is moving relative to the cosmic microwave background at 600 km/s in the general direction of the Centaurus cluster—some 1.3 million miles per hour. Four nearby clusters of galaxies and an enormous supercluster called the *Shapley concentration* are pulling us in that direction.

17-4 A period of vigorous inflation followed the Big Bang

The cosmic background radiation provides compelling support for the Big Bang. Yet its near-isotropy was, until recently, a major problem for the theory. The opposite sides of our observable universe may be as much as 26 billion light-years apart, twice as far as light can have traveled since the Big Bang. *Why, then, do these unrelated parts of the universe have the same temperature?* How could each region "know" the temperature of the other if they have not had time to "communicate" since the beginning of the universe?

In the early 1980s, Alan Guth offered a remarkable solution to this long-standing dilemma. Guth analyzed the suggestion that the universe experienced extremely rapid expansion during the first second of its existence (Figure 17-7). During this brief **inflationary epoch,** the universe ballooned outward in all directions to become many billions of times its original size. In theory, pressure briefly built up and literally inflated the universe.

Inflation took an initially tiny region of space and enlarged it so much that it became our entire universe. Because this tiny region was already at a single temperature, the cosmic microwave radiation throughout it is still uniform today. Thus, when we examine microwaves from opposite directions in the sky, we are seeing radiation from regions of the universe that were originally in intimate contact. There may be many other "universes" that we cannot see, each at a different temperature.

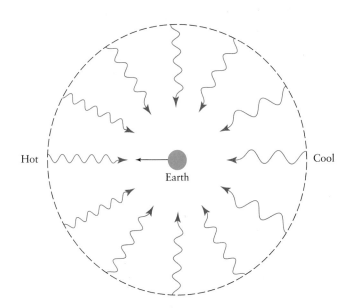

Figure 17-6 **Our Motion Through the Microwave Background** Because of the Doppler effect, the microwave background is slightly warmer in that part of the sky toward which we are moving. Recent measurements indicate that our Galaxy, along with the rest of the Local Group, is moving in the general direction of the Centaurus cluster.

Figure 17-7 **Size of the Universe With and Without Inflation** According to the inflationary model, the universe expanded by a factor of about 10^{50} shortly after the Big Bang. This sudden growth in the size of the observable universe probably occurred during a very brief interval, as indicated by the vertical shaded area on the graph. For comparison, the projected size of the universe without inflation is also shown.

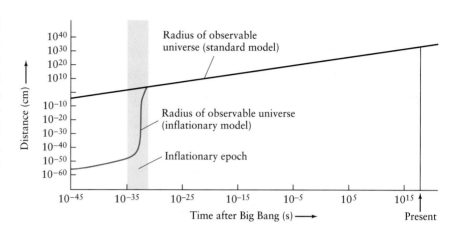

A Brief History of Matter

17-5 All four forces in nature initially had the same strength

The composition of the early universe was profoundly different than it is today. To understand why, we need to briefly expand on the nature of the four physical forces: gravity, electromagnetism, and the strong and weak nuclear forces. We are all familiar with the force of gravity. The electromagnetic force, which holds electrons in orbit about the nuclei in atoms, also acts over long distances. Over large volumes of space, the net effects of electromagnetism cancel each other, because there is a negative electric charge for every positive charge and a south magnetic pole for every north magnetic pole. Gravity, on the other hand, has no "negative mass" to cancel itself out. Therefore gravity dominates the universe at astronomical distances.

The **strong nuclear force** holds protons and neutrons together inside the nuclei of atoms. Without this force, nuclei would disintegrate because of the electromagnetic repulsion of the positively charged protons. Thus, the strong nuclear force overpowers the electromagnetic force inside nuclei. The **weak nuclear force** is at work in certain kinds of radioactive decay, such as the transformation of a neutron into a proton. It is so weak that it does not hold anything together. The strong and the weak nuclear forces are both said to be short range because their influences extend only over distances less than about 10^{-15} m.

Protons and neutrons are composed of more basic particles called **quarks.** A proton is composed of two "up" quarks and one "down" quark, whereas a neutron is made of two down quarks and one up quark. The weak nuclear force is at work whenever a quark changes from one variety to another. For example, in radioactive decay, a neutron transforms into a proton, and one of the neutron's down quarks changes into an up quark.

To examine details of the physical forces, scientists use particle accelerators that hurl high-speed electrons and protons at targets. In such experiments, physicists find that the different forces begin to behave the same way as the particles' speeds approach the speed of light. In fact, during experiments at the CERN accelerator in Europe in the 1980s, particles were slammed together with such violence that the electromagnetic force and the weak nuclear force had equal strength. It seems that all four forces would have the same strength if particles were slammed together with energy trillions of times greater than that achievable in the CERN accelerator.

Physicists have no hope of building accelerators that powerful. However, immediately after the Big Bang, the universe was so hot that particles in it were moving with tremendous speeds. The earliest moments of the universe thus provide a laboratory in which scientists can explore some of the most elegant ideas in physics. Figure 17-8 summarizes the connections between particle physics and cosmology. Figure 17-9 shows that during the beginning moments of the universe (from 0 to 10^{-43} s, called the **Planck time**), the four forces were unified, meaning there was only one force in nature.

By the end of the Planck time, the energy of particles in the universe had declined to the extent that gravity was no longer unified with the other three forces. We can therefore say that at 10^{-43} s, when the temperature of the universe was 10^{32} K, gravity "froze out" of the otherwise unified hot soup that filled all space.

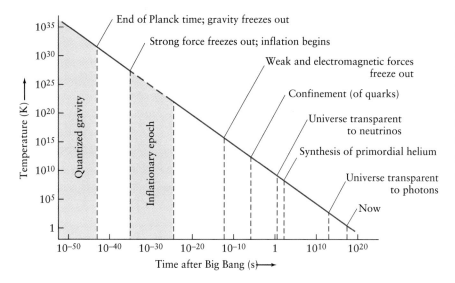

Figure 17-8 **The Early History of the Universe** As the universe cooled, the four forces "froze out" of their initial unified state. The inflationary epoch lasted from 10^{-35} s to 10^{-24} s after the Big Bang. Neutrons and protons froze out one-millionth of a second after the Big Bang. The universe became transparent to light (that is, photons decoupled from matter) when the universe was 300,000 years old.

By 10^{-35} s, the temperature of the universe had fallen to 10^{27} K, and the energy of particles had declined to the extent that the strong nuclear force was no longer unified with the electromagnetic and weak nuclear forces. At this point, the strong nuclear force emerged, distinct from the other two forces. Calculations suggest that the inflationary epoch lasted from 10^{-35} s to about 10^{-24} s, during which time the universe increased in size enormously, perhaps by a factor of 10^{50}.

At 10^{-12} s, when the temperature of the universe had dropped to 10^{15} K, a final freeze-out separated the electromagnetic force from the weak nuclear force. From that moment on, all four forces interacted with particles essentially as they do today.

The next significant event occurred at 10^{-6} s, when the universe's temperature was 10^{13} K. Prior to this moment, particles collided so violently that individual protons and neutrons could not exist, because they were constantly being fragmented into quarks. After this moment, appropriately called **confinement**, quarks could finally stick together to form individual protons and neutrons.

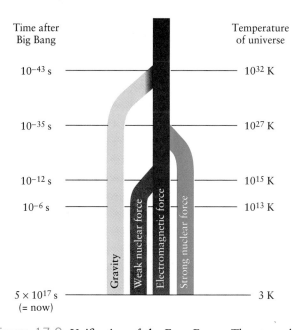

Figure 17-9 **Unification of the Four Forces** The strength of the four physical forces depends on the speed or energy with which these particles interact. As shown in this diagram, the higher the temperature of the universe, the more the forces resemble each other. Also included here is the age of the universe at which the various forces were equal.

17-6 During the first second, most of the matter and antimatter in the universe annihilated each other

Early in the first second, the universe was so hot that photons possessed incredibly high energies. This energy was enough to create matter and antimatter according to Einstein's equation $E = mc^2$. As we saw in Chapter 9, to make a particle of mass m, you need an amount of energy E at least as great as mc^2, where c is the speed of light.

Figure 17-10 **Pair Production and Annihilation** A particle and an antiparticle can be created when high-energy photons collide. Conversely, a particle and an antiparticle can annihilate each other by giving up energy in the form of gamma rays.

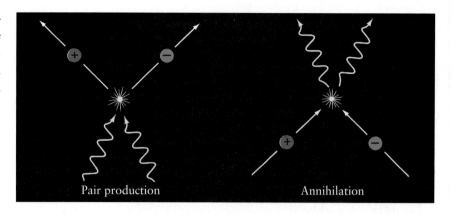

Pair production   Annihilation

The creation of matter from energy is routinely observed in laboratory experiments involving high-energy gamma rays. When a highly energetic photon collides with a second photon, both vanish and matter appears in their place, as sketched in Figure 17-10. This process, called **pair production,** always creates one ordinary particle and one so-called antiparticle. For instance, if one of the particles is an electron, the other is an antielectron (Figure 17-11). (Recall from Chapter 13 that pairs of particles can also be produced near a black hole.)

There is nothing mysterious about antimatter except that it has an opposite electric charge to matter. An antielectron (usually called a *positron*) is an ordinary electron except that it has a positive charge. At times earlier than 1 s in the life of the universe, pair production was continually occurring. In other words, shortly after the Big Bang all space was chock full of protons, neutrons, electrons, and their antiparticles, which were immersed in an inconceivably hot bath of high-energy photons.

After that brief second, pair production ceased. As the universe expanded, the temperature declined until the photons could no longer create pairs of particles and antiparticles. A particle and its antiparticle would often collide, annihilating each other and converting their mass back into high-energy photons. But nothing could replenish the supply of particles and antiparticles.

If particles and antiparticles were always created or destroyed in pairs, why aren't they found in equal numbers? For every proton created, there should have been an antiproton; for every electron there should have been an antielectron. Stranger still, had particles kept on annihilating their antiparticles, no matter would be left at all.

Obviously this did not happen. Physicists therefore say that a "symmetry breaking" occurred. This means that the number of particles was slightly greater than the number of antiparticles. For every billion antiprotons, perhaps a billion plus one protons formed; for every billion antielectrons, a billion plus one electrons formed. As far as we can tell, all the antiparticles annihilated with normal particles shortly after they all formed. The universe was left with the slight excess of normal matter.

As the universe expanded further, the remaining protons, neutrons, and electrons began colliding and fusing together. For the most part they were quickly separated again by the gamma rays in which they were bathed. However, during the first three minutes, enough fusions occurred that were not subsequently unfused to create most of the helium and the trace elements of lithium and beryllium that exist today. Hydrogen, helium, lithium, and beryllium are the four lowest-mass elements. After a few minutes, the universe was too cool to allow fusion to create more massive elements such as carbon and iron. All elements other than the four lowest-mass ones formed later as a result of stellar evolution.

17-7 With the formation of neutral atoms, matter came to dominate the universe

For the first 300,000 years, photons prevented neutral atoms from forming. Electrons were ionized as quickly as they came to orbit a nucleus. That early time, called the **radiation-dominated universe,** lasted until the universe was so large and photons were so spread out that neutral atoms could form without being ionized. Thereafter the behavior of matter was determined by gravitational forces. We live in this **matter-dominated universe.**

Although the universe is dominated by matter today, the number of photons left over from the Big Bang is

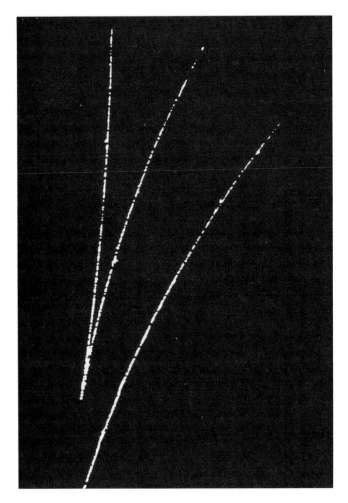

The transition from a radiation-dominated universe to a matter-dominated universe, called **decoupling**, occurred about 300,000 years after the Big Bang. We can use Wien's law to calculate the temperature of the cosmic background radiation at this time. The peak wavelength of 0.001 mm existing at decoupling corresponds to 3000 K. The temperature history of the universe is graphed in Figure 17-12.

In the first 300,000 years, the universe was completely filled with a shimmering expanse of high-energy photons colliding vigorously with protons and electrons. This state of matter, called *plasma*, is opaque, just as the glowing gases inside a neon advertising sign or fluorescent light bulb are opaque. The term **primordial fireball** describes the universe during this time.

After 300,000 years, when the temperature of the radiation field fell below 3000 K, the photons no longer had enough energy to keep the protons and electrons apart. Protons and electrons everywhere began combining to form hydrogen atoms. Because hydrogen gas is clear, the universe suddenly became transparent! The photons that just a few seconds earlier had been vigorously colliding with charged particles could now stream unimpeded across space. Today we see these same photons as the microwave background.

Figure 17-11 **Creation of Matter from Energy** This photograph shows the conversion of a gamma ray into matter inside a bubble chamber, a device filled with liquid hydrogen that is designed to make the path of a charged particle visible as a long row of tiny bubbles. The path of the gamma ray is not visible because photons are electrically neutral. Near the bottom of the photograph, the energy carried by the gamma ray is converted into an electron and an antielectron. Because of a magnetic field surrounding the bubble chamber, the electron is deflected to the right while the antielectron veers toward the left. The path of a stray electron is also seen at the right.

very large. From the physics of blackbody radiation, astronomers calculate that there are now 550 million cosmic microwave photons in every cubic meter of space. In contrast, if all the visible matter in the universe were uniformly spread throughout space, there would be roughly one hydrogen atom in every three cubic meters of space. In other words, photons outnumber atoms by roughly a billion to one.

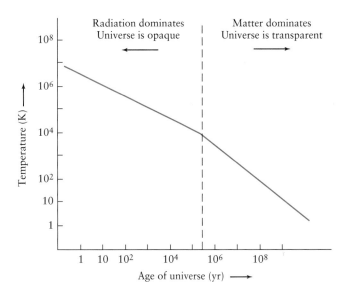

Figure 17-12 **The Declining Temperature of the Universe** As the universe expands, the photons in the radiation background become increasingly redshifted and the temperature of the radiation field falls. Roughly 300,000 years after the Big Bang, when the temperature fell below 3000 K, hydrogen atoms formed and the universe became transparent.

This dramatic moment, when the universe went from being opaque to being transparent, is referred to as the **era of recombination.** Because the universe was opaque in its first 300,000 years, we cannot see any further into the past than the era of recombination. The microwave background contains the most ancient photons we shall ever be able to observe. This microwave background is today a ghostly relic of its former dazzling splendor.

The Big Bang explains the existence of spacetime, matter, and energy. It accounts for the hydrogen and most of the helium that exists today. But so far we cannot use it to explain the existence of superclusters of galaxies, clusters of galaxies, or even individual galaxies. Indeed, inflation suggests a universe far too uniform for these large features to evolve.

Happily, large-scale structure can still occur despite inflation—if there are bumps in the density of the universe. Quantum mechanics predicts just such tiny density bumps at the beginning of the universe. Inflation expanded these tiny inhomogeneities to the colossal sizes of superclusters of galaxies.

All these bumps throughout the inflated universe must have been only small density enhancements for the microwave background to appear as uniform as it does now. Part of the mission of the COBE satellite was to find the bumps, and find them it did! They are only about 1 part in 10,000 denser than the average density of matter in the universe (see the opening figure for this chapter). COBE's images are taken from radiation emitted at the time of decoupling.

Wherever the universe had a tiny extra density of matter after inflation, the gravity there would have kept all the mass in that region bound together. This material would then have formed one supercluster and, inside it, any number of clusters. Galaxies would then have formed in individual clusters, depending on the even-smaller bumps found in each cluster.

Galaxy Formation

17-8 Galaxies formed from huge clouds of primordial gas

Astronomers can probe the past to gain important clues about galactic evolution simply by looking deep into space. The more distant a galaxy is, the longer its light takes to reach us. Consequently, *as we examine galaxies at increasing distances from Earth, we are seeing them at increasingly earlier stages of their lives.*

By observing remote galaxies, astronomers have discovered that galaxies were bluer and brighter in the past than they are today. These changes in color and brightness suggest that newly formed galaxies had an over-abundance of young, bright, hot, massive stars (recall that these are also the stars that highlight spiral arms). As galaxies aged, these blue O and B stars became red supergiants and eventually died off. Therefore, galaxies became somewhat redder and dimmer. This was especially true for elliptical galaxies, which formed nearly all their stars in one vigorous burst of activity that lasted for only about a billion years.

In contrast, spiral galaxies have been forming stars for at least the past 10 billion years, although at a gradually decreasing rate. There is still plenty of interstellar hydrogen in the disks of spiral galaxies like our Milky Way to fuel star formation today. Figure 17-13 compares the rates at which spiral and elliptical galaxies form stars.

Galaxies formed from huge clouds of hydrogen and helium that comprised virtually all the matter in the early universe. Based on observations of the most ancient, metal-poor stars we have seen and on our model of stellar evolution, the galaxies formed 12 to 18 billion years ago. Under the action of gravity, these pregalactic clouds of gas started to contract and form *protogalaxies* stud-

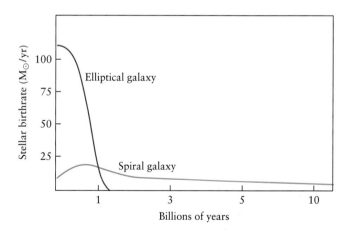

Figure 17-13 **Stellar Birth Rate in Galaxies** Most of the stars in an elliptical galaxy are created in a brief burst of star formation when the galaxy is very young. In spiral galaxies, stars form at a more leisurely pace that extends over billions of years.

ded with the first generation of stars. Astronomers are just now beginning to observe protogalaxies.

17-9 Star formation determines a galaxy's initial structure

The rate of star formation in a protogalaxy determines whether it initially becomes a spiral or an elliptical. If the rate is low, then the gas from which the galaxy formed has plenty of time to settle by collision with other in-falling gas into a flattened disk. A flattened disk is the natural consequence of the rotation of the original gas cloud that formed the entire galaxy. It is quite analogous to the formation of the disk of the solar system. Star formation continues in the protogalactic disk because it contains an ample supply of hydrogen, and thus a spiral galaxy is created.

If, however, the initial stellar birthrate is high in the protogalaxy, virtually all of the pregalactic gas is used up in the creation of stars before a disk can form. In this case, an elliptical galaxy is created. Figure 17-14 depicts these contrasting scenarios.

Figure 17-14 **The Creation of Spiral and Elliptical Galaxies** A galaxy begins as a huge cloud of primordial gas that collapses gravitationally. (a) If the rate of star birth is low, then much of the gas collapses to form a disk and a spiral galaxy is created. (b) If the rate of star birth is high, then the gas is converted into stars before a disk can form, resulting in an elliptical galaxy.

Galactic evolution is a difficult and often controversial subject in which many questions remain. For instance, what happened in the early universe to cause the primordial hydrogen and helium to gather in clouds destined to evolve into galaxies instead of becoming objects a million times bigger or smaller?

Even more troublesome is the puzzle of dark matter. The observable stars, gas, and dust in a galaxy account for only about 10% of its mass. We have very little idea of what the remaining 90% looks like, how it's distributed in space, or what it's made of. With this level of ignorance, any discussion of galactic evolution must be woefully incomplete. Because of the great distances involved, observing the early universe is among the greatest challenges facing astronomers.

The Fate of the Universe

We now turn to the fate of the universe. Will it last forever? Or will it someday stop expanding and collapse?

17-10 The future of the universe is determined by the average density of matter

During the 1920s, Alexandre Friedmann in Russia, Georges Lemaître in France, Willem de Sitter in the Netherlands, and, of course, Einstein himself applied the general theory of relativity to cosmology. They found that gravity is slowing the cosmological expansion. We say that the universe is *decelerating.*

Consider a cannonball shot upward from the surface of the Earth. The Earth's gravitational force slows the ball's ascent. If the cannonball's speed upward is less than the escape velocity from the Earth's gravity (about 11 km/s), it will fall back to Earth. If the cannonball's speed equals the escape velocity, it will just barely escape falling back to Earth, coming to a stop an infinite distance away. And if the ball's speed exceeds the escape velocity, it can easily leave the Earth, despite the relentless pull of the Earth's gravity.

The universe obeys a similar set of rules. The mutual attraction of the receding superclusters on each other is slowing the recession. *The universe is expanding, but more slowly all the time.* The universe is expanding today only because of the momentum given matter by the Big Bang and the inflationary period that followed; no

external force is pushing the superclusters apart. Whether the expansion ceases is critical in determining the fate of the universe.

If the average density of matter throughout space is sufficiently low, gravity will be too weak to stop the universe's expansion. In that case, the universe will continue to expand forever, like the cannonball that escapes from the Earth's gravitational clutches. The superclusters of galaxies will continue to rush away from each other forever. Under such circumstances, we say that the universe is **unbound.**

Conversely, if the average density of matter across space is sufficiently high, then gravity will eventually halt the expansion. Like the cannonball that falls back down, the universe will begin contracting. In this case gravity will pull the superclusters of galaxies back toward each other. If this is the case, the universe is **bound.**

If the average density of matter is at the *critical density,* the superclusters will just barely manage to keep moving away from each other forever. The future history of the universe depends on two things: its present expansion rate and its average mass density. Estimates of the critical density depend on the Hubble constant. Using a Hubble constant of 75 km/s/Mpc, we find that the critical density is 2.4×10^{-26} kg/m^3, which is equivalent to a density of about 14 hydrogen atoms per cubic meter of space.

Present-day estimates of the density are not accurate enough to tell us whether the universe is bound or unbound. The average density of luminous matter that we see in the sky seems to be about 5×10^{-28} kg/m^3, which is about $\frac{1}{50}$ of the critical density. As discussed in Chapter 15, we can see only a small fraction of all the matter in the universe. The rest remains to be discovered. Some physicists argue that this dark matter is probably not composed of particles like protons, neutrons, or electrons; otherwise, we would have already detected it. Instead, they believe that the dark matter is vastly different from anything we have ever encountered.

17-11 The deceleration of the universe is related to the redshifts of extremely distant galaxies

The deceleration of the universe shows up as a deviation from the straight-line relationship predicted by the Hubble law. Consider galaxies several billion light-years from Earth. The light from these galaxies has taken billions of years to get to our telescopes, so redshift measurements reveal how fast the universe was expanding billions of years ago. Because the universe was expanding faster in the past than it is today, our data deviate slightly from the straight-line Hubble law.

Figure 17-15 displays the relationship between the deceleration of the universe and the Hubble law for great distances. Astronomers call the amount of deceleration the **deceleration parameter,** designated q_0. Appropriately, $q_0 = 0$ corresponds to no deceleration at all. This is possible only if the universe is completely empty and thus there is no gravity to slow down the expansion. As sketched in Figure 17-15, if q_0 is 0, the universe will expand forever at a constant rate.

If $q_0 = \frac{1}{2}$, matter is at the critical density. Such a universe just barely manages to expand forever. If q_0 is between 0 and $\frac{1}{2}$, the universe is unbound and will continue to expand forever. Such a universe contains matter at less than the critical density. If q_0 is greater than $\frac{1}{2}$, the universe is bound and is filled with matter of a density greater than the critical density. This universe is doomed to collapse upon itself and ultimately end in a **Big Crunch** in the extremely distant future.

During the Planck time, in the first 10^{-43} s, the universe was so dense and particles were interacting so vio-

lently that no present physical theory can properly describe what happened then. However, immediately after the Planck time, the fate of the universe became very sensitive to the density of matter. Calculations demonstrate that the slightest deviation from the precise critical density would have mushroomed very rapidly, doubling itself every 10^{-35} s. If the density were even slightly less than the critical density near the beginning of time, the universe would soon have become wide open and virtually empty. If, on the other hand, the density were slightly greater than the critical density, the universe would soon have become tightly closed and so packed with matter that the entire cosmos would have rapidly collapsed in a Big Crunch. In other words, immediately after the Big Bang, the fate of the universe hung in the balance, like a pencil teetering on its point, so that the tiniest deviation from critical density would have rapidly propelled the universe away from the special case of $q_0 = \frac{1}{2}$.

Ever since Edwin Hubble discovered that the universe is expanding, astronomers have struggled to determine the deceleration parameter. In principle it should be possible to determine q_0 by measuring the redshifts and distances of many remote galaxies and then plotting the data on a Hubble diagram. If the data points fall above the $q_0 = \frac{1}{2}$ line, the universe is bound. If the data points fall between the $q_0 = 0$ and $q_0 = \frac{1}{2}$ lines, the universe is unbound.

Unfortunately, such observations are extremely difficult to make. The deceleration of the universal expansion is so slight that galaxies nearer than a billion light-years are of no help in determining q_0. And beyond a billion light-years, uncertainties cloud determinations of distance. For example, Figure 17-16 shows data obtained by Allan Sandage with the 200-in. telescope at the Palomar Observatory. Note how the data points are scattered about the $q_0 = \frac{1}{2}$ line. This scatter prevents us from determining conclusively whether the universe is bound or unbound. Nevertheless, because the data lies close to the $q_0 = \frac{1}{2}$ line, many astronomers think that the universe is not far from being marginally bound.

During the 1960s and 1970s, various teams of astronomers reported various values for q_0, some slightly larger than $\frac{1}{2}$ and some slightly smaller. Because $q_0 = \frac{1}{2}$ is the special case separating a bound cosmological model from an unbound model, the predicted fate of the universe swung back and forth. According to some data, the universe seems to be just barely unbound and infinite, whereas other data indicate that the universe is just barely bound and ultimately doomed to collapse.

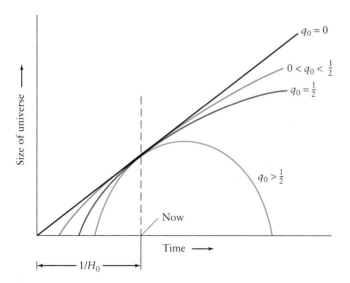

Figure 17-15 **Deceleration and the Size of the Universe** This graph shows the possible evolution of the universe. The case $q_0 = 0$ is an empty universe. The case $q_0 = \frac{1}{2}$ is a marginally bound universe. If q_0 is between these two values, then the universe is unbound and will expand forever. If q_0 is greater than $\frac{1}{2}$, the universe is bound and will some day collapse.

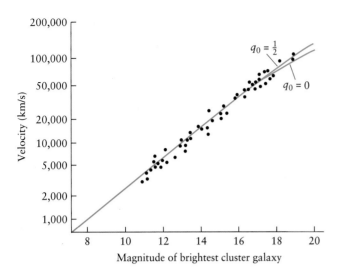

Figure 17-16 **The Hubble Diagram** This graph shows the Hubble diagram extended to include extremely remote galaxies. The magnitude of the brightest galaxy in a cluster is directly correlated with the distance to the cluster. If the data points fall between the curves marked $q_0 = 0$ and $q_0 = \frac{1}{2}$, then the universe is unbound. If the data points fall above the curve marked $q_0 = \frac{1}{2}$, then the universe is bound.

17-12 The shape of the universe is governed by the density of matter

We can also relate the fate of the universe to the fabric of spacetime. Our present understanding of the universe is based on Einstein's general theory of relativity, which explains that gravity curves the fabric of spacetime. By measuring this curvature, we should be able to discover whether the universe is bound or unbound and thus discover its ultimate fate.

To see what astronomers mean by the curvature of the universe, imagine shining two powerful laser beams out into space. Suppose that we can align these two beams so that they are perfectly parallel as they leave the Earth. Finally, suppose that nothing gets in the way of these two beams, so that we can follow them for billions of light-years across the universe, across the spacetime whose curvature we wish to detect.

With this arrangement, there are only three possibilities. First, we might find that our two beams of light remain perfectly parallel, even after traversing billions of light-years. In this case, we would say that space is not curved: The universe has *zero curvature* and space is *flat*. This is the $q_0 = \frac{1}{2}$ universe. As we have seen, such a universe will (barely) last forever.

Alternatively, we might find that our two beams of light gradually get closer and closer together as they move across the universe, eventually intersecting at some enormous distance from Earth. In this case, space would not be flat. Recall that lines of longitude on the Earth's surface are parallel at the equator but intersect at the poles. Thus, in this case, the shape of the universe would be analogous to the surface of a sphere. We would then say that space is *spherical* and the universe has *positive curvature*. Such a universe is said to be **closed**. In this case, $q_0 > \frac{1}{2}$. The universe would expand to a finite size and then collapse back to the Big Crunch.

The third and final possibility is that the two parallel beams of light would gradually diverge, becoming farther and farther apart as they moved across the universe. In this case, we would say that the universe has negative curvature. A horse's saddle is a good example of a negatively curved surface. Parallel lines drawn on a saddle always diverge. Mathematicians say that saddle-shaped surfaces are hyperbolic. Thus, in a negatively curved universe, we would describe space as *hyperbolic*. In this case $q_0 < \frac{1}{2}$. Such a universe is said to be **open** and will last forever.

Figure 17-17 illustrates these cases with two-dimensional analogies: a sphere, a plane, and a saddle. Of course, real space is three-dimensional. These relationships are summarized in Table 17-1. *Both the flat and the hyperbolic universes are infinite*. They extend forever in all directions.

17-13 Inflation accounts for the flatness of the universe

Since observations reveal that q_0 is today approximately $\frac{1}{2}$, the density of the universe immediately after the Big Bang must have been equal to the critical density to an incredibly high degree of precision. Calculations demonstrate that, in order for q_0 to be even roughly $\frac{1}{2}$ today, density right after the Big Bang must have been equal to the critical density *to better than 50 decimal places!*

What could have happened immediately after the Planck time to ensure that the density was critical to such an astounding degree of accuracy? Because critical density means that space is flat, this enigma is called the **flatness problem.**

Inflation accounts for the flatness of the observable universe. The newborn universe might still have begun with a highly curved shape, because its mass might have strongly curved spacetime, just as the mass of a black

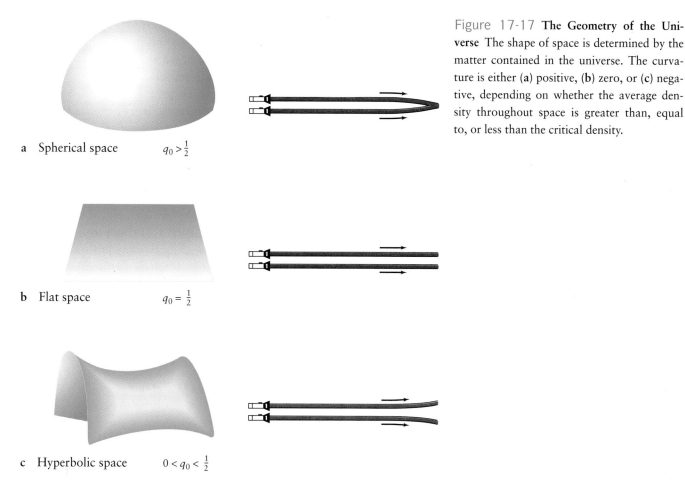

a Spherical space $q_0 > \frac{1}{2}$

b Flat space $q_0 = \frac{1}{2}$

c Hyperbolic space $0 < q_0 < \frac{1}{2}$

Figure 17-17 **The Geometry of the Universe** The shape of space is determined by the matter contained in the universe. The curvature is either (**a**) positive, (**b**) zero, or (**c**) negative, depending on whether the average density throughout space is greater than, equal to, or less than the critical density.

hole curves spacetime today. Nonetheless, inflation expanded the universe so much that the portion we can see today looks quite flat. The observable universe, which extends outward from Earth roughly 13 billion light-years in all directions, is such a tiny fraction of the entire inflated universe that any overall curvature is virtually undetectable. To see why, think about a small portion of the Earth's surface, such as your backyard. For all practical purposes, it is impossible to detect the Earth's curvature over such a small area, so your backyard looks very

TABLE 17-1

Shape and Fate of the Universe

Geometry of space	Curvature of space	Average density throughout space	Deceleration parameter (q_0)	Type of universe	Ultimate future of the universe
Spherical	Positive	Greater than critical density	Greater than $\frac{1}{2}$	Closed	Eventual collapse
Flat	Zero	Exactly equal to critical density	Exactly equal to $\frac{1}{2}$	Flat	Perpetual expansion (just barely)
Hyperbolic	Negative	Less than critical density	Between 0 and $\frac{1}{2}$	Open	Perpetual expansion

flat. Like your backyard, the segment of space within our horizon looks flat.

If q_0 is indeed nearly equal to $\frac{1}{2}$, the average density of matter must be almost exactly equal to the critical density. Suppose, then, that during the Big Bang the average density of matter was slightly larger (or smaller) than the critical density. How would this deviation grow (or decrease) as the universe evolved?

The answer hinges on the age of the universe and q_0. If $q_0 = 0$, then the universe is empty. As we saw earlier, its age is then $1/H_0$. For a Hubble constant of 75 km/s/Mpc, that turns out to be 13 billion years. If the universe is marginally bounded, $q_0 = \frac{1}{2}$, then the age of the universe equals $\frac{2}{3}(1/H_0)$, or about 9 billion years. If the universe is unbounded, it has an age between these two values. If the universe is bounded, then its age is less than 9 billion years. To see how $q_0 = 0$ governs the age of the universe, note where the different lines begin in Figure 17-15.

All of these estimates for the age of the universe are uncomfortably low. It is impossible for the universe to be younger than the stars it contains; as we mentioned before, some stars seem to be 14 billion years old. Perhaps, then, our understanding of the ages of stars is not correct. Perhaps H_0 is significantly less than 75 km/s/Mpc. Or maybe a straightforward cosmology involving only two constants (H_0 and q_0) does not adequately describe the universe.

As we have seen, H_0 may be as low as 55 km/s/Mpc, making the age of the universe 12 billion years old if $q_0 = \frac{1}{2}$. Although still low, this figure is much closer to the ages of the oldest known stars.

So what is the fate of the universe? To put it bluntly, we are not yet sure. But we have some very important clues. To date astronomers have not observed enough matter to conclude that the universe is closed. Going strictly by the known mass, we would conclude that the q_0 is less than $\frac{1}{2}$ and the universe will expand forever. However, the amount of matter that has been identified is not thousands, millions, or trillions of times too little to close the universe and make it recollapse. Rather, the known mass is already a few percent of the critical density. This amount is too close to critical density to be a coincidence.

Because the inflationary model of the Big Bang *requires* that the universe have just the critical mass, there is growing evidence that $q_0 = \frac{1}{2}$. If this turns out to be true, then the universe is going to expand forever, more and more slowly all the time. In such a universe, the stars will eventually use up all the available hydrogen and helium fuel. They will wink out and all that will be left are black holes, stellar remnants, planets, photons, and slowly expanding spacetime. If the inflationary theory of the universe is correct, the universe is fated to become forever cold and dark.

WHAT DID YOU THINK?

1 *What is the universe?* It is all matter, energy, and spacetime.

2 *Did the universe have a beginning?* Yes, it probably occurred between 12 and 18 billion years ago in the Big Bang.

3 *Will the universe end?* Present observations suggest that it will expand forever.

Key Words

Big Bang	cosmological redshift	inflationary epoch	quark
Big Crunch	cosmology	isotropy	radiation-dominated
bound universe	deceleration parameter	matter-dominated	universe
closed universe	(q_0)	universe	strong nuclear force
confinement	decoupling	open universe	unbound universe
cosmic microwave	era of recombination	pair production	weak nuclear force
background	expanding universe	Planck time	
cosmic particle horizon	flatness problem	primordial fireball	

Key Ideas

• The universe began as an infinitely dense cosmic singularity that expanded explosively in an event called the Big Bang.

  The Hubble law describes the ongoing expansion of the universe; the distance between widely separated galaxies is growing ever larger.

  The observable universe extends about 13 billion light-years in every direction from the Earth to what is called the cosmic particle horizon. We cannot see objects beyond the cosmic particle horizon because light from these objects has not had enough time to reach us.

• According to the theory of inflation, much of the material originally near our location moved far beyond the limits of our observable universe during the inflationary period. The observable universe today is thus expanding into space containing matter and radiation that was in close contact with our matter and radiation during the first instant after the Big Bang. This explains the isotropic appearance of the universe.

• Four basic forces—gravity, electromagnetism, the strong nuclear force, and the weak nuclear force—explain all of the interactions observed in the universe.

  According to current theory, all four forces were identical just after the Big Bang.

  At the end of the Planck time, gravity "froze out" to become a distinctive force. A short time later, the strong nuclear force became a distinctive force. A final "freeze-out" separated the electromagnetic force from the weak nuclear force.

• A Hubble constant of 75 km/s/Mpc implies that the age of the universe ($1/H_0$) is 13 billion years or less. This conclusion is distressing because the oldest stars seem to have ages of about 14 billion years.

• Before the Planck time (about 10^{-43} s after the Big Bang), the universe was so dense that known laws of physics do not properly describe the behavior of space, time, and matter.

• The cosmic microwave background radiation is the greatly redshifted remnant of the very hot universe that existed about 300,000 years after the Big Bang.

During the first 300,000 years of the universe, matter and energy formed an opaque plasma called the primordial fireball.

By 300,000 years after the Big Bang, expansion caused the temperature of the universe to fall below 3000 K, enabling protons and electrons to combine to form hydrogen atoms; this event is called the era of recombination.

• The universe became transparent during the era of recombination, meaning that the microwave background radiation contains the oldest photons in the universe.

• The universe today is matter-dominated, but during the first 300,000 years, before the background radiation became greatly redshifted, the universe was radiation-dominated.

• The average density of matter in the universe determines the curvature of space and the ultimate fate of the universe.

  If the average density of matter in the universe is greater than the critical density, then space is spherical (with positive curvature), the deceleration parameter q_0 has a value greater than $\frac{1}{2}$, and the universe is closed (bound) and will ultimately collapse.

  If the average density of matter in the universe is less than the critical density, then space is hyperbolic (with negative curvature), q_0 has a value less than $\frac{1}{2}$, and the universe is open (unbound) and will continue to expand forever.

  If the average density of matter in the universe is exactly equal to the critical density, then space is flat (with zero curvature), q_0 is exactly equal to $\frac{1}{2}$, and the universe is marginally bound, and expansion will just barely continue forever.

• That the universe is nearly flat and that the cosmic microwave background is almost perfectly isotropic may be the result of a brief period of very rapid expansion (the inflationary epoch).

Review Questions

1 What does it mean when astronomers say that we live in an expanding universe?

2 Explain the difference between a Doppler shift and a cosmological redshift. Why is it incorrect to think of the redshifts of remote galaxies and quasars as being a result of the Doppler effect?

3 In what ways are the fate of the universe, the geometry of the universe, and the average density of the universe related?

4 Suppose that the universe will expand forever. What will eventually become of the microwave background radiation?

5 What does it mean to say that the universe is matter-dominated? When was the universe radiation-dominated?

6 Explain the difference between an electron and an anti-electron?

7 Where did most of the photons in the cosmic microwave background come from?

8 Describe an example of each of the four basic forces in the physical universe.

9 What is the observational evidence for (a) the Big Bang, (b) the inflationary epoch, and (c) the confinement of quarks?

Advanced Questions

* 10 How old would a $q_0 = \frac{1}{2}$ universe be if $H_0 = 60$ km/s/Mpc?

11 Explain why the cosmic microwave background is a major blow to the steady state theory.

12 The separation of nuclei by energetic photons in the early universe was caused by the same mechanism as what other transformation we studied earlier?

Discussion Questions

13 Suppose we were living in a radiation-dominated universe. How would such a universe be different from what we now observe?

14 Discuss the theological implications of the idea that we cannot use science to tell us what existed before the Big Bang.

18 ▶ The Search for Extraterrestrial Life

IN THIS CHAPTER

You will explore the question of whether we are alone in the universe. Could other planets orbiting other stars support complex life? You will begin here on Earth, with a look at the requirements for life. Then you will look at the instruments used to search for life beyond our own solar system—and at the results of those searches.

The Prevalence of Life This painting, entitled *DNA Embraces the Planets*, artistically expresses the suspicion of many scientists that carbon-based life may be a common phenomenon in the universe. Other scientists argue, however, that we may be unique and no intelligent alien civilizations exist. In either case, humanity has the clear mandate to preserve and protect the abundance of life forms with which we share our planet.

WHAT DO YOU THINK?

1   What do people who search for extraterrestrial intelligence look for?

••••••••••••••••••••••

THE HEAVENS INSPIRE US to contemplate—from the formation of the Earth to the nature of the stars and the creation of the universe. Perhaps no subject is as compelling, however, as the question of extraterrestrial life. Are we alone? Is there life elsewhere in the universe? What are the chances that we might someday make contact with an alien civilization?

There are not yet any firm answers to these questions. Furthermore, our probes to other worlds have failed to detect life on any of them. And despite all the alleged UFO sightings and personal encounters, no one has produced convincing evidence that an extraterrestrial life form has ever visited Earth. However, we cannot ever rule out the possibility of life elsewhere, and so the search continues.

18-1   Our planet is special to us, but not unique in its possibilities for life

As you have seen throughout this book, one of the great lessons of modern astronomy is that our circumstances are quite typical. We inhabit one of nine planets orbiting an ordinary star—just one among billions of stars in the Milky Way Galaxy alone. There is growing evidence of planets around various stars, and the elements necessary to make life as we know it are ubiquitous throughout our Galaxy. Is it possible that we are also biologically commonplace? The question is at the heart of the search for extraterrestrial intelligence (also known by its acronym, **SETI**).

All terrestrial life is based on the unique properties of the element carbon. Carbon atoms form chemical bonds that can combine to form especially long, complex molecules. These molecules can be further linked in elaborate chains, lattices, and fibers. No wonder carbon-based compounds, called **organic molecules,** are the stuff of life. The variety and stability of living organisms depends on the complex, self-regulating chemical reactions of organic molecules.

Carbon is also among the most abundant elements in the universe. So are other constituents of organic molecules, such as hydrogen, nitrogen, oxygen, sulfur, and phosphorus. With the exception of hydrogen, all of these

elements were synthesized during stellar evolution. In interstellar clouds, carbon atoms have combined with other elements to produce an impressive variety of organic compounds. Since the 1960s radio astronomers have detected telltale microwave emission lines from interstellar clouds that help identify dozens of these carbon-based chemicals. Examples include ethyl alcohol (CH_3CH_2OH), formaldehyde (H_2CO), methyl cyanoacetylene (CH_3C_3N), and acetaldehyde (CH_3CHO). The unique versatility and abundance of carbon suggest that extraterrestrial biology would also be based on organic chemistry.

Although no life has been discovered on any other body in the solar system, further evidence of extraterrestrial organic molecules comes from newly fallen meteorites called carbonaceous chondrites (Figure 18-1). Many of these contain a variety of organic substances. As noted in Chapter 8, carbonaceous chondrites are ancient meteorites dating from the formation of the solar system. So it seems reasonable to conclude that, even from their earliest days, the planets have been continually bombarded with organic compounds.

Interstellar space is not the only source of organic material. In a classic experiment performed in 1952, American chemists Stanley Miller and Harold Urey demonstrated that simple chemicals can combine to form prebiological compounds under supposedly primitive Earthlike conditions. In a closed container, they subjected a mixture of hydrogen, ammonia, methane, and

Figure 18-1   **A Carbonaceous Chondrite** Carbonaceous chondrites are ancient meteorites that date back to the formation of the solar system. Chemical analysis of newly fallen specimens discloses that they are rich in the chemical building blocks of life. This sample is a piece of the large Allende carbonaceous chondrite that fell in Mexico in 1969.

water vapor to an electric arc (to simulate lightning bolts) for a week. At the end of this period, the inside of the container had become coated with a reddish-brown substance rich in compounds essential to life.

Because Earth's primordial atmosphere was more likely to have contained carbon dioxide, nitrogen, and water vapor outgassed from volcanoes along with some hydrogen, modern versions of the Miller–Urey experiment have used these common gases (Figure 18-2). Again, organic compounds were produced. Could these compounds be forming today elsewhere in the universe as well?

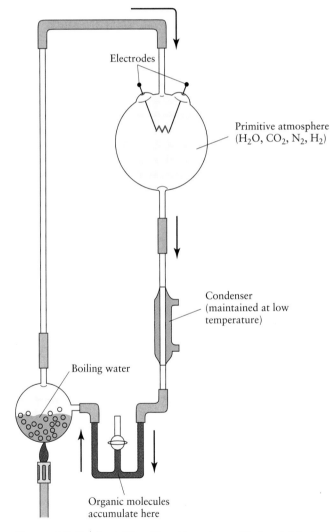

Figure 18-2 **The Miller–Urey Experiment Updated** Modern versions of this classic experiment prove that numerous organic compounds important to life can be synthesized from gases that were present in Earth's primordial atmosphere. This experiment supports the hypothesis that life on Earth arose as a result of ordinary chemical reactions.

It is important to emphasize that scientists have not created life in a test tube. Biologists have yet to figure out, among other things, how these simple organic molecules gathered into cell-like arrangements and developed systems for self-replication. Nevertheless, because so many chemical components of life are so easily synthesized under conditions that simulate the primordial Earth, it seems reasonable to suppose that life could have originated on Earthlike planets elsewhere.

A widespread abundance of organic precursors does not guarantee that life is commonplace throughout the universe. If a planet is too close to a star, the temperature and ultraviolet radiation levels will be too high for life to evolve. If the planet is too far away, it will be too cold. If a planet's environment is hostile, life may either never get started or quickly become extinct. If the star is too massive, it will explode before life evolves very far on planets orbiting it. If the star is too small, the planet would have to be so close to it to be warm that tidal forces from the star would lock the planet in synchronous rotation, making most of its surface uninhabitable.

While the two planets in orbit around the pulsar PSR B1257+12 (see Chapter 12) are not expected to harbor any life, astronomers have been actively searching for Earthlike planets in orbit around stars similar to the Sun. They measure stellar Doppler shifts, searching for stars in binary systems with companions of less than $0.08 \, M_\odot$ (that is, brown dwarfs or smaller). The first such extraterrestrial planet orbiting a main sequence star was discovered in 1995, near 51 Pegasus, a G5 star located some 42 ly from Earth in the constellation Pegasus.

The star 51 Pegasus is barely visible to the naked eye (but quite clear through binoculars), located halfway between the two western stars in the great square of Pegasus. From the star's Doppler shift, astronomers conclude that the planet completes an orbit every 4.2 days. This puts the planet, estimated to be about half the mass of Jupiter, $\frac{1}{20}$ AU from the star. If the planet originally formed at that distance, the star's heat and ultraviolet radiation assures that it has no hydrogen and helium envelope, nor any liquid water; it is a parched ball of rock and metal. However, there is evidence for a second planet orbiting farther away from the same star.

18-2 SETI uses radio telescopes

How might we ascertain whether life-supporting worlds exist, given the tremendous distances that separate us from them? Many astronomers hope to learn about

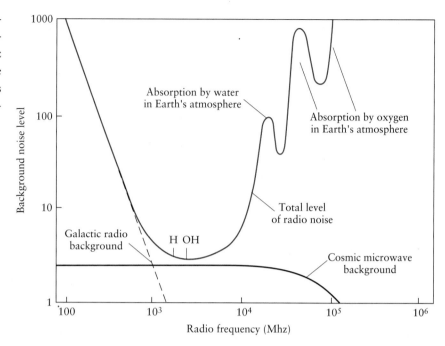

Figure 18-3 **The "Water Hole"** The so-called water hole is a range of radio frequencies from about 1000 to 10,000 megahertz (Mhz) that happens to have relatively little cosmic noise. Some scientists suggest that this noise-free region would be well suited for interstellar communication.

extraterrestrial civilizations by detecting radio transmissions from them. As we have seen, radio waves can travel immense distances without being significantly altered by the gas and dust through which they pass. Because of this ability to penetrate the interstellar medium, radio waves are a logical choice for interstellar communication.

If extraterrestrial beings were purposefully sending messages into space, it seems reasonable that they might choose a frequency that is fairly free of emissions from extraneous sources. SETI pioneer Bernard Oliver has pointed out that a range of relatively noise-free frequencies exists in the neighborhood of the microwave emission lines of hydrogen and the hydroxyl radical (OH) (Figure 18-3). This region of the microwave part of the radio spectrum is called the **water hole,** a humorous reference to the H and OH lines being so close together. Another plausible wavelength that aliens might choose is 21 cm, because astronomers studying the distribution of hydrogen around the Galaxy would already have their radio telescopes tuned to that wavelength (recall Figure 14-7).

Even if there are only a few alien civilizations scattered across the Galaxy, we have the technology to detect radio transmissions from them. One of the first searches was made in 1960, when Frank Drake used a radio telescope at the National Radio Astronomy Observatory in West Virginia to "listen" to two Sunlike stars, τ Ceti and ε Eridani, without success.

NASA funded a SETI program called the Microwave Observing Project, which began operations in October

1992. Continuing now with private donations, scientists are using radio telescopes along with sophisticated electronic equipment and powerful computers to conduct both a targeted search and an all-sky survey. The targeted search will examine some 800 nearby solar-type stars over a frequency range that covers the long-wavelength side of the water hole. The largest radio telescopes are being used in order to achieve the highest possible sensitivity. Specialized equipment for processing digital signals enables astronomers to study tens of millions of frequency channels simultaneously. This equipment can automatically detect continuous waves or pulses, whether they remain constant in frequency or drift slowly because of some relative motion between the transmitter and the receiver.

The all-sky survey uses several dish-shaped antennas of NASA's Deep Space Network (Figure 18-4) to extend the search over the entire sky. Some sensitivity is sacrificed, but the survey does cover the entire water hole. The project is almost completely automated in order to sift through incoming data at a rate far beyond human capabilities.

The detection of a message from an alien civilization would be one of the greatest events in human history. Such a message could dramatically change the course of civilization through the sharing of scientific information or an awakening of social or humanistic enlightenment. In only a few years our technology, industry, and social structure could advance the equivalent of centuries into the future. Such changes would touch every person on

Figure 18-4   **A Deep-Space Tracking Antenna** This dish-shaped antenna, located in the Mojave Desert in California, was originally built by NASA to track interplanetary spacecraft. It is was recently used in an all-sky survey, along with an antenna at the Arecibo Observatory in Puerto Rico, to search for extraterrestrial intelligence. In 1996 an antenna in Canberra, Australia, joined the network.

Earth. Mindful of these profound implications, scientists push ahead with the search for extraterrestrial communication despite 40 unsuccessful searches to date using radio telescopes around the world.

18-3 The Drake equation predicts how many civilizations are in the Milky Way

The growing evidence that planets exist around other stars raises the question of just how many of them are likely to harbor complex life. The first person to tackle this question quantitatively was Frank Drake. Drake proposed that the number of technologically advanced civilizations in the Galaxy (designated by the letter N) could be estimated by the **Drake equation:**

$$N = R_* f_p n_e f_l f_i f_c L$$

where

R_* = rate at which solar-type stars form in the Galaxy

f_p = fraction of stars that have planets

n_e = number of planets per star system suitable for life

f_l = fraction of those habitable planets on which life actually arises

f_i = fraction of those life forms that evolve into intelligent species

f_c = fraction of those species that develop adequate technology and then choose to send messages out into space

L = lifetime of that technologically advanced civilization

The Drake equation is enlightening because it expresses the number of extraterrestrial civilizations as a product of terms, some of which can be estimated from what we know about stars and stellar evolution. For instance, the first two terms, R_* and f_p, can be determined by observation. In estimating R_*, we should probably exclude massive stars (those larger than about 1.5 $M_\odot$), because they have main sequence lifetimes shorter than the time it took for intelligent life to develop here on Earth. Life on Earth originated some 3.5 to 4.0 billion

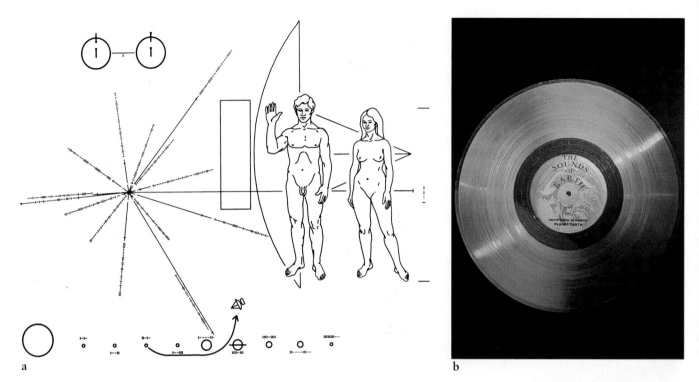

Figure 18-5 **Human Memorabilia on Voyager** The Voyager 1 and 2 spacecraft, now in interstellar space, each carry a plaque (**a**), with images of a man and a woman, as well as a phonograph record (**b**). There are also instructions for playing the record, which contains information about our biology, our technology, and our knowledge base. Each record also contains the sounds of children's voices. It is remotely possible that another race might someday discover the spacecraft.

years ago. If that is typical of the time needed to evolve higher life forms, then a massive star probably becomes a red giant or a supernova before intelligent creatures can appear on any of its planets.

Although low-mass stars have much longer lifetimes, they, too, seem unsuited for life because they are so cool. Only planets very near a low-mass star would be sufficiently warm for water to be a liquid, an essential ingredient for life to evolve as we know it. However, a planet that close would become tidally coupled to the star, with one side continuously facing the star while the other is in perpetual, frigid darkness.

This leaves us with main sequence stars like the Sun, those with spectral types between F5 and M0. Based on statistical studies of star formation in the Milky Way, some astronomers estimate that roughly one of these Sunlike stars forms in the Galaxy each year, thus yielding a value for R_* of 1 per year.

We learned in Foundations II that the planets in our solar system formed as a natural consequence of the birth of the Sun. We have also seen evidence that similar processes of planetary formation may be commonplace around isolated stars. Many astronomers therefore give

f_p a value of 1, meaning they believe it likely that most Sunlike stars have planets.

Unfortunately, the rest of the terms in the Drake equation are very uncertain. Let's play with some hypothetical values. The chances that a planetary system has an Earthlike world are not known. Were we to consider our own solar system as representative, we could put n_e at 1. Let's be more conservative, however, and suppose that 1 in 10 solar-type stars is orbited by a habitable planet, making $n_e = 0.1$.

From what we know about the evolution of life on Earth, we might assume that, given appropriate conditions, the development of life is a certainty, which would make $f_l = 1$. This is, of course, an area of intense interest to biologists. For the sake of argument, we might also assume that evolution naturally leads to the development of intelligence (a conjecture that is hotly debated) and also make $f_i = 1$. It's anyone's guess as to whether these intelligent extraterrestrial beings would attempt communication with other civilizations in the Galaxy, but were we to assume they would, f_c would be put at 1 also.

The last variable, L, the longevity of technological civilization, is the most uncertain of all and certainly cannot be subjected to testing! Looking at our own example, we see a planet whose atmosphere and oceans are increasingly polluted, thereby potentially destroying the food chain. When we add in how close we have come to destroying ourselves with weapons of mass destruction, it may be that we humans are among the lucky few technological civilizations to squeak through its first years. In other words, L may be as short as 100 years. Putting all these numbers together, we arrive at

$$N = 1 \times 1 \times 0.1 \times 1 \times 1 \times 1 \times 100 = 10$$

Therefore, out of the hundreds of billions of stars in the Galaxy, there may be only ten technologically advanced civilizations with whom we might communicate.

A wide range of values have been proposed for the terms in the Drake equation, and these various guesses produce vastly different estimates of N. Some scientists argue that there is exactly one advanced civilization in the Galaxy and that we are it. Others speculate that there may be tens of millions of planets inhabited by intelligent creatures. We just don't know.

18-4 We have been sending signals into space for a century

We have been doing more than passive searching. For a century now we have been transmitting messages about ourselves. Sometimes we have signaled our presence intentionally, but most often our transmissions into space have been inadvertent. It all began when radio signals were first broadcast. Some of those waves have been traveling outward from the Earth for a hundred years. Space within 75 ly of the solar system is filled with signals from radio and television shows. If there are other advanced civilizations within that sphere, they could very well be watching radio or television shows from decades past, trying to make sense out of our species.

Astronomers have also intentionally broadcast well-focused signals through radio telescopes toward likely star systems. In a more pedestrian vein, the two Voyager spacecraft, now making their way into interstellar space, each have on them a plaque with information about who and where we are, as well as phonograph records (remember them?) with human voices (Figure 18-5). Because space is so vast and the spacecraft are so small, the likelihood of their being discovered is truly remote. And yet millions, perhaps billions, of years from now, one of them might be detected by another race of intelligent creatures. From its path and the information on board, they might be able to determine where it came from. If their travel budgets are sufficiently large, they might even decide to come and visit us. At least, they could send us radio messages. The question is: Will our descendants be here to receive them?

WHAT DID YOU THINK?

1 *What do people who search for extraterrestrial intelligence look for?* They search for radio signals from other advanced civilizations.

Key Words

Drake equation   organic molecule   SETI   water hole

Key Ideas

• The chemical building blocks of life exist throughout the Milky Way Galaxy.

Organic molecules have been discovered in interstellar clouds and in some meteorites.

• Astronomers are using radio telescopes to search for signals from other advanced life in the Galaxy. This effort is called the search for extraterrestrial intelligence, or SETI. So far, these searches have not detected any life off Earth.

SETI is primarily done at frequencies where radio waves pass most easily through the interstellar medium.

• The Drake equation is used to estimate the number of technologically advanced civilizations there might be in the Galaxy.

• Everyday radio and television transmissions from Earth, along with intentional broadcasts into space, may have been detected by other life forms.

Discussion Questions

1 What information would you have put on the record sent out on the Voyager spacecraft?

2 What arguments could you make against sending messages via radio or spacecraft into interstellar space?

3 Try using the Drake equation with values that you find reasonable. How many civilizations do you estimate there are in our Galaxy?

4 Discuss the possible biological effects of our probes visiting other life-sustaining worlds. What social effects might probes from other worlds have on us?

5 List some of the pros and cons in the argument that alien spacecraft have visited the Earth.

6 Why would most of a planet locked in synchronous rotation be uninhabitable?

WHAT IF...

The Earth Were Closer to the Center of the Galaxy?

I T IS NIGHT. The Sun and Moon are down, and stars by the thousands twinkle serenely against the ebony darkness of space. Overhead the soft white of the Milky Way catches your attention and you try to see individual stars in the glowing haze. Most of the 6000 stars visible throughout the year are within 300 ly of the Earth. The rest of the Galaxy's 200 billion or so stars are too dim or too obscured by interstellar gas and dust to be easily observed. What if the solar system were one-third of its present distance from the center of our Galaxy? At that distance we would still be in the realm of the spiral arms, extremely close to the nuclear bulge, and none of the stars we see now would be visible.

Out in the galactic suburbs, where we are today, there is about one star per 300 cubic light-years. The displaced Earth would be surrounded by nearly five times as many stars, or one per 60 cubic light-years. The solar system would pass much more frequently through the dust-rich spiral arms of the Galaxy. Scattering starlight, this matter would glow as diaphanous wisps throughout our sky. Furthermore, whenever the solar system was actually in a spiral arm, several of the nearby stars would be of the high-mass, high-luminosity variety. The combined light from these stars and the shimmering clouds would be so great that for millions of years at a time there would never be night.

Recall from Chapter 12 that high-mass stars evolve much more rapidly than the Sun. If we were closer to the center of the Milky Way and frequently passing through the spiral arms of the Galaxy. massive stars would explode near us much more frequently than they do today. Those nearby supernovae, occurring within 50 ly of

Earth, would deposit lethal radiation, damage life, and cause mass extinctions. As a result, the direction of evolution would change more frequently.

Today, the closest star, Proxima Centauri, is over 3 ly from Earth. Would there be a danger of colliding with a star if the solar system were closer to the center of the Galaxy? There would certainly be several stars much closer than Proxima Centauri is now. However, stars are so small compared to the vastness of a galaxy that the likelihood of a collision would still be nil. Much more likely would be the passage of a star so close to the Sun that the Earth's orbit would be disturbed. If the Earth's orbit became more elliptical, the change of seasons would be noticeably affected. With the seasonal effect of the Earth's tilt compounded by greater changes in distance between the Earth and Sun, one hemisphere of the Earth would have much more extreme temperatures than it does today, while the other would have less variation. This would affect the evolution of life, of course, and the distribution of life forms on the planet.

Earth-evolved life at our new location would be more likely to encounter sentient beings on other planets. Today, after 75 years of broadcasting radio and television signals, there is a 150-ly-diameter sphere of space centered on the Earth filled with such signals. There are about 6300 stars in that sphere. At our new location there would be 31,500 stars in the same volume and many more stars with life-supporting planets orbiting them. Perhaps one of the reasons that we have been able to develop so long undisturbed is because the solar system exists near the fringe of the Galaxy.

Appendix

· ·

TABLE A-1

The Planets: Orbital Data

Planet	Semimajor axis (AÚ)	Semimajor axis (10⁶ km)	Sidereal period (yr)	Sidereal period (d)	Synodic period (d)	Mean orbital speed (km/s)	Orbital eccentricity	Inclination of orbit to ecliptic (°)
Mercury	0.3871	57.9	0.2408	87.97	115.88	47.9	0.206	7.00
Venus	0.7233	108.2	0.6152	224.70	583.92	35.0	0.007	3.39
Earth	1.0000	149.6	1.0000	365.26	—	29.8	0.017	0.00
Mars	1.5237	227.9	1.8809	686.98	779.94	24.1	0.093	1.85
(Ceres)	2.7656	413.7	4.603	1,681.3	466.6	17.9	0.097	10.61
Jupiter	5.2028	778.3	11.862	4,332.7	398.9	13.1	0.048	1.31
Saturn	9.5388	1427.0	29.458	10,760	378.1	9.6	0.056	2.49
Uranus	19.1914	2871.0	84.01	30,685	369.7	6.8	0.046	0.77
Neptune	30.0611	4497.1	164.79	60,082	367.5	5.4	0.010	1.77
Pluto	39.5294	5913.5	248.5	90,767	366.7	4.7	0.248	17.15

· ·

TABLE A-2

The Planets: Physical Data

Planet	Equatorial diameter (km)	Equatorial diameter (Earth = 1)	Mass (Earth = 1)	Mean density (kg/m³)	Rotation period* (d)	Inclination of equator to orbit (°)	Surface gravity (Earth = 1)	Albedo	Brightest visual magnitude	Escape velocity (km/s)
Mercury	4,878	0.38	0.055	5430	58.65	2(?)	0.39	0.106	−1.9	4.3
Venus	12,102	0.95	0.815	5250	−243.01	177.3	0.88	0.65	−4.4	10.4
Earth	12,756	1.00	1.000	5520	0.997	23.4	1.00	0.37	—	11.2
Mars	6,786	0.53	0.107	3950	1.026	25.2	0.38	0.15	−2.0	5.0
Jupiter	142,984	11.21	317.94	1330	0.410	3.1	2.34	0.52	−2.7	59.6
Saturn	120,536	9.45	95.18	690	0.426	26.7	0.93	0.47	+0.7	35.5
Uranus	51,118	4.01	14.53	1290	−0.746	97.9	0.79	0.50	+5.5	21.3
Neptune	49,528	3.88	17.14	1640	0.800	29.6	1.12	0.5	+7.8	23.3
Pluto	2,300	0.18	0.002	2030	−6.387	122.5	0.04	0.5	+15.1	1.1

*Negative values indicate retrograde rotation.

TABLE A-3

Satellites of the Planets

Planet	Satellite	Discoverer(s)	Mean distance from planet (km)	Sidereal period* (d)	Orbital eccentricity	Diameter of satellite† (km)	Approximate magnitude at opposition
Earth	Moon	——	384,400	27.322	0.05	3476	−13
Mars	Phobos	Hall (1877)	9,380	0.319	0.01	28 × 23 × 20	+11
	Deimos	Hall (1877)	23,460	1.263	0.00	16 × 12 × 10	12
Jupiter	Metis	Synnott (1979)	127,960	0.295	0.00	(40)	+18
	Adrastea	Jewitt et al. (1979)	128,980	0.298	0(?)	24 × 16 × 20	19
	Amalthea	Barnard (1892)	181,300	0.498	0.00	270 × 200 × 155	14
	Thebe	Synnott (1979)	221,900	0.675	0.01	(100)	16
	Io	Galileo (1610)	421,600	1.769	0.00	3630	5
	Europa	Galileo (1610)	670,900	3.551	0.01	3138	5
	Ganymede	Galileo (1610)	1,070,000	7.155	0.00	5262	5
	Callisto	Galileo (1610)	1,883,000	16.689	0.01	4800	6
	Leda	Kowal (1974)	11,094,000	238.72	0.15	(16)	20
	Himalia	Perrine (1904)	11,480,000	250.57	0.16	(180)	15
	Lysithea	Nicholson (1938)	11,720,000	259.22	0.11	(40)	18
	Elara	Perrine (1905)	11,737,000	259.65	0.21	(80)	17
	Ananke	Nicholson (1951)	21,200,000	631r	0.17	(30)	19
	Carme	Nicholson (1938)	22,600,000	692r	0.21	(44)	18
	Pasiphae	Melotte (1908)	23,500,000	735r	0.38	(70)	17
	Sinope	Nicholson (1914)	23,700,000	758r	0.28	(40)	18
Saturn	Pan	Showalter (1990)	133,570	0.573	0.00	20	+19
	Atlas	Terrile (1980)	137,640	0.602	0(?)	40 × 30 × 30	18
	Prometheus	Collins et al. (1980)	139,350	0.613	0.00	140 × 80 × 100	16
	Pandora	Collins et al. (1980)	141,700	0.629	0.00	110 × 70 × 100	16
	Epimethius	Walker (1966)	151,422	0.694	0.01	140 × 100 × 100	16
	Janus	Dolfus (1966)	151,472	0.695	0.01	220 × 160 × 200	14
	Mimas	Herschel (1789)	185,520	0.942	0.02	390	13
	Enceladus	Herschel (1789)	238,020	1.370	0.00	500	12
	Tethys	Cassini (1684)	294,660	1.888	0.00	1050	10
	Telesto	Smith et al. (1980)	294,660	1.888	0(?)	(24)	19
	Calypso	Smith et al. (1980)	294,660	1.888	0(?)	30 × 20 × 25	19
	Dione	Cassini (1684)	377,400	2.737	0.00	1120	10

Planet	Satellite	Discoverer(s)	Mean distance from planet (km)	Sidereal period* (d)	Orbital eccentricity	Diameter of satellite† (km)	Approximate magnitude at opposition
	Helene	Laques et al. (1980)	377,400	2.737	0.01	40 × 30 × 30	18
	Rhea	Cassini (1672)	527,040	4.518	0.00	1530	10
	Titan	Huygens (1655)	1,221,850	15.945	0.03	5150	8
	Hyperion	Bond (1848)	1,481,000	21.277	0.10	410 × 260 × 220	14
	Iapetus	Cassini (1671)	3,561,000	79.331	0.03	1440	11
	Phoebe	Pickering (1898)	12,952,000	550.48	0.16	220	16
Uranus	Cordelia	Voyager 2 (1986)	49,750	0.335	0(?)	(30)	+24
	Ophelia	Voyager 2 (1986)	53,760	0.376	0(?)	(30)	24
	Bianca	Voyager 2 (1986)	59,160	0.435	0(?)	(40)	23
	Cressida	Voyager 2 (1986)	61,770	0.464	0(?)	(70)	22
	Desdemona	Voyager 2 (1986)	62,660	0.474	0(?)	(60)	22
	Juliet	Voyager 2 (1986)	64,360	0.493	0(?)	(80)	22
	Portia	Voyager 2 (1986)	66,100	0.513	0(?)	(110)	21
	Rosalind	Voyager 2 (1986)	69,930	0.558	0(?)	(60)	22
	Belinda	Voyager 2 (1986)	75,260	0.624	0(?)	(70)	22
	Puck	Voyager 2 (1986)	86,010	0.762	0(?)	150	20
	Miranda	Kuiper (1948)	129,780	1.414	0.00	470	16
	Ariel	Lassell (1851)	191,240	2.520	0.00	1160	14
	Umbriel	Lassell (1851)	265,970	4.144	0.00	1170	15
	Titania	Herschel (1787)	435,840	8.706	0.00	1580	14
	Oberon	Herschel (1787)	582,600	13.463	0.00	1520	14
Neptune	Naiad	Voyager 2 (1989)	48,230	0.296	0(?)	(50)	+25
	Thalassa	Voyager 2 (1989)	50,070	0.312	0(?)	60	24
	Despoina	Voyager 2 (1989)	52,530	0.333	0(?)	80	23
	Galatea	Voyager 2 (1989)	61,950	0.429	0(?)	180	23
	Larissa	Voyager 2 (1989)	73,550	0.554	0(?)	150	21
	Proteus	Voyager 2 (1989)	117,640	1,121	0(?)	415	20
	Triton	Lassell (1846)	354,800	5.877r	0.00	2700	14
	Nereid	Kuiper (1949)	5,513,400	359.16	0.75	(340)	19
Pluto	Charon	Christy (1978)	19,640	6.387r	0.00	1190	+17

*"r" indicates a retrograde orbit.
†A diameter given in parentheses is estimated from the amount of sunlight it reflects.

TABLE A-4

The Nearest Stars

Name	Parallax (arc sec)	Distance (ly)	Spectral type*	Radial velocity (km/s)	Proper motion (arcsec/yr)	Apparent visual magnitude	Luminosity (Sun = 1.0)
Sun			G2 V			−26.7	1.0
Proxima Centauri	0.772	4.2	M5e	−16	3.85	11.05	0.00006
α Centauri A	0.750	4.3	G2 V	−22	3.68	−0.01	1.6
α Centauri B			K0 V			1.33	0.45
Barnard's star	0.545	5.9	M5 V	−108	10.31	9.54	0.00045
Wolf 359	0.421	7.6	M8e	+13	4.70	13.53	0.00002
BD +36°2147	0.397	8.1	M2 V	−84	4.78	7.50	0.0055
Luyten 726-8A	0.387	8.4	M6e	+29	3.36	12.52	0.00006
Luyten 726-8B (UV Ceti)			M6e	+32		13.02	0.00004
Sirius A	0.377	8.6	A1 V	−8	1.33	−1.46	23.5
Sirius B			wd			8.3	0.003
Ross 154	0.345	9.4	M5e	−4	0.72	10.45	0.00048
Ross 248	0.314	10.3	M6e	−81	1.60	12.29	0.00011
ε Eridani	0.303	10.7	K2 V	+16	0.98	3.73	0.30
Ross 128	0.298	10.8	M5	−13	1.38	11.10	0.00036
61 Cygni A	0.294	11.2	K5 V	−64	5.22	5.22	0.082
61 Cygni B			K7 V			6.03	0.039
ε Indi	0.291	11.2	K5 V	−40	4.70	4.68	0.14
BD +43°44A	0.290	11.2	M1 V	+13	2.90	8.08	0.0061
BD +43°44B			M6 V	+20		11.06	0.00039
Luyten 789-6	0.290	11.2	M7e	−60	3.26	12.18	0.00014
Procyon A	0.285	11.4	F5 IV–V	−3	1.25	0.37	7.65
Procyon B			wd			10.7	0.00055
BD +59°1915A	0.282	11.5	M4	0	2.29	8.90	0.0030
BD +59°1915B			M5	+10	2.27	9.69	0.0015
CD −36°1668	0.279	11.7	M2 V	+10	6.90	7.35	0.0013
G51-15	0.278	11.7			1.27	14.81	0.00001
τ Ceti	0.277	11.9	G8 V	−16	1.92	3.50	0.45
BD +5°1668	0.266	12.2	M5	+26	3.77	9.82	0.0015
Luyten 725-32 (YZ Ceti)	0.261	12.4	M5e	+28	1.32	12.04	0.0002
CD −39°14192	0.260	12.5	M0 V	+21	3.46	6.66	0.028
Kapteyn's star	0.256	12.7	M0 V	+245	8.72	8.84	0.0039
Kruger 60 A	0.253	12.8	M3	−26	0.86	9.85	0.0016
Kruger 60 B			M5e			11.3	0.0004

Note: In the case of double stars, like Sirius A and B or Procyon A and B, both members have the same parallax, distance, etc. These redundant values are not repeated for the second (dimmer) member.

*"wd" stands for white dwarf; "e" means that the star's spectrum contains emission lines.

TABLE A-5

The Visually Brightest Stars

Star	Name	Apparent visual magnitude	Spectral type	Absolute magnitude	Distance (ly)	Radial velocity (km/s)	Proper motion (arcsec/yr)
α CMa A	Sirius	−1.46	A1 V	+1.4	9	−8	1.324
α Car	Canopus	−0.72	F0 I	−3.1	98	+21	0.025
α Boo	Arcturus	−0.04	K2 III	−0.2	36	−5	2.284
α Cen A	Rigel Kents	0.00	G2 V	+4.4	4	−25	3.676
α Lyr	Vega	0.03	A0 V	+0.5	26	−14	0.345
α Aur	Capella	0.08	G8 III	−0.5	42	+30	0.435
β Ori A	Rigel	0.12	B8 Ia	−7.1	910	+21	0.001
α CMi A	Procyon	0.38	F5 IV	+2.6	11	−3	1.250
α Eri	Achernar	0.46	B3 IV	−1.6	85	+19	0.098
α Ori	Betelgeuse	0.50	M2 Iab	−5.6	510	+21	0.028
β Cen	Hadar	0.61	B1 II	−5.1	460	−12	0.035
α Aql	Altair	0.77	A7 IV–V	+2.2	17	−26	0.658
α Tau A	Aldebaran	0.85	K5 III	−0.3	68	+54	0.202
α Sco A	Antares	0.96	M1 Ib	−4.7	330	−3	0.029
α Vir	Spica	0.98	B1 V	−3.5	260	+1	0.054
β Gem	Pollux	1.14	K0 III	+0.2	36	+3	0.625
α PsA	Fomalhaut	1.16	A3 V	+2.0	22	+7	0.367
α Cyg	Deneb	1.25	A2 Ia	−7.5	1800	−5	0.003
β Cru	Mimosa	1.25	B0.5 III	−5.0	420	+20	0.049
α Leo A	Regulus	1.35	B7 V	−0.6	85	+4	0.248

Note: Acrux, the brightest star in Crux (the Southern Cross), appears to the naked eye as a star of magnitude +0.87, which suggests that it should be in this table. A small telescope, however, reveals that Acrux is actually a double star whose blue-white components have visual magnitudes of 1.4 and 1.9, and so they are dimmer than any of the stars listed here.

TABLE A-6

Some Important Astronomical Quantities

Astronomical unit:	$1 \text{ AU} = 1.496 \times 10^{11} \text{ m}$
Light-year:	$1 \text{ ly} = 9.461 \times 10^{15} \text{ m} = 63{,}240 \text{ AU}$
Parsec:	$1 \text{ pc} = 3.086 \times 10^{16} \text{ m} = 3.262 \text{ ly}$
Solar mass:	$1 \text{ M}_\odot = 1.989 \times 10^{30} \text{ kg}$
Solar radius:	$1 \text{ R}_\odot = 6.960 \times 10^{8} \text{ m}$
Solar luminosity:	$1 \text{ L}_\odot = 3.827 \times 10^{26} \text{ W}$
Earth's mass:	$1 \text{ M}_\oplus = 5.974 \times 10^{24} \text{ kg}$
Earth's equatorial radius:	$1 \text{ R}_\oplus = 6.378 \times 10^{6} \text{ m}$
Moon's mass:	$1 \text{ M}_\text{☽} = 7.348 \times 10^{22} \text{ kg}$
Moon's equatorial radius:	$1 \text{ R}_\text{☽} = 1.738 \times 10^{6} \text{ m}$

TABLE A-7

Some Important Physical Constants

Speed of light:	$c = 2.998 \times 10^{8} \text{ m/s}$
Gravitational constant:	$G = 6.668 \times 10^{-11} \text{ N m}^2 \text{ kg}^{-2}$
Planck constant:	$h = 6.625 \times 10^{-34} \text{ J s}$
	$= 4.136 \times 10^{-15} \text{ eV s}$
Boltzmann constant:	$k = 1.380 \times 10^{-23} \text{ J K}^{-1}$
	$= 8.617 \times 10^{-5} \text{ eV K}^{-1}$
Stefan–Boltzmann constant:	$\sigma = 5.669 \times 10^{-8} \text{ W m}^{-2} \text{ K}^{-4}$
Mass of electron:	$m_\text{e} = 9.108 \times 10^{-31} \text{ kg}$
Mass of ^{1}H atom:	$m_\text{H} = 1.673 \times 10^{-27} \text{ kg}$

Temperatures and Temperature Scales

Three temperature scales are in common use. Throughout most of the world, temperatures are expressed in degrees Celsius (°C), named in honor of the Swedish astronomer Anders Celsius, who proposed it in 1742. The Celsius temperature scale (also known as the "centigrade scale") is based on the behavior of water, which freezes at 0°C and boils at 100°C at sea level on Earth.

Scientists usually prefer the Kelvin scale, named after the British physicist Lord Kelvin (William Thomson), who made many important contributions to our knowledge about heat and temperature. On the Kelvin temperature scale, water freezes at 273 K and boils at 373 K. Note that we do not use the degree symbol with the Kelvin temperature scale. For instance, room temperature is roughly 295 kelvins, or 295 K.

Because water must be heated 100 K or 100°C to go from its freezing point to its boiling point, you can see that the size of a kelvin is the same as the size of a degree Celsius. When considering temperature *changes,* measurements in kelvin and in degrees Celsius lead to the same number.

A temperature expressed in kelvins is always equal to the temperature in degrees Celsius plus 273. Scientists prefer the Kelvin scale because it is closely related to the physical meaning of temperature. All substances are made of atoms, which are very tiny (a typical atom has a diameter of about 10^{-10} m) and constantly in motion. The temperature of a substance is directly related to the average speed of its atoms. If something is hot, its atoms are moving at high speeds. If a substance is cold, its atoms are moving much more slowly.

The coldest possible temperature is the temperature at which atoms move as slowly as possible (they can never quite stop completely). This minimum possible temperature, called absolute zero, is the starting point for the Kelvin scale. Absolute zero is 0 K, or −273°C. Since it is impossible for anything to be colder than 0 K, there are no negative temperatures on the Kelvin scale.

In the United States, many people still use the now-archaic Fahrenheit scale, which expresses temperatures in degrees Fahrenheit (°F). When the German physicist Gabriel Fahrenheit introduced this scale in the early 1700s, he intended 0°F to represent the coldest temperature then achievable (with a mixture of ice and saltwater) and 100°F to represent the temperature of a healthy human body. On the Fahrenheit scale, water freezes at 32°F and boils at 212°F. Because there are 180 degrees Fahrenheit between the freezing and boiling points of water, a degree Fahrenheit is only $\frac{100}{180}$ ($= \frac{5}{9}$) as large as either a degree Celsius or a kelvin.

The following equation converts from degrees Celsius to degrees Fahrenheit:

$$T_F = \frac{9}{5} T_C + 32$$

where (T_F) is the temperature in degrees Fahrenheit and (T_C) is the temperature in degrees Celsius. To convert from Fahrenheit to Celsius, a simple rearrangement of terms gives the relationship

$$T_C = \frac{5}{9} (T_F - 32)$$

Example: Consider a typical room temperature of 72°F. Using the second equation, we can convert this measurement to the Celsius scale as follows:

$$T_C = \frac{5}{9} (72 - 32) \approx 22°C$$

To arrive at the Kelvin scale, we simply add 273 degrees to the value in degrees Celsius. Thus, 72°F ≈ 22°C = 295 K.

The diagram below displays the relationships between these three temperature scales.

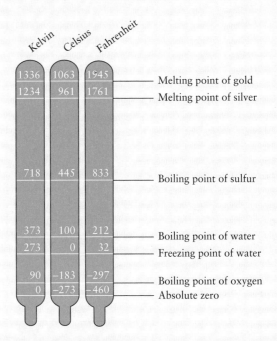

Kelvin	Celsius	Fahrenheit	
1336	1063	1945	— Melting point of gold
1234	961	1761	— Melting point of silver
718	445	833	— Boiling point of sulfur
373	100	212	— Boiling point of water
273	0	32	— Freezing point of water
90	−183	−297	— Boiling point of oxygen
0	−273	−460	— Absolute zero

Astronomy on the Internet

The Internet—an informal, worldwide network of computers—can provide you with an extraordinary range of astronomical resources. You will find news of important discoveries, video clips, and libraries of essential information. Most of the planetarium software mentioned in Chapter 1 can be downloaded from the Net; so can the latest images from the Hubble Space Telescope and major observatories. If you have a question about your new interest in the heavens, someone is out there to discuss it with you. Like the night sky itself, it is all waiting for you to explore.

Internet resources are furnished by thousands of scientists, engineers, government agencies, universities, amateur astronomers, and even students. They have set up electronic "locations" where outsiders can view what they have stored on their computers. Getting there is very simple: The Internet has been around long enough that you will find ample user-friendly software. Most colleges and universities offer free Internet access to students and staff, along with help in getting started. The Internet provides four basic services:

The World Wide Web (WWW) offers access to up-to-the-minute images and facts at different home pages.

Gopher and ftp sites are major libraries of images and facts that you can download to your own computer.

Newsgroups give you a forum to ask questions and receive answers about astronomy.

Through subscriptions to mailing lists, correspondence is sent to you automatically. You can also correspond personally by e-mail with other individuals.

An Internet Starter Kit: A Few Excellent Home Pages

Discovering the Universe (special information designed for this book)	http://www.whfreeman.com—and check out the home page built into the CD-ROM based on this book.
Space Telescope Electronic Information Service (Hubble Space Telescope)	http://stsci.edu/top.html
Astronomy and Space Science (U.S. Geological Survey)	http://info.er.usgs.gov/network/science/astronomy/index.html
NASA Information Services	http://www.gsfc.nasa.gov/NASA_homepage.html
The Astronomical Society of the Pacific (A major organization of amateur and professional astronomers)	http://www.physics.sfsu.edu/asp/asp.html
Students for the Exploration and Development of Space (SEDS)	http://seds.lpl.arizona.edu/
NASA Jet Propulsion Laboratory (JPL)	http://www.jpl.nasa.gov/
Webstars (Astrophysics in Cyberspace)	http://guinan.gsfc.nasa.gov/WebStars.html
JPL Space Calendar	http://newproducts.jpl.nasa.gov:80/calendar/
Virtual Library: Astronomy and Astrophysics	http://www.w3.org/hypertext/DataSources/bySubject/astro/astro.html
Comet Observation	http://encke.jpl.nasa.gov/
Search for Extraterrestrial Intelligence	http://www.seti-inst.edu/
Access to many search engines	http://cuiwww.unige.ch/meta-index.html

EYES ON . . .   Getting Connected

To get started on the information superhighway, you need to get connected. That means a computer, a modem (or network hardware), some software, and an Internet account. If you are new at this, check first to see if your college or university provides you with free access. If not, computer clubs and stores in your area can help you find other ways to get connected.

Suppose you want to know the latest progress of the Galileo spacecraft investigating Jupiter, but you don't know where to look. Do what I did: Choose a search engine, and pick the terms you will use to limit the search. I first used Yahoo, with the two words *Galileo* and *Jupiter.* Yahoo was too busy to reply, but another tool, called Search the Web at EINet Galaxy, turned up 72 *hits*—Web locations where the words *Galileo* and *Jupiter* are found together. Most of these were not relevant to astronomy. I clicked on the first astronomical one, "Space Calendar (JPL)."

As its name implies, the Space Calendar lists significant astronomical events. Using the Netscape "Find"

command, I looked down the home page for the word Galileo and found "Galileo Probe Separation from Orbiter." Clicking on this title, I viewed the press release from the Jet Propulsion Lab (JPL) on this subject. At the bottom of the press release page was a link to "Galileo Project Home Page (JPL)" with images, news, and links to other resources about Galileo. I had found it!

So that I would not have to go through the same search again, I used the "Bookmark" command to save the address of the Galileo home page. Now I can go there directly whenever I want.

Sometimes searches take longer, with many dead ends. Sometimes you find a legitimate address, but it doesn't take you exactly where you want to go. Keep in mind that connections on the Internet are sufficiently numerous that eventually you can find virtually anything you look for. Keep searching!

You can easily find yourself browsing dozens of home pages, pursuing the built-in links from one page to another. You can use gopher to access a wealth of data or have ftp (which stands for "file transfer protocol") send it to you. You can exchange ideas very informally in newsgroups. Use mailing lists with discretion, however, since they can inundate you with e-mail that you are not interested in. And there are other resources, such as the Wide Area Information System (WAIS), that deserve mention. Commercial services independent of the Internet—CompuServe, America On-Line, and Prodigy—also offer information on astronomy.

We discuss mostly the Internet's World Wide Web, because it is the fastest-growing resource as well as the most attractive for newcomers. For information about other Internet resources, check out the informative article in the August 1995 issue of *Sky & Telescope* magazine.

The World Wide Web allows you to look at images, movies, and text directly on your computer. Some home pages also contain sound clips. First, you "bring up" your *browser,* or interface software, such as Mosaic or Netscape. Then you type in a computer address (called a URL, for "uniform resource locator") or click on a

bookmark that you saved from a previous visit. Addresses are entered by choosing File/Open URL on Mosaic or File/Open Location on Netscape. The address you enter connects your computer with another one on which astronomical information is available to you. An address often contains slashes (/), because a single computer usually offers many different home pages. Finding the one you want may mean moving up or down through several "levels." If you know what topic you are interested in but don't know how to find it on the Web, *search engines* can help you locate the appropriate address. Popular ones include Yahoo and WebCrawler. (Search engines for ftp and other Internet resources include Veronica, Archie, and Jughead. Do you detect a pattern?) At least one search engine is probably built into your browser, but there are a good dozen more, and most are free.

From any home page, you can use your mouse to click on points of interest. Many images can be enlarged just by double-clicking on them. Highlighted *hot links* will take you to other parts of the home page or to other home pages. One well-designed home page can become your door to the entire Web. You might begin with the home page included in the CD-ROM based on this book.

There are many, many ways to connect to most locations. For example, you can get to NASA's home page by typing in its address directly. (Be sure to type each address *exactly* as given.) However, you can also get there from the Space Telescope Science Institute's home page, from the U.S. Geological Survey's Astronomy and Space Science page, and from any of several hundred other pages on the Internet. (Other Internet resources, such as gopher sites, are similarly linked.)

Your browser can quickly retrace your steps during a session, so that your computer does not spend time searching the Web for the same home page twice. However, you do not have to "back out" of one location through all the other connections you have made to arrive at a new place. You can zip directly from any address to any other.

The table on page 398 lists a few of the more exciting, information-filled pages on the Internet. It is very important to keep in mind that owners of publicly accessible pages often change addresses! Although all the addresses listed here were active at press time, some may have changed or been discontinued since then. Still others are being created every day.

For Further Reading

Chapter 1

Allen, D., and Allen, C. *Eclipse*. Allen & Unwin, 1987.

Allen, R. *Star Names: Their Lore and Meaning.* Originally published 1899; Dover, 1963.

Astronomy magazine. Kalmbach Publishing, published monthly.

Berry, R. *Discover the Stars*. Harmony, 1987.

Brown, R. H. *Man and the Stars*. Oxford University Press, 1978.

Burnham, R., Jr. *Burnham's Celestial Handbook,* 3 vols. Dover, 1978.

Chartrand, M. *Skyguide*. Western Publications, 1982.

Gallant, R. *The Constellations: How They Came To Be.* Four Winds Press, 1979.

Gingerich, O. "The Origin of the Zodiac," *Sky & Telescope,* March 1984.

Kundu, M. "Observing the Sun during Eclipses," *Mercury,* July/August 1981.

Kunitzsch, P. "How We Got Our Arabic Star Names," *Sky & Telescope,* January 1983.

Menzel, D., and Pasachoff, J. "Solar Eclipse: Nature's Superspectacular," *National Geographic,* August 1970.

Motz, L., and Nathanson, C. *The Constellations: An Enthusiast's Guide to the Night Sky.* Doubleday, 1988.

Ridpath, I., and Tirion, W. *Universe Guide to Stars and Planets*. Universe, 1985.

Sagan, C. "The Shores of the Cosmic Ocean," in *Cosmos*. Random House, 1980.

Sky & Telescope magazine. Sky Publishing, published monthly.

Stephenson, F. "Historical Eclipses," *Scientific American,* October 1982.

Chapter 2

Christianson, G. *This Wild Abyss*. Free Press, 1978.

——. "Newton's Principia: A Retrospective," *Sky & Telescope,* June 1987.

Cohen, I. B. "Newton's Discovery of Gravity," *Scientific American,* March 1981.

Drake, S. "Newton's Apple and Galileo's Dialogue," *Scientific American,* August 1980.

Gingerich, O. "Copernicus and Tycho," *Scientific American,* December 1973.

——. "The Galileo Affair," *Scientific American,* August 1982.

——. "How Galileo Changed the Rules of Science," *Sky & Telescope,* March 1993.

Lerner, L., and Gosselin, E. "Galileo and the Specter of Bruno," *Scientific American,* November 1986.

Wilson, C. "How Did Kepler Discover His First Two Laws?" *Scientific American,* March 1972.

Chapter 3

Brunier, S. "Temples in the Sky," *Sky & Telescope,* February, June, and December 1993.

Burns, J., et al. "Observatories on the Moon," *Scientific American,* March 1990.

Chaisson, E. *The Hubble Wars*. HarperCollins Publishers, 1994.

Cohen, M. *In Quest of Telescopes*. Sky Publishing and Cambridge University Press, 1980.

Cole, S. "Astronomy on the Edge: Using the Space Telescope," *Sky & Telescope,* October 1992.

Giacconi, R. "The Einstein X-Ray Observatory," *Scientific American,* February 1980.

Harrington, S. "Selecting Your First Telescope," *Mercury,* July/August 1982.

Henbest, N., and Marten, M. *The New Astronomy.* Cambridge University Press, 1983.

Janesick, J., and Blouke, M. "Sky on a Chip: The Fabulous CCD," *Sky & Telescope,* September 1987.

Jastrow, R., and Baliunas, S. "Mount Wilson: America's Observatory," *Sky & Telescope,* March 1993.

Kondo, Y., Wamsteker, W., and Stickland, D. "IUE: 15 Years and Counting," *Sky & Telescope,* September 1993.

Krisciunas, K. *Astronomical Centers of the World.* Cambridge University Press, 1988.

Kristian, J., and Blouke, M. "Charge-Coupled Devices in Astronomy," *Scientific American,* October 1982.

Learner, R. *Astronomy through the Telescope*. Van Nostrand Reinhold, 1981.

MacRobert, A. "Astronomy with a $5 Telescope," *Sky & Telescope,* April 1990.

Powell, C. "Mirroring the Cosmos," *Scientific American,* November 1991.

Preston, R. *First Light: The Search for the Edge of the Universe.* Atlantic Monthly Press, 1987.

Readhead, A. "Radio Astronomy by Very-Long-Baseline Interferometry," *Scientific American,* June 1982.

Robinson, L. "The Frigid World of IRAS," *Sky & Telescope,* January 1984.

Robinson, L., and Murray, J. "The Gemini Project: Twins in Trouble?" *Sky & Telescope,* May 1993. See also the follow-up note in the March 1994 issue of *Sky & Telescope* (pp. 10–11).

Schorn, R. A., "Listening to the Universe," *Sky & Telescope,* November, 1988.

Sinnott, R., and Nyren, K. "The World's Largest Telescopes," *Sky & Telescope,* July 1993.

Sullivan, W. "Radio Astronomy's Golden Anniversary," *Sky & Telescope,* December 1982.

Tucker, W., and Tucker, K. *The Cosmic Inquirers.* Harvard University Press, 1986.

Verschuur, G. *The Invisible Universe Revealed.* Springer Verlag, 1987.

Chapter 4

Achinstein, P. *Particles and Waves: Historical Essays in the Philosophy of Science.* Oxford University Press, 1991.

Bova, B. *The Beauty of Light.* Wiley, 1988.

Cline, B. *Men Who Made a New Physics.* Signet, 1965.

Davies, J. K. "The Extreme Ultraviolet: A Promising New Window on the Universe," *Astronomy,* July 1987.

Feynman, R. *QED.* Princeton University Press, 1985.

Field, G. B., and Chaisson, E. J. *The Invisible Universe.* Vintage Books, 1987.

Gingerich, O. "Unlocking the Chemical Secrets of the Cosmos," *Sky & Telescope,* July 1981.

Gribbin, J. *In Search of Schroedinger's Cat.* Bantam, 1984.

Griffin, R. "The Radial-Velocity Revolution," *Sky & Telescope,* September 1989.

Sobel, M. *Light.* University of Chicago Press, 1987.

Snow, C. *The Physicists.* Macmillan, 1981.

van Heel, A., and Velzel, C. *What Is Light?* McGraw-Hill, 1968.

Weinberg, S. *Subatomic Particles.* Scientific American Books, 1983.

Chapter 5

Beatty, J. K. "The Making of a Better Moon," *Sky & Telescope,* December 1986.

Broecker, W. S. *How To Build A Habitable Planet.* Eldigio Press, 1985.

Burnham, R. "Watching the Lunar Phases," *Astronomy,* July 1989.

Cattermole, P., and Moore, P. *The Story of the Earth.* Cambridge University Press, 1985.

Comins, N. F. *What If The Moon Didn't Exist?* HarperCollins, 1993.

Davis, N. *The Aurora Watcher's Handbook.* University of Alaska Press, 1992.

French, B. "What's New On the Moon?" *Sky & Telescope,* March and April, 1977.

Hartmann, W. "The Moon's Early History," *Astronomy,* September 1976.

Kelley, K., ed. *The Home Planet.* Addison-Wesley and Mir, 1988.

Kitt, M. T. "Eight Lunar Wonders," *Astronomy,* March 1989.

——. "One Day at Copernicus Crater," *Astronomy,* September 1988.

Moore, P. *The Moon.* Rand McNally, 1981.

Murphy, J. B., and Nance, R. D. "Mountain Building and the Supercontinent Cycle," *Scientific American,* April 1992.

Van Andel, T. *New Views on an Old Planet.* Cambridge University Press, 1985.

Chapter 6

Bazilevskiy, A. T. "The Planet Next Door," *Sky & Telescope,* April 1989.

Bolt, B. A. *Earthquakes and Geological Discovery.* Scientific American Library, 1993.

Burgess, E. *Venus: An Arrant Twin.* Columbia University Press, 1985.

Burnham, R. "What Makes Venus Go?" *Astronomy,* January 1993.

Chaikin, A. "Four Faces of Mars," *Sky & Telescope,* July 1992.

Chapman, C. R. "Mercury's Heart of Iron," *Astronomy,* November 1988.

Cooper, H. *The Search for Life on Mars.* Holt, Rinehart & Winston, 1980.

Davies, M., et al., eds. *Atlas of Mercury.* NASA SP-423, 1978.

Eicher, D. J. "*Magellan* Scores at Venus," *Astronomy,* January 1991.

Goldman, S. J. "Venus Unveiled," *Sky & Telescope,* March 1992.

Haberle, R. M. "The Climate of Mars," *Scientific American,* May 1986.

Hartmann, W. "The Significance of the Planet Mercury," *Sky & Telescope,* May 1976.

——. "What's New on Mars?" *Sky & Telescope,* May 1989.

Horowitz, N. H. *To Utopia and Back.* W. H. Freeman and Company, 1986.

Murray, B., and Burgess, E. *Flight to Mercury.* Columbia University Press, 1977.

Powell, C. S. "Venus Revealed," *Scientific American,* January 1992.

Saunders, R. S. "The Surface of Venus," *Scientific American,* December 1990

Strom, R. *Mercury, the Elusive Planet,* Cambridge University Press, 1987.

——. "Mercury: The Forgotten Planet," *Sky & Telescope,* September 1990.

Toon, O. B. "How Climate Evolved on the Terrestrial Planets," *Scientific American,* February 1988.

Chapter 7

Beatty, J. "A Place Called Uranus," *Sky & Telescope,* April 1986.

——. "Getting to Know Neptune," *Sky & Telescope,* February 1990.

——. "Pluto and Charon: The Dance Goes On," *Sky & Telescope,* September 1987.

——. "Report on the *Voyager* Encounters with Saturn," *Sky & Telescope,* January, October, and November 1981.

Beebe, R. F. "Queen of the Giant Storms," *Sky & Telescope,* October 1990.

Bennett, J. "The Discovery of Uranus," *Sky & Telescope,* March 1981.

Berry, R. "Neptune Revealed," *Astronomy,* December 1989.

Binzel, R. "Pluto," *Scientific American,* June 1990.

Brown, R., and Cruikshank, D. "The Moons of Uranus, Neptune, and Pluto," *Scientific American,* July 1985.

Burgess, E. *By Jupiter.* Columbia University Press, 1982.

Burnham, R. "The Saturnian Satellites," *Astronomy,* November 1981.

Carroll, M. "Project Galileo: The Phoenix Rises," *Sky & Telescope,* April 1987.

Cuzzi, J. N., and Esposito, L. W. "The Rings of Uranus," *Scientific American,* July 1987.

Elliot, J., et al. "Discovering the Rings of Uranus," *Sky & Telescope,* June 1977.

Ingersoll, A. "Jupiter and Saturn," *Scientific American,* December 1981.

——. "Uranus," *Scientific American,* January 1987.

Kinoshita, J. "Neptune," *Scientific American,* November 1989.

MacRobert, A. M. "Hunting the Moons and Saturn," *Sky & Telescope,* July 1991.

Morrison, D. *Exploring Planetary Worlds.* Scientific American Library, 1993.

——. "The Enigma Called Io," *Sky & Telescope,* March 1985.

——. *Voyages to Saturn.* NASA SP-451, 1982.

Morrison, D., and Samz, J. *Voyage to Jupiter.* NASA SP-439, 1980.

Pollack, J., and Cuzzi, J. "Rings in the Solar System," *Scientific American,* November 1981.

Soderblom, L. A. "The Galilean Moons of Jupiter," *Scientific American,* January 1980.

"*Voyager* at Jupiter," *Astronomy,* special issue, May 1979.

Washburn, M. *Distant Encounters: The Exploration of Jupiter and Saturn.* Harcourt Brace Jovanovich, 1983.

Chapter 8

Alvarez, W., and Asaro, F. "What Caused the Mass Extinction? An Extraterrestrial Impact," *Scientific American* October 1990.

Beatty, J. K. "An Inside Look at Halley's Comet," *Sky & Telescope,* May 1986.

——. "Killer Crater in the Yucatán?" *Sky & Telescope,* July 1991.

Binzel, R. P., Barucci, M. A., and Fulchignoni, M. "The Origins of the Asteroids," *Scientific American,* October 1991.

Brandt, J. C., and Chapman, R. D. *Rendezvous in Space: The Science of Comets.* W. H. Freeman, 1992.

Cunningham, C. J. "Giuseppe Piazzi and the "Missing Planet,'" *Sky & Telescope,* September 1992.

Dodd, R. *Thunderstones and Shooting Stars: The Meaning of Meteorites.* Harvard University Press, 1986.

Durda, D. "All in the Family," *Astronomy,* February 1993.

Hartmann, W. "The Smaller Bodies of the Solar System," *Scientific American,* September 1975.

Knacke, R. "Sampling the Stuff of a Comet," *Sky & Telescope,* March 1987.

Kowal, C. *Asteroids.* Ellis Horwood/John Wiley, 1988.

Levy, D. H. "How to Discover a Comet," *Astronomy,* December 1987.

Marsden, B. G., "Comet Swift-Tuttle: Does it Threaten Earth?" in *Sky & Telescope,* January 1993.

Morrison, D., and Chapman, C. R. "Target Earth: It *Will* Happen," *Sky & Telescope,* March 1990.

Olson, R. "Giotto's Portrait of Halley's Comet," *Scientific American,* July 1979.

Russell, D. "The Mass Extinctions of the Late Mesozoic," *Scientific American,* January 1982.

Sagan, C., and Druyan, A. *Comet.* Random House, 1985.

Weissman, P. "Are Periodic Bombardments Real?" *Sky & Telescope,* March 1990.

——. "Realm of the Comet," *Sky & Telescope,* March 1987.

Whipple, F. "The Black Heart of Comet Halley," *Sky & Telescope*, March 1987.

—. *The Mystery of Comets*. Smithsonian Institution Press, 1985.

—. "The Nature of Comets," *Scientific American*, February 1974.

—. "The Spin of Comets," *Scientific American*, March 1980.

Chapter 9

Bahcall, J. "Where Are the Solar Neutrinos?" *Astronomy*, March 1990.

Eddy, J. *A New Sun*. NASA SP-402, 1979.

—. "The Case of the Missing Sunspots," *Scientific American*, May 1977.

Foukal, P. "The Variable Sun," *Scientific American*, February 1990.

Frazier, K. *Our Turbulent Sun*. Prentice-Hall, 1983.

Friedman, H. *Sun and Earth*. Scientific American Library, 1986.

Giovanelli, R. *Secrets of the Sun*. Cambridge University Press, 1984.

Harvey, J., et al. "GONG: To See inside Our Sun," *Sky & Telescope*, November 1987.

Leibacher, J. W., et al. "Helioseismology," *Scientific American*, September 1985.

Levine, R. "The New Sun," in J. Cornell and P. Gorenstein, eds., *Astronomy from Space: Sputnik to Space Telescope*. MIT Press, 1983.

Mitton, S. *Daytime Star*. Scribner, 1981.

Nichols, R. "Solar Max: 1980–89," *Sky & Telescope*, December 1989.

Noyes, R. *The Sun, Our Star*. Harvard University Press, 1982.

Robinson, L. J. "The Disquieting Sun: How Big, How Steady?" *Sky & Telescope*, April 1982.

—. "The Sunspot Cycle: Tip of the Iceberg," *Sky & Telescope*, June 1987.

Taylor, M. "Observing from the South Pole," *Sky & Telescope*, October 1988.

Wallenhorst, S. "Sunspot Numbers and Solar Cycles," *Sky & Telescope*, September 1982.

Wentzel, D. *The Restless Sun*. Smithsonian Institution Press, 1989.

Wolfson, R. "The Active Solar Corona," *Scientific American*, February 1983.

Chapter 10

Ashbrook, J. "Visual Double Stars for the Amateur," *Sky & Telescope*, November 1980.

Evans, D., et al. "Measuring Diameters of Stars," *Sky & Telescope*, August 1979.

Gingerich, O. "A Search for Russell's Original Diagram," *Sky & Telescope*, July 1982.

Griffin, R. "The Radial-Velocity Revolution," *Sky & Telescope*, September 1989.

Kaler, J. "Origins of the Spectral Sequence," *Sky & Telescope*, February 1986.

—. *Stars and Their Spectra*. Cambridge University Press, 1989.

Mitton, J., and MacRobert, A. "Colored Stars," *Sky & Telescope*, February 1989

Nielsen, A. "E. Hertzsprung—Measurer of Stars," *Sky & Telescope*, January 1968

Phillip, A., and Green, L. "Henry N. Russell and the H–R Diagram," *Sky & Telescope*, April 1978, May 1978

Tomkin, J., and Lambert, D. "The Strange Case of Beta Lyrae," *Sky & Telescope*, October 1987

Upgren, A. "New Parallaxes for Old: Coming Improvements in the Distance Scale of the Universe," *Mercury*, November/December 1980.

Chapter 11

Blitz, L. "Giant Molecular Cloud Complexes in the Galaxy," *Scientific American*, April 1982.

Cohen, M. *In Darkness Born: The Story of Star Formation*. Cambridge University Press, 1988.

Fortier, E. "Touring the Stellar Cycle," *Astronomy*, March 1987.

Hartley, K. "How a Star is Born," *Astronomy*, December 1989.

Johnson, B. "Red Giant Stars," *Astronomy*, December 1976.

Kaler, J. B. "Cousins of Our Sun: The G Stars," *Sky & Telescope*, November 1986.

—. "Journeys on the H–R diagram," *Sky & Telescope*, May 1988.

—. "The K Stars: Orange Giants and Dwarfs," *Sky & Telescope*, August 1986.

—. "The M Stars: Supergiants to Dwarfs," *Sky & Telescope*, May 1986.

—. "The Spectacular O Stars," *Sky & Telescope*, November 1987.

—. "The Temperate F Stars," in *Sky & Telescope*, February 1987.

—. *Stars*. Scientific American Library, 1992.

Kippenhahn, R. *100 Billion Suns: The Birth, Life and Death of Stars*. Basic Books, 1983.

Lada, C. "Energetic Outflows from Young Stars," *Scientific American*, July 1982.

Marschall, L. *The Supernova Story*. Plenum Press, 1988.

Robinson, L. "Orion's Stellar Nursery," *Sky & Telescope*, November 1982.

Stahler, S. W. "The Early Life of Stars," *Scientific American*, July 1991.

Stahler, S., and Comins, N. "The Difficult Birth of Sunlike Stars," *Astronomy*, September 1988.

Verschuur, G. *Interstellar Matters*. Springer-Verlag, 1988.

Chapter 12

Bethe, H., and Brown, G. "How a Supernova Explodes," *Scientific American*, May 1985.

Clark, D. *Superstars*. McGraw-Hill, 1984.

Greenstein, G. *Frozen Star*. Freundlich Books, 1984.

Kwok, S. "Not with a Bang But a Whimper," *Sky & Telescope*, May 1982.

Lattimer, J., and Burrows, A. "Neutrinos from Supernova 1987A," *Sky & Telescope*, October 1988.

Margon, B. "The Bizarre Spectrum of SS433," *Scientific American*, October 1980.

Marschall, L. *The Supernova Story*. Plenum, 1988.

Marschall, L., and Brecher, K. "Will Supernova 1987A Shine Again?" *Astronomy*, February 1992.

White, N. "New Wave Pulsars," *Sky & Telescope*, January 1987.

Woosley, S., and Weaver, T. "The Great Supernova of 1987," *Scientific American*, August 1989.

Chapter 13

Chaisson, E. *Relatively Speaking: Relativity, Black Holes and the Fate of the Universe*. W. W. Norton, 1988.

Croswell, K. "The Best Black Hole in the Galaxy," *Astronomy*, March 1992.

Greenstein, G. *Frozen Star*. Freundlich Books, 1983.

McClintock, J. "Do Black Holes Exist?" *Sky & Telescope*, January 1988.

Rees, M. J. "Black Holes in Galactic Centers," *Scientific American*, November 1990.

Thorne, K. *Black Holes and Time Warps*. W. W. Norton, 1994.

——. "The Search for Black Holes," *Scientific American*, December 1974.

Wheeler, J. A. *A Journey Into Gravity and Spacetime. Scientific American Library*, 1990.

Chapter 14

Bok, B. "A Bigger and Better Milky Way," *Astronomy*, January 1984.

——. "The Milky Way Galaxy," *Scientific American*, March 1981.

Bok, B., and Bok, P. *The Milky Way*, 5th ed. Harvard University Press, 1981.

Chaisson, E. "Journey to the Center of the Galaxy," *Astronomy*, August 1980.

Geballe, T. "The Central Parsec of the Galaxy," *Scientific American*, July 1979.

Kaufmann, W. J. "Our Galaxy, Part I," *Mercury*, May/June 1989.

——. "Our Galaxy, Part 2," *Mercury* July–August 1989.

Kraus, J. "The Center of Our Galaxy," *Sky & Telescope*, January 1983.

Palmer, E. "Unveiling the Hidden Milky Way," in *Astronomy*, November 1989.

Weaver, H. "Steps toward Understanding the Large-Scale Structure of the Milky Way," *Mercury*, September/October 1975, November/December 1975, January/February 1976.

Chapter 15

Barrow, J. D. and Silk, J. *The Left Hand of Creation*. Basic Books 1983.

Christianson, G. E. *Edwin Hubble: Mariner of the Nebulae*. Farrar, Straus, & Giroux, 1995.

de Vaucouleurs, G. "The Distance Scale of the Universe," *Sky & Telescope*, December 1983.

——. "M31's Spiral Shape," *Sky & Telescope*, December 1987.

Ferris, T. *Galaxies*. Stewart, Tabori, and Chang, 1982.

Field, G. "The Hidden Mass in Galaxies," *Mercury*, May/June 1982.

Freeman, M. "Galaxies," *Smithsonian*, January 1989.

Gorenstein, P., and Tucker, W. "Rich Clusters of Galaxies," *Scientific American*, November 1978.

Hartley, K. "Elliptical Galaxies Forged by Collision," *Astronomy*, May 1989.

Hodge, P. *Galaxies*. Harvard University Press, 1986.

Keel, W. "Crashing Galaxies, Cosmic Fireworks," *Sky & Telescope*, January 1989.

Osterbrock, D., Brashear, R., and Gwinn, J. "Young Edwin Hubble," *Mercury*, January/February 1990.

Rubin, V. "Dark Matter in Spiral Galaxies," *Scientific American*, June 1983.

Silk, J. "Formation of the Galaxies," *Sky & Telescope*, December 1986.

Smith, R. *The Expanding Universe: Astronomy's Great Debate*. Cambridge University Press, 1982.

Tully, R. "Unscrambling the Local Supercluster," *Sky & Telescope*, June 1982.

Wray, J. *The Color Atlas of Galaxies*. Cambridge University Press, 1988.

Chapter 16

Blandford, R., et al. "Cosmic Jets," *Scientific American*, May 1982.

Burns, J., and Price, R. "Centaurus A: The Nearest Active Galaxy," *Scientific American*, November 1983.

McCarthy, P. "Measuring Distances to Remote Galaxies and Quasars," *Mercury*, January/February 1988.

Mood, J. "Star Hopping to a Quasar," *Astronomy*, May 1987.

Preston, R. *First Light*. Atlantic Monthly Press, 1987.

Shipman, H. *Black Holes, Quasars, and the Universe*, 2nd ed. Houghton Mifflin, 1980.

Verschuur, G. *The Invisible Universe Revealed*. Springer-Verlag, 1987.

Wilkes, B. "The Emerging Picture of Quasars," *Astronomy*, December 1991.

Chapter 17

Barrow, J., and Silk, J. *The Left Hand of Creation: Origin and Evolution of the Universe*. Basic Books, 1983.

Davies, P. *The Forces of Nature*, 2nd ed. Cambridge University Press, 1986.

——. *Superforce*. Simon & Schuster, 1984.

Discus, D., et al. "The Future of the Universe," *Scientific American*, March 1983.

Disney, M. *The Hidden Universe*. Macmillan, 1984.

Dressler, A. "The Large-Scale Streaming of Galaxies," *Scientific American*, September 1987.

Gribbin, J. *In Search of the Big Bang*. Bantam, 1986.

Guth, A., and Steinhardt, P. "The Inflationary Universe," *Scientific American*, May 1984.

Harrison, E. *Darkness at Night*. Harvard University Press, 1987.

Hawking, S. *A Brief History of Time*. Bantam 1988. (Also on CD-ROM from W. H. Freeman.)

Henbest, N., and Couper, H. *The Guide to the Galaxies*. Cambridge University Press 1994.

Islam, J. N. *The Ultimate Fate of the Universe*. Cambridge University Press, 1983.

Kanipe, J. "Beyond the Big Bang," *Astronomy*, April 1992.

Monda, R. "Shedding Light on Dark Matter," *Astronomy*, February 1992.

Riordan, M., and Schramm, D. N. *The Shadows of Creation: Dark Matter and the Structure of the Universe*. W. H. Freeman, 1991.

Rowan-Robinson, M. *Our Universe: An Armchair Guide*. W. H. Freeman, 1990.

Silk, J. *A Short History of the Universe*. Scientific American Library, 1994.

——. *The Big Bang*, 2nd ed., W. H. Freeman, 1989.

Smoot, G., and Davidson, K. *Wrinkles in Time*. William Morrow, 1993.

Weinberg, S. *The First Three Minutes*, updated. HarperCollins, 1988.

Chapter 18

Baugher, J. *On Civilized Stars: The Search for Intelligent Life in Outer Space*, Prentice-Hall, 1985.

Beatty, J. K. "The New, Improved SETI," in *Sky & Telescope*, May 1983.

Davies, P. *The Cosmic Blueprint*, Simon & Schuster, 1988.

Dawkins, R. *The Blind Watchmaker*. Norton, 1987.

Feinberg, G., and Shapiro, R. *Life beyond Earth: The Intelligent Earthling's Guide to Life in the Universe*. Morrow, 1980.

Finney, B., and Jones, E. *Interstellar Migration and the Human Experience*. University of California Press, 1985.

Goldsmith, D., ed. *The Quest for Extraterrestrial Life: A Book of Readings*. University Science Books, 1980.

Goldsmith, D., and Owen, T. *The Search for Life in the Universe*. Benjamin/Cummings, 1980.

Gould, S. *Wonderful Life*. Scribner, 1990.

Harrison, E. *Masks of the Universe*. MacMillan, 1985.

Hart, M., and Zuckerman, B., eds. *Extraterrestrials Where Are They?* Pergamon Press, 1982.

Horowitz, N. *To Utopia and Back: The Search for Life in the Solar System*, W. H. Freeman, 1986.

Kutter, G. S. *The Universe and Life*, Jones and Bartlett, 1987.

Marx, G., ed. *Bioastronomy: The Next Steps*. Kluwer, 1988.

Papagiannis, M. D. "Bioastronomy: The Search for Extraterrestrial Life," in *Sky & Telescope*, June 1984.

Reeves, H. *The Hour of Our Delight: Cosmic Evolution, Order, and Complexity*. W. H. Freeman, 1991.

Regis, E. *Extraterrestrials: Science and Alien Intelligence*. Cambridge University Press, 1985.

Rood, R., and Trefil, J. *Are We Alone?* Scribner, 1981.

Sagan, C., and Drake, F. "The Search for Extraterrestrial Intelligence," *Scientific American*, May 1975.

Schorn, R. A. "Extraterrestrial Beings Don't Exist," *Sky & Telescope*, September 1981.

Shklovskii, I. S., and Sagan, C. *Intelligent Life in the Universe*. Holden-Day, 1966.

Wilson, A. C. "The Molecular Basis of Evolution," *Scientific American*, October 1985.

Glossary

A ring The outermost ring of Saturn visible from Earth; it is located just beyond Cassini's division.

absolute magnitude The apparent magnitude that a star would have if it were 10 parsecs from Earth.

absorption line spectrum Dark lines superimposed on a continuous spectrum.

acceleration A change in the direction or magnitude of a velocity.

accretion The gradual accumulation of matter by an astronomical body, usually caused by gravity.

accretion disk An orbiting disk of matter spiraling in toward a star or black hole.

achromatic lens A compound lens designed to minimize the effect of chromatic aberration.

active galaxy A very luminous galaxy, often containing an active galactic nucleus.

adaptive optics Primary telescope mirrors that are continuously and automatically adjusted to compensate for the distortion of starlight due to the motion of the Earth's atmosphere.

AGB star *See* asymptotic giant branch (AGB) star.

albedo The fraction of sunlight that a planet, asteroid, or satellite reflects directly back into space.

amino acid A class of chemical compounds that are the building blocks of proteins.

angle The opening between two straight lines that meet at a point.

angular diameter (angular size) The arc angle across an object.

angular resolution The angular size of the smallest detail of an astronomical object that can be distinguished with a telescope.

annular eclipse An eclipse of the Sun in which the Moon is too distant to cover the Sun completely so that a ring of sunlight is seen around the Moon at mid-eclipse.

anorthosite A light-colored rock found throughout the lunar highlands and in some very old mountains on Earth.

aphelion The point in its orbit where a planet or other solar system body is farthest from the Sun.

Apollo asteriod An asteroid that is sometimes closer to the Sun than is the Earth.

apparent magnitude A measure of the brightness of light from a star or other object as seen from Earth.

arc angle The measurement of the angle between two objects or two parts of the same object.

asteroid (minor planet) Any of the rocky objects larger than a few hundred meters in diameter (and not classified as a planet or moon) that orbits the Sun.

asteroid belt A $1\frac{1}{2}$-astronomical-unit-wide region between the orbits of Mars and Jupiter in which most of the asteroids are found.

astronomical unit (AU) The average distance between the Earth and the Sun: 1.5×10^8 kilometers $\approx$ 93 million miles.

astronomy The branch of science dealing with objects and phenomena that lie beyond the Earth's atmosphere.

astrophysics That part of astronomy dealing with the physics of astronomical objects and phenomena.

asymptotic giant branch (AGB) star A red giant star that has completed core helium fusion and has re-expanded for a second time.

atom The smallest particle of an element that has the properties characterizing that element.

atomic number The number of protons in the nucleus of an atom.

aurora (*plural* aurorae) Light radiated by atoms and ions in the Earth's upper atmosphere; seen mostly in the polar regions.

autumnal equinox The intersection of the ecliptic and the celestial equator where the Sun crosses the equator moving from north to south.

average density The mass of an object divided by its volume.

B ring The brightest of the three rings of Saturn visible from Earth; it lies just inside of Cassini's division.

barred spiral galaxy   A spiral galaxy in which the spiral arms begin from the ends of a bar running through the nuclear bulge.

belt asteriod   An asteroid whose orbit lies in the asteroid belt.

belts (on Jupiter)   Dark, reddish bands in Jupiter's cloud cover.

Big Bang   An explosion roughly 15 billion years ago creating all space, matter, and energy in which the universe emerged.

Big Crunch   The gravitational collapse of the universe; the ultimate fate of a bound universe.

binary star   Two stars revolving about each other; a double star.

birth line   A line on the Hertzsprung–Russell diagram corresponding to where stars with different masses transform from protostars to pre-main-sequence stars.

BL Lacertae (BL Lac) object   A type of active galaxy.

black hole   An object whose gravity is so strong that the escape velocity from it exceeds the speed of light.

blackbody   A hypothetical perfect radiator that absorbs and re-emits all radiation falling upon it.

blackbody curve   The curve obtained when the intensity of radiation from a blackbody at a particular temperature is plotted against wavelength.

blackbody radiation   Electromagnetic radiation emitted by a blackbody.

blueshift   A shift of all spectral features toward shorter wavelengths; the Doppler shift of light from an approaching source.

Bode's law   A numerical sequence that gives the approximate distances of the planets from the Sun in astronomical units.

Bohr atom   A model of the atom, described by Niels Bohr, in which electrons revolve about the nucleus in various circular orbits.

bound universe   A universe that expands, reaches a maximum size, and then contracts; *see also* closed universe.

breccia   A rock formed by the sudden amalgamation of various rock fragments under pressure.

brown dwarf   Any of the planetlike bodies with less than 0.08 $M_\odot$; such bodies do not have enough mass to sustain fusion in their cores.

burster   A nonperiodic ray source that emits powerful bursts of x rays or gamma rays.

C ring   The faint, inner portion of Saturn's main ring system.

caldera   The crater at the summit of a volcano.

Callisto   One of the four Galilean satellites of Jupiter.

capture theory   The idea that the Moon was created at a different location in the solar system and subsequently captured by Earth's gravity.

carbon fusion   The thermonuclear fusion of carbon nuclei to produce nitrogen.

carbonaceous chondrite   A class of extremely ancient, carbon-rich meteorites.

Cassegrain focus   An optical arrangement in a reflecting telescope in which light rays are reflected by a secondary mirror through a hole in the primary mirror.

Cassini division   A prominent gap between Saturn's A and B rings discovered in 1675 by J. D. Cassini.

celestial equator   A great circle on the celestial sphere 90° from the celestial poles.

celestial poles   Points about which the celestial sphere appears to rotate.

celestial sphere   A sphere of very large radius centered on the observer; the apparent sphere of the night sky.

Celsius scale   *See* temperature (Celsius).

Cepheid variable star   One of two types of yellow, supergiant, pulsating stars.

Cerenkov radiation   Radiation produced by particles traveling through a substance faster than light can.

Ceres   The largest known asteroid and the first to be discovered.

Chandrasekhar limit   The maximum mass of a white dwarf, about 1.4$M_\odot$.

charge-coupled device (CCD)   A type of solid-state silicon wafer designed to detect photons.

chromatic aberration   An optical property whereby different colors of light passing through a lens are focused at different distances from it.

chromosphere   The layer in the solar atmosphere between the photosphere and the corona.

circumpolar stars   All the stars that never set at a given latitude; all the stars between Polaris and the northern horizon.

close binary   A binary star whose members are separated by a few stellar diameters.

closed universe   A universe that contains enough matter to cause it to recollapse. It is spherically shaped and has no "outside."

cluster (of galaxies)   A collection of a few hundred to a few thousand galaxies bound by gravity.

cocreation theory   The theory that the Moon formed simultaneously with the Earth and in orbit around it.

collision–ejection theory   The theory that the Moon was created by the impact of a planet-sized object with the Earth; presently considered the most plausible theory of the Moon's formation.

color index   The difference in the magnitudes of a star's brightness measured in two separate wavelength bands.

color–magnitude diagram   A plot of the surface temperatures (colors) versus the absolute magnitudes of stars.

coma (of a comet)   The nearly spherical, diffuse gas surrounding the nucleus of a comet near the Sun.

comet   A small body of ice and dust in orbit about the Sun. While passing near the Sun, a comet's vaporized ices give rise to a coma and tails.

conduction   The transfer of heat by passing energy directly from atom to atom.

configuration (of a planet)   A particular geometric arrangement of the Earth, a planet, and the Sun.

confinement   The moment shortly after the Big Bang when quarks bound together to form particles like protons and neutrons.

conic section   The curve of intersection between a circular cone and a plane. This curve can be a circle, ellipse, parabola, or hyperbola.

conjunction   The alignment of two bodies in the solar system so that they appear in the same part of the sky as seen from Earth.

conservation of angular momentum   The law of physics stating that the total amount of angular momentum in an isolated system remains constant.

constellation   Any of the 88 contiguous regions that cover the entire celestial sphere, including all the objects in each region; also, a configuration of stars often named after an object, a person, or an animal.

contact binary   A close binary system in which both stars fill or overflow their Roche lobes.

continental drift   The gradual movement of the continents over the surface of the Earth due to plate tectonics.

continuous spectrum   A spectrum of light over a range of wavelengths without any spectral lines.

continuum   *See* continuous spectrum.

convection   The transfer of energy by moving currents of fluid or gas containing that energy.

convective zone   A layer in a star where energy is transported outward by means of convection; also known as the convective envelope or convection zone.

core   The central portion of any astronomical object.

core helium fusion   The fusion of helium to form carbon and oxygen at the center of a star.

core hydrogen fusion   The fusion of hydrogen to form helium at the center of a star.

corona   The Sun's outer atmosphere.

coronagraph   A specially designed telescope with a baffle that blocks out the solar disk so that the corona can be photographed.

coronal hole   A dark region of the Sun's inner corona as seen at x-ray wavelengths.

cosmic censorship   The belief that the only connection between a black hole and the universe is the black hole's event horizon.

cosmic microwave background   Photons from every part of the sky with a blackbody spectrum at 2.73 K; the cooled-off radiation from the primordial fireball that originally filled all space.

cosmic particle horizon   A sphere, centered on the Earth, whose radius equals the distance traveled by light since the Big Bang.

cosmogony   The study of the formation and evolution of the solar system.

cosmological constant   A number sometimes inserted in the equations of general relativity that represents a pressure that opposes gravity throughout the universe.

cosmological redshift   An increase in wavelength from distant galaxies and quasars caused by the expansion of the universe.

cosmology   The study of the formation, organization, and evolution of the universe.

coudé focus   A reflecting telescope in which a series of mirrors direct light to a remote focus away from the moving parts of the telescope.

crater   A circular depression on a planet or moon caused by the impact of a meteoroid, asteroid, or comet, or by a volcano.

crescent Moon   A lunar phase during which the Moon appears less than half full.

critical density   The average density throughout the universe for which space is flat and galaxies just barely continue to recede from each other infinitely far into the future.

critical surface   An imaginary double-teardrop-shaped surface surrounding the stars in a binary that delineates the gravitational domain of each star; Roche lobes.

crust   The solid surface layer of some astronomical bodies including the terrestrial planets, the moons, the asteroids, and some stellar remnants.

dark matter (missing mass)   The as-yet-undetected matter in the universe that is underluminous and probably quite different from ordinary matter.

dark nebula   A cloud of interstellar gas and dust that obscures the light of more distant stars.

deceleration parameter (q_0)   A number that specifies how fast the expansion of the universe is slowing down.

declination   The coordinate on the celestial sphere exactly analogous to latitude on Earth; measured north and south of the celestial equator.

decoupling   The epoch in the early universe when electrons and ions first combined to create stable atoms; the time when electromagnetic radiation ceased to dominate over matter.

deferent   A fixed circle in the Earth-centered universe along which a smaller circle (an epicycle) moves carrying a planet, the Sun, or Moon.

degeneracy   The condition in which all the lower energy states for particles (electrons or neutrons) in a gas have been filled, thereby causing the gas to behave differently than an ordinary gas.

degree (°)   A unit of angular measure or of temperature measure.

dense core   Any of the regions of interstellar gas clouds that are slightly denser than normal and destined to collapse to form one or a few stars.

density   The ratio of the mass of an object to its volume.

density wave   A spiral-shaped compression of the gas and dust in a spiral galaxy.

density-wave theory   An explanation of spiral arms in galaxies elaborated by C. C. Lin and F. Shu.

detached binary   A binary system in which the surfaces of both stars are inside their Roche lobes.

differential rotation   The rotation of a nonrigid object in which parts at different latitudes or different radial distances move at different speeds.

differentiation (chemical)   The separation of different kinds of material into different layers inside a planet.

direct motion   The gradual, eastward apparent motion of a planet against the background stars as seen from Earth.

disk (of a galaxy)   A flattened assemblage of stars, gas, and dust in a spiral galaxy like the Milky Way.

diurnal motion   Cyclic motion with a one-day period.

Doppler effect (or **shift**)   The change in wavelength of radiation due to relative motion between the source and the observer along the line of sight.

double-line spectroscopic binary   A spectroscopic binary whose spectrum exhibits spectral lines of both stars.

double radio source   An extragalactic radio source characterized by two large lobes of radio emission, often located on either side of an active galaxy.

Drake equation   A mathematical equation used to estimate the number of extraterrestrial civilizations that may exist in our Galaxy.

dust tail (of a comet)   A comet tail caused by dust particles escaping from the comet's nucleus.

dwarf elliptical galaxy   A small elliptical galaxy with far fewer stars than a typical galaxy.

dwarf star   Any star smaller than a giant, such as a main sequence star or a white dwarf.

dynamo theory   The generation of a magnetic field by circulating electric charges.

eccentricity   *See* orbital eccentricity.

eclipse   The blocking of part or all of the light from one celestial object by another.

eclipse path   The track of the tip of the Moon's shadow along the Earth's surface during a total or annular solar eclipse.

eclipsing binary   A double star system in which stars periodically pass in front of each other as seen from Earth.

ecliptic   The annual path of the Sun on the celestial sphere; the plane of the Earth's orbit around the Sun.

Einstein cross   The appearance of four images of the same galaxy or quasar due to gravitational lensing by an intervening galaxy.

Einstein ring   The circular or arc-shaped image of a distant galaxy or quasar created by gravitational lensing by an intervening galaxy.

ejecta blanket   The ring of material surrounding a

crater that was ejected during the crater-forming impact.

electromagnetic radiation Radiation consisting of oscillating electric and magnetic fields such as gamma rays, x rays, visible light, ultraviolet and infrared radiation, and radio waves.

electromagnetic spectrum The entire array or family of electromagnetic radiation.

electron A negatively charged subatomic particle usually found in orbit about the nucleus of an atom.

electron degeneracy pressure A powerful pressure produced by repulsion of closely packed (i.e., degenerate) electrons.

element A substance that cannot be decomposed by chemical means into simpler substances.

ellipse A closed curve obtained by cutting completely through a circular cone with a plane; the shape of planetary orbits.

elliptical galaxy A galaxy with an elliptical shape, little interstellar matter, and no spiral arms.

elongation The angle between a planet and the Sun as seen from Earth.

emission line spectrum A spectrum that contains only bright emission lines.

emission nebula A glowing gaseous nebula whose light comes from fluorescence caused by a nearby star.

Encke division A thin gap in Saturn's A ring, possibly first seen by J. F. Encke in 1838.

energy The ability to do work.

energy flux The amount of energy emitted from each square meter of an object's surface per second.

energy level (in an atom) A particular amount of energy possessed by an electron in orbit around a nucleus.

epicycle In the Earth-centered universe, a moving circle about which planets revolve.

equations of stellar structure A set of relationships that describe the interactions of matter, energy, and gravity inside a star.

equinox Either of the two days of the year when the Sun crosses the celestial equator and is therefore directly over the Earth's equator; *see also* autumnal equinox *and* vernal equinox.

era of recombination The time, roughly 300 thousand years after the Big Bang, when the universe became transparent.

ergoregion The region of space immediately outside the event horizon of a rotating black hole where it is impossible to remain at rest.

escape velocity The speed needed by one body to just escape the gravitational attraction of another body and move into free space.

Europa One of the Galilean satellites of Jupiter.

event horizon The location around a black hole where the escape velocity equals the speed of light; the boundary of a black hole.

evolutionary track On the Hertzsprung–Russell diagram, the path followed by a point representing an evolving star.

excitation The process of imparting energy to an electron in an atom or ion.

expanding universe The motion of the superclusters of galaxies away from each other.

eyepiece lens A magnifying lens used to view the image produced at the focus of a telescope.

F ring A thin ring just beyond the outer edge of Saturn's main ring system.

Fahrenheit scale *See* temperature (Fahrenheit).

filament A dark curve seen above the Sun's photosphere that is the top view of a solar prominence.

first quarter Moon A phase of the waxing Moon when Earth-based observers see half of the Moon's illuminated hemisphere.

fission theory The theory that the Moon formed from matter flung off the Earth because the planet was rotating extremely fast.

flare A sudden, temporary outburst of photons and particles from an extended region of the solar surface.

flatness problem The quandary associated with the improbable fact that space throughout the universe seems to be essentially flat.

flocculent spiral galaxy A spiral galaxy whose spiral arms are broad, fuzzy, and poorly demarcated.

focal length The distance from a lens or concave mirror to where converging light rays meet.

focal plane The plane at the focal length of a lens or concave mirror on which an extended object is focused.

focus (of a lens or concave mirror) The place at the focal length where light rays from a point object (i.e., one that is too distant or tiny to resolve) are converged by a lens or concave mirror.

focus (*plural* **foci**) (of an ellipse)   The two points inside an ellipse, the sum of whose distances from any point on the ellipse is constant.

force   That which can change the momentum of an object.

frequency   The number of waves that cross a given point per unit time; the number of vibrations per unit time.

full Moon   A phase of the Moon during which its full daylight hemisphere can be seen from Earth.

galactic cannibalism   A collision between two galaxies of unequal mass and size in which the smaller galaxy is absorbed by the larger galaxy.

galactic merger   A collision and subsequent merger of two roughly equal-sized galaxies.

galactic nucleus   The center of a galaxy; the center of the Milky Way Galaxy.

galaxy   A large assemblage of stars, gas, and dust bound together by their mutual gravitational attraction.

Galilean satellite (or **moon**)   Any one of the four large moons of Jupiter that are visible from Earth through even a small telescope.

gamma ray   The most energetic form of electromagnetic radiation.

gamma-ray burster   An object that emits a short burst of gamma rays; most gamma-ray bursters are believed to be outside our Galaxy.

Ganymede   One of the Galilean satellites of Jupiter.

gas (ion) tail   The relatively straight tail of a comet produced by the solar wind acting on ions in a comet's coma.

general theory of relativity   A description of spacetime formulated by Einstein explaining how gravity affects the geometry of space and the flow of time.

geocentric cosmology (or **cosmogony**)   The belief that the Earth is at the center of the universe.

giant elliptical galaxy   A very large, extremely massive elliptical galaxy, usually located near the center of a rich cluster of galaxies.

giant molecular cloud   A large interstellar cloud of cool gas and dust in a galaxy.

giant star   A star whose diameter is roughly 10 to 100 times that of the Sun.

gibbous Moon   A phase of the Moon in which more than half, but not all, of the Moon's daylight hemisphere is visible from Earth.

globular cluster   A large spherical cluster of gravitationally bound stars usually found in the outlying regions of a galaxy.

grand-design spiral galaxy   A spiral galaxy whose spiral arms are thin, graceful, and well-defined.

granulation   The rice-grain-like structure of the solar photosphere due to convection of solar gases.

granules   Lightly colored convection features about 1000 kilometers in diameter seen constantly in the solar photosphere.

grating   An optical device consisting of closely spaced lines ruled on a piece of glass that is used like a prism to disperse light into a spectrum.

gravitation (**gravity**)   The tendency of all matter to attract all other matter.

gravitational lensing   The distortion of the appearance of an object by a source of gravity between it and the observer.

gravitational redshift   The redshift of photons leaving the gravitational field of any massive object, such as a star or black hole.

gravitational waves (**gravitational radiation**)   Ripples in the overall geometry of space produced by moving objects.

gravity   *See* gravitation.

Great Dark Spot   A large, dark, oval-shaped storm that used to be in Neptune's southern hemisphere.

Great Red Spot   A large, red-orange, oval-shaped storm in Jupiter's southern hemisphere.

Great Wall   A huge arc of galaxies between two voids in the cosmos.

greatest elongation   The largest possible angle between the Sun and an inferior planet.

greenhouse effect   The trapping of infrared radiation near a planet's surface by the planet's atmosphere.

ground state   The lowest energy level of an atom.

H I region   A region of neutral hydrogen in interstellar space.

H II region   A region of ionized hydrogen in interstellar space.

halo (of a galaxy)   A spherical distribution of globular clusters, isolated stars, and possibly dark matter that surrounds a galaxy.

Hawking process The formation of real particles from virtual ones just outside a black hole's event horizon; the means by which black holes evaporate.

head–tail source A radio galaxy whose radio emission is deflected from the galaxy.

heliocentric cosmogony A theory of the formation and evolution of the solar system with the Sun at the center.

helioseismology The study of vibrations of the solar surface.

helium flash The explosive ignition of helium fusion in the core of a low-mass, red giant star.

helium fusion The thermonuclear fusion of helium to produce carbon.

helium shell flash The explosive ignition of helium fusion in a thin shell surrounding the core of a low-mass star.

Hertzsprung–Russell (H–R) diagram A plot of the absolute magnitude or luminosity of stars versus their surface temperatures or spectral classes.

highlands (lunar) Heavily cratered, mountainous regions of the lunar surface.

horizontal branch stars A group of post-helium-flash stars near the main sequence on the Hertzsprung–Russell diagram of a typical globular cluster.

hot-spot volcanism The creation of volcanoes on a planet's surface caused by a reservoir of hot magma in the planet's mantle under a thin part of the crust.

Hubble classification A system of classifying galaxies according to their appearance into one of four broad categories: spirals, barred spirals, ellipticals, and irregulars.

Hubble constant The constant of proportionality in the relation between the recessional velocities of remote galaxies and their distances; the correct number will determine the age of the universe.

Hubble flow The recession of the galaxies caused by the expansion of the universe.

Hubble law The relationship which states that the redshifts of remote galaxies are directly proportional to their distances from Earth.

hydrocarbon A molecule based on hydrogen and carbon.

hydrogen envelope An extremely large, tenuous sphere of hydrogen gas surrounding the head of a comet.

hydrogen fusion (hydrogen burning) The thermonuclear fusion of hydrogen to produce helium.

hydrostatic equilibrium A balance between the weight of a layer in a star and the pressure that supports it.

hyperbola An open curve obtained by cutting a cone with a plane.

impact breccia A rock consisting of various fragments cemented together by the impact of a meteoroid.

impact crater A crater on the surface of a planet or moon produced by the impact of an asteroid, meteoroid, or comet.

inferior conjunction The configuration when Mercury or Venus is directly between the Sun and the Earth.

inflationary epoch A brief period shortly after the Big Bang during which the scale of the universe increased very rapidly.

infrared radiation Electromagnetic radiation of a wavelength longer than visible light yet shorter than radio waves.

instability strip A region on the Hertzsprung–Russell diagram occupied by pulsating stars.

interferometry A method of increasing resolving power by combining electromagnetic radiation obtained by two or more telescopes.

intergalactic gas Gas located between the galaxies within a galactic cluster.

interstellar dust Microscopic solid grains of various compounds in interstellar space.

interstellar gas Sparse gas in interstellar space.

interstellar medium Interstellar gas and dust.

inverse-square law The gravitational attraction between two objects and the apparent brightness of a light source are both inversely proportional to the square of its distance.

Io One of the Galilean satellites of Jupiter.

ion An atom that has become electrically charged due to the loss or addition of one or more electrons.

ionization The process by which an atom loses or gains electrons.

iron meteorite A meteorite composed of iron with a small admixture of nickel; also called an iron.

irregular cluster (of galaxies) A unevenly distributed group of galaxies bound together by their mutual gravitational attraction.

irregular galaxy   An asymmetrical galaxy having neither spiral arms nor an elliptical shape.

isotope   Any of several forms of the same chemical element whose nuclei all have the same number of protons but different numbers of neutrons.

isotropy   The fact that the average number of galaxies at different distances from Earth is the same in all directions; also, the fact that the temperature of the cosmic microwave background is essentially the same in all directions.

Jeans instability   The condition under which gravitational forces overcome thermal forces to cause part of an interstellar cloud to collapse.

kelvin   *See* temperature (Kelvin).

Kepler's laws   Three statements, formulated by Johannes Kepler, that describe the motions of the planets.

Kerr black hole   Any rotating, uncharged black hole.

kiloparsec (kpc)   One thousand parsecs; about 3260 light-years.

Kirchhoff's laws   Three statements formulated by Gustav Kirchhoff describing spectra and spectral analysis.

Kirkwood gaps   Gaps in the spacing of asteroid orbits discovered by Daniel Kirkwood.

Kuiper belt   A doughnut-shaped ring of space around the Sun beyond Pluto containing many frozen comet bodies, some of which are occasionally deflected toward the inner solar system.

Large Magellanic Cloud (LMC)   A companion galaxy to the Milky Way.

last quarter Moon   A phase of the waning Moon when Earth-based observers see half of the Moon's illuminated hemisphere.

law of equal areas   Kepler's second law.

law of inertia   The physical law that an object will stay at rest or move at a constant speed unless acted upon by an outside force.

laws of physics   Basic principles that govern the behavior of physical reality.

lenticular galaxy   A disk-shaped galaxy without spiral arms.

light   Electromagnetic radiation, which travels in packets called photons.

light curve   A graph that displays variations in the brightness of a star or other astronomical object over time.

light-gathering power   A measure of how much light a telescope intercepts and brings to a focus.

light-year (ly)   The distance that light travels in a vacuum in one year.

lighthouse model   The explanation that a pulsar pulses by rotating and funneling energy outward via magnetic fields that are not aligned with the rotation axis.

limb (of Sun)   The apparent edge of the Sun as seen in the sky.

limb darkening   The phenomenon whereby the Sun is darker near its limb than near the center of its disk.

line of nodes   The line along which the plane of the Moon's orbit intersects the plane of the ecliptic.

liquid metallic hydrogen   A metallike form of hydrogen that is produced under extreme pressure.

Local Group   The cluster of galaxies of which our own Galaxy is a member.

long-period comet   A comet that takes tens of thousands of years or more to orbit the Sun once.

luminosity   The rate at which electromagnetic radiation is emitted from a star or other object.

luminosity class   The classification of a star of a given spectral type according to its luminosity; the classes are supergiant, bright giant, giant, subgiant, and main sequence.

lunar   Referring to the Moon.

lunar eclipse   An eclipse during which the Earth blocks light that would have struck the Moon.

lunar month   *See* synodic month.

lunar phases   The names given to the apparent shapes of the Moon.

Lyman series   A series of spectral lines of hydrogen produced by electron transitions to and from the lowest energy state of the hydrogen atom.

magnetic dynamo   A theory that explains phenomena of the solar cycle as a result of periodic winding and unwinding of the Sun's magnetic field in the solar atmosphere.

magnetic field   A region of space near a magnetized body within which magnetic forces can be detected.

magnetosphere The region around a planet occupied by its magnetic field.

magnification (magnifying power) The number of times larger in angular diameter an object appears through a telescope than as seen by the naked eye.

magnitude A measure of the amount of light received from a star or other luminous object.

magnitude scale The system of denoting magnitudes.

main sequence A grouping of stars on the Hertzsprung–Russell diagram extending diagonally across the graph from the hottest, brightest stars to the dimmest, coolest stars.

main sequence star A star, fusing hydrogen to helium in its core, whose surface temperature and luminosity place it on the main sequence on the Hertzsprung–Russell diagram.

mantle (of a planet) That portion of a terrestrial planet located between its crust and core.

mare (*plural* **maria**) Latin for "sea;" a large, relatively crater-free plain on the Moon.

mare basalt Dark, solidified lava that covers the lunar maria.

marginally bound universe A universe that just barely manages to expand forever.

mass A measure of the total amount of material in an object.

mass–luminosity relation The linear relationship between the masses and luminosities of main sequence stars.

matter-dominated universe A universe in which the radiation field that fills all space is unable to prevent the existence of neutral atoms.

megaparsec (Mpc) One million parsecs.

mesosphere The layer in the Earth's atmosphere above the stratosphere.

metal-poor star A star whose abundance of heavy elements is significantly less than that of the Sun.

metal-rich star A star whose abundance of heavy elements is comparable to that of the Sun.

meteor The streak of light seen when any space debris vaporizes in the Earth's atmosphere; a "shooting star."

meteor shower Frequent meteors that seem to originate from a common point in the sky.

meteorite A fragment of space debris that has survived passage through the Earth's atmosphere.

meteroid A small rock in interplanetary space.

Milky Way Galaxy The galaxy in which our solar system resides.

minor planet *See* asteroid.

minute of arc (arcmin) One-sixtieth of a degree of arc.

missing mass *See* dark matter.

model A hypothesis that has withstood observational or experimental tests.

molecule A bound combination of two or more atoms.

momentum A measure of the inertia of an object; an object's mass multiplied by its velocity.

neap tide The least change from high to low tide during a day; it occurs during first and third quarter phases of the Moon.

nebula (*plural* **nebulae**) A cloud of interstellar gas and dust.

neon fusion The thermonuclear fusion of neon nuclei.

neutrino A subatomic particle with no electric charge and little or no mass yet which is important in many nuclear reactions and in supernovae.

neutron A nuclear particle with no electric charge and with a mass nearly equal to that of the proton.

neutron degeneracy pressure A powerful pressure produced by degenerate neutrons.

neutron star A very compact, dense stellar remnant composed almost entirely of neutrons.

New General Catalogue (**NGC**) A catalogue of star clusters, nebulae, and galaxies, first published in 1888.

new Moon The phase of the Moon when it is nearest the Sun in the sky.

Newtonian laws of motion A branch of physics based on Newton's laws of mechanics and gravitation.

Newtonian reflector An optical arrangement in a reflecting telescope in which a small, flat mirror reflects converging light rays to a focus on one side of the telescope tube.

nonthermal radiation Radiation emitted by charged particles moving through a magnetic field; synchrotron radiation.

north celestial pole The location on the celestial sphere directly above the Earth's northern rotation pole.

northern lights (aurora borealis) Light radiated by atoms and ions in the Earth's upper atmosphere due to high-energy particles from the Sun and seen mostly in the northern polar regions.

nova (*plural* **novae**)   A star that experiences a sudden outburst of radiant energy, temporarily increasing its luminosity by a factor of between 10^4 and 10^6.

nuclear   Referring to the nucleus of an atom.

nuclear bulge   A distribution of stars in the shape of a flattened sphere that surrounds the nucleus of a spiral galaxy like the Milky Way.

nuclear density   The density of matter in the nucleus of an atom; about 10^{17} kilograms per cubic meter (kg/m^3).

nucleus (of an atom)   The massive part of an atom, composed of protons and neutrons, about which electrons revolve.

nucleus (of a comet)   A collection of ices and dust that constitute the solid part of a comet.

nucleus (of a galaxy)   The concentration of stars and dust at the very center of a galaxy.

OB association   An unbound group of very young, massive stars predominantly of spectral types O and B.

OBAFGKM sequence   The sequence of stellar spectral classifications from hottest to coolest stars.

objective lens   The principal lens of a telescope.

observable universe   All space that is nearer to us than the distance traveled by light since the time of the Big Bang.

occultation   The eclipsing of an astronomical object by the Moon or a planet.

Occam's razor   The principle of choosing the simplest scientific theory that correctly explains any phenomenon.

Oort cloud   A spherical region of the solar system beyond the Kuiper belt where most comets are believed to spend most of their time.

open cluster   A loose association of young stars in the disk of the galaxy; a galactic cluster.

open universe   A universe that lacks the mass necessary to stop its expansion; a universe that will expand forever.

opposition   The configuration of a planet when it is at an elongation of 180° and thus appears opposite the Sun in the sky.

optical double   A pair of stars that appear to be near each other but which are unbound and at very different distances from Earth.

optics   The branch of physics dealing with the behavior and properties of light.

orbit   The path of an object that is moving about a second object.

orbital eccentricity   A measure between 0 and 1 indicating how close to circular a planet's orbit is (the eccentricity of a circular orbit is 0).

organic molecule   A carbon-based compound.

oxygen fusion   The thermonuclear fusion of oxygen nuclei.

ozone layer   The lower stratosphere, where most of the ozone in the air exists.

pair production   The creation of a particle and an antiparticle from energetic photons.

Pallas   The second asteroid to be discovered.

parabola   An open curve formed by cutting a circular cone at an angle parallel to the sides of the cone.

parallax   The apparent displacement of an object caused by viewing it from different locations.

parsec (pc)   A unit of distance equal to 3.26 light-years.

partial eclipse   A lunar or solar eclipse in which the eclipsed object does not appear completely covered.

Pauli exclusion principle   A principle of quantum mechanics that says that two identical particles cannot simultaneously have the same position and momentum.

peculiar galaxy   Any Hubble class of galaxy that appears to be blowing apart.

penumbra   The portion of a shadow in which only part of the light source is covered by the shadow-making body.

penumbral eclipse   A lunar eclipse in which the Moon passes only through the Earth's penumbra.

perihelion   The point in its orbit where a planet is nearest the Sun.

period   The interval of time between successive repetitions of a periodic phenomenon.

period–luminosity relation   A relationship between the period and average luminosity of a pulsating star.

periodic table   A listing of the chemical elements according to their properties; invented by D. Mendeleev.

phase (of the Moon)   The appearance of the Moon at different points in its orbit of the Earth.

photodisintegration   The breakup of nuclei in the core

of a massive star due to the effects of energetic gamma rays.

photometry The measurement of light intensities.

photon A discrete unit of electromagnetic energy.

photon pressure The force per unit area exerted by photons on stellar or interstellar gas.

photosphere The region in the solar atmosphere from which most of the visible light escapes into space.

pixel A contraction of the term "picture element;" usually refers to one square of a grid into which the light-sensitive component of a charge-coupled device is divided.

plage A bright spot on the Sun associated with an emerging magnetic field.

Planck time The brief interval of time, about 10^{-43} second, immediately after the Big Bang, when all four forces had the same strength.

Planck's law A relationship between the energy carried by a photon and its wavelength.

planetary differentiation The process early in the life of each planet whereby denser elements sank inward and lighter ones rose.

planetary nebula A luminous shell of gas ejected from an old, low-mass star.

planetesimal Primordial asteroidlike object from which the planets accreted.

plasma A hot, ionized gas.

plate tectonics The motions of large segments (plates) of the Earth's surface over the underlying mantle.

polymer A long molecule composed of many smaller molecules.

population I star A star whose spectrum exhibits spectral lines of many elements heavier than helium; a metal-rich star.

population II star A star whose spectrum exhibits comparatively few spectral lines of elements heavier than helium; a metal-poor star.

positron An electron with a positive rather than negative electric charge; an antielectron.

powers of ten A convenient method of writing large and small numbers that uses a number between 1 and 10 multiplied by a power of 10.

pre-main-sequence star The stage of star formation just before the main sequence; it involves slow contraction of the young star.

precession (of the Earth) A slow, conical motion of the Earth's axis of rotation caused by the gravitational pull of the Moon and Sun on the Earth's equatorial bulge.

precession of the equinoxes The slow westward motion of the equinoxes along the ecliptic because of the Earth's precession.

primary mirror The large, concave, light-gathering mirror in a reflecting telescope, analogous to the objective lens on a refracting telescope.

prime focus The point in a reflecting telescope where the primary mirror focuses light.

primordial black hole A relatively low-mass black hole formed at the beginning of the universe.

primordial fireball The extremely hot gas that filled the universe immediately following the Big Bang.

prism A wedge-shaped piece of glass used to disperse white light into a spectrum.

prominence Flamelike protrusion seen near the limb of the Sun and extending into the solar corona.

proper motion The change in the location of a star on the celestial sphere.

proton A heavy, positively charged nuclear particle.

protoplanet The embryonic stage of a planet when it is growing due to collisions with planetesimals.

protostar The earliest stage of a star's life before fusion commences and when gas is rapidly falling onto it.

protosun The Sun prior to the time when hydrogen fusion began in its core.

pulsar A pulsating radio source associated with a rapidly rotating neutron star with an off-axis magnetic field.

quantum mechanics The branch of physics dealing with the structure and behavior of atoms and their interactions with each other and with light.

quark A particle that is a building block of the heavy nuclear particles such as protons and neutrons.

quarter Moon A phase of the Moon when it is located 90° from the Sun in the sky.

quasar (quasi-stellar radio source) A starlike object with a very large redshift.

quasi-stellar object (QSO) A quasar.

radial velocity That portion of an object's velocity parallel to the line of sight.

radial-velocity curve   A plot showing the variation of radial velocity with time for a binary star or variable star.

radiation   Electromagnetic energy; photons.

radiation (photon) pressure   The transfer of momentum carried by radiation to an object on which the radiation falls.

radiation-dominated universe   The time at the beginning of the universe when the electromagnetic radiation prevented ions and electrons from combining to make neutral atoms.

radiative zone   A region inside a star where energy is transported outward by the movement of photons through a gas from a hot location to a cooler one.

radio astronomy   That branch of astronomy dealing with observations at radio wavelengths.

radio galaxy   A galaxy that emits an unusually large amount of radio waves.

radio lobes   Vast regions of radio emission on opposite sides of a radio galaxy.

radio telescope   A telescope designed to detect radio waves.

radio wave   Long-wavelength electromagnetic radiation.

radioactivity   The process whereby certain atomic nuclei naturally decompose by spontaneously emitting particles.

red giant   A large, cool star of high luminosity.

red supergiant   An extremely large, cool star of high luminosity; a star in the upper right corner of the Hertzsprung–Russell diagram.

redshift   The shifting to longer wavelengths of the light from remote galaxies and quasars; the Doppler shift of light from any receding source.

reflecting telescope (reflector)   A telescope in which the principal light-gathering component is a concave mirror.

reflection   The rebounding of light rays off a smooth surface.

reflection nebula   A comparatively dense cloud of gas and dust in interstellar space that is illuminated by a star between it and the Earth.

refracting telescope (refractor)   A telescope in which the principal light-gathering component is a lens.

refraction   The bending of light rays upon passing from one transparent medium to another.

regolith   The powdery, lifeless material on the surface of a moon or planet.

regular cluster (of galaxies)   An evenly distributed group of galaxies bound together by mutual gravitational attraction.

resolution   The degree to which fine details in an optical image can be distinguished.

resolving power   A measure of the ability of an optical system to distinguish fine details in the image it produces.

resonance   The large response of an object to a small periodic gravitational tug from another object.

retrograde motion   The occasional backward (i.e., westward) apparent motion of a planet against the background stars as seen from Earth. Retrograde motion is an optical illusion.

retrograde rotation   The rotation of a planet opposite to its direction of revolution around the Sun. Only Pluto, Uranus, and Venus have retrograde rotation.

revolution   The orbit of one body about another.

right ascension   The celestial coordinate analogous to longitude on Earth and measured around the celestial equator from the vernal equinox.

rille   A winding crack or depression in the lunar surface.

ringlet   Any one of numerous, closely spaced, thin bands of particles in Saturn's ring system.

Roche lobe   The teardrop-shaped regions around each star in a binary star system inside of which gas is gravitationally bound to that star.

rotation   The spinning of a body about an axis passing through it.

rotation curve (of a galaxy)   A graph showing how the orbital speed of material in a galaxy depends on the distance from the galaxy's center.

RR Lyrae variable star   A type of pulsating star with a period less than one day.

Sa, Sb, Sc   Categories of spiral galaxies determined by the sizes of their nuclear bulges or how tightly wound their spiral arms are; Sa is the most tightly wound.

Sagittarius A   The strong radio source associated with the nucleus of the Milky Way galaxy.

satellite   A body that revolves about a larger one.

SBa, SBb, SBc   Categories of barred spiral galaxies

determined by how tightly wound their spiral arms are; SBa are the most tightly wound.

scarp   A cliff on Mercury believed to have formed when the planet cooled and shrank.

Schmidt corrector plate   A specially shaped lens used with spherical mirrors that corrects for spherical aberration and provides an especially wide field of view.

Schwarzschild black hole   Any nonrotating, uncharged black hole.

Schwarzschild radius   The distance from the center to the event horizon in a black hole.

scientific method   The method of doing science based on observation, experimentation, and the formulation of hypotheses (theories) that can be tested.

scientific theory   An idea about the natural world that is subject to testing and refinement.

seafloor spreading   The process whereby magma upwelling along rifts in the ocean floor causes adjacent segments of the Earth's crust to separate.

second of arc (arcsec)   One-sixtieth of a minute of arc.

secondary mirror   A relatively small mirror used in reflecting telescopes to guide the light out the side or bottom of the telescope.

seeing disk   The size that a star appears to have on a photographic or charge-coupled device image as a result of the changing refraction of the starlight passing through the Earth's atmosphere.

seismic waves   Vibrations traveling through or around an astronomical body usually associated with earthquakelike phenomena.

seismograph   A device used to record and measure seismic waves, such as those produced by earthquakes.

seismology   The study of earthquakes and related phenomena.

self-propagating star formation   The process whereby the birth of stars in one part of a galaxy stimulates star formation in a neighboring region of that galaxy.

semidetached binary   A close binary system in which one star fills or is overflowing its Roche lobe.

semimajor axis (of an ellipse)   Half of the longest dimension of an ellipse.

SETI   The search for extraterrestrial intelligence.

Seyfert galaxy   A spiral galaxy with a bright nucleus whose spectrum exhibits emission lines.

Shapley–Curtis debate   An inconclusive debate between Harlow Shapley and Heber Curtis in 1920 about whether certain nebulae were beyond the Milky Way.

shell helium fusion   Helium fusion that occurs in a thin shell surrounding the core of a star.

shell hydrogen fusion   Hydrogen fusion that occurs in a thin shell surrounding the core of a star.

shepherd satellite   A small satellite whose gravitational tug is responsible for maintaining a sharply defined ring of matter around a planet such as Saturn or Uranus.

shock wave   An abrupt, localized region of compressed gas caused by an object traveling through the gas at a speed greater than the speed of sound.

shooting star   *See* meteor.

short-period comet   A comet that orbits the Sun in the vicinity of the planets, thereby reappearing with tails every 200 years or less.

sidereal month   The period of the Moon's revolution about the Earth measured with respect to the Moon's location among the stars.

sidereal period   The orbital period of one object about another measured with respect to the stars.

single-line spectroscopic binary   A spectroscopic binary whose periodically varying spectrum exhibits the spectral lines of only one of its two stars.

singularity   A place of infinite curvature of spacetime in a black hole.

Small Magellanic Cloud (SMC)   An irregular galaxy that is a companion to the Milky Way.

solar corona   The Sun's outer atmosphere.

solar cycle   A 22-year cycle during which the Sun's magnetic field reverses its polarity.

solar day   From noontime to the next noontime; for Earth it is 24 hours.

solar eclipse   An eclipse during which the Moon blocks the Sun.

solar flare   A violent eruption on the Sun's surface.

solar model   A set of equations that describe the internal structure and energy generation of the Sun.

solar nebula   The cloud of gas and dust from which the Sun and the rest of the solar system formed.

solar seismology   The study of the Sun's interior from observations of vibrations of its surface.

solar system   The Sun, planets, their satellites, asteroids, comets, and related objects that orbit the Sun.

solar wind   A radial flow of particles (mostly electrons and protons) from the Sun.

solstice   Either of two points along the ecliptic at which the Sun reaches its maximum distance north or south of the celestial equator.

south celestial pole   The location on the celestial sphere directly above the Earth's south rotation pole.

southern lights (aurora australis)   Light radiated by atoms and ions in the Earth's upper atmosphere due to high-energy particles from the Sun; seen mostly in the southern polar regions.

spacetime   The concept from special relativity that space and time are both essential in describing the position, motion, and action of any object or event.

special theory of relativity   A description of mechanics and electromagnetic theory formulated by Einstein according to which measurements of distance, time, and mass are affected by the observer's motion.

spectral analysis   The identification of chemicals by the appearance of their spectra.

spectral lines   A dark or bright line at a specific wavelength in a spectrum.

spectral type   A classification of stars according to the appearance of their spectra.

spectrogram   The photograph of a spectrum.

spectrograph   A device for photographing a spectrum.

spectroscope   A device for directly viewing a spectrum.

spectroscopic binary star   A double star whose binary nature can be deduced from the periodic Doppler shifting of lines in its spectrum.

spectroscopic parallax   A method of determining a star's distance from the Earth by measuring its surface temperature, luminosity, and apparent magnitude.

spectroscopy   The study of spectra.

spectrum (*plural* **spectra**)   The result of electromagnetic radiation passing through a prism or grating so that different wavelengths are separated.

speed   The rate at which an object moves.

spherical aberration   An optical property whereby different portions of a spherical lens or spherical, concave mirror have slightly different focal lengths.

spicule   A narrow jet of rising gas in the solar chromosphere.

spin (of an electron or proton)   A small, well-defined amount of angular momentum possessed by electrons, protons, and other particles.

spiral arms   Lanes of interstellar gas, dust, and young stars that wind outward in a plane from the central regions of some galaxies.

spiral density wave   A spiral-shaped pressure wave that orbits the disk of a spiral galaxy and induces new star formation.

spiral galaxy   A flattened, rotating galaxy with pinwheel-like spiral arms winding outward from the galaxy's nuclear bulge.

spoke   A moving dark region of Saturn's rings.

spring tide   The greatest daily difference between high and low tide, occurring when the Moon is new or full.

stable Lagrange points   Locations throughout the solar system where the gravitational forces from the Sun and a planet keep space debris trapped.

standard candle   An object whose known luminosity can be used to deduce the distance to a galaxy.

star   A self-luminous sphere of gas.

starburst galaxy   A galaxy where there is an exceptionally high rate of star formation.

Stefan–Boltzmann law   A relationship between the temperature of a blackbody and the rate at which it radiates energy.

stellar evolution   The changes in size, luminosity, temperature, and chemical composition that occur as a star ages.

stellar model   The result of theoretical calculations that give details of physical conditions inside a star.

stellar parallax   The apparent shift in a nearby star's position on the celestial sphere resulting from the Earth's orbit around the Sun.

stellar spectroscopy   The study of the properties of stars encoded in their spectra.

stony meteorite   A meteorite composed of rock with very little iron; also called a stone.

stony-iron meteorite   A meteorite composed of roughly equal amounts of rock and iron.

stratosphere   The second layer in the Earth's atmosphere, directly above the troposphere.

strong nuclear force   The force that binds protons and neutrons together in nuclei.

subduction zone   A location where colliding tectonic plates cause the Earth's crust to be pulled down into the mantle.

subgiant   A star whose luminosity is between that of

main sequence stars and normal giants of the same spectral type.

summer solstice The point on the ecliptic where the Sun is farthest north of the celestial equator.

Sun The star about which the Earth and other planets revolve.

sunspot A temporary cool region in the solar photosphere created by protruding magnetic fields.

sunspot cycle The semiregular 11-year period with which the number of sunspots fluctuates.

sunspot maximum The time during the solar cycle when the number of sunspots is exceptionally high.

sunspot minimum The time during the solar cycle when the number of sunspots is exceptionally low.

supercluster (of galaxies) A gravitationally bound collection of many clusters of galaxies.

supergiant A star of very high luminosity.

supergranule A large convective cell in the Sun's chromosphere containing many granules.

superior conjunction The configuration when a planet is behind the Sun as seen from Earth.

supermassive black hole A black hole whose mass exceeds a thousand solar masses.

supernova A stellar outburst during which a star suddenly increases its brightness roughly a millionfold.

supernova remnant A nebula left over after a supernova detonates.

synchronous rotation The condition when a moon's rotation rate and revolution rate are equal.

synchrotron radiation The radiation emitted by charged particles moving through a magnetic field; nonthermal radiation.

synodic month (lunar month) The period of revolution of the Moon with respect to the Sun; the length of one cycle of lunar phases.

synodic period The interval between successive occurrences of the same configuration of a planet as seen from Earth.

T Tauri stars Young, variable stars associated with interstellar matter that show erratic changes in luminosity.

tail (of a comet) Gas and dust particles from a comet's nucleus that have been swept away from the comet's nucleus by the radiation pressure of sunlight and impact of the solar wind.

telescope An instrument for viewing remote objects.

temperature (Celsius) Temperature measured on a scale where water freezes at 0° and boils at 100°.

temperature (Fahrenheit) Temperature measured on a scale where water freezes at 32° and boils at 212°.

temperature (Kelvin) Absolute temperature measured in Celsius degree intervals. Water freezes at 273 K and boils at 373 K.

terminator The line dividing day and night on the surface of the Moon or a planet; the line of sunset or sunrise.

terrestrial planet Any of the planets Mercury, Venus, Earth, or Mars; a planet with a composition and density similar to Earth's.

theory A hypothesis that has been demonstrated to describe a range of phenomena accurately.

thermal energy The energy associated with the motions of atoms or molecules in a substance.

thermal equilibrium A balance between the input and outflow of heat in a system.

thermonuclear fusion A reaction in which the nuclei of atoms are fused together at a high temperature.

thermosphere A layer high in the Earth's atmosphere above the mesosphere.

tidal force A gravitational force whose strength and/or direction varies over a body and thus tends to deform the body.

time zone One of 24 divisions of the Earth's surface separated by 15° along lines of constant longitude (with allowances for some political boundaries).

total eclipse A solar eclipse during which the Sun is completely hidden by the Moon, or a lunar eclipse during which the Moon is completely immersed in the Earth's umbra.

trailing-arm spiral A spiral arm pointing away from the direction of rotation; characteristic of all spiral galaxies.

transition (electronic) The change in energy and orbit of an electron around an atom or molecule.

Trojan asteroid One of several asteroids at stable Lagrange points that share Jupiter's orbit about the Sun.

troposphere The lowest level of the Earth's atmosphere.

Tully–Fisher relation A correlation between the width of the 21-centimeter line of a spiral galaxy and its absolute magnitude.

turbulence   Random motions in a gas or liquid.

turnoff point   The location of the brightest main sequence stars on the Hertzsprung–Russell diagram of a globular cluster.

21-cm radiation   Radio emission from a hydrogen atom caused by the flip of the electron's spin orientation.

twinkling   The apparent change in a star's brightness, position, or color due to the motion of gases in the Earth's atmosphere.

Type I Cepheid   Population I Cepheid variable star found in the disks of spiral galaxies.

Type I supernova   A supernova occurring after a white dwarf accretes enough mass from a companion star to exceed the Chandrasekhar limit.

Type II Cepheid   Population II Cepheid variable star found in elliptical galaxies and in the halos of disk galaxies that is 1.5 magnitudes dimmer than a Type I Cepheid.

Type II supernova   A supernova occurring after a massive star's core is converted to iron.

UBV filters   A system of filters that yield stellar magnitudes at different wavelengths in the ultraviolet, blue, and visible (yellow) spectral regions.

ultraviolet (UV) radiation   Electromagnetic radiation of wavelengths shorter than those of visible light but longer than those of x rays.

umbra   The central, completely dark portion of a shadow.

unbound universe   A universe with so little mass that it expands forever.

universal constant of gravitation   The constant of proportionality in Newton's law of gravitation.

universal law of gravitation   Newton's law of gravitation, which describes how the gravitational force between two bodies depends on their masses and separation.

universe   All space along with all the matter and radiation in space.

Van Allen radiation belts   Two flattened, doughnut-shaped regions around the Earth where many charged particles (mostly protons and electrons) are trapped by the Earth's magnetic field.

variable star   A star whose luminosity varies.

velocity   A quantity that specifies both direction and speed of an object.

verification   The testing of a scientific theory.

vernal equinox   The point on the ecliptic where the Sun crosses the celestial equator from south to north.

very-long-baseline interferometry (VLBI)   a method of connecting widely separated radio telescopes to make observations of very high resolution.

visual binary star   A double star in which the two components can be resolved through a telescope.

void   A huge, roughly spherical region of the universe where exceptionally few galaxies are found.

water hole   The part of the electromagnetic spectrum at a few thousand megahertz where there is very little background noise from space.

wavelength   The distance between two successive peaks in a wave.

waning   An adjective that means "decreasing," as in the "waning crescent Moon" or the "waning gibbous Moon."

waxing   An adjective that means "increasing," as in the "waxing crescent Moon" or the "waxing gibbous Moon."

weak nuclear force   A nuclear interaction involved in certain kinds of radioactive decay.

weight   The force with which a body presses down on the surface of the Earth.

white dwarf   A low-mass stellar remnant that has exhausted all its thermonuclear fuel and contracted to a size roughly equal to the size of the Earth.

Widmanstätten patterns   Crystalline structure seen inside iron meteorites.

Wien's law   A relationship between the temperature of a blackbody and the wavelength at which it emits the greatest intensity of radiation.

winter solstice   The point on the ecliptic where the Sun reaches its greatest distance south of the celestial equator.

wormhole   A hypothetical connecting passage between black holes and other places in the universe.

x ray   Electromagnetic radiation whose wavelength is between that of ultraviolet light and gamma rays.

x-ray burster   A neutron star in a binary star system that accretes mass, undergoes thermonuclear fusion on its surface, and therefore emits short bursts of x rays.

year The sidereal period of revolution of the Earth about the Sun.

Zeeman effect A splitting or broadening of spectral lines in the presence of a magnetic field.

zenith The point on the celestial sphere directly overhead.

zero-age main sequence (ZAMS) The positions of stars on the Hertzsprung–Russell diagram that have just begun to fuse hydrogen in their cores.

zodiac A band of 13 constellations around the sky through which the Sun appears to move throughout the year.

zones (on Jupiter) Light-colored band in Jupiter's cloud cover.

Answers to Computation Questions

Foundations I

2.a. 5.974×10^{24} kg
  b. 1.989999×10^{30} kg
  c. 6.9599×10^5 km
  d. 3.1558×10^7 s

Chapter 1

35. 1 more sidereal month than synodic month

Chapter 2

12.a. 5.2 square AU in 1995, 26 square AU in 5 years
13. $a = 100$ AU, maximum distance is almost 200 AU
14. 2.8 yr
15. 100 times weaker than at present
16. use $P^2 = a^3$

Chapter 3

 4. 9 times more
17. ½s
18. Palomar gathers 10^6 times more light than the human eye.
20.a. 222×
  b. 100×
  c. 36×

Chapter 4

10. $7\frac{1}{2}$ times
11. 238 nm
12. 6520 K
13. 500 nm, 4 times brighter at same distance from Earth
14. approaching, 13 km/s
15. receding, 21 km/s
16. about 90,000 km/s

Chapter 8

16. 5.0×10^{14} tons

Chapter 9

 3. next maximum in 2002, next minimum in 1997
14. 1400 kg/m^3
15. 4.8%
16. 500 nm = visible light; 58 nm = ultraviolet; 1.9 nm = x ray

Foundations III

 7. 4.3 pc

Chapter 11

21. 2000
22. 200 times longer

Chapter 12

22. about 7465 years ago

Chapter 13

 8. 8.9 km, 89 km

Chapter 14

13. 20 times
15. about once every 750 years

Chapter 15

16. 293 Mpc
17. 7500 km/s

Chapter 17

10. 11 billion years

Illustration Credits

Foundations I Facing p. 1:NASA. Fig I-1:Dennis L. Mammana. Fig. I-2:Scala/Art Resource, NY; NASA; NOAO; Mexico State University; C. Holmes. Fig. I-3a:NASA. p. 5:Scientific American Books, NASA, and the Anglo-Australian Observatory. Fig. I-3b:L. Golub, Naval Observatory; IBM Research and SAO. Fig. I-4a:NASA. Fig. I-4b:NOAO. Fig. I-4c:European Southern Observatory. Fig. I-4d:C. R. O'Dell, Rice University, John Bally, University of Colorado, Ralph Sutherland, University of Colorado. Fig. I-4e:Copyright © R. J. Dufour, Rice University. Fig. I-4f:F. N. Owen and J. J. Puschell/ NRAO/AUI.

Chapter 1 p. 10:Roger Ressmeyer © 1989 Corbis. Fig. 1-1a: Robert C. Mitchell, Central Washington University. Fig. 1-1b:Drawing from Elijah Burrit's Atlas, Janus Publications, Wichita, Kansas. Fig. 1-7:NOAO. Fig. 1-9:Greenwich Observatory, England. Fig. 1-13:Dennis di Cicco. Fig. 1-20: UCO/Lick Observatory. Fig. 1-21:Copyright © Art Wolfe. Fig. 1-27:Mike Harms. Fig. 1-28:William R. Dellinges. Fig. 1-30:NASA. Fig. 1-31:Dennis di Cicco.

Chapter 2 p. 36:NASA. Fig. 2-8:New Mexico State University Observatory. Fig. 2-10:Clifford Holmes. Fig. 2-11:Yerkes Observatory. p. 48:(left and middle) Giraudon/Art Resource, NY; (right) Erich Lessing/Art Resource, NY. p. 49:Art Resource, NY. Fig. 2-13:Copyright © 1986 Jack B. Marling, Lumicon. Fig. 2-14:UCO/Lick Observatory Photographs.

Chapter 3 p. 54:NOAO. Fig. 3-13:Yerkes Observatory. Fig. 3-20:NASA/ESA. Fig. 3-21:Kitt Peak National Observatory. Fig. 3-22:Roger Ressmeyer © 1994 Corbis. Fig. 3-23: European South Observatory. Fig. 3-24:IRFA, University of Hawaii, and Roger Ressmeyer © 1993 Corbis. Fig. 3-25a–c: Patrick Seitzer, NOAO. p. 72:Meade Instruments Corporation. Figs. 3-26, 3-27:National Radio Astronomy Observatory. Fig. 3-28a:NASA. Fig. 3-28b:Copyright © 1982 Associated Universities, Inc., under contract with the National Science Foundation (VLA observations by Imke de Pater, J. R. Dickel). Fig. 3-29:NASA. Fig. 3-30a:George R. Carruthers, NRL. Fig. 3-30b:NASA. Fig. 3-30c:Robert C. Mitchell, Central Washington University. Fig. 3-31:NASA/ESA. Figs. 3-32, 3-33:NASA.

Chapter 4 p. 80:University of Michigan. Fig. 4-4:NOAO. Figs. 4-7, 4-13:The Observatories of the Carnegie Institution of Washington. Fig. 4-19:Yerkes Observatory. p. 94:NOAO.

Foundations II p. 98:NASA/JPL. M. Buie, K. Horne and D. Tholen. Figs. II-1 and II-2:Copyright © 1980 and 1984 Anglo-Australian Observatory. Fig. II-4:University of Arizona and Jet Propulsion Laboratory. Fig. II-5:Adapted from a computer simulation by George W. Wetherill. Figs. II-6, II-10, II-11:NASA.

Chapter 5 p. 110 and Fig. 5-1:NASA. Fig. 5-2:Adapted from P. M. Hurley. Fig. 5-3:Courtesy of Marie Tharp and Bruce Heezen. Figs. 5-6, 5-7:NASA. Fig. 5-10a:PSSC Physics, 3d ed, Lexington, Mass., D. C. Heath, 1971. Fig. 5-12:Courtesy of S. I. Akasofu, Geophysical Institute, University of Alaska.

Figs. 5-14, 5-15, 5-16, 5-17, 5-18:NASA. Fig. 5-19:Carnegie Observatories. Figs. 5-20, 5-21, 5-22, 5-23, 5-24:NASA. Fig. 5-26a–f:Courtesy of W. Benz. Fig. 5-31:Randall L. Ricklefs, McDonald Observatory, University of Texas at Austin.

Chapter 6 p. 133:NASA, USGS. Fig. 6-1a,b:New Mexico State University. Fig. 6-2:NASA, UCO/Lick Observatory Photograph. Figs. 6-3, 6-4, 6-5, 6-6:NASA. Fig. 6-8:Courtesy of W. Benz, A. G. W. Cameron, and W. Slattery. Fig. 6-9:NASA. Fig. 6-13a:Courtesy of C. M. Pieters and the U.S.S.R. Academy of Sciences. Fig. 6-14:NASA. Fig. 6-15:NASA/JPL. Fig. 6-16:NASA. Fig. 6-17:NASA/JPL. Fig. 6-18:Courtesy of S. M. Larson Fig. 6-19:NASA. Figs. 6-20, 6-21:A. S. McEwen, USGS. Figs. 6-22, 6-23:NASA. Fig. 6-24:NASA/JPL. Fig. 6-25: NASA. Fig. 6-26:Photo Researchers, Inc. © Terranova International. Fig. 6-27:NASA, Johnson Space Flight Center. Figs. 6-28, 6-29, 6-30, 6-31a,b:NASA.

Chapter 7 p. 156:Stephen P. Meszaros, NASA. M. Buie, K. Horne, and D. Tholen. Fig. 7-1:Courtesy of S. Larson. Figs. 7-2, 7-3, 7-4a,b, 7-5:NASA. Figs. 7-8, 7-9:STSI/ NASA. Figs. 7-10, 7-11, 7-12, 7-13a,b, 7-14, 7-15, 7-16, 7-17, 7-18, 7-19, 7-20, 7-21,7-22:NASA. Fig. 7-23:STSI/NASA. Figs. 7-25, 7-26, 7-27, 7-28, 7-29, 7-30:NASA. Fig. 7-31:Peter Smith and Mark Lemmon, University of Arizona, courtesy of Space Telescope Science Institute. Figs. 7-32, 7-37, 7-38, 7-39, 7-40, 7-41, 7-42, 7-43:NASA. Fig. 7-44:UCO/Lick Observatory. Fig. 7-45:Courtesy, U.S. Naval Observatory. Fig. 7-46:NASA.

Chapter 8 p. 185:Copyright © 1985 Anglo-Australian Observatory. Fig. 8-3:Yerkes Observatory. Fig. 8-5a,b:NASA. Fig. 8-7:Courtesy of Jim Scotti, Spacematch on Kitt Peak. Fig. 8-8:Art adapted from STSI/NASA. Fig. 8-9a,b:Alan Fitzsimmons, Queen's University of Belfast. Fig. 8-10:Max Planck Institut fur Aeronomie. Fig. 8-11:Johns Hopkins University and the Naval Research Laboratory. Fig. 8-12: Hans Vehrenberg. Fig. 8-14:Palomar Observatory. Figs. 8-16, 8-17:UCO/Lick Observatory. Fig. 8-19:New Mexico State University Observatory. Fig. 8-20:A. H. Weaver and P. D. Sellman, STSI/NASA. Fig. 8-21:Ronald A. Oriti. Fig. 8-22:Meteor Crater Enterprises, Arizona. Fig. 8-23: Jim Pryal. Figs. 8-24, 8-25, 8-26, 8-27:Ronald A. Oriti. Fig. 8-28:Chip Clark. Fig. 8-29:Courtesy of Sovfoto. Fig. 8-30: Courtesy of J. A. Woods. Fig. 8-31:Courtesy of W. Alvarez.

Chapter 9 p. 205:Leon Golub, Center for Astrophysics, and Serge Koutchmy, Centre National da la Recherche Scientifique. Fig. 9-1:Celestron International. Fig. 9-2:E. C. Olson, Mt. Wilson and Las Campanos Observatories. Fig. 9-4a:Gary McDonald. Fig. 9-4b:Richard Berry. Figs. 9-5, 9-7:NOAO. Fig. 9-8:Roland and Margorie Christen; Astro-Physics Inc. Figs. 9-9, 9-10:National Solar Observatory, Sacremento Peak, Sunspot, New Mexico. Fig. 9-12a,b:NOAO. Fig. 9-13:The Observatories of the Carnegie Institution of Washington. Fig. 9-15a,b:NOAO. Fig. 9-17:Copyright © Sacramento Peak Observatory Association of Universities for Research in Astronomy, Inc. Fig. 9-18:Naval Research Laboratory.

Fig. 9-19:NOAO. Fig. 9-20:NASA and Harvard College Observatory. Fig. 9-24:National Solar Observatory. Fig. 9-25: Kenneth G. Libbrecht, Big Bear Solar Observatory. Fig. 9-26: Raymond Davis, Brookhaven National Laboratory.

Foundations III p. 224:Okiro Fujii/ L' Astronomia.

Chapter 10 p. 232:Copyright © Anglo-Australian Observatory, photography by David Malin. Fig. 10-4:N. Houk, N. Irvine, and D. Rosenbush. Fig. 10-9:Yerkes Observatory. Figs. 10-12, 10-14:UCO/Lick Observatory.

Chapter 11 p. 248:Copyright © 1981 Anglo-Australian Observatory. Fig. 11-1:Royal Observatory, Edinburgh. Fig. 11-2: Anglo-Australian Observatory. Fig. 11-3a,b:Courtesy of Ronald J. Maddalena, Mark Morris, J. Moscowitz, and P. Thaddeus. Fig. 11-4a:Copyright © European Southern Observatory. Fig. 11-4b:NASA. Fig. 11-5:Copyright © 1984 Anglo-Australian Observatory. Fig. 11-6a,b:NOAO. Fig. 11-8: STSI/NASA. Fig. 11-9:Copyright © 1980 Anglo-Australian Observatory. Fig. 11-10:Anglo-Australian Observatory. Fig. 11-11:Jeff Hester and Paul Scowen (Arizona State University), and NASA. Figs. 11-13, 11-14a:Copyright © 1981 Anglo-Australian Observatory. Fig. 11-15:Anglo-Australian Observatory. Fig. 11-16:Copyright © 1980 Anglo-Australian Observatory. Figs. 11-17, 11-19:U.S. Naval Observatory. Fig. 11-22:UCO/Lick Observatory.

Chapter 12 p. 270:J. P. Harrington and K. J. Borkowski (University of Maryland) and NASA. Fig. 12-3:Copyright © 1995 Anglo-Australian Observatory. Fig. 12-5:Courtesy of R. B. Minton. Fig. 12-7:Copyright © 1979 Anglo-Australian Observatory. Fig. 12-8:European Southern Observatory. Fig. 12-9a,b:Copyright © 1987 Anglo-Australian Observatory. Fig. 12-10:Courtesy of K. S. Luttrell. Fig. 12-11:Dr. Christopher Burrows, ESA/STSI/NASA. Fig. 12-13:Royal Observatory, Edinburgh. Fig. 12-14a:Courtesy of ESA/NASA-HEASARC. Fig. 12-14b:Smithsonian Institution and the Very Large Array. Fig. 12-16:Courtesy of David Malin and Jay Pasachoff, Caltech 1992. Fig. 12-18a,b:F. R. Harnden Jr., Smithsonian Astrophysical Observatory. Fig. 12-22:NRAO, Very Large Array. Fig. 12-23a,b:UCO/Lick Observatory Photographs.

Chapter 13 p. 292:Wood, Ronsaville, Harlin, Inc. Fig. 13-6: Courtesy of J. Kristian, The Observatories of the Carnegie Institution of Washington. Fig. 13-7a:NASA/ESA. Fig. 13-7b:NASA.

Foundations IV p. 304:Copyright © 1994 Smithsonian Institution/Corbis "Over the Great Wall," digital image opener by Margaret Geller and Enhanced by Roger Ressmeyer. Fig. IV-1a: Courtesy of Mary Lea Shane Archives of the UCO/Lick Observatory. Fig. IV-1b:Courtesy of Lund Humpries. Figs. IV-2:NOAO. Figs. IV-3, IV-4:Palomar Observatory, copyright © California Institute of Technology.

Chapter 14 p. 311:Dennis di Cicco. Fig. 14-1:Steward Observatory. Fig. 14-2:Harvard Observatory. Fig. 14-3:NOAO. Fig. 14-5:U.S. Naval Observatory. Fig. 14-8:G. Westerhout. Fig. 14-9a,b:Anglo-Australian Observatory, VLA, NRAO. Fig. 14-10b:Palomar/Caltech. Fig. 14-13a,b:NASA. Fig. 14-13c:A. Eckart, R. Genzel, R. Hofmann, B. J. Sams, and L. E. Tacconi-Garman; ESO. Fig. 14-14a,b:VLA, NRAO; K. Y. Lo and N. Kileen, VLA NCSA.

Chapter 15 p. 323 and Fig. 15-1:Copyright © 1980 and © 1987 Anglo-Australian Observatory. Fig. 15-2:(Sa) Dr. Rudy Schild, Smithsonian Astrophysical Observatory; (Sb) NOAO; (Sc) copyright © 1992 Anglo-Australian Observatory. Fig. 15-3a:NOAO. Fig. 15-3b:Anglo-Australian Observatory, photo by David Malin. Fig. 15-3c:Dr. Rudy Schild, Smithsonian Astrophysical Observatory. Fig. 15-4a:Anglo-Australian Observatory, photo by David Malin. Fig. 15-4b:P. Seiden, D. Elmegreen, B. Elmegreen, and A. Mobarak, IBM. Figs. 15-7, 15-8:J. D. Wray; Mcdonald Observatory. Fig. 15-9:Dr. Rudy Schild, Smithsonian Astrophysical Observatory. Fig. 15-10:Copyright © 1987 Anglo-Australian Observatory. Figs. 15-12, 15-13, 15-14:Anglo-Australian Observatory, photos by David Malin, copyright © 1984 Royal Observatory, Edinburgh. Fig. 15-15:S. J. Maddox, W. J. Sutherland, G. P. Efstathuou, and J. Loveday, Oxford Astrophysics. Fig. 15-16:Adapted from V. de Lapparent, M. Geller, and J. Huchra. Fig. 15-18a:R. Ibata, University of British Columbia and G. Gilmore, M. Irwin, Royal Greenwich Observatory. Fig. 15-18b:R. C. Krann-Korteweg/Photo Researchers. Fig. 15-19:Anglo-Australian Observatory, photos by David Malin, Copyright © 1987 Royal Observatory, Edinburgh. Fig. 15-20:Hubble Space Telescope WFPC Team, Caltech and NASA. Fig. 15-21:NOAO. Fig. 15-22:Kirk Borne (Space Telescope Science Institute) and NASA. Fig. 15-23: Lick Observatory. Fig. 15-24a:M. Yun and P. Ho, Harvard-Smithsonian Center for Astrophysics. Fig. 15-24b:Copyright © 1960 National Geographic Society—Palomar Sky Survey, reproduced by permission of the California Institute of Technology. Fig. 15-25:Joshua Barnes, Canadian Institute for Theoretical Astrophysics. Fig. 15-26:Palomar Observatory. Fig. 15-27:W. C. Keel, University of Alabama. Fig. 15-28: Courtesy of Lars Hernquist, Institute for Advanced Study with simulations performed at the Pittsburgh Supercomputing Center. Fig. 15-30:The Observatories of the Carnegie Institution of Washington. Fig. 15-32:UCO/Lick Observatory.

Chapter 16 p. 346:Drawing by Joh Kagaya for Makoto Inoue, National Astronomical Observatory of Japan, © Hoshi No Techou. Fig. 16-1:Palomar Observatory. Fig. 16-2:Alex G. Smith, Rosemary Hill Observatory, University of Florida. Fig. 16-3:Geneva Observatory. Fig. 16-7:Dr. Rudy Schild, Smithsonian Astrophysical Observatory. Fig. 16-8:J. Holtzman/NASA. Fig. 16-9a:NOAO. Fig. 16-9b:NRAO/AUI. Fig. 16-9c:Harvard Smithsonian Center for Astrophysics. Fig. 16-10:Courtesy of T. D. Kinman, Kitt Peak National Observatory. Fig. 16-11:R. A. Perley, J. W. Dreher, J. J. Cowan; NRAO. Fig. 16-12:NRAO. Fig. 16-13a:Palomar Observatory, Copyright © California Institute of Technology. Fig. 16-13b:NASA/ESA. Fig. 16-14: NOAO. Fig. 16-15a:Copyright © 1987 Anglo-Australian Observatory. Fig. 16-15b:NASA. Fig. 16-16a:NASA/ESO. Fig. 16-16b:W. Couch (University of New South Wales), R. Ellis (Cambridge University), and NASA. Fig. 16-18: NASA/ESA. Fig. 16-20:NASA.

Chapter 17 p. 361:Goddard Space Flight Center, NASA. Fig. 17-2:Bell Telephone Laboratories. Fig. 17-3:John Mather, NASA. Fig. 17-5:NASA. Fig. 17-11:Lawrence Berkeley Laboratory.

Chapter 18 p. 381:J. Lomberg. Fig. 18-1:From the collection of Ronald A. Oriti, Santa Rosa Junior College. Fig. 18-4:Jet Propulsion Laboratory. Fig. 18-5a,b: NASA.

Index

THE NIGHT SKY IN MARCH

Chart time (Local Standard Time):

10 pm...First of March
9 pm...Middle of March
8 pm...Last of March

NORTHERN HORIZON

EASTERN HORIZON

WESTERN HORIZON

CASSIOPEIA
CEPHEUS
DENEB
CYGNUS
LYRA
VEGA
DRACO
POLARIS "NORTH STAR"
URSA MINOR "LITTLE DIPPER"
CAPELLA
AURIGA
GEMINI
CASTOR
POLLUX
HERCULES
CORONA BOREALIS
BOOTES
URSA MAJOR "BIG DIPPER"
CANCER
CANIS MINOR
PROCYON
OPHIUCHUS
SERPENS
ARCTURUS
LEO
REGULUS
VIRGO
LIBRA
SPICA
CORVUS
HYDRA
SCORPIUS
ANTARES

SOUTHERN HORIZON

THE NIGHT SKY IN MAY

Chart time (Daylight Savings Time):

11 pm...First of May
10 pm...Middle of May
9 pm...Last of May